# TISSUE ENGINEERING FOR TISSUE AND ORGAN REGENERATION

Edited by **Daniel Eberli**

**Tissue Engineering for Tissue and Organ Regeneration**
Edited by Daniel Eberli

**Published by InTech**
Janeza Trdine 9, 51000 Rijeka, Croatia



**Publishing Process Manager** Romina Krebel

**Technical Editor** Teodora Smiljanic

**Cover Designer** InTech Design Team

Additional hard copies can be obtained from orders@intechopen.com

Tissue Engineering for Tissue and Organ Regeneration, Edited by Daniel Eberli
p. cm.
ISBN 978-953-307-688-1

For EU product safety concerns:
IN TECH d.o.o., Prolaz Marije Krucifikse Kozulić 3, 51000 Rijeka, Croatia,
info@intechopen.com or visit our website at intechopen.com.

# Meet the editor

Daniel Eberli MD. Ph.D. is a scientific physician working in the translational field of urologic tissue engineering. He has a medical degree from the Medical School in Zurich, Switzerland, and a Ph.D. in Molecular Medicine from Wake Forest University, Winston Salem, NC. He currently has a faculty position at the Department of Urology at the University Hospital Zurich, where he devotes half of his time to patient care. He is a lecturer at the Medical School of Zurich and the Swiss Federal Institute of Technology. Together with his research team, he is working on novel biomaterials for bladder reconstruction, improving autonomic innervation, cellular treatment of incontinence and tracking of stem cells.

# Contents

# Preface

Over the last decade Tissue Engineering progressed rapidly and first biological substitutes were developed for several tissues in the body. Today, Tissue Engineering is one of the major approaches of Regenerative Medicine and represents a growing and exciting field of research. With the understanding and application of new knowledge of structure, biology, physiology and cell culture techniques, Tissue Engineering may offer new treatment alternatives for organ replacement or repair deteriorated organs. Among the clinical applications of Tissue Engineering are the production of artificial skin for burn patients, tissue engineered trachea, cartilage for knee-replacement procedures, urinary bladder replacement, urethra substitutes and cellular therapies for the treatment of urinary incontinence.

The classical principle of Tissue Engineering is to dissociate cells from a tissue biopsy, to expand them in culture, and to seed them onto a scaffold material in vitro in order to generate a viable tissue construct prior to re-implantation into the recipient's organism. In the appropriate biochemical and biomechanical environment these tissues will unfold their full functional potential and serve as native tissue equivalents. Tissue Engineering products may be fully functional at the time of treatment, or have potential to integrate and evolve into the expected functional tissue after implantation. While these steps may seem logical and easy to understand, the underlying biology is far more complicated and more profound questions have to be answered before the engineering of tissue and organs becomes a routine practice.

Even so, the Tissue Engineering approach has major advantages over traditional organ transplantation and circumvents the problem of organ shortage. Tissues reconstructed from readily available biopsy material implicate only minimal or no immunogenicity when reimplanted in the patient. This eventually conquers several limitations encountered in tissue transplantation approaches.

This book is aimed at anyone interested in the application of Tissue Engineering in different organ systems. With a colorful mix of topics which explain the obstacles and possible solutions, it offers insights into a wide variety of strategies applying the principles of Tissue Engineering to tissue and organ regeneration. As more and more applications move toward clinical application, a reliable preclinical model system to evaluate the developed techniques becomes crucial. Several animal models and Tissue Engineering approaches for a variety of organ systems are presented in this book.

Finally, I would like to thank all the authors who have supported this book with their contributions.

**Daniel Eberli**
University Zürich
Switzerland

# Part 1

## Cardiac Muscle

# 1

# Myocardial Tissue Engineering

Tatsuya Shimizu
*Institute of Advanced Biomedical Engineering & Science*
*Tokyo Women's Medical University*
*Japan*

## 1. Introduction

Many lives are lost due to heart diseases including myocardial infarction and cardiomyopathy. Recent reports have demonstrated that regenerative medicine has promising potential for recovering severe heart failure. Regenerative therapies for heart failure include cytokine, gene and cell therapy. Because many types of cardiovascular stem cells have been identified and their clinical potentials have been demonstrated for the past decade, cell injection therapy has most attracted both researchers and clinicians (Wollert 2008). On the other hand, significant cell loss due to washing out and cell death has become problematic in cell injection technique. So, as next generation of regenerative therapy for impaired heart, transplantation of myocardial patches fabricated by tissue engineering technology are emerging and are clinically applied. Furthermore, several challenges for fabricating functional myocardial tissues/organs, which are electrically communicated, pulsate synchronously and evoke contraction power, have also started (Zimmermann, Didie et al. 2006). These ambitious challenges may lead to reconstruction of malformed hearts and become alternative therapy for heart transplantation.

Heart tissues are composed of high-dense cylindrical cardiomyocytes and fibroblasts with abundant vascular network and collagen-based extracellular matrix (ECM). Cardiomyocytes pulsate via sodium and calcium ion transient through cell membrane. They are also electrically coupled by gap junctions composed of connexion 43 and rapid electrical propagation realizes simultaneous beating as a whole. Continuous blood flow supplies oxygen and nutrition, and withdraw the waste for high metabolic demand of heart tissues. These structure and function produce mechanical contractions as a blood pump. Therefore the researchers should take into account high density culture of cardiomyocyte and surrounding cells, sufficient micro blood vessel fabrication, cell/ECM orientation and proper cell-to-cell coupling for engineering heart tissues/organs.

Here, previous and current status of cell injection therapy, myocardial patch transplantation and pulsatile myocardial tissue fabrication is described with some future views.

## 2. Cell injection therapy

Cell injection therapy for damaged heart has been researched since the early 1990's. Many researchers have demonstrated the therapeutic potential of isolated cell transplantation into myocardium using various types of cell sources both in animal models and in some clinical trials (Puceat 2008). The mechanism of myocardial tissue regeneration has not been completely cleared, but most researchers have agreed that transplanted cells secrete several cytokines which promote neovascularization, prohibit fibrosis, decrease cell death and recruit stem cells, leading to heart function improvement. It has been also asserted that some of injected cells differentiate into functional cardiomyocytes and may directly contribute to heart contraction improvement. Although some differences may exist in according to cell types, multifactorial mechanisms seem to relate with myocardial tissue regeneration.

In addition to cell sourcing, different routes are used for cell administration. Systematic intravenous infusion is performed through a central or peripheral vein. This method is simple and less invasive, however widespread distribution cause low ratio of cell engraftment. Most popular approach is intracoronary cell infusion via a balloon-catheter. Injected cells are reached directly in the target myocardial region, however, cells have to transmigrate across endothelium wall. Intracardiac injection is performed via pericardium during open heart surgery and via endocardium by a catheter with a 3-D electromechanical mapping system (NOGA mapping system). These methods realize relatively targeted delivery, but myocardial damage and arrhythmia induction are problematic. Future clarification will be needed to decide which is the best approach for cell injection.

### 2.1 Skeletal myoblasts

Skeletal myoblasts were the first cell source to enter the clinical application for heart tissue repair. They lie in a quiescent state on the basal membrane of myofibers and have the potential to start to proliferate and differentiate into functional skeletal muscle in response to muscle damage. They can be isolated autologously and be expanded from a single biopsy. In addition, skeketal myoblasts are relatively resistant to ischemia. Menasche and colleagues first applied skeletal myoblast injection via epicardium for patients undergoing coronary artery bypass grafting (CABG) (Menasche, Hagege et al. 2001). The phase I clinical study (MAGIC I) have shown the feasibility of skeletal myoblast implantation, however, increased risk of ventricular arrhythmias after the operation. Then, MAGIC II trial was performed to clarify the safety and efficacy, in which all patients received preventive medication and an implantable cardioverter-defibrillator for rescuing critical ventricular arrhythmias. In result, skeletal myoblast injection failed to significantly improve heart function, leading to sample size reduction (Menasche, Alfieri et al. 2008). On the other hand, the trial indicated the possibility that high dose cell injection might recover left ventricular dilatation. In addition, the other clinical trials of catheter-based myoblast implantation via endocardium have revealed functional efficacy (Opie and Dib 2006). According to these results, not the regenerative potential of myoblasts themselves but the amount of injected cells and delivery system may affect the efficacy. Therefore, it seems that skeletal myoblasts should not be excluded as a cell source for heart tissue repair. More optimization of cell delivery and comparison of cell sources can address these critical issues.

### 2.2 Bone marrow-derived cells

Bone marrow-derived cells are the most used cells in clinical trials for myocardial tissue repair (Wollert 2008). The discovery of circulating progenitor cells originated from human bone marrow has stimulated research and clinical use of bone marrow-derived cells

(Asahara, Murohara et al. 1997). Bone marrow cells contain different stem and progenitor cells which will differentiate into various types of cells including endothelial cells, smooth muscle cells and cardiomyocytes. Bone marrow mononuclear cells (BMNCs), which can be isolated simply by gradient sedimentation after bone marrow aspiration without culture expansion, have been clinically injected via coronary artery from the first. BMNCs include heterogeneous cell population of monocytes, hematopoietic stem cells and endothelial progenitor cells (EPCs). Therefore, some groups have used BMNCs selected by surface markers ($CD34^+$, $CD133^+$) and demonstrated more efficacy of their injection. As another cell population, mesenchymal stem cells (MSCs) have been researched and clinically used. Although, MSCs represent between 0.01 and 0.001% of all nucleated cells in bone marrow, they can be readily expanded in culture. MSCs have the potential to differentiate into various types of cells and injected MSCs in heart seem to differentiate into myocardial composing cells. Recent studies have revealed rare happening of cardiomyocyte differentiation, therefore MSCs seem to recover heart function via their cytokine secretion and partial differentiation into vascular cells. As a unique feature, MSCs have the potential to escape from immune detection due to the direct inflammatory inhibition and the lack of cell-surface molecules. This property has realized allogenic mesenchymal stem cell transplantation in clinic and has given high impact on cell therapy research field.

Recent randomized controlled trials of bone marrow-derived cell injection revealed overall feasibility and safety. However the data has revealed only marginal increases of ejection fraction (EF) even in positive studies (0-5%) (Martin-Rendon, Brunskill et al. 2008). For establishing more effective bone marrow-derived cell therapy, optimization of cell source, cell dose, delivery method and deliver timing will be needed.

### 2.3 Adipose-derived stem cells

In addition to bone marrow-derived MSCs, stem cells isolated from the stroma of adipose tissues have represented regenerative potential for heart tissues (Psaltis, Zannettino et al. 2008). Adipose tissue-derived stem cells (ASCs) display features similar to that of bone marrow-derived MSCs and their angiogenic potential have been reported. Some studies have also revealed cardiomyocyte differentiation from ASCs. It has not been clarified which mesenchymal stem cells are superior to other cell types, however, reatively easy isolation of adipose tissue may push the clinical application of ASCs.

### 2.4 Cardiac stem cells

Cardiac stem cells (CSCs) are also possible cell source for myocardial tissue regeneration. Two groups first reported CSC existence in 2003 (Beltrami, Barlucchi et al. 2003; Oh, Bradfute et al. 2003). Until then, it was common knowledge that heart was a post mitotic organ, but those reports accelerated the researches for identifying surface marker of CSCs and culturing them. Islet-1, Sca-1 and c-kit have been known as CSC markers. Recently, it has been also confirmed that heart has renewal ability at normal state and the annual rate of turning over is 1% at the age of 25 (Bergmann, Bhardwaj et al. 2009). Although the ability of CSCs may increase after heart injury, newly formed cardyomyocytes are not sufficient for replacing damaged muscle tissues. Therefore isolation and expansion of CSCs have been extensively examined. Some groups have used a different approach to make cardiospheres from biopsied myocardium, which lead to efficient CSC expansion (Lee, White et al. 2011).

Clinical trials for injection therapy of autologous CSCs isolated from biopsy sample are now on going.

### 2.5 Embryonic stem cells

Although abundant studies demonstrated that MSCs, ASCs and CSCs have the potential of cardiomyocyte differentiation regarding gene and protein expression, there are no studies clearly showing beating cardiomyocytes differentiated from those stem cells. On the other hand, many researchers have confirmed that embryonic stem cells (ESCs) can differentiate into beating cardiomyocytes in vitro and implantation of ESC-derived cardiomyocytes improves damaged heart function. Several signal pathways for cardiac differentiation have been already clarified and various molecules have been reported as its promoters. For example, noggin increased cardiac differentiation efficacy via regulation of Bone morphogenetic protein (BMP) signalling pathway (Yuasa, Itabashi et al. 2005) and insulin-like growth-factor-binding protein 4 (IGFBP4) promotes cardiogenesis by inhibitor of canonical Wnt signalling (Zhu, Shiojima et al. 2008). In addition, fibloblast growth factor (FGF), retinoic acid, ascorbic acid and cyclosporine A have been reported to have the potential to enhance cardiac differentiation from ESCs. The important issue as well as cardiac differentiation is purification of cardiomyocytes from heterogeneous cell mixture, because contamination of immature cells leads to teratoma formation. Although gene-modified ESCs harboring neomycin resistance gene or green fluorescent protein (GFP) gene in the cardiac-specific gene locus are very useful in non-clinical experiments, safe and efficient isolation technologies will be needed for clinical application. Culture media control focusing on the differences of cell metabolism may be useful for safe cell selection. Moreover immune response of the host is another critical issue. Nucler transfer or cell banking is possible approach avoiding immunoreaction.

Electrical communication and simultaneous beating of implanted ESC-derived cardiomyocytes should be also requested for improving damaged heart function without arrhythmia. In vivo electrophisiological analyses and the transplantation technology for synchronization will be essential for clinical application of these cells.

### 2.6 Induced pluripotent stem cells

Induced pluripotent stem cells (iPSCs) also hold great promise for myocardial tissue engineering (Vunjak-Novakovic, Tandon et al. 2010). Terminally differentiated cells can be reprogrammed to have the same potential as ESCs by introducing 3 or 4 transcriptional factor genes. Furthemore none-gene transfer technologies have been developed in the world. The superiority of iPSCs to ESCs is autologous cells, which do not cause immune response. Cardiac differentiation of human iPSCs has been reported in the same manner with ESCs.

Several critical issues must be clarified for clinical use, but ESCs/iPSCs-derived cardiomyocytes should contribute to myocardial tissue engineering in the view point of their pulsatile function and scaling-up.

### 2.7 Problems of cell injection therapy

Cell injection therapies for heart failure are now world-widely performed. While moderate success of direct cell injection has been observed, the efficacies seem not to reach the level that general clinicians think cell therapy a reliable treatment for heart failure. More optimization of cell source, cell preparation process, injection route, injection timing and patient population may increase the effectiveness; however one of the essential issues is cell delivery methodology. Cell injection therapy has significant difficulties about cell retention in the target tissue. The shape, size, and position of the grafted cells are often uncontrollable and large amount of the cells are washed-out. Moreover, once retaining cells die due to necrosis and apoptosis. Time course quantification with TUNEL assay demonstrated that a large number of the grafted cells die within a few days after injection in rat models (Zhang, Methot et al. 2001). In the clinical trial using bone marrow-derived cells, it has been also demonstrated that only 1-3% of the cells infused via coronary arteries could be detected by 3D positron emission tomography (PET) imaging of the patient heart. In this study, a large percentage of cells were found in the liver and spleen immediately after the procedures (Hofmann, Wollert et al. 2005). To clear the problem of cell loss, hydrogel-cell mixture injection has been pursued. Fibrin, collagen and alginate hydrogels are now used. Hydrogels with cells are injected as a liquid phase through syringe or catheter, then, they are polymerized and fixed in the target tissues (Kofidis, de Bruin et al. 2004). In hydrogel-cell mixture injection therapy, local tissue damage due to space occupation of hydrogel itself and inflammatory reaction due to hydrogel biodegradation are problematic.

Therefore, more advanced cell delivery systems have been requested to spread the regenerative therapy as one of the reliable treatments for heart failure.

## 3. Tissue engineering

Recent advance of tissue engineering technologies have realized the transplantation of tissue-engineered construct "myocardial patch" covering over damaged heart surface instead of simple cell injection into myocardium. Grafted cells within myocardial patches can survive more and secrete more cytokines, resulting in more heart function improvement. Furthermore pulsatile myocardial tissues have been successfully engineered by using cardiomyocytes as a seeding cell source. These tissues may directly help heart contraction and total heart wall replacement may be possible in future. There are several contexts of tissue engineering.

### 3.1 Scaffold-based tissue engineering

Most popular technology of tissue engineering is to seed cells into 3-D pre-fabricated biodegradable scaffolds which are made from synthetic polymer and biological material. Hydrogel formation after mixing cells and scaffold solution is another approach. Decellularized tissues have been also used as scaffolds. These scaffolds play as alternatives for extra cellular matrix (ECM), therefore, their cell-adhesiveness and porosity affect survived cell amount and engineered tissue quality. Scaffold modification can control its biodegradation and tissue formation. Growth factor linkage leads to accelerating tissue formation. Now these scaffold-based tissue engineering has been widely applied to cardiovascular tissue regeneration as well as other tissue repair (Vunjak-Novakovic, Tandon et al. 2010).

### 3.2 Cell sheet-based tissue engineering

In contrast to scaffold-based tissue engineering, our group have developed unique technique involving cell sheet stacking to fabricate 3-D tissues (Shimizu, Yamato et al. 2003).

Cell sheets are 2-D connecting pure cells without any scaffolds, therefore cell-dense 3-D tissues can be fabricated by stacking cell sheets. Cell sheets are harvested from intelligent culture surface "temperature-responsive culture surface", which are covalently grafted with temperature-responsive polymer, poly (*N*-isopropylacrylamide) (PIPAAm) (Okano, Yamada et al. 1993). The surfaces are slightly hydrophobic and cell-adhesive at 37°C, on the other hand, the surface changes to hydrophilic and not cell adhesive below 32°C. Confluently cultured cells on the surface can detach as a contiguous cell sheet simply by reducing temperature. Furthermore, biological molecules underneath cell sheets are also preserved and play a critical role as an adhesive agent during cell sheet stacking. Cell sheet-based tissue engineering has been applied for a wide range of regenerative medicine including corneal epithelial replacement, heart tissue repair, pneumothorax repair, liver tissue repair and so on.

According to the spread of the concept fabricating 3-D tissues from 2-D confluent cells, several other technologies using this concept have emerged. Cell sheet fabrication techniques using fibrin coated dishes or nanofibrous polycaprolactone meshes have been reported (Shin, Ishii et al. 2004; Itabashi, Miyoshi et al. 2005). Cell sheet-like constructs have been also engineered using magnetite nanoparticles (Ito, Hibino et al. 2005). Magnetically labelled cells are attached on culture materials by magnetic force and confluent cells are harvested as a cell sheet by magnetic force release. Thus, cell sheet-based tissue engineering has now spread in the world as scaffold-free tissue engineering.

## 4. Myocardial patch transplantation

Both scaffold-based and cell sheet-based tissue engineering have been used for myocardial patch fabrication. Not only cardiomyocytes but also other types of cells have been used for creating myocardial patches and some myocardial patches using non-cardiomyocytes have been already clinically transplanted over damaged hearts. (Fig. 1.)

### 4.1 Scaffold-based mayocardial patch

In myocardial patch fabrication, synthetic polymer, biological material and decellularized tissue have been used as prefabricated scaffolds. Li and colleagues, who were one of the pioneer groups of myocardial tissue engineering, first demonstrated that gelatine sponges seeded with cardiac cells have therapeutic potentials for cryoinjured rat hearts (Li, Jia et al. 1999). Leor and colleagues reported that bioengineered heart grafts using porous alginate scaffolds attenuated left ventricular dilatation and heart function deterioration in infarction model (Leor, Aboulafia-Etzion et al. 2000). Eschenhagen and Zimmermann's group have developed innovative myocardial tissue engineering approach (Zimmermann, Schneiderbanger et al. 2002). They have fabricated 3-D tissues by gelling mixture of cardiac cells and collagen solution. The constructs induced systolic wall thickening of the left ventricle infracted area and improved fractional shortening of damaged hearts in rat myocardial infarction model (Zimmermann, Melnychenko et al. 2006). Small intestinal submucosa (SIS) has also been used as a scaffold for myocardial patch. MSC-seeded SIS improved heart contraction in rabbit infarction model (Tan, Zhi et al. 2009). There have been various types of myocardial patches using different scaffolds and different cell sources. Although implantable human myocardial patches using beating cardiomyocytes have not been established now, clinical trials of collagen-based myocardial patch with bone marrow cells (MAGNUM trial) (Chachques, Trainini et al. 2007) and vicryl mesh-based myocardial patches with fibroblasts (Anginera) ((Mirsadraee, Wilcox et al. 2006)) have revealed feasibility and safety of myocardial patch transplantation.

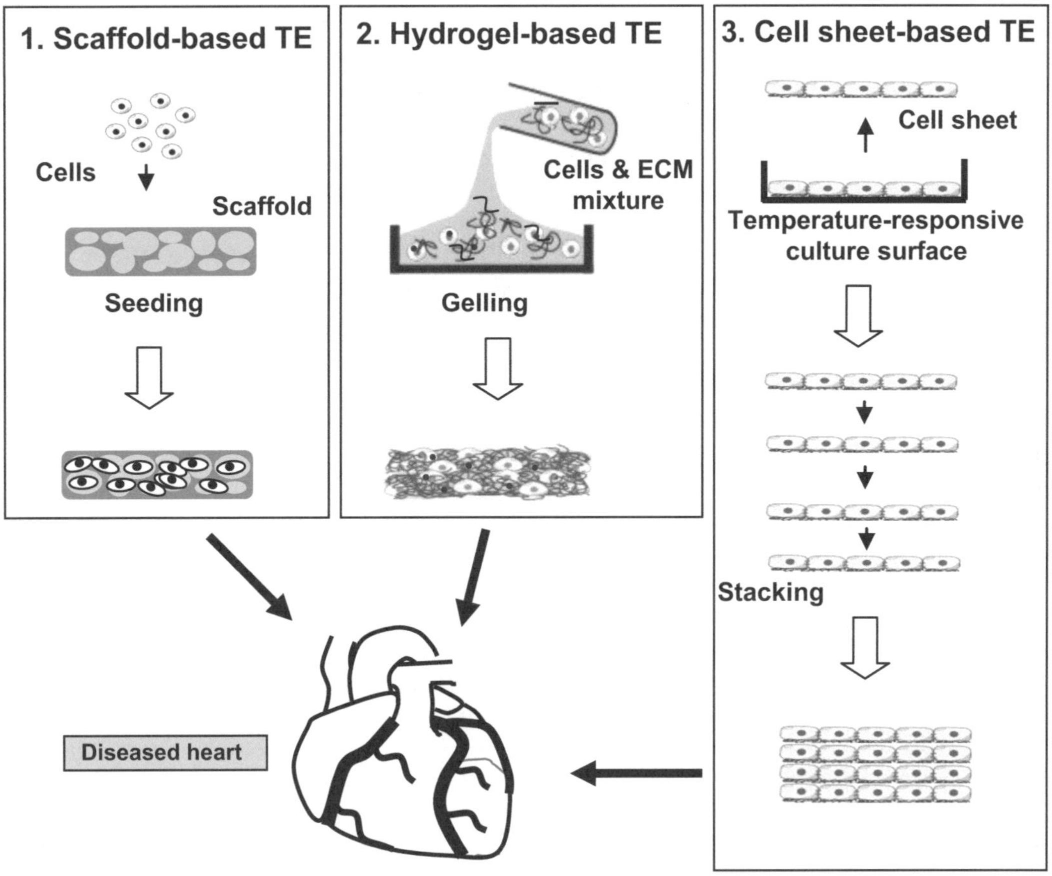


Fig. 1. Tissue engineering (TE) strategies for myocardial patch fabrication

### 4.2 Cell sheet-based myocardial patch

Many types of cell sheets have been reported to improve impaired heart function (Shimizu, Sekine et al. 2009). Cell sheets are transplanted onto heart surface directly via open heart surgery and cells can be more effectively delivered as thin, but large-area cell-dense grafts than isolated cell injection. Scaffold-based myocardial patches are usually transplanted on myocardium with suture, on the other hand, cell sheets are transplanted with no suture because biological adhesive proteins underneath cell sheets promote the attachment. When neonatal rat cardiac cell sheets were transplanted onto infracted rat hearts, grafted cardiomyocytes communicated with host myocardium via gap junctions and blood vessels formed within the graft, resulting in significant improvement of heart function (Miyagawa, Sawa et al. 2005; Sekine, Shimizu et al. 2006).

Sawa and colleagues have started to use skeletal myoblasts for cell sheet fabrication, because myoblasts can be isolated autologously and are relatively resistant to ischemic condition.

The recovery of heart function by skeletal myoblast transplantation has been confirmed in rat ischemic model, in dilated cardiomyopathy hamster model, in pacing-induced canine heart failure model and in pig infarction model (Memon, Sawa et al. 2005; Hata, Matsumiya et al. 2006; Kondoh, Sawa et al. 2006; Miyagawa, Saito et al. 2010). Regarding stacking cell sheet number, 3-5 layers are optimal and more layering cause primary necrosis of the constructs (Sekiya, Matsumiya et al. 2009). They have demonstrated more hematopoietic stem cells and less fibrosis in cell sheet transplantation than in isolated cell injection in accordance with more expression of stromal-derived factor 1 (SDF-1), hepatocyte growth factor (HGF), and vascular endothelial growth factor (VEGF). Based on these results, clinical trial of autologous myoblast sheet transplantation for severe heart failure has started and the detailed results will appear soon.

In the same manner with cell injection therapy, MSCs are used as a candidate cell source for human implantable cell sheet. Adipose tissue-derived MSCs and menstrual blood-derived MSCs have improved damaged heart function in rat infarction model (Miyahara, Nagaya et al. 2006; Hida, Nishiyama et al. 2008). MSCs can gradually grow to form a thick stratum containing newly formed blood vessels and some cells seem to differentiate into cardiomyocytes at least by histological analyses. Further studies will be needed to confirm the differentiation into functional beating cardiomyocytes and possibilities to differentiate into unexpected cell types.

As emerging cell source, cell sheets of stem cell antigen 1-positive (Sca-1-positive) CSCs ameliorates cardiac dysfunction in mouse infarction model through cardiomyocyte differentiation and paracrine mechanisms mediated via soluble vascular cell adhesion molecule 1 (VCAM-1)/very late antigen-4 (VLA-4) signaling pathway (Matsuura, Honda et al. 2009). In addition, cardiac cell sheets originated from ESCs/iPSCs have been successfully fabricated and their transplantation into animal models is now ongoing.

For enhancing the efficacy of cell sheet transplantation, gene-modified cell sheets have been examined. Bcl-2 expressed myoblast sheets prolonged survival, increased production of proangiogenic paracrine mediators, and enhanced the therapeutic efficacy (Kitabayashi, Siltanen et al. 2010). HGF overexpression in myoblast sheets enhances their angiogenic potential in rat chronic heart failure model (Siltanen, Kitabayashi et al. 2011). As another concept, cell sheets co-cultured with endothelial cell sources have been transplanted in rat infarction models. Transplantation of EPC co-cultured fibroblast sheet improved heart function more than only fibroblast sheet implantation or EPC injection (Kobayashi, Shimizu et al. 2008). Furthermore, endothelial cell co-culture within cardiomyocyte sheets induced more neovascularization and more improvement of cardiac function than only cardiomyocyte sheets (Sekine, Shimizu et al. 2008). These studies indicate advanced strategies of cell sheet transplantation.

As mentioned previously, it is considered that the main mechanism of heart function improvement is neovascularization, fibrosis inhibition, apoptosis inhibition and stem cell recruitment due to various cytokines secreted from grafted cells. In comparison with cell injection approach, increase of cell survival within myocardial patches leads to more cytokine secretion, then, resulting in more function improvement. In addition to cytokine secretion, myocardial patches may have girdling effect and prohibit heart dilatation. Therefore, myocardial patch transplantation is quite different cell delivery method from cell injection and has more potential to rescue diseased hearts. In the case of myocardial patches using beating cardiomyocytes, direct enhancement of contraction power is additionally expected, however, electrical synchronization between host hearts and transplanted patches is a critical issue to be clarified.

## 5. Engineering plusatile myocardial tisssue

Beyond myocardial patch fabrication, several research groups have challenged to fabricate pulsatile myocardial tissues by their original tissue engineering strategies. Bioengineered contractile myocardial tissues may realize new therapeutics for severe heart diseases and be useful as alternatives for animal models.

### 5.1 Pre-fabricated scaffold-based myocardial tissue fabrication

The first approach for engineering functional myocardial tissue is seeding cardiomyocytes into synthetic or biological 3-D scaffolds. Vunjak-Novakovic and colleagues first reported that seeding primary cultured cardiomyocytes onto disc-shaped polyglycolic acid (PGA) scaffolds in rotating bioreactor system resulted in spontaneously pulsatile myocardial tissues (Papadaki, Bursac et al. 2001). Optimization of cell population, serum concentration and scaffold coating improved electrical conduction velocity of engineered constructs. Radisc and colleagues seeded rat cardiomyocytes in Matrigel onto collagen sponges and stimulated the constructs electrically. The stimulation improved the conductive and contractile properties in accordance with increased expression of myosin heavy chain and connexion 43. Furthermore, cardiomyocytes in the electrically stimulated constructs were more aligned and elongated as same as those in native heart tissue (Radisic, Park et al. 2007).

Following these studies, many research groups have started to engineer myocardial tissue in vitro by using various types of scaffolds. Scaffold porosity is one of the critical factors for pre-fabricated scaffold-based tissue engineering. High porosity increases seeded cell number and facilitates mass transport. Surface modyfication is also important for cell attachment and survival. Laminin coating improved cardiomyocyte adhesiveness. In addition, scaffold elasticity and degradability affect contraction property of engineered myocardium. Further studies are ongoing to development appropriate scaffold materials for myocardial tissue engineering.

### 5.2 Hydrogel-based myocardial tissue fabrication

The second approach is to form 3-D tissues by gelling of cardiac cell and matrix solution mixture. Eschenhagen and Zimmermann have continuously developed this strategy using collagen gel and successfully engineered macroscopically beating cardiac tissues (Zimmermann and Cesnjevar 2009). First, neonatal rat cardiomyocytes were suspended in collagen I solution and the mixture was poured into the mold. After gelling, the constructs were unidirectionally stretched with the mechanical devise. They have also realized contraction force measurement. Cyclic stretch introduced cell alignment along the stretching direction and increased mitochondrial density, leading to native heart-like tissue. The contraction force of engineered myocardium was comparative with native heart tissue and responded to pharmacological agents properly. Ring-shaped myocardial tissues were also fabricated and combined 5 constructs were transplanted onto infarcted rat hearts. Interestingly, the constructs synchronized to each other and improved damaged heart function. They have also confirmed that co-culture constructs including cardiomyocytes, fibroblasts and endothelial cells were superior to cardiomyocyte rich constructs in morphology and function. Recently, they have also started to utilize human cardiomyocytes differentiated from ESC/iPSC as cell source and challenged to create human myocardial tissues (Zimmermann 2011). In contrast to pre-fabricated scaffold usage, relatively homogeneous myocardial tissues are engineered by hydrogel-based approach. Therefore collagen gel-based myocardial tissue engineering has now become popular in the world.

### 5.3 Cell sheet-based myocardial tissue fabrication

The third approach is to engineer 3-D pulsatile myocardial tissues by stacking cardiac cell sheets. As mentioned previously, 2-D cell sheets can be harvested from temperature-responsive culture dishes only by lowering temperature and do not include any materials. 3-D tissues are constructed by layering cell sheets. Because 2-D confluent cells are directly stacked without any scaffolds, resulting constructs are cell-dense 3-D tissues. It is well-known that 2-D confluent cardiomyocytes connect to each other electrically via gap junctions resulting in synchronized beating. Cardiac cell sheets harvested from temperature-responsive culture dishes maintain this synchronized pulsation (Shimizu, Yamato et al. 2002). For creating 3-D functional heart tissues by layering cardiac cell sheets, morphological and electrical communications between cell sheets are critical. Multiple-electrode extracellular recording system revealed that double-layer rat cardiac cell sheets coupled electrically about one hour after layering and histological analysis showed the existence of connexin 43 between two cardiac cell sheets. Adhesive proteins deposited on cell sheet surface are considered to promote these rapid electrical communications (Haraguchi, Shimizu et al. 2006). Stacked cardiac cell sheets beat synchronously in macroscopic view and the constructs transplanted into rat subcutaneous tissues also pulsated continuously at least up to one year and eight months after implantation. Morphological analyses showed elongated cardiomyocytes, well-differentiated sarcomeres, gap junctions and multiple blood vessels, which were characteristic structure of native heart tissue (Shimizu, Yamato et al. 2002). Long-term observation revealed that their size, conduction velocity, and contractile force increased in proportion to the host growth (Shimizu, Sekine et al. 2006). Recently, fabrication of cardiac cell sheets using ESC-originated cardiomyocytes have just started and human cardiac cell sheets will appear in near future.

### 5.4 Fabrication of vascularized myocardial tissue

One of the major obstacles in myocardial tissue engineering is scaling-up of the constructs. Insufficient supply of oxygen and nutrient, and waste accumulation limit their thickness. Actually, cells are sparse in the central area, on the other hand, cells are dense in the outer surface (100-200μm) area in scaffold-based myocardial tissue engineering. In the case of cell sheet-based myocardial tissue engineering, thickness limit is approximately 80μm (3 layers) (Shimizu, Sekine et al. 2006). Several approaches have been examined in the point of view overcoming diffusion limit. Perfusion of culture media through the constructs using porous scaffolds is one possible approach. Media penetration increased cell migration depth and improved cell metabolism. However shear stress due to media flow may prohibit tight cell attachment on the scaffold material. Media perfusion with oxygen carrier, perfluorocarbon (PFC) has been also examined for improving oxygen transport. PFC usage increased cell proliferation and improved pulsatile function. Media penetration is useful to some extent, however, it becomes more difficult as cell density increases.

To overcome this problem, it has been requested to develop new technologies for introducing vasculature or vascular-like structure into engineered tissues. Several researchers have tried to generate microchannel network within porous 3-D scaffolds by microfabrication techniques including $CO_2$ laser ablation. The technology has not reached to mimicking native micro capillary network. On the other hand, recent studies have revealed that co-cultured endothelial cells within cardiac constructs can spontaneously form vascular-like network in vitro and tubular formation has been found in some parts. It has been also confirmed that this pre-vascular structure connected to host blood vessels immediately after transplantation and the newly developed vessels within the constructs were blood-supplied within a few days (Sekiya, Shimizu et al. 2006). We have already demonstrated that the tissue thickness of cardiac cell sheets co-cultured with endothelial cells were just twice as the thickness of cardiac cell sheets without endothelial cells (Sekine, Shimizu et al. 2008). Although endothelial cell co-culture is helpful for accelerating blood vessel formation, more scaling-up is still limited due to primary ischemia until sufficient vascularization.

One possible idea for scaling-up is utilizing in vivo vascularization power. Our group has reported that triple-layer cardiac cell sheets were repeatedly implanted after waiting enough vascular formation within previously implanted tissues. In result, synchronously beating thick myocardial tissues with sufficient micro capillaries were successfully fabricated and 10-times transplantation of triple-layer constructs (totally 30 sheets) formed 1-mm thick, pulsatile myocardial tissues. Furthermore, when triple-layer grafts were transplanted repeatedly over a surgically connectable artery and vein in leg, the multilayer constructs were blood-supplied from the thick artery and vein. The constructs were successfully resected with the connectable blood vessels and were ectopically transplanted in neck with direct vessel anastomoses (Shimizu, Sekine et al. 2006). Recently several groups have also utilized in vivo power for myocardial tissue engineering. Cardiomyocytes, ECM alternatives and native blood vessels were packed in the special chamber and incubated in vivo. Vascularized heart-like tissues were created in the body (Morritt, Bortolotto et al. 2007; Birla, Dhawan et al. 2009).

Furthemore, next challenge is now in vitro fabrication of vascularized myocardial tissues. Kofidis and colleagues have constructed fibrin gel-based myocardial tissues containing rat aortas (1-2mm), through which culture media was perfused (Kofidis, Lenz et al. 2003). Cell survival and metabolism were improved, however formation of functional blood vessels connecting with central aortas were not clear. We are now trying to promote endothelial cell tubular formation within in vitro engineered cardiac tissues and to perfuse culture media through the newly formed vessels using perfusion bioreactors. Further studies will be needed to break through the obstacles for in vitro scaling up.

### 5.5 From tissue engineering to organ engineering

For future organ engineering, some groups have challenged to engineer myocardial constructs with pumping function. Ott and colleagues have used decellularized organ as a scaffold. They decellularized rat whole hearts and re-seeded cardiac cells into decellularized hearts. Heart contraction was recovered and pump function was generated (Ott, Matthiesen et al. 2008). Zimmermann's group developed pouch-like myocardial tissue by their technology as previously described and covered heart with pouch-like constructs (Yildirim, Naito et al. 2007).

Regarding cell sheet technology, myocardial tubes have been fabricated by wrapping rat cardiac cell sheets around fibrin tubes and rat resected aortas. The engineered myocardial tubes revealed spontaneous, synchronized pulsation and small but significant inner pressure changes (about 0.1 mmHg) in vitro (Kubo, Shimizu et al. 2007). On the other hand, resected rat aortas wrapped with cardiac cell sheets were micro surgically transplanted in place of the abdominal aorta. After 1 month, in vivo myocardial tubes demonstrated spontaneous beating and evoked independent blood pressures (about 6 mm Hg). The value of in vivo myocardial tubes was much bigger than in vitro myocardial tubes (Sekine, Shimizu et al. 2006). Comparing in vitro and in vivo, it was considered that pulsation due to host blood flow has induced cardiomyocyte hypertrophy, leading to improvement of pumping function. Therefore pulsatile perfusion bioreactors may improve pumping function of in vitro engineered myocardial tubes.

Thus, small size myocardial constructs evoking pumping function have been realized. Expansion and selection of cardiomyocytes, and sufficient blood vessel formation for scaling-up are now critical issues for organ engineering.

## 6. Conclusions

As the first generation of cardiac regenerative therapy, many clinical trials of cell injection therapy have been already performed. The controversial arguments about its effectiveness will be settled in next several years. Tissue engineered myocardial patches have now emerged as the second generation and previous studies indicate promising potential for rescuing damaged heart. As the third generation, tissue-engineered pulsatile myocardial tissues should support heart contraction physically. Furthermore, future development of cell sourcing and scaling-up technologies may realize "bioengineered hearts".

## 7. Acknowledgment

This work is granted by the Japan Society for the Promotion of Science (JSPS) through the "Funding Program for World-Leading Innovative R&D on Science and Technology (FIRST Program)," initiated by the Council for Science and Technology Policy (CSTP).

## 8. References

Asahara, T., T. Murohara, et al. (1997). Isolation of putative progenitor endothelial cells for angiogenesis. *Science* 275(5302): 964-967.

Beltrami, A. P., L. Barlucchi, et al. (2003). Adult cardiac stem cells are multipotent and support myocardial regeneration. *Cell* 114(6): 763-776.

Bergmann, O., R. D. Bhardwaj, et al. (2009). Evidence for cardiomyocyte renewal in humans. *Science* 324(5923): 98-102.

Birla, R. K., V. Dhawan, et al. (2009). Cardiac cells implanted into a cylindrical, vascularized chamber in vivo: pressure generation and morphology. *Biotechnol Lett* 31(2): 191-201.

Chachques, J. C., J. C. Trainini, et al. (2007). Myocardial assistance by grafting a new bioartificial upgraded myocardium (MAGNUM clinical trial): one year follow-up. *Cell Transplant* 16(9): 927-934.

Haraguchi, Y., T. Shimizu, et al. (2006). Electrical coupling of cardiomyocyte sheets occurs rapidly via functional gap junction formation. *Biomaterials* 27(27): 4765-4774.

Hata, H., G. Matsumiya, et al. (2006). Grafted skeletal myoblast sheets attenuate myocardial remodeling in pacing-induced canine heart failure model. *J Thorac Cardiovasc Surg* 132(4): 918-924.

Hida, N., N. Nishiyama, et al. (2008). Novel cardiac precursor-like cells from human menstrual blood-derived mesenchymal cells. *Stem Cells* 26(7): 1695-1704.

Hofmann, M., K. C. Wollert, et al. (2005). Monitoring of bone marrow cell homing into the infarcted human myocardium. *Circulation* 111(17): 2198-2202.

Itabashi, Y., S. Miyoshi, et al. (2005). A new method for manufacturing cardiac cell sheets using fibrin-coated dishes and its electrophysiological studies by optical mapping. *Artif Organs* 29(2): 95-103.

Ito, A., E. Hibino, et al. (2005). Construction and delivery of tissue-engineered human retinal pigment epithelial cell sheets, using magnetite nanoparticles and magnetic force. *Tissue Eng* 11(3-4): 489-496.

Kitabayashi, K., A. Siltanen, et al. (2010). Bcl-2 expression enhances myoblast sheet transplantation therapy for acute myocardial infarction. *Cell Transplant* 19(5): 573-588.

Kobayashi, H., T. Shimizu, et al. (2008). Fibroblast sheets co-cultured with endothelial progenitor cells improve cardiac function of infarcted hearts. *J Artif Organs* 11(3): 141-147.

Kofidis, T., J. L. de Bruin, et al. (2004). Injectable bioartificial myocardial tissue for large-scale intramural cell transfer and functional recovery of injured heart muscle. *J Thorac Cardiovasc Surg* 128(4): 571-578.

Kofidis, T., A. Lenz, et al. (2003). Pulsatile perfusion and cardiomyocyte viability in a solid three-dimensional matrix. *Biomaterials* 24(27): 5009-5014.

Kondoh, H., Y. Sawa, et al. (2006). Longer preservation of cardiac performance by sheet-shaped myoblast implantation in dilated cardiomyopathic hamsters. *Cardiovasc Res* 69(2): 466-475.

Kubo, H., T. Shimizu, et al. (2007). Creation of myocardial tubes using cardiomyocyte sheets and an in vitro cell sheet-wrapping device. *Biomaterials* 28(24): 3508-3516.

Lee, S. T., A. J. White, et al. (2011). Intramyocardial injection of autologous cardiospheres or cardiosphere-derived cells preserves function and minimizes adverse ventricular remodeling in pigs with heart failure post-myocardial infarction. *J Am Coll Cardiol* 57(4): 455-465.

Leor, J., S. Aboulafia-Etzion, et al. (2000). Bioengineered cardiac grafts: A new approach to repair the infarcted myocardium? *Circulation* 102(19 Suppl 3): III56-61.

Li, R. K., Z. Q. Jia, et al. (1999). Survival and function of bioengineered cardiac grafts. *Circulation* 100(19 Suppl): II63-69.

Martin-Rendon, E., S. J. Brunskill, et al. (2008). Autologous bone marrow stem cells to treat acute myocardial infarction: a systematic review. *Eur Heart J* 29(15): 1807-1818.

Matsuura, K., A. Honda, et al. (2009). Transplantation of cardiac progenitor cells ameliorates cardiac dysfunction after myocardial infarction in mice. *J Clin Invest* 119(8): 2204-2217.

Memon, I. A., Y. Sawa, et al. (2005). Repair of impaired myocardium by means of implantation of engineered autologous myoblast sheets. *J Thorac Cardiovasc Surg* 130(5): 1333-1341.

Menasche, P., O. Alfieri, et al. (2008). The Myoblast Autologous Grafting in Ischemic Cardiomyopathy (MAGIC) trial: first randomized placebo-controlled study of myoblast transplantation. *Circulation* 117(9): 1189-1200.

Menasche, P., A. A. Hagege, et al. (2001). Myoblast transplantation for heart failure. *Lancet* 357(9252): 279-280.

Mirsadraee, S., H. E. Wilcox, et al. (2006). Development and characterization of an acellular human pericardial matrix for tissue engineering. *Tissue Eng* 12(4): 763-773.

Miyagawa, S., A. Saito, et al. (2010). Impaired myocardium regeneration with skeletal cell sheets--a preclinical trial for tissue-engineered regeneration therapy. *Transplantation* 90(4): 364-372.

Miyagawa, S., Y. Sawa, et al. (2005). Tissue cardiomyoplasty using bioengineered contractile cardiomyocyte sheets to repair damaged myocardium: their integration with recipient myocardium. *Transplantation* 80(11): 1586-1595.

Miyahara, Y., N. Nagaya, et al. (2006). Monolayered mesenchymal stem cells repair scarred myocardium after myocardial infarction. *Nat Med* 12(4): 459-465.

Morritt, A. N., S. K. Bortolotto, et al. (2007). Cardiac tissue engineering in an in vivo vascularized chamber. *Circulation* 115(3): 353-360.

Oh, H., S. B. Bradfute, et al. (2003). Cardiac progenitor cells from adult myocardium: homing, differentiation, and fusion after infarction. *Proc Natl Acad Sci U S A* 100(21): 12313-12318.

Okano, T., N. Yamada, et al. (1993). A novel recovery system for cultured cells using plasma-treated polystyrene dishes grafted with poly(N-isopropylacrylamide). *J Biomed Mater Res* 27(10): 1243-1251.

Opie, S. R. and N. Dib (2006). Surgical and catheter delivery of autologous myoblasts in patients with congestive heart failure. *Nat Clin Pract Cardiovasc Med* 3 Suppl 1: S42-45.

Ott, H. C., T. S. Matthiesen, et al. (2008). Perfusion-decellularized matrix: using nature's platform to engineer a bioartificial heart. *Nat Med* 14(2): 213-221.

Papadaki, M., N. Bursac, et al. (2001). Tissue engineering of functional cardiac muscle: molecular, structural, and electrophysiological studies. A*m J Physiol Heart Circ Physiol* 280(1): H168-178.

Psaltis, P. J., A. C. Zannettino, et al. (2008). Concise review: mesenchymal stromal cells: potential for cardiovascular repair. *Stem Cells* 26(9): 2201-2210.

Puceat, M. (2008). Pharmacological approaches to regenerative strategies for the treatment of cardiovascular diseases. *Curr Opin Pharmacol* 8(2): 189-192.

Radisic, M., H. Park, et al. (2007). Biomimetic approach to cardiac tissue engineering. *Philos Trans R Soc Lond B Biol Sci* 362(1484): 1357-1368.

Sekine, H., T. Shimizu, et al. (2008). Endothelial cell coculture within tissue-engineered cardiomyocyte sheets enhances neovascularization and improves cardiac function of ischemic hearts. *Circulation* 118(14 Suppl): S145-152.

Sekine, H., T. Shimizu, et al. (2006). Cardiomyocyte bridging between hearts and bioengineered myocardial tissues with mesenchymal transition of mesothelial cells. *J Heart Lung Transplant* 25(3): 324-332.

Sekine, H., T. Shimizu, et al. (2006). Pulsatile myocardial tubes fabricated with cell sheet engineering. *Circulation* 114(1 Suppl): I87-93.

Sekiya, N., G. Matsumiya, et al. (2009). Layered implantation of myoblast sheets attenuates adverse cardiac remodeling of the infarcted heart. *J Thorac Cardiovasc Surg* 138(4): 985-993.

Sekiya, S., T. Shimizu, et al. (2006). Bioengineered cardiac cell sheet grafts have intrinsic angiogenic potential. *Biochem Biophys Res Commun* 341(2): 573-582.

Shimizu, T., H. Sekine, et al. (2006). Long-term survival and growth of pulsatile myocardial tissue grafts engineered by the layering of cardiomyocyte sheets. *Tissue Eng* 12(3): 499-507.

Shimizu, T., H. Sekine, et al. (2009). Cell sheet-based myocardial tissue engineering: new hope for damaged heart rescue. *Curr Pharm Des* 15(24): 2807-2814.

Shimizu, T., H. Sekine, et al. (2006). Polysurgery of cell sheet grafts overcomes diffusion limits to produce thick, vascularized myocardial tissues. *FASEB J* 20(6): 708-710.

Shimizu, T., M. Yamato, et al. (2002). Electrically communicating three-dimensional cardiac tissue mimic fabricated by layered cultured cardiomyocyte sheets. *J Biomed Mater Res* 60(1): 110-117.

Shimizu, T., M. Yamato, et al. (2002). Fabrication of pulsatile cardiac tissue grafts using a novel 3-dimensional cell sheet manipulation technique and temperature-responsive cell culture surfaces. *Circ Res* 90(3): e40.

Shimizu, T., M. Yamato, et al. (2003). Cell sheet engineering for myocardial tissue reconstruction. *Biomaterials* 24(13): 2309-2316.

Shin, M., O. Ishii, et al. (2004). Contractile cardiac grafts using a novel nanofibrous mesh. *Biomaterials* 25(17): 3717-3723.

Siltanen, A., K. Kitabayashi, et al. (2011). hHGF Overexpression in Myoblast Sheets Enhances Their Angiogenic Potential in Rat Chronic Heart Failure. *PLoS One* 6(4): e19161.

Tan, M. Y., W. Zhi, et al. (2009). Repair of infarcted myocardium using mesenchymal stem cell seeded small intestinal submucosa in rabbits. *Biomaterials* 30(19): 3234-3240.

Vunjak-Novakovic, G., N. Tandon, et al. (2010). Challenges in cardiac tissue engineering. *Tissue Eng Part B* Rev 16(2): 169-187.

Wollert, K. C. (2008). Cell therapy for acute myocardial infarction. *Curr Opin Pharmaco*l 8(2): 202-210.

Yildirim, Y., H. Naito, et al. (2007). Development of a biological ventricular assist device: preliminary data from a small animal model. *Circulation* 116(11 Suppl): I16-23.

Yuasa, S., Y. Itabashi, et al. (2005). Transient inhibition of BMP signaling by Noggin induces cardiomyocyte differentiation of mouse embryonic stem cells. *Nat Biotechnol* 23(5): 607-611.

Zhang, M., D. Methot, et al. (2001). Cardiomyocyte grafting for cardiac repair: graft cell death and anti-death strategies. *J Mol Cell Cardiol* 33(5): 907-921.

Zhu, W., I. Shiojima, et al. (2008). IGFBP-4 is an inhibitor of canonical Wnt signalling required for cardiogenesis. *Nature* 454(7202): 345-349.

Zimmermann, W. H. (2011). Embryonic and embryonic-like stem cells in heart muscle engineering. *J Mol Cell Cardiol* 50(2): 320-326.

Zimmermann, W. H. and R. Cesnjevar (2009). Cardiac tissue engineering: implications for pediatric heart surgery. *Pediatr Cardiol* 30(5): 716-723.

Zimmermann, W. H., M. Didie, et al. (2006). Heart muscle engineering: an update on cardiac muscle replacement therapy. *Cardiovasc Res* 71(3): 419-429.
Zimmermann, W. H., I. Melnychenko, et al. (2006). Engineered heart tissue grafts improve systolic and diastolic function in infarcted rat hearts. *Nat Med* 12(4): 452-458.
Zimmermann, W. H., K. Schneiderbanger, et al. (2002). Tissue engineering of a differentiated cardiac muscle construct. *Circ Res* 90(2): 223-230.

# 2

# Cardiac Muscle Engineering: Strategies to Deliver Stem Cells to the Damaged Site

Giancarlo Forte[1], Stefania Pagliari[2], Francesca Pagliari[2], Paolo Di Nardo[2] and Takao Aoyagi[1]
[1]*Biomaterials Center, International Center for Materials Nanoarchitectonics (MANA), National Institute for Materials Science (NIMS), Tsukuba,*
[2]*Laboratorio di Cardiologia Molecolare e Cellulare, Dipartimento di Medicina Interna, Università di "Tor Vergata", Roma,*
[1]*Japan*
[2]*Italy*

## 1. Introduction

In healthy human hearts, only 10-20% of the total cells are contractile cardiomyocytes and, at the age of 25 years, no more than 1% of them are annually substituted by progenitor cells, this percentage reducing to less than 0.5% at the age of 75. In total, less than 50% of cardiomyocytes are renewed during a normal human life span [1]. For this reason, the topic of cardiac repair is among the major challenges for the tissue engineers worldwide. In fact, cardiac diseases are a predominant cause of mortality and morbidity in industrialized countries, despite the recent advancements achieved in pharmacological treatment and interventional cardiology procedures. Nonetheless, end-stage heart failure management still relies on organ transplantation as unique approach, and, notwithstanding the use of massive immunosuppressive drugs, still a percentage falling within 20%-40% of patients encounters immune rejection during the first year post-transplant [2]. Among the patients not facing severe immune rejection, almost 70% is forced to retire or reduce their working activity, their survival rate falling below 70% during the first five years post organ transplantation [3]. Last, but not least, the economic impact of cardiovascular diseases and stroke has been estimated in 2010 at $503.2 billion [4].

Currently, post-infarction myocardial revascularization protocols include the administration of raw bone marrow stem cells, while a number of clinical trials have been performed or are currently in progress in which different cell subsets are implanted in the damaged tissue by means of surgical techniques. The results of such trials are still controversial. In fact, when autologous skeletal myoblasts were injected into the heart of patients suffering from ischemic cardiomyopathy, the modest functional improvement obtained was impaired by the arising of arrhythmia events, thus requiring the adoption of a pacemaker [5]. On the other side, intracoronary administration of bone marrow mesenchymal stem cells resulted in minimal improvements in cardiac contractile function in patients with dilated cardiomyopathy [6]. These mild results were mostly ascribed to a paracrine effect exerted on host tissue, rather than to a direct contribution of stem cells to the contractile activity.

Thus, among the criticisms to be challenged before efficient cell therapy protocols for cardiac diseases can be setup, the choice of the appropriate cell subset to generate new vessels and contractile cardiomyocytes, as well as the route of cell delivery remain key steps. The solution of such problems requires additional efforts in basic research to clarify the processes leading to stem cell differentiation as well as technological advancements to setup efficient protocols to implant the cells.

In principle, adult stem cells could be extracted from patient's own tissues and expanded in culture by means of well-known techniques (Figure 1).

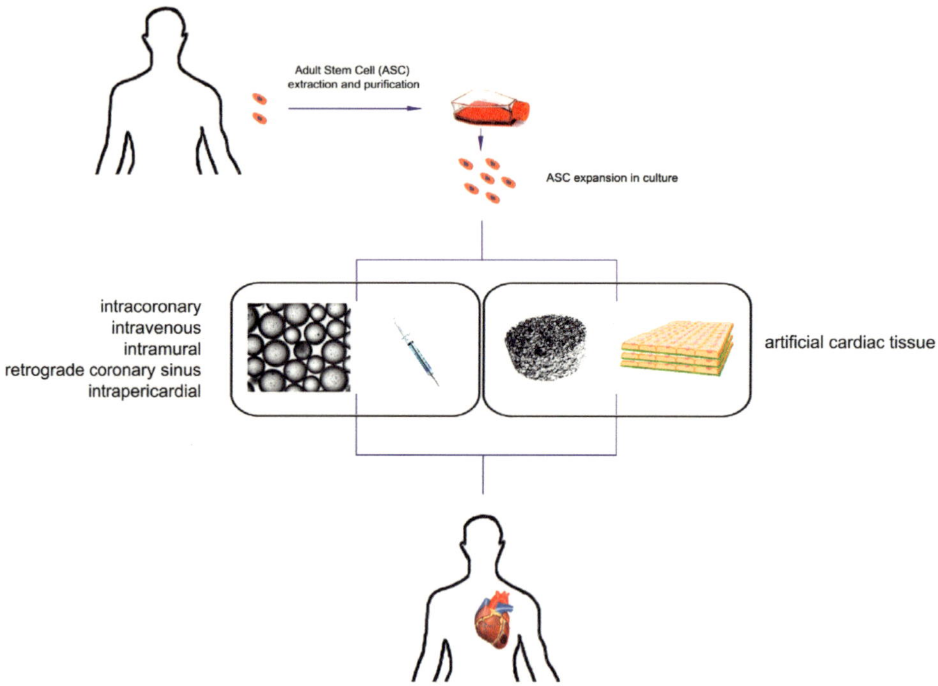


Fig. 1. Cardiac Tissue Engineering paradigm. Adult stem cells can be harvested, purified from the patient and expanded in culture. Such cells can be delivered to the injured heart by injection (intramural or through bloodstream with or without injectable carriers), or in the form of solid bio-constructs. Stem cell-derived bio-constructs can be obtained by culturing the cells on scaffolds or by scaffold-free technology

Nonetheless, a number of issues should be challenged before safe procedures to manipulate stem cells *in vitro* for cardiac transplant can be setup. In fact, stem cells should be amplified *in vitro* to reach a critical number (Figure 1). During this passage, malignant transformation is likely to occur in *ex vivo* cells when standard culture conditions are adopted to expand stem cells [7, 8]. On the other side, stem cells could encounter senescence after a short number of passages *in vitro* [9]. Moreover, the use of animal-derived supplements during the phase of cell expansion would hinder the use of stem cells for cardiac cell therapy.

The employment of autologous stem cells would avoid the problem of immune rejection and the need for immune-suppressive drugs, while, in the treatment of pathologies for which a genetic basis is suspected the use of autologous cells is hampered. As far as the use of autologous cells is concerned, the possibility that a significant patient-to-patient variability in stem cell quality exists should be taken into account [10]. Finally, the use of cellular and tissue-based products in human disease therapy is subjected to regulations issued by the European Union and Food and Drug Administration (FDA) aimed at establishing classification criteria for advanced therapy medicinal products (ATMP). In particular, the European Regulation states that human cells to be used in cell therapy have to comply with the principles of Good Manufacturing Practice (GMP) protocols [11, 12].

## 2. Adult stem cells for cardiac repair

A number of stem cells and progenitors have been so far proposed for cardiac repair, due to the inability of cardiomyocytes to proliferate after birth [1]. Among the cell sources challenged for the possibility to produce new cardiomyocytes, skeletal myoblasts have proven to be able to acquire a contractile phenotype *in vitro* [13]. Moreover, when implanted *in vivo* in a canine model of dilated cardiomyopathy (DCM), they attenuated cardiac remodeling [14]. This result is likely to be due to the fusion of skeletal myoblasts with the surrounding myocardium rather than to direct cell differentiation, as suggested by *in vitro* experiments [15]. As discussed in the following section, clinical trials demonstrated that skeletal myoblasts are not able to couple electrically with host tissue, leading to arrhythmia events [5].

The role of hematopoietic stem cells (HSC) in cardiac repair has been investigated by several research groups and their contribution to cardiac regeneration *in vivo* has been heavily debated, being the ability of HSC to transdifferentiate to other lineages still questionable. Indeed, evidence of the ability of bone marrow-derived c-kit+ HSC to help cardiac tissue healing has been given using two different approaches: c-kit+ cells were (i) either delivered to the infarcted site by intramural injection [16] or (ii) mobilized from bone marrow through growth factor administration [17]. More recently, elegant experiments compellingly clarified that HSC are not able to acquire contractile phenotype *in vivo* [18-20]. Nonetheless, a subset of bone marrow hematopoietic precursors expressing CD34 and CD133 has been proven to contain endothelial progenitors. Thus, they have been tested for revascularization protocols in hind limb ischemic animals and could be proposed for cardiac infarction therapy [21]. On the other hand, the results obtained in preliminary investigations in which another bone marrow-derived stem cell subset, mesenchymal stem cells (BM-MSC or MSC) were challenged as a candidate for cellular cardiomyoplasty, raised great enthusiasm for such a cell subpopulation. Recent studies clarified that the direct contribution of MSC to cardiac repair in terms of production of new contractile cells is minimal if any, while a paracrine effect on the diseased tissue of such cells is universally recognized [22]. Such cells are also appealing for their ability to induce a certain degree of immune tolerance [23].

The presence of a small reservoir of cardiac resident progenitor cells (CPC or CSC) has been recently demonstrated in human as well as in other mammals' heart [24]. Such tissue-resident cells participate in myocardial homeostasis and retain a limited regenerative capacity throughout organism lifespan [1]. All the subsets so far identified through the expression of stemness markers (c-kit+, Sca-1+, Islet-1+) demonstrated the ability to give birth to new contractile cells *in vitro*, while only c-kit+, Sca-1+ progenitors were shown to be involved in post-natal cardiac tissue homeostasis *in vivo* [25]. In fact, the presence of Islet-1+ cells appears to be limited to fetal life and their contribution to the endogenous program of cardiovascular repair is still unknown; on the other hand, the very low number of c-kit+ and Sca-1+ cells in the myocardium is considered the limiting factor of cardiac regeneration [26]. Furthermore, among the adult stem cells, a novel "artificial" subset can be recognized: induced pluripotent stem cells (iPSC, Figure 2). This cell type can be produced *in vitro* by transducing somatic cells with a combination of transcription factors able to induce the nuclear reprogramming of differentiated cells. These cells, which display the functional features of pluripotent embryonic stem cells, have been credited of the ability to produce new cardiomyocytes. They could thus be the source of autologous, although genetically modified, patient-specific contractile cells [27]. Moreover, the possibility to directly obtain functional cardiomyocytes by the genetic reprogramming of postnatal cardiac or dermal fibroblasts has been demonstrated [28]. Such a result was firstly obtained *in vitro* but also when the cells were transplanted into mouse hearts one day after transduction of transcription factors (GATA-4, MEF-2c, Tbx-5) known to be involved in cardiac muscle development. Nonetheless, the reprogramming and differentiation efficiency of these cells appears to be really low, thus requiring an efficient purification step before they can be implanted *in vivo*. Additionally, safety concerns due to the use of genetically modified cells and / or viral vectors remain.

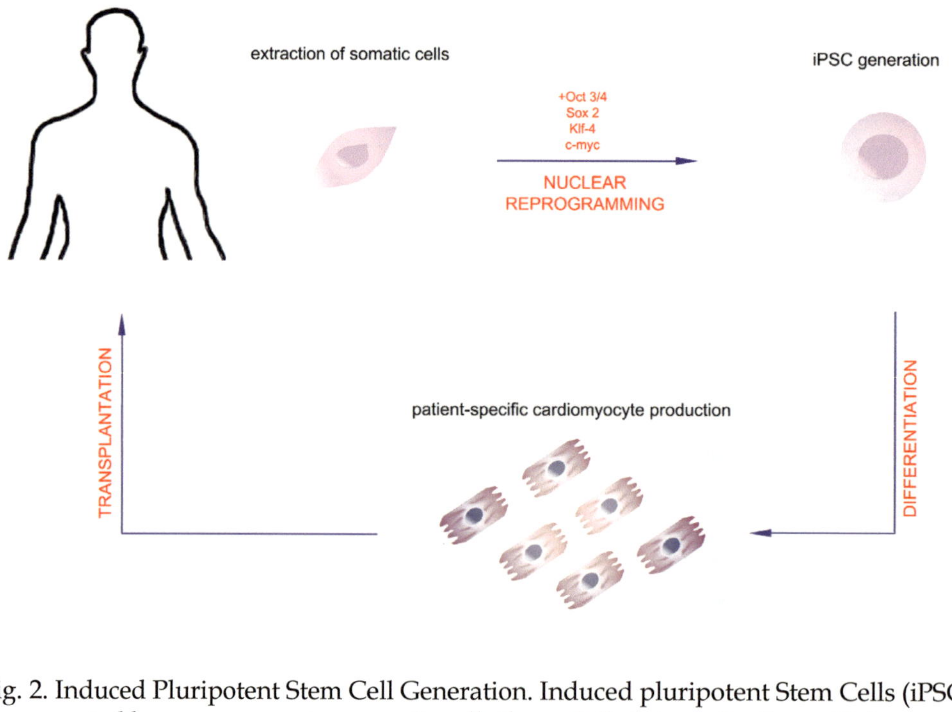


Fig. 2. Induced Pluripotent Stem Cell Generation. Induced pluripotent Stem Cells (iPSC) can be generated by reprogramming somatic cells through their transduction with four transcription factors. iPSC share functional similarities with Embryonic Stem Cells (ESC) and can be differentiated towards cardiomyocytes, thus representing an autologous source of contractile cells

## 3. Stem cell delivery to the injured heart

As previously said, cell route of delivery to damaged heart represents the major topic in the setup of efficient, minimally invasive techniques to treat cardiac pathologies. Recently, a number of techniques to deliver stem cells to the injured site have been proposed but questions remain regarding the optimal approach able to favor high cell retention, differentiation rate and clinically relevant improvement in cardiac performance.

**a) Direct injection**

Stem cell direct *intramural injection*, including trans-epicardial and trans-endocardial cell injection, is the elective strategy for patients with severe occlusion of coronary vessels. In particular, trans-epicardial approach consists in the direct injection of a high number of cells into the infarcted area or around the border zone. Endocardial stem cell injection is performed using catheters such as MyoStar™ injection catheter (Biosense Webster) integrated with imaging systems like NOGA® system (Cordis Corp., Warren, NJ, USA), which allows real-time three-dimensional reconstruction of left ventricle as well as the targeting and functional assessment of specific myocardial area [29]. Such procedures are highly invasive since they require open-heart surgery and gave contrasting results so far. For example, pre-clinical studies performed on experimental animals demonstrated that, although a certain extent of cardiac repair was achieved when bone marrow Stro-3+ perivascular cells are implanted *in vivo*, the cells vanished from the application site within few days [30]. In other reports, when Sca-1+ cardiac resident stem cells were injected in infarction border zone, a modest but significant improvement in cardiac function was reported, with evidence of cell engraftment and differentiation [31]. Finally, in another pre-clinical study, bone marrow-derived c-kit+ cells were shown to repair entire ventricular areas while massively engrafting and differentiating in contractile and vascular figures *in vivo* [32]. Of interest, independent groups already demonstrated that c-kit+ bone marrow-derived hematopoietic stem cells fail to acquire contractile phenotype when implanted in diseased myocardium [19, 20]. Such discrepancies are not surprising since different stem cell subsets or preparation protocols were probably used in these studies.

Stem cells can be delivered *intravenously* to the heart, through *coronary arteries* or even through *retrograde coronary sinus*. The major drawback of stem cells being infused through peripheral venous system seems to be the low retention of cells into infarcted area. Results obtained in pre-clinical animal models showed that this minimally invasive approach results in a significant percentage of injected cells being sequestered in lungs, liver or spleen, due to blood flow [33]. On the other hand, intracoronary or retrograde coronary sinus infusion of the cells are mainly performed after acute myocardial infarction using an angioplasty balloon and high pressure to deliver cells to heart muscle [34]. The coronary route was proven to be free of stem cell systemic delivery, while a limited number of cells could be found in the infarcted area [35].

Finally, an interesting attempt with stem cells being injected into the *pericardial cavity* has been proposed. By this means, a higher number of cells could be deposited and retained in the pericardial cavity, while migration across the visceral pericardium is required (Table 1).

| DELIVERY METHOD | ADVANTAGES | DRAWBACKS |
|---|---|---|
| intravenous | No invasive technique | Cells can be sequestered in lung, liver, spleen |
| intracoronary | No risk of systemic delivery<br>Direct delivery to the target site | Few cells delivered |
| intramyocardial | Direct delivery | Risk of perforation |
| retrograde coronary sinus | Homogeneous cell delivery | Endothelial wall transmigration required |
| intrapericardial | Large number of cells delivered | Visceral pericardium transmigration required |

Table 1. Advantages and disadvantages of injecting stem cells by intravenous, intracoronary, intramyocardial, retrograde coronary sinus or intra-pericardial route

**b) Injectable scaffolds**

Injectable scaffolds are defined as materials offering the unique solution of replacing damaged myocardial ECM and/or delivering cells directly to the infarcted region while holding the potential for minimally invasive delivery [36]. Such scaffolds can be composed of biocompatible microspheres or in situ gelling materials having reasonable dimensions as to surpass capillary barrier. They are considered a promising tool for stem cell delivery to damaged myocardium. In situ gelling materials are generally made of components of extracellular matrix (ECM), which are induced to a transition after being implanted *in situ*. Complex injectable gelling materials have been prepared by decellularization technique out of ventricular or epicardial ECM, thus possibly avoiding animal-derived components and paving the way to the definition of patient-specific treatments.

The use of injectable, synthetic microspheres has already been proven promising in the treatment of neurological diseases *in vivo* [37]. Recently the possibility of using injectable scaffolds in cardiac cell therapy has been explored by interfacing murine mesenchymal (mMSC) and cardiac stem cell (mCSC) lines with poly-lactic acid (PLA) microspheres having a diameter of 30 and 100 μm. Preliminary *in vitro* experiments demonstrated that such cells can be grown onto PLA microspheres while preserving their phenotype, but the formation of cell clumps can hamper the application of this technique [38]. The use of dynamic seeding techniques (i.e. bioreactors) would favor a more homogeneous distribution of the cells. An interesting approach has been recently proposed to deliver human mesenchymal stem cells to the injured myocardium: RGD-modified alginate microsphere (diameter: 200-700 μm) encapsulation of hMSC was setup. *In vitro* experiments showed that hMSC could survive, proliferate and migrate through the porous material. When intramyocardially injected in a rat model of myocardial infarction by left anterior descendant coronary (LAD) ligation, cell-loaded alginate microspheres promoted angiogenesis and prevented LV negative remodeling [39]. Nonetheless, few human cells were found in the injection area after few days, while microbead remains were still present within host myocardium 10 weeks after the injection. The aspect of microbead resorption should thus be addressed before clinical perspectives could be foreseen.

**c) Scaffold-based technology**

The possibility of using biocompatible scaffolds to deliver stem cells to the injured heart has been explored by a number of independent research groups so far. The scaffolds proposed are natural of synthetic but when designing cardiac-specific constructs, a number of requirements should be fulfilled. For example, it cannot be neglected that myocardial contractile function relies on the transmission of electrical and mechanical forces throughout a functional syncytium. So, the integrity of the tissue has to be preserved. For this reason, a cardiac-specific scaffold should comply with tissue architecture and thus be deformable enough to indulge and, if possible sustain cardiac contraction. Moreover, as far as stem cell engraftment is concerned, scaffolds should be able to start at least cell alignment and commitment to favor stem cell electromechanical coupling with host tissue. In this respect, the work of Mandoli and collaborators using Cerium Oxyde nanoparticles to affect poly-lactic acid film surface and obtain a controlled nanorugosity appears intriguing [40]. In fact, far from being a noxious compound for stem cells, ceria was able to induce cardiac stem cell alignment and growth. Nonetheless, cardiac tissue is extremely complex and highly demanding in terms of blood supply and catabolite removal, so that porous scaffolds that could allow microvascular branches formation and oxygen perfusion are to be preferred. To fulfill such requirements, the first attempts were performed by the group of Thomas Eschenhagen. Neonatal cardiomyocytes were seeded in Collagen I + Matrigel to produce Engineered Heart Tissue (EHT). Continuous contractile activity up to 1 week *in vitro* as well as cell survival and integration *in vivo* in singenic rat hearts were reported [41]. In another attempt, anisotropic accordion-like honeycomb scaffolds were prepared by excimer laser microablation using poly(glycerol sebacate) as an elastomeric tool to mimic anisotropic cardiac muscle stiffness distribution [42]. Although the authors demonstrated that such scaffolds promote neonatal rat cardiomyocyte alignment and contraction, *in vivo* testing has not been performed so far. The same material has been utilized to produce elastomeric patches on which human embryonic stem cell-derived cardiomyocytes were grown, showing that it is indeed possible to observe spontaneous beating activity *in vitro* up to 3 months [43]. Such patches were shown to be suitable as delivery systems and, when sutured in the absence of cells onto healthy rat left ventricle, they did not affect cardiac contractile activity. More basic studies were also conducted to study the ability of stem cells to interface with different synthetic and natural materials. In this respect, few research groups focused on the possibility to drive a certain extent of stem cell commitment through tailoring scaffold physical and chemical properties, independently of biological cues. In this respect, a common agreement on the ability of stem cells to sense substrate rugosity and elasticity has been reached [44]. Thus, in order to rule out the occurrence of spontaneous events of differentiation in implanted cells, the possibility to induce *in vitro* stem cell commitment on scaffolds towards a desired phenotype is being investigated. Indeed, Engler and collaborators compellingly demonstrated that the possibility to affect stem cell fate determination by simply tuning substrate elasticity as to match tissue-specific stiffness, exists. Recently, this concept has been corroborated by other research groups, showing that cardiac resident progenitors (Sca-1+ CPC) can be committed to cardiac phenotype by the physico-chemical signals arising from matrix, but biological factors are needed to complete the differentiation process [45, 46].

**d) Preparation of thick cardiac substitutes by Scaffold-free technology**

To overcome the problem of poor cell retention reported in cell injection experiments in the heart [30] and avoid the release of possibly harmful scaffold byproducts, scaffold-free technology has been developed, in which cells are grown in a monolayer onto thermo-responsive surfaces and easily detached in the form of cell sheet by lowering the temperature [47]. Such technology takes advantage of the ability of polymers like poly-N-isopropylacrylamide (PNIPAAm) to shift between hydrophobic and hydrophilic status when the temperature ranges from 37°C to 32°C. Cell sheets can be serially stacked to obtain multilayered scaffoldless constructs (Figure 3). Such an approach has already been applied to obtain cell sheets composed of rodent [48, 49] and human [50] cells. Given the need for thick cardiac substitutes suited to comply with cardiac muscle continuous contractility, thermo-responsive technology has been envisaged as a possible answer to the lack of heart donors. Pre-clinical trials performed onto experimentally infarcted animals demonstrated that when a murine adipose-derived monolayer sheet is leant onto injured myocardium, it can be retained and help tissue repair [48]. Similarly, striking results are obtained when a Sca-1+ cardiac progenitor cell-derived sheet is used [49]. Finally, an interesting approach has been recently proposed to deliver cardiac stem cells cultured in the form of cardiospheres to the injured heart: cardiospheres were embedded into a cardiac stromal cell-derived sheet obtained by using poly-lysine/ collagen IV-coated dishes [51]. The formation of mature vessels as well as new cardiomyocytes *in vivo* was reported after 3 weeks.

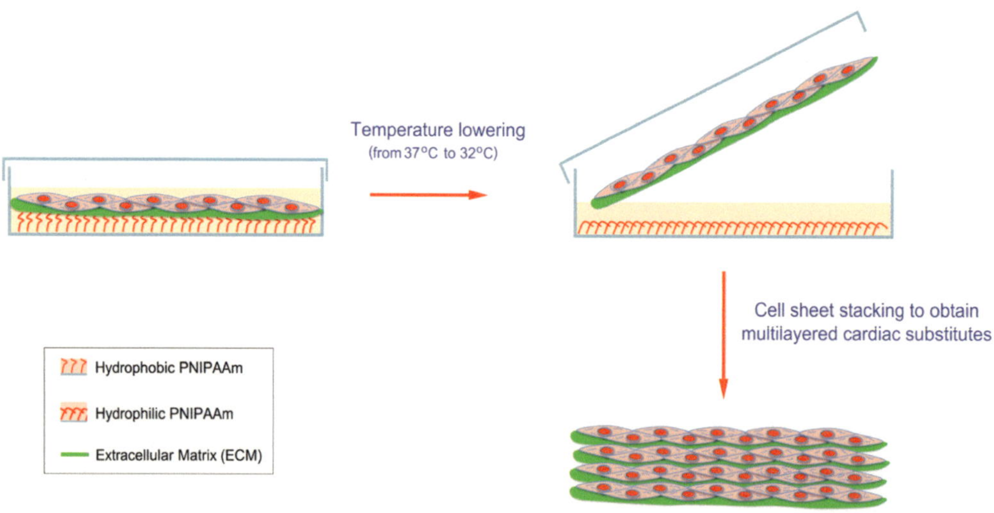


Fig. 3. Generation of scaffoldless multilayered bio-constructs by means of thermo-responsive technology: cells grown in a monolayer onto thermo-responsive poly-N-isopropilacrylamide (PNIPAAm)-coated dishes can be detached by lowering the temperature below 32°C. At 37°C the surface is highly hydrophobic and allows cell adhesion. When the temperature is lowered, PNIPAAm becomes hydrophilic, the cell sheet is detached and extracellular matrix (ECM) preserved. Multilayered cell sheets can be obtained by serially stacking monolayered sheets

## 4. Clinical trials

In the attempt to transfer bench experience to bedside, a number of clinical trials in which different stem cell or progenitor subsets are used have been approved (see http://www.clinicaltrials.gov). Most of them are still in the recruitment phase while some already gave indications and preliminary results. Since most of the ongoing trials are based on the injection of raw stem cell preparations (mostly bone marrow-derived cells), the time and route of cell application remain the key problems to be addressed before proceeding to routine clinical practice. In this respect, recent animal experiments demonstrated that the acute phase of myocardial infarction is probably not suitable for stem cell engraftment and differentiation [52]. Therefore, the right moment in which stem cells should be delivered is to be studied. An overview on some of the ongoing clinical trials is given below.

1. **MAGIC (Myoblast Autologous Grafting in Ischemic Cardiomyopathy).** In one of the first phase II clinical trials setup to study the possibility to use stem cells to treat cardiac pathologies, ninety-seven (97) patients undergoing coronary artery bypass grafting (CABG) were enrolled. 400-800 X $10^6$ autologous myoblasts harvested from patient muscle biopsy were implanted in the akinetic area of ventricular wall 21 days after *in vitro* culture. The follow-up after 30 days and 6 months demonstrated the arising of arrhythmia events, thus requiring the implantation of pacemaker. Moreover, no cardiac function improvement was reported. Such negative results were ascribed to the inability of skeletal myoblasts to balance cell death and achieve complete electromechanical integration with the recipient myocardium. Finally, skeletal myoblast administration was reported to determine no enhancement in major cardiac adverse events and mild effects on left ventricular remodeling process [53, 54]. More recently, final results from **SEISMIC** [Safety and Effects of Implanted (Autologous) Skeletal Myoblasts (MyoCell) Using an Injection Catheter] Trial, a phase II-a study encompassing 40 patients experiencing congestive heart failure and receiving percutaneous intramyocardial injection of autologous skeletal myoblasts, reported the feasibility and safety of this procedure without significant arrhythmogenic events recorded at 6-month follow-up with respect to control groups, although left ventricular ejection fraction did not result significantly improve. These encouraging results suggest that myoblast cell therapy could be considered as a potential effective treatment when associated with standard medical therapy in patients with previously implanted cardiac defibrillators [55].
2. **TOPCARE-CHD, -AMI, -DCM (Transplantation of Progenitor Cells and Regeneration Enhancement in Acute Myocardial Infarction, Chronic Stable Ischemic Heart disease or Dilated Cardiomyopathy)**. In this complex clinical trial, a total of 346 patients were classified to CHD, AMI or DCM pathologies and infused either with bone marrow cells (BMCs), blood-derived stem cells, or no infusion. In TOPCARE-CHD, 121 patients (mean age: 59) with chronic stable ischemic heart disease (CHD) were treated. Although complications occurred in 21% of the patients during 3 months follow-up, BMC intracoronary administration was related with a reduction of both brain and atrial natriuretic peptide (NTP) serum levels (indicators of LV remodelling process) in the remaining population (79%), especially in patients with higher NTP levels at baseline and receiving a greater BMC number with a high functional capacity. Moreover, these results were also correlated with a left ventricular ejection fraction (LVEF) increase and better survival during the further follow-up, suggesting that cell therapy could be

associated with cardiac function enhancements in patients with advanced chronic post-infarction heart failure [56]. Similarly, two hundred and four (204) patients were treated using bone-marrow-derived progenitor cells directly into the infarct artery three to seven days after an acute myocardial infarction (AMI). A statistically significant 2.5% improvement in left ventricular ejection fraction at four months was reported for patients randomized to the bone marrow injection [57]. Finally, intracoronary infusion of bone marrow cells was performed in 33 patients with dilated cardiomyopathy (DCM) by using an over-the-wire balloon catheter. Three month follow-up demonstrated an improvement in left ventricular pump function while a modest improvement in Brain Natriuretic Peptide (BNP) levels was reported after 1 year [6]. Importantly, the conditions chosen in the present clinical trial were representative of different conditions (acute, chronic phase) encountered in the clinic. Unfortunately, no clear indication on stem cell characterization or on their actual ability to regenerate contractile cells is available.

3. **TRACIA STUDY (Intracoronary Autologous Stem Cell Transplantation in ST Elevation Myocardial Infarction).** The phase II/ III clinical trial aimed at evaluating the effects of intracoronary administration of adult stem cells on LV ejection fraction and major adverse cardiovascular events (MACE) after 6 months follow-up. For this reason, 1-2 million CD34+ cells were injected through the infarct-related artery few days after post-infarct angioplasty using an "over-the-wire" catheter in 80 patients aging from 20 to 75 years. The results of this study are still to be published.
4. **Combined CABG and Stem-Cell Transplantation for Heart Failure.** Intramyocardial delivery of autologous bone marrow cells extracted from iliac crest and purified by Ficoll centrifugation, during cardiac surgery for CABG intervention in 30 patients, as compared to 30 patients undergoing CABG without cell infusion. Although information on the number and characteristics of cells to be injected has not been given, the trial is currently ongoing and the follow-up is scheduled in 6-12 months (http://clinicaltrials.gov).
5. **POSEIDON-Pilot Study (The Percutaneous Stem Cell Injection Delivery Effects on Neomyogenesis Pilot Study)** Poseidon-pilot Study is a phase I/ II multi-center trial in which the trans-endocardial injection of autologous Mesenchymal Stem Cells (20-, 100-, 200 X $10^6$) is compared to autologous non-purified bone marrow cells and to allogeneic human Mesenchymal Stem Cells. The implant is performed during cardiac catheterization using the Biocardia Helical Infusion Catheter in fifty (50) patients suffering from chronic ischemic left ventricular dysfunction secondary to myocardial infarction. The data collection is currently ongoing.
6. **SCIPIO (Cardiac Stem Cell Infusion in Patients With Ischemic CardiOmyopathy).** This phase I clinical trial is aimed at assessing the safety and effectiveness of intracoronary autologous cardiac stem cell therapy. As such, forty (40) patients suffering from ischemic cardiomyopathy are exposed to intracoronary injection of cardiac resident stem cells (CSC). Cardiac stem cells are harvested from right atrial appendages and selected for c-kit expression, cultured and expanded in vitro prior to injecting them via intracoronary route, three to five months after CABG surgery. The hypothesis is that CSC infused into nonviable myocardial segments will regenerate infarcted myocardium by differentiating into cardiomyocytes and vascular cells. The preliminary results are encouraging: in the nine patients treated at four months after

CSC infusion, LVEF increased from 31.3 + 2.5 percent before CSC infusion to 38.8 + 3.2 percent four months after CSC infusion. Moreover, in the five patients in whom data are available at 12 months after stem cell infusion, the improvement in LVEF observed at four months was even greater, averaging 15% at 12 months. The follow-up is scheduled in 1,5 years.

7. **ALCADIA (AutoLogous Human CArdiac-Derived Stem Cell to Treat Ischemic cArdiomyopathy).** In this phase I, multicenter clinical trial, a rather different approach is followed. In fact, patients' own cardiac stem cells obtained by endo-myocardial biopsies are delivered by a single intramyocardial injection. The cells injected are 0.5 million cells/kg (patient body weight) and their engraftment should be favored by the concomitant implantation of gelatin hydrogel sheet releasing human recombinant beta Fibroblast Growth Factor (bFGF), during CABG surgery. The study has been designed to treat refractory heart failure, ischemic cardiomyopathy or ventricular dysfunction cases. Importantly, this is the first clinical trial, to our knowledge, in which a human recombinant growth factor is used. Unfortunately, the number of enrolled patients is limited to six (6).
8. **REGEN-IHD (Bone Marrow Derived Adult Stem Cells for Chronic Heart Failure).** In this phase II/ III study, granulocyte-colony stimulating factor (G-CSF) is subcutaneously administered for 5 days to patients with heart failure secondary to ischemic heart disease to mobilize CD34+ bone marrow stem cells. A concomitant intracoronary or intramyocardial administration of bone marrow derived stem cells is performed. The number of enrolled patients is high (165) and the aim of the study is to compare the effects of G-CSF and autologous bone marrow progenitor cell infusion on the quality of life and left ventricular function in the patients. The follow-up timepoint is scheduled in 6-12 months.

A number of papers reporting statistical analyses and comparisons among the clinical trials in which stem and progenitor cells have been adopted are currently available. [For further information, please refer to www.clinicaltrials.gov].

## 5. Conclusions

The possibility to treat cardiac diseases by cell therapy techniques is an extraordinary promise. While a number of different approaches has been so far proposed to setup minimally invasive techniques for cardiac repair, few of them being already in the clinical experimental phase, basic questions still need to be addressed. In fact, the molecular processes leading to cardiac differentiation still need to be fully clarified, while the impact of novel, genetically modified cell types obtained from adult differentiated cells on cardiac microenvironment deserve further investigations. More importantly, the seek to identify suitable delivery systems (i.e. scaffolds) able to foster stem cell survival, growth and differentiation, while degrading without negative effects as the formation of new tissue occurs is still open. A look at the literature reveals that an impressive effort to translate the information obtained by *in vitro* and pre-clinical studies to the bedside is being produced. In particular, a number of stem cell subsets, which have been previously tested *in vitro* and in animal models, are currently being tested in phase I, II clinical trials. As expected, the predominant delivery system used in the ongoing clinical trials is intracoronary or intramural injection of stem cells. The possibility to adopt tissue engineering techniques to design patient-specific cardiac substitutes containing synthetic or natural scaffolds is still far from being taken into consideration for clinical application, since any single formulation will have to be approved before clinical testing.

## 6. References


[1] Bergmann O, Bhardwaj RD, Bernard S, Zdunek S, Barnabé-Heider F, Walsh S, Zupicich J, Alkass K, Buchholz BA, Druid H, Jovinge S, Frisén J. Evidence for Cardiomyocyte Renewal in Humans. *Science* 2009: 324; 98-102.

[2] Patel JK, Kobashigawa JA. Should we be doing routine biopsy after heart transplantation in a new era of anti-rejection? Curr Opin Cardiol 2006; 21: 127-131.

[3] Hertz MI, Aurora P, Christie JD, Dobbels F, Edwards LB, Kirk R, Kucheryavaya AY, Rahmel AO, Rowe AW, Stehlik J, Taylor DO. Scientific Registry of the International Society for Heart and Lung Transplantation. *J Heart Lung Transplant* 2009: 28; 989-1049.

[4] American Heart Association. Heart disease and stroke statistics-2010 update. Dallas, Texas: *American Heart Association*; 2010. © 2010, American Heart Association.

[5] Menasché P, Alfieri O, Janssens S, McKenna W, Reichenspurner H, Trinquart L, Vilquin JT, Marolleau JP, Seymour B, Larghero J, Lake S, Chatellier G, Solomon S, Desnos M, Hagège AA. The Myoblast Autologous Grafting in Ischemic Cardiomyopathy (MAGIC) trial: first randomized placebo-controlled study of myoblast transplantation. *Circulation*. 2008; 117: 1189-1200.

[6] Fischer-Rasokat U, Assmus B, Assmus B, Seeger FH, Honold J, Leistner D, Fichtlscherer S, Schächinger V, Tonn T, Martin H, Dimmeler S, Zeiher MA. A pilot trial to assess potential effects of selective intracoronary bone marrow-derived progenitor cell infusion in patients with nonischemic dilated cardiomyopathy: final 1-year results of the transplantation of progenitor cells and functional regeneration enhancement pilot trial in patients with nonischemic dilated cardiomyopathy. *Circ Heart Fail* 2009; 2: 417-423.

[7] Foudah D, Redaelli S, Donzelli E, Bentivegna A, Miloso M, Dalprà L, Tredici G. Monitoring the genomic stability of in vitro cultured rat bone-marrow-derived mesenchymal stem cells. *Chromosome Res*. 2009; 17: 1025-1039.

[8] Momin EN, Vela G, Zaidi HA, Quiñones-Hinojosa A. The Oncogenic Potential of Mesenchymal Stem Cells in the Treatment of Cancer: Directions for Future Research. *Curr Immunol Rev*. 2010; 6:137-148

[9] Vacanti V, Kong E, Suzuki G, Sato K, Canty JM, Lee T. Phenotypic changes of adult porcine mesenchymal stem cells induced by prolonged passaging in culture. *J Cell Physiol* 2005: 194-201.

[10] Itzhaki-Alfia A, Leor J, Raanani E, Sternik L, Spiegelstein D, Netser S, Holbova R, Pevsner-Fischer M, Lavee J, Barbash IM. Patient characteristics and cell source determine the number of isolated human cardiac progenitor cells. *Circulation* 2009; 120: 2559-2566.

[11] Regulation (EC) No 1394/2007 of the European Parliament and of the Council of 13 November 2007 on advanced therapy medicinal products and amending Directive 2001/83/EC and Regulation (EC) No 726/2004.

[12] Food and Drug Administration 21 CFR 1271 (2006).

[13] Formigli L, Francini F, Tani A, Squecco R, Nosi D, Polidori L, Nistri S, Chiappini L, Cesati V, Pacini A, Perna AM, Orlandini GE, Zecchi Orlandini S, Bani D. Morphofunctional integration between skeletal myoblasts and adult cardiomyocytes in coculture is favoured by direct cell-cell contacts and relaxin treatment. *Am J Physiol Cell Physiol* 2005; 288: C795-804.

[14] Hata H, Matsumiya G, Miyagawa S, Kondoh H, Kawaguchi N, Matsuura N, Shimizu T, Okano T, Matsuda H, Sawa H. Grafted skeletal myoblasts sheets attenuate myocardial remodelling in pacing-induced canine heart failure model. *J Thorac Cardiovasc Surg* 2009; 138: 460-467.

[15] Reinecke H, Minami E, Poppa V, Murry CE. Evidence for fusion between cardiac and skeletal muscle cells. *Circ Res* 2004; 94: e56-e60.

[16] Orlic D, Kajstura J, Chimenti S, Jakoniuk I, Anderson SM, Li B, Pickel J, McKay R, Nadal-Ginard B, Bodine DM, Leri A, Anversa P. Bone marrow cells regenerate infarcted myocardium. *Nature* 2001; 410: 221-229.

[17] Orlic D, Kajstura J, Chimenti S, Limana F, Jakoniuk I, Quaini F, Nadal-Ginard B, Bodine DM, Leri A, Anversa P. Mobilized bone marrow cells repair the infarcted heart, improving function and survival. *Proc Natl Acad USA* 2001; 98: 10344-10349

[18] Wagers AJ, Sherwood RI, Christensen JL, Weissman IL. Little evidence for developmental plasticity of adult hematopoietic stem cells. *Science* 2002; 297: 2256-2259.

[19] Murry CE, Soonpaa MH, Reinecke H, Nakajima H, Nakajima HO, Rubart M, Pasumarthi KB, Virag JI, Bartelmez SH, Poppa V, Bradford G, Dowell JD, Williams DA, Field LJ. Haematopoietic stem cells do not transdifferentiate into cardiac myocytes in myocardial infarcts. *Nature.* 2004; 428: 664-668.

[20] Balsam LB, Wagers AJ, Christensen JL, Kofidis T, Weissman IL, Robbins RC. Haematopoietic stem cells adopt mature haematopoietic fates in ishaemic myocardium. *Nature* 2004; 428: 668-673.

[21] Asahara T, Murohara T, Sullivan A, Silver M, van der Zee R, Li T, Witzenbichler B, Schatteman G, Isner JM. Isolation of putative progenitor cells for angiogenesis. *Science* 1997; 275: 964-967.

[22] Nesselmann C, Ma N, Bieback K, Wagner W, Ho A, Konttinen YT, Zhang H, Hinescu ME, Steinhoff G. Mesenchymal stem cells and cardiac repair. *J Cell Mol Med* 2008; 12: 1795-1810.

[23] Amado L, Saliaris A, Schuleri K, St. John M, Xie JS, Cattaneo S, Durand DJ, Fitton T, Kuang JQ, Stewart G, Lehrke S, Baumgartner WW, Martin BJ, Heldman AW, Hare JM. Cardiac repair with intramyocardial injection of allogenic mesenchymal stem cells after myocardial infarction. *Proc Natl Acad Sci USA* 2005; 102: 11474-11479.

[24] Quaini, F., Urbanek, K., Beltrami, AP, Finato, N, Beltrami, CA, Nadal-Ginard, B, Kajstura, J, Leri, A, and Anversa, P. Chimerism of the transplanted heart. *N Engl J Med* 2002; 346: 5-15.

[25] Bearzi C, Rota M, Hosoda T, Tillmanns J, Nascimbene A, De Angelis A, Yasuzawa-Amano S, Trofimova I, Siggins RW, Lecapitaine N, Cascapera S, Beltrami AP, D'Alessandro DA, Zias E, Quaini F, Urbanek K, Michler RE, Bolli R, Kajstura J, Leri A, Anversa P. Human cardiac stem cells. *Proc Natl Acad Sci U S A.* 2007; 104: 14068 - 14073.

[26] Di Nardo P, Forte G, Ahluwalia A, Minieri M. Cardiac progenitor cells: Potency and control. *J. Cell. Physiol.* 2010; 224: 590–600.
[27] Yamanaka S, Blau HM. Nuclear reprogramming to a pluripotent state by three approaches. *Nature* 2010; 465: 704-712.
[28] Ieda M, Fu JD, Delgado-Olguin P, Vedantham V, Hayashi Y, Bruneau BG, Srivastava D. Direct reprogramming of fibroblasts into functional cardiomyocytes by defined factors. *Cell.* 2010; 142: 375-386.
[29] Perin EC and Lopez J. Methods in stem cell delivery in cardiac diseases. *Nat Clinic Pract Cardiovasc Med* 2006: 3 S1.
[30] Dixon JA, Gorman RC, Stroud RE, Bouges S, Hirotsugu H, Gorman JH 3rd, Martens TP, Itescu S, Schuster MD, Plappert T, St John-Sutton MG, Spinale FG. Mesenchymal cell transplantation and myocardial remodeling after myocardial infarction. *Circulation.* 2009; 120: S220-S229.
[31] Smits AM, van Vliet P, Metz CH, Korfage T, Sluijter JPG, Doevendans PA, Goumans MJ. Human cardiomyocyte progenitor cells differentiate into functional mature cardiomyocytes: an in vitro model for studying human cardiac physiology and pathophysiology. *Nat Protoc* 2009; 4: 232-243.
[32] Rota M, Kajstura J, Hosoda T, Bearzi C, Vitale S, Esposito G, Iaffaldano G, Padin-Iruegas ME, Gonzalez A, Rizzi R, Small N, Muraski J, Alvarez R, Chen X, Urbanek K, Bolli R, Houser SR, Leri A, Sussman MA, Anversa P: Bone marrow cells adopt the cardiomyogenic fate in vivo. *Proc Natl Acad Sci U S A* 2007, 104: 17783-17788.
[33] Gao J, Dennis JE, Muzic RF, Lundberg M, Caplan L. The dynamic *in vivo* distribution of bone marrow-derived mesenchymal stem cells after infusion. *Cells Tissues Organs* 2001; 169: 12-20.
[34] Bui QT, Gertz ZM, Wilensky RL. Intracoronary delivery of bone-marrow-derived stem cells. *Stem Cell Res Ther.* 2010; 1:29-35.
[35] Bartunek J, Vanderheyden M, Vandekerckhove B, Mansour S, De Bruyne B, De Bondt P, Van Haute I, Lootens N, Heyndrickx G, Wijns W. Intracoronary injection of CD133-positive enriched bone marrow progenitor cells promotes cardiac recovery after recent myocardial infarction: feasibility and safety. *Circulation* 2005; 112: 178–183.
[36] Singelyn JM, Christman KL. Injectable materials for the treatment of myocardial infarction and heart failure: the promise of decellularized matrices. *J Cardiovasc Transl Res.* 2010; 3: 478-486.
[37] Menei P, Montero-Menei C, Venier MC, Benoit JP. Drug delivery into the brain using poly(lactide-co-glycolide) microspheres. *Expert Opin Drug Deliv.* 2005 2: 363-376.
[38] Forte G, Franzese O, Pagliari S, Pagliari F, Cossa P, Laudisi A, Di Francesco AM, Fiaccavento R, Carotenuto F, Bonmassar E, Fiaccavento R, Minieri M, Di Nardo P. Interfacing Sca-1$^{pos}$ Mesenchymal Stem Cells with Biocompatible Scaffolds with Different Chemical Composition and Geometry. *J Biomed Biotechnol* 2009; doi: 10.1155/2009/910610.
[39] Yu J, Du KT, Fang Q, Gu Y, Mihardja SS, Sievers RE, Wu JC, Lee RJ. The use of human mesenchymal stem cells encapsulated in RGD modified alginate microspheres in the repair of myocardial infarction in the rat. *Biomaterials.* 2010; 31: 7012-7020
[40] Mandoli C, Pagliari F, Pagliari S, Forte G, Di Nardo P, Licoccia S, Traversa E. Stem cell aligned growth induced by CeO2 nanoparticles in PLGA scaffolds with improved bioactivity for regenerative medicine. *Adv Funct Mater* 2010; 20: 1617-1624.

[41] Zimmermann WH, Melnychenko I, Wasmeier G, Didié M, Naito H, Nixdorff U, Hess A, Budinsky L, Brune K, Michaelis B, Dhein S, Schwoerer A, Ehmke H, Eschenhagen T. Engineered heart tissue grafts improve systolic and diastolic function in infarcted rat hearts. *Nat Med.* 2006; 12: 452-458.

[42] Engelmayr GC Jr, Cheng M, Bettinger CJ, Borenstein JT, Langer R, Freed LE. Accordion-like honeycombs for tissue engineering of cardiac anisotropy. *Nat Mater.* 2008; 7: 1003-1010.

[43] Chen QZ, Ishii H, Thouas GA, Lyon AR, Wright JS, Blaker JJ, Chrzanowski W, Boccaccini AR, Ali NN, Knowles JC, Harding SE. An elastomeric patch derived from poly(glycerol sebacate) for delivery of embryonic stem cells to the heart. *Biomaterials.* 2010; 31: 3885-3893.

[44] Engler AJ, Sen S, Sweeney HL, Discher DE. Matrix elasticity directs stem cell lineage specification. *Cell.* 2006; 126: 677-689.

[45] Forte G, Carotenuto F, Pagliari F, Pagliari S, Cossa P, Fiaccavento R, Ahluwalia A, Vozzi G, Vinci B, Serafino A, Rinaldi A, Traversa E, Carosella L, Minieri M, Di Nardo P. Criticatility of the biological and physical stimuli array inducing resident stem cell determination. *Stem Cells* 2008; 26: 2093-2103.

[46] Pagliari S, Vilela-Silva AC, Forte G, Pagliari F, Mandoli C, Vozzi G, Pietronave S, Prat M, Licoccia S, Ahluwalia A, Traversa E, Minieri M, Di Nardo P. Cooperation of Biological and Mechanical Signals in Cardiac Progenitor Cell Differentiation. *Adv Mater* 2010; 23: 514-518.

[47] Masuda S, Shimizu T, Yamato M, Okano T. Cell sheet engineering for heart tissue repair. *Adv Drug Deliv Rev.* 2008; 60: 277-285.

[48] Miyahara Y, Nagaya N, Kataoka M, Yanagawa B, Tanaka K, Hao H, Ishino, Ishida H, Shimizu T, Kangawa K, Sano S, Okano T, Kitamura S, Mori H. Monolayered mesenchymal stem cells repair scarred myocardium after myocardial infarction. *Nat Med* 2006; 12: 459-465.

[49] Matsuura K, Honda A, Nagai T, Fukushima N, Iwanaga K, Tokunaga M, Shimizu T, Okano T, Kasanuki H, Hagiwara N, Komuro I. Transplantation of cardiac progenitor cells ameliorates cardiac dysfunction after myocardial infarction in mice. *J Clin Invest.* 2009; 119: 2204–2217.

[50] Arauchi A, Shimizu T, Yamato M, Obara T, Okano T. Tissue-engineered thyroid cell sheet rescued hypothyroidism in rat models after receiving total thyroidectomy comparing with nontransplantation models. *Tissue Eng* Part A. 2009; 15: 3943-3949.

[51] Zakharova L, Mastroeni D, Mutlu N, Molina M, Goldman S, Diethrich E, Gaballa MA. Transplantation of cardiac progenitor cell sheet onto infarcted heart promotes cardiogenesis and improves function. *Cardiovasc Res.* 2010; 87: 40-49.

[52] Chen YR, Li Y, Chen L, Yang XC, Su PX, Cai J. The infarcted myocardium does not selectively promote embryonic stem cell differentiation into cardiomyocytes. *Cardiovasc. Pathol.* 2010; doi:10.1016/j.carpath.2009.12.003.

[53] Menasché P. Stem cell therapy for heart failure: are arrhythmias a real safety concern? *Circulation.* 2009; 119:2735-2740.

[54] Zenovich AG, Davis BH, Taylor DA. Comparison of intracardiac cell transplantation: autologous skeletal myoblasts versus bone marrow cells. *Handb Exp Pharmacol.* 2007; 180: 117-165.

[55] Duckers HJ, Houtgraaf J, Hehrlein C, Schofer J, Waltenberger J, Gershlick A, Bartunek J, Nienaber C, Macaya C, Peters N, Smits P, Siminiak T, van Mieghem W, Legrand V, Serruys PW. Final results of a phase IIa, randomised, open-label trial to evaluate the percutaneous intramyocardial transplantation of autologous skeletal myoblasts in congestive heart failure patients: the SEISMIC trial. *EuroIntervention.* 2011; 6: 805-812.

[56] Assmus B, Fischer-Rasokat U, Honold J, Seeger FH, Fichtlscherer S, Tonn T, Seilfried E, Schaechinger V, Dimmeller S, Zeiher AM; TOPCARE-CHD Registry. Transcoronary transplantation of functionally competent BMCs is associated with a decrease in natriuretic peptide serum levels and improved survival of patients with chronic postinfarction heart failure: results of the TOPCARE-CHD Registry. *Circ Res* 2007; 100: 1234-1241.

[57] Schaechinger V, Assmus B, Britten MB, Honold J, Lehmann R, Teupe C, Abolmaali ND, Vogl TJ, Hofmann WK, Martin H, Dimmeler S, Zeiher AM. Transplantation of progenitor cells and regeneration enhancement in acute myocardial infarction: final one-year results of the TOPCARE-AMI trial. *J Am Coll Cardiol* 2004; 44: 1690-1698.

3

# Cardiovascular Tissue Engineering Based on Fibrin-Gel-Scaffolds

Stefan Jockenhoevel[1] and Thomas C. Flanagan[2]
[1]*Department of Tissue Engineering & Textile Implants, AME-Helmholtz Institute for Biomedical Engineering, RWTH Aachen University, Aachen,*
[2]*School of Medicine & Medical Science, Health Sciences Centre, University College Dublin, Dublin*
[1]*Germany*
[2]*Ireland*

## 1. Introduction

Cardiovascular disease is a major cause of death in the Western World. Novel drugs and innovative devices have enhanced the quality of life for patients with cardiovascular disease, but such treatments are not without limitations and complications. The major constraint with these current treatments is the inability for growth, repair and remodeling of the structure. The emergence of tissue engineering as an alternative therapy for cardiovascular disease has generated an intensity of research into the development of many components of the cardiovascular system, including heart valves, small-calibre vascular grafts and biological stent materials. The composition of the biomaterial used as a support for the developing cardiovascular structure is a key mediator of cell behaviour and function in the tissue, and the ideal scaffold biomaterial for development of a successful end-product continues to be a matter of debate. Fibrin, a major structural protein involved in wound healing, represents an ideal scaffold for the rapid synthesis of autologous tissue-engineered cardiovascular grafts, as its primary scaffold constituents (fibrinogen and thrombin) can be isolated directly from a blood sample of the patient requiring the graft. Fibrin gel scaffolds offer immediate high cell seeding efficiency and homogenous cell distribution by gelation entrapment, and have a degradation rate that can be controlled by protease inhibitors, e.g. tranexamic acid or aprotinin. Fibrin is also known to stimulate the secretion of reinforcing extracellular matrix (ECM) proteins by seeded cells. The potential to control the fibrin polymerisation process also offers the opportunity to produce complex 3-D structures, like heart valve prostheses and to embed porous, textile or metal (stent) structures. This book chapter reviews the properties of fibrin that make it an ideal scaffold candidate for applications in the area of cardiovascular tissue engineering, and documents the successful development of fibrin-based heart valves, vascular grafts and biostents for clinical application.

## 2. Scaffold materials

Scaffolds play a central role in cardiovascular tissue engineering. Essential requirements for the ideal cardiovascular scaffold are easy handling properties and the ability to mould complex 3-D structures from the material, such as aortic roots or vessels with complex side branches. The scaffold material should neither be toxic, nor elicit any immunological side effects. The diffusion barrier of the scaffold material should have the lowest possible resistivity in order to guarantee an optimal nutrition supply in thicker tissues. Furthermore, both the mechanical and the chemical properties (e.g. the integration of growth factors) of the scaffold material should be modifiable. Controllable degradation of the material is also important in order to adapt the structural support of the scaffold with regard to the developing tissue.

A multitude of scaffolds are currently employed in the field of tissue engineering, e.g. synthetic polymers (polyurethanes, polyglyolic acid, polylactic acid, polyhydroxybutyrate, copolymers of lactic and glycolic acids, polyanhydrides, polyorthoesters) and natural polymers (chitosan, glycosaminoglycans, collagen), or biological scaffolds such as acellularised porcine aortic conduits (Bader et al., 1998; Chevallay & Herbage, 2000; Flanagan et al., 2006; Freed et al., 1994; Grande et al., 1997). Scaffold-related problems including cytotoxic degradation products, fixed degradation times, limited mechanical properties and the absence of growth modulation, etc. necessitate further extensive investigations in developing the ideal cardiovascular scaffold.

## 3. Fibrin as scaffold material?

Based on the assumption that successful tissue engineering should mimic the process of tissue regeneration, and that regeneration is closely related to haemostatis, fibrin (gel) seems to be an ideal candidate as a tissue engineering scaffold by virtue of its role as a "physiological scaffold" in tissue regeneration. Several influences of fibrin gel on tissue development have been described in the literature: it is known that fibrin gel is one of the major ligands for ß3 integrins, which leads to cell migration into a wound/tissue-engineered construct (Ikari et al., 2000; Nomura et al., 1999). Thrombin, fibrinogen, fibrin monomers and fibrinopepide B all increase DNA synthesis in smooth muscle cells (SMCs) and consequently the proliferation of the cell (Pakala et al., 2001).

### 3.1 Physiology of fibrin

Fibrin is the end-product of the coagulation cascade following the conversion of fibrinogen in the presence of thrombin and calcium (Figure 1). Fibrinogen is a soluble plasma glycoprotein, which is produced by the liver. Fibrinogen is an acute phase protein with a normal blood concentration of 1.4 – 3.5 g/l. The fibrinogen molecule has a length of 45 nm, a molecular weight of 340 KDa and consists of 2 subunits and 3 polypeptides chains - $\alpha$, $\beta$ and $\gamma$. During the polymerisation process, the fibrinopeptide A of the $\alpha$-chain and the fibrinopeptide B of the ß-chain are cleaved by thrombin. The exposed N-terminal positions of the fibrinopeptides bind to the $\gamma$-chain of the fibrinogen and produce the so-called *proteofibrils*. In the subsequent step, the *lateral association* leads to apposition of the proteofibrils to form a 3-D fibrin network structure (Meyer, 2004). FXIIIa stabilises fibrin further by incorporation of the fibrinolysis receptors, alpha-2-antiplasmin and TAFI (thrombin activatable fibrinolysis inhibitor, procarboxypeptidase B), and binding to several adhesive proteins of various cells (Muszbek et al., 2008).

The polymerised fibrin gel matrix is a hydro-gel, which contains ~95-98% water. The water content can easily be exchanged against a buffer solution or a cell culture medium, allowing an optimal nutrition supply of the cells that are embedded after the gelation process.

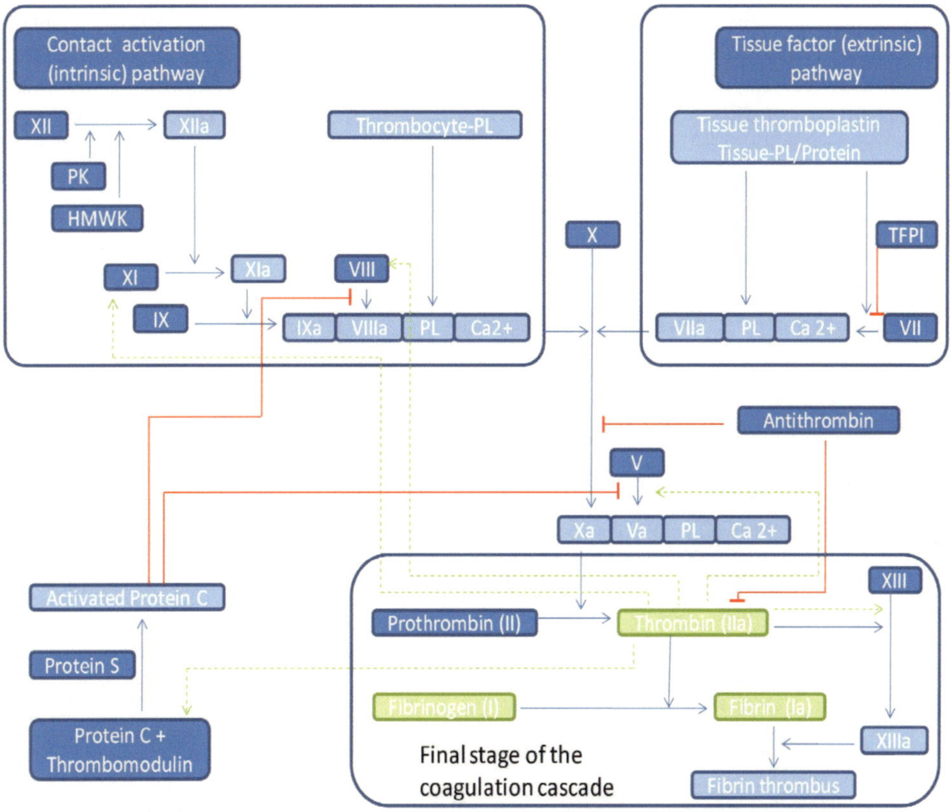


| | | | |
|---|---|---|---|
| I | Fibrinogen | XI | Plasma thromboplastin antecedent (PTC) Rosenthal factor |
| II | Prothrombin | XII | Hageman factor |
| V | Proaccelerin | XIII | Fibrin stabilizing factor (FSF) |
| VII | Proconvertin | PK | Prekallikrein, Fletcher factor |
| VIII | anti-hemophilic factor (AHF) | HMWK | High Molecular Weight Kininogen, Fitzgerald factor |
| IX | Plasma Thromboplastic Component (PTC) | PL | Phospholipid |
| X | Stuart-(Power-) Factor | a | activated |

Fig. 1. Coagulation cascade: the conversion of fibrinogen into fibrin is triggered by thrombin and calcium

### 3.2 Production of autologous fibrin

The classical approach for production of autologous fibrin is the cryoprecipitation method: after the production of platelet-poor plasma (PPP), the plasma is frozen at -80°C and thawed overnight at +4°C. The precipitate formed contains ~60-70% of fibrinogen. After centrifugation, the supernatant is decanted and the precipitate is subsequently washed twice in rinsed water. After the precipitate is dissolved in water, overnight dialysis against calcium-free TRIS buffer solution is necessary to provide optimal conditions for the embedded cells.

The cryoprecipitation method has two major disadvantages: (1) the efficiency of fibrinogen isolation is relatively low with only ~20-25% of the total fibrinogen content removed, and (2) the production process is time-consuming (~2 days). The low isolation efficiency is particularly problematic regarding the use of autologous fibrin gel scaffolds in paediatric patients, as the volume of sampled blood needs to be kept to an absolute minimum. Therefore, the use of alternative precipitation methods with different chemicals has been evaluated: (1) ethanol (Kjaergard et al., 1992; Weis-Fogh, 1988), (2) ammonium sulphate alone, and (3) in combination with the cryoprecipitation method (Wolf, 1983), (4) albumin plus cryoprecipitation, and (5) polyethylene glycol (PEG) plus cryoprecipitation (Epstein et al., 1986). Heselhaus investigated each of these different precipitation methods with regard to their efficiency and their use in the development of fibrin scaffold materials for cardiovascular applications (Heselhaus, 2011):

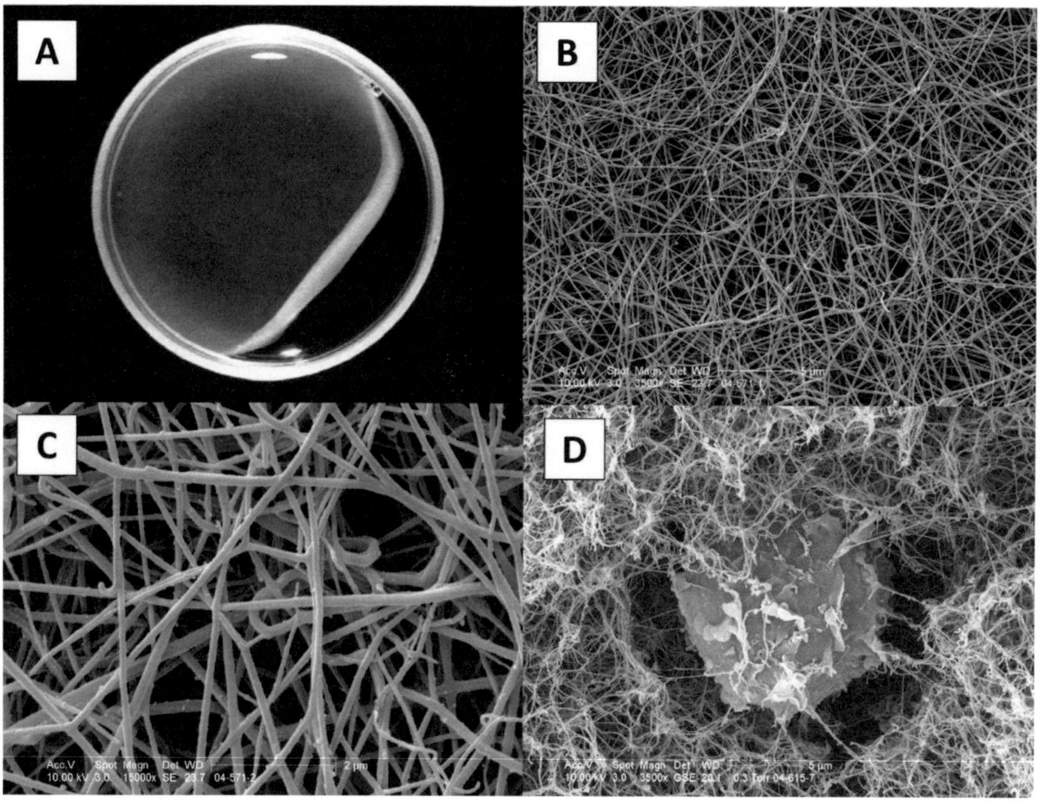


Fig. 2. [A] Fibrin gel after polymerisation in a 6-well plate. [B-D] Scanning electron microscopy (SEM) images demonstrating the nano-fibre network structure of the fibrin, which enables the gentle embedding of cells, with a vascular smooth muscle cell (SMC) shown in [D] immediately after the gelation of the fibrin within a web-like network surrounded by cell culture medium

Figure 3 demonstrates that all of the reported alternative methods are more efficient than the standard cryoprecipitation method. Here, the technique using ethanol as the precipitation reagent is observed as the most efficient method, with an isolation efficiency of ~80%, ~4-times higher than the efficiency of the standard method. The technique applying both albumin and cryoprecipitation indicates a false positive high result due to contamination of the precipitate with albumin (shown by a significant band in SDS gel electrophoreses) (Heselhaus, 2011).

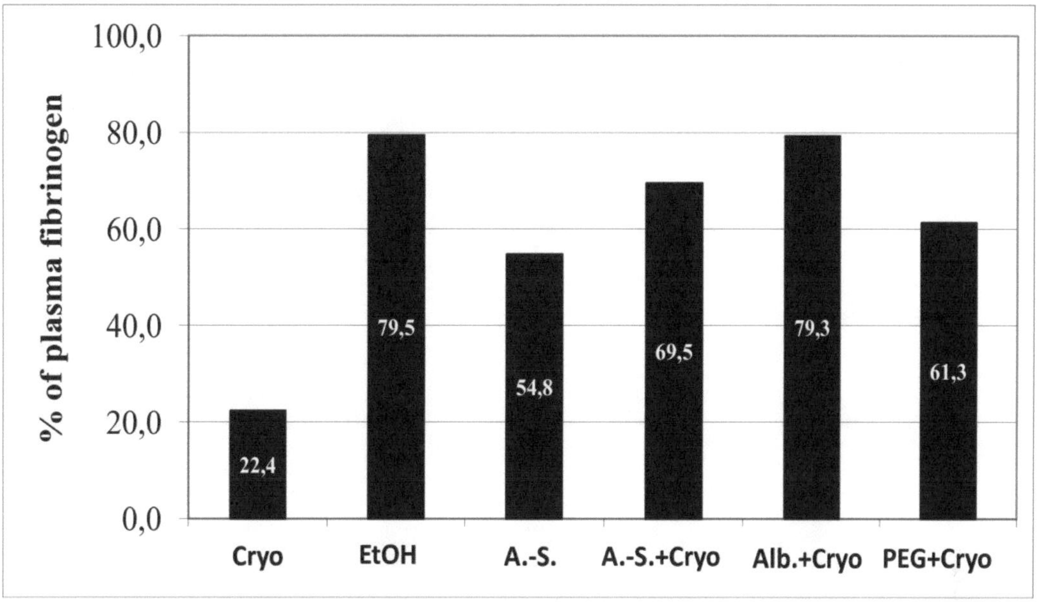


Fig. 3. Comparison of different precipitation methods for the isolation of fibrinogen (Cryo - cryoprecipitation; EtOH - ethanol; A.-S. - ammonium sulphate; Alb.–albumin; PEG - polyethylene glycol)

### 3.3 Control of scaffold degradation (aprotinin vs. tranexamic acid)

In nature, fibrin production and fibrinolysis are finely balanced. Figure 4 demonstrates the process of fibrinolysis induced by plasmin and the role of different regulatory factors, which stimulate and inhibit the fibrinolysis, respectively.

In the context of *in vitro* cultivation, both of these processes are restricted by the limited concentration of substrates. This offers the opportunity to adapt the degradation of the fibrin scaffold material to the individual need during the tissue maturation process. Until recently, the most commonly used substrate to control the degradation of fibrin gel was aprotinin with a concentration of 130 K.I.U. per mL medium (Ye et al., 2000). Aprotinin is a polypeptide serine protease inhibitor, which stops fibrinolysis by inhibiting kallikrein, plasmin and platelet-activation factors. Aprotinin was used as anti-fibrinolyticum for many years in the clinic, but negative side-effects in the post-operative recovery phase of cardiothoracic patients led to the suspension of aprotinin in the worldwide market.

Therefore, tranexamic acid (trans-4-aminomethely-cyclohexane-1-carboxylic acid; *t*AMCA) was evaluated as alternative drug to control fibrin gel degradation *in vitro* and as far as possible *in vivo*. Tranexamic acid is clinically approved and competitively inhibits the conversion of plasminogen into plasmin via reversibly binding to the lysine-binding site on plasminogen.

Cholewinski et al. (2009) demonstrated that tranexamic acid at a concentration of 160 mg per mL medium has a comparable inhibition effect on fibrinolysis in comparison with aprotinin. Furthermore, no negative side-effects with regard to proliferation, apoptosis, necrosis or the burst strength of the produced fibrin gel scaffolds could be demonstrated (Cholewinski et al., 2009).

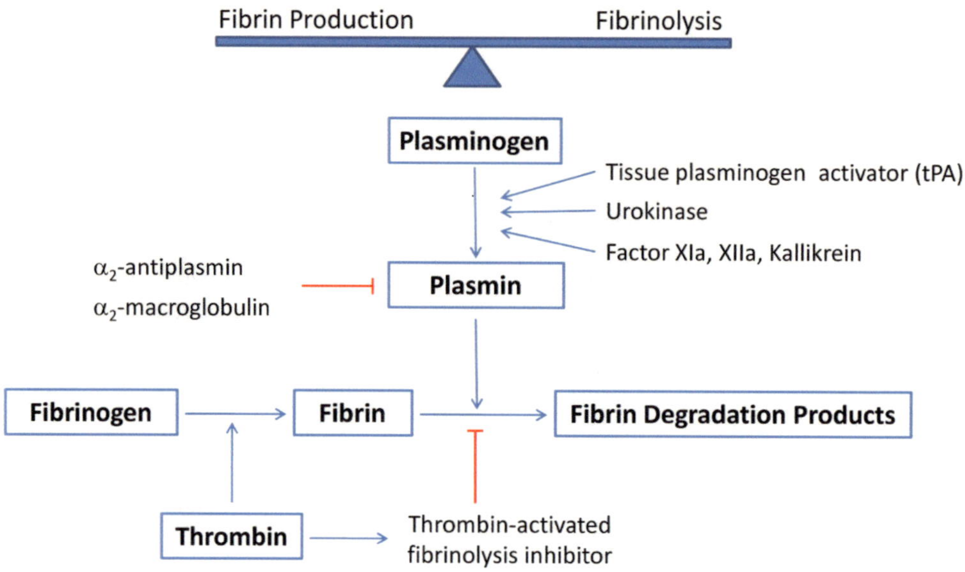


Fig. 4. Balance between fibrin production and fibrinolyis ( →positive and --| negative effect)

### 3.4 Fibrin as autologous growth factor delivery system (PDGF)

The autologous source of fibrin, whole blood, contains a number of important growth factors. One major growth factor is the platelet-derived growth factor (PDGF), a cytokine produced in megacaryocytes and stored in the α-granules (Cianciolo et al., 1999; Ross et al., 1986). PDGF has been described as one particular factor that positively influences proliferation, migration and ECM secretion of fibroblasts and SMCs (Claesson-Welsh, 1996). PDGF is released in conjunction with the platelet release reaction (Witte et al., 1978). Platelet activation induces the release of a variety of low-molecular weight growth factors and more than 60 of these growth factors are directly involved in tissue repair mechanisms such as chemotaxis, cell proliferation, angiogenesis, ECM deposition and remodelling (Borzini & Mazzucco, 2007). Platelets are easily isolated from blood, and can be concentrated in a low volume of plasma known as platelet-rich plasma (PRP). Thus, PRP is a storage vehicle for platelet growth factors. In addition to the abundant mixture of growth factors, PRP also contains proteins known to act as cell adhesion molecules and matrix for bone, connective tissue and epithelial migration, namely fibrin, fibronectin and vitronectin (Marx, 2004).

PRP can be used as a fibrin-based scaffold material by inducing gel polymerisation. The use of PRP as an autologous scaffold material and growth factor delivery vehicle has attracted the attention of researchers in the field of tissue engineering as a new possibility to optimise the composition of the "ideal" autologous scaffold.

To further optimise fibrin gel scaffolds, the use of PRP as a basis for autologous gel scaffolds instead of platelet-poor plasma (PPP) or pure fibrinogen solution has been evaluated. It was postulated that the use of PRP in contrast to PPP or fibrinogen solution as a basis for human autologous fibrin gels leads to an increased release of autologous PDGF-AB, which may have a consequent positive effect on tissue development. Therefore, a protocol for plasma preparation and subsequent plasma gel production was developed and the release kinetics of PDGF-AB from autologous plasma gels were investigated. Wirz and colleagues (2011)

analysed (i) the concentration of PDGF-AB in the PRP vs. PPP during gel preparation, (ii) the influence of plasma gels and particularly PDGF-AB (autologous and recombinant) on growth behaviour and cell proliferation, and (iii) the secretion of ECM by human umbilical cord artery smooth muscle cells (HUASMCs) in both plasma-based and pure fibrin gels. The study demonstrated that the use of PPP leads to almost complete loss of PDGF, whereas the use of PRP in combination with a concentration step (by a factor of two) retains almost serum levels of PDGF. The subsequent steps in the protocol allow the removal of all platelets without losing the PDGF in the concentrate (Wirz et al., 2011).
But is the absence of platelets in the fibrin-based scaffolds good or not? To answer this question, Wirz et al. compared PDGF release from such fibrin-based scaffolds with and without incorporated platelets. Whereas the release of PDGF of platelet-rich and platelet-free fibrin gels were similar during the first 3 days, the platelet-rich gels showed a significantly higher level of PDGF in the supernatant after the first medium exchange, indicating that the platelets with the α-granules continue to work as a 'natural' reservoir of the growth factor PDGF (Wirz et al., 2011).
Nevertheless, against all expectations, the significantly higher level of PDGF-AB in fibrin gels was shown to have neither a significant influence on cell proliferation nor on the ECM synthesis of vascular SMCs. Although the vascular SMCs are the basic cell source for cardiovascular tissue engineering, these cells do not express the receptor for PDGF-A and lose the receptor for PDGF-B at a very early stage of *in vitro* primary cell culture (P0 to P4) (Wirz et al., 2011). Therefore, PDGF-rich fibrin gels may ultimately have a positive effect on fibroblasts and mesenchymal stem cells, for example, but not on primary vascular SMCs.

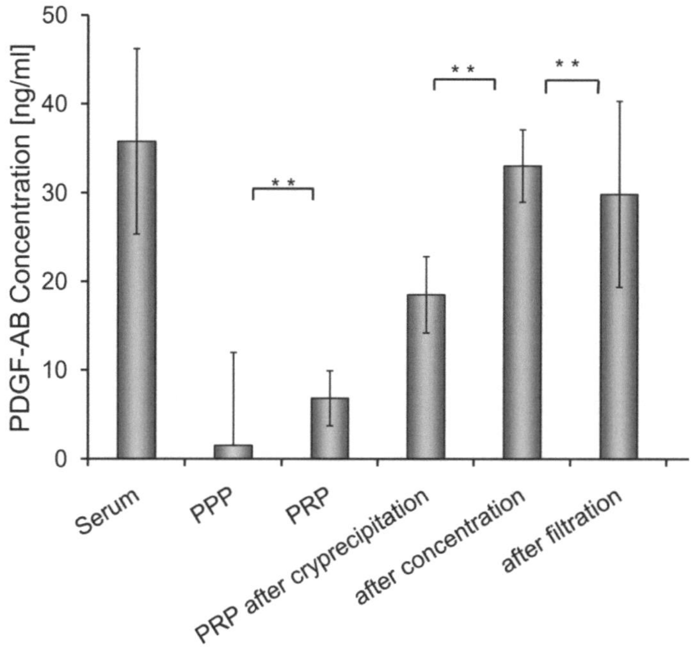


Fig. 5. PDGF-AB release of platelet poor plasma (PPP), platelet rich plasma (PRP) before and after cryoprecipitation, after concentration of the plasma volume (by a factor of two), and after sterile filtration process (0.22μm filter)

## 4. Examples of fibrin-based cardiovascular tissue engineering

### 4.1 Heart valve

The principles of fibrin-based cardiovascular tissue engineering can be demonstrated using the example of a completely autologous heart valve prosthesis as shown in Figure 6 (Jockenhoevel et al., 2001b). For paediatric application, the umbilical cord is the optimal cell source. The umbilical cord contains the myofibroblasts/vascular SMCs, which are responsible for the structural and mechanical stability of the neo-tissue. Furthermore, the endothelial cells of the umbilical cord vein are an ideal cell source for the luminal coating of cardiovascular structures. The endothelial cells are mainly responsible for the haemocompatibility of the tissue-engineered graft. The third component of the umbilical cord that is applicable to tissue engineering is the blood, from which the compounds for autologous fibrin-gel synthesis can be extracted.

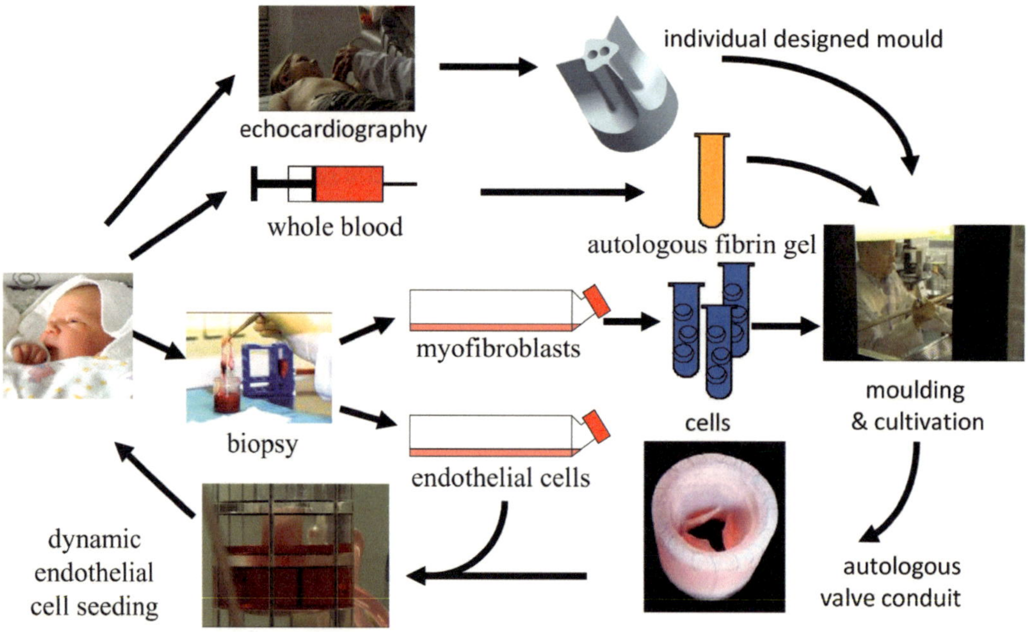


Fig. 6. Principles of fibrin-based tissue engineering using the example of a completely autologous heart valve prosthesis

Based on clinical imaging methods such as echocardiography, it is possible to manufacture a mould that is customised for the individual. This allows the surgical correction of complex congenital heart defects with a shape-optimised implant.

The production of the basic heart valve structure is realised using an injection moulding technique: at the outset, the first of two essential components is produced, and consists of myofibroblasts suspended in a buffer solution that is supplemented with calcium and thrombin. The second component consists of the autologous fibrinogen solution. With a 2-piston-dual syringe system, both components are injected into the customised mould, which consists of a negative and positive stamp. The gelation process begins immediately and homogenously entraps the myofibroblasts in the fibrin gel. After the gel polymerisation is

complete, the newly moulded heart valve conduit is decast from the mould and transferred into a bioreactor system. A suitable nutrition supply and biomechanical stimulation are essential for the maturation of these tissue-engineered structures, as they are too fragile at the outset for direct implantation (Flanagan et al., 2007).
After preconditioning in the bioreactor system, the mechanically stable heart valve prosthesis has been developed. The lining of the luminal surface with autologous endothelial cells is the final production step and is important for complete immunological integrity and physiological haemostasis. At the end of the production chain, a completely autologous heart valve implant is realised, which has the ability to withstand the mechanical properties in the low pressure/pulmonary circulation, as demonstrated previously (Flanagan et al., 2009).

### 4.2 Vascular graft

Autologous vascular grafts based on a fibrin scaffold have also been developed for applications in a number of settings, including coronary artery and peripheral artery bypass procedures, arteriovenous access grafts for haemodialysis patients, or in the paediatric setting for congenital pulmonary artery reconstruction. Fibrin alone does not possess sufficient mechanical properties for direct implantation into the high pressure setting of small-calibre arteries (< 6 mm); therefore, the fibrin-based grafts are supported by a bioabsorbable, macroporous mesh (poly(L/D)lactide 96/4; PLA) with a controlled degradation profile, which acts as a temporary supporting system upon implantation until the implanted cells have remodelled the fibrin matrix into an autologous tissue (Tschoeke et al., 2009).

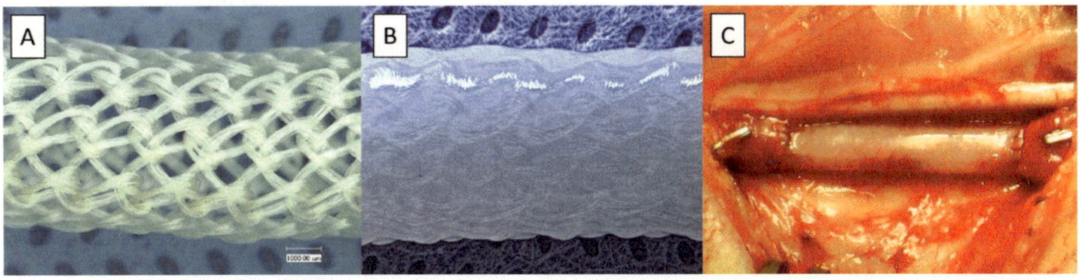


Fig. 7. Bioresorbable, macroporous mesh with a pore size of ~ 1mm before [A] and after [B] the embedding into the fibrin/cell matrix. [C] Fibrin-based vascular graft after implantation in the arterial circulation (ovine carotid model)

The synthesis of the fibrin-based vascular graft is based on similar principles to those used in developing the autologous fibrin-based heart valve (i.e. injection moulding technique). Once the vascular grafts are constructed in a tubular mould, however, the mould acts as the bioreactor system and is connected to a flow support that imparts pulsatile luminal flow to the graft. We have shown for the first time that fibrin-based vascular grafts attain a supraphysiological burst strength sufficient for arterial implantation after just 21 days of mechanical conditioning in such a bioreactor system (Tschoeke et al., 2008). We followed up this report with a preclinical study in a large animal model, which presented data on the first series of fibrin-based grafts to be implanted in the arterial circulation (ovine carotid model) (Koch et al., 2010). In this model, the grafts showed no evidence for thrombus formation, aneurysm, calcification or infection, and remained patent for at least 6 months *in vivo*. The grafts maintained a functional endothelial lining *in vivo*, and the fibrin-based scaffold was completely replaced by autologous connective tissue elements (e.g. collagen, elastin) after 6 months of implantation (Koch et al., 2010).

### 4.3 BioStent

During recent years, percutaneous stent angioplasty has become well established in the treatment of peripheral and coronary atherosclerosis. Nevertheless, the patency rates in small-calibre vessels, particularly those of the femoral and femoropopliteal region, are not satisfactory (Cejna et al., 2001; Grenacher et al., 2004; Schillinger et al., 2006).The incessant problem of *instent restenosis* is a result of: (1) the proliferation stimulus resulting from the pressure trauma of stent implantation (>10 bar in non-self-expanding stents), which leads to (2) an ingrowth of myofibroblasts through the gap of the stent structure, and (3) an increased production of ECM proteins; (4) finally, acute thrombotic occlusion can occur in the stented segment of the vessel. Occlusion rates of more than 20% after 6 months could be overcome with a viable stent prosthesis that includes a confluent, functionally active endothelial cell layer from the outset.

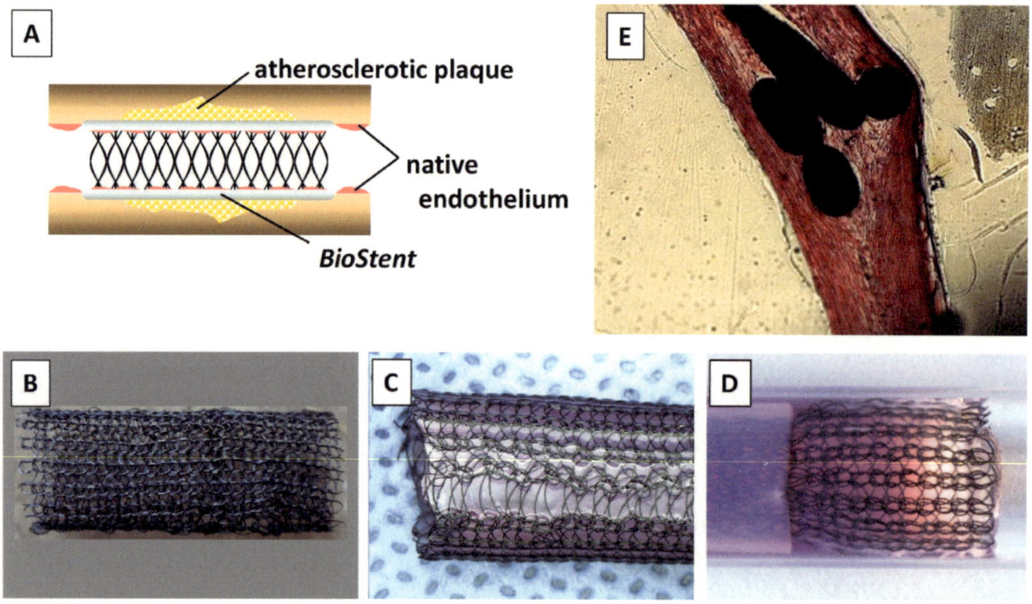


Fig. 8. Concept of the BioStent [A]: the knitted, self-expanding Nitinol-Stent [B] is embedded into a fibrin-cell matrix [C] and can be deployed without tissue destruction [D]. H&E staining demonstrates a complete coating of the Nitinol stent structure with viable tissue [E]

The *BioStent* concept merges the principles of self-expanding stent technology with those of vascular tissue engineering: the moulding process of vascular grafts based on a fibrin gel scaffold allows the complete integration of a self-expanding stent structure within the tissue-engineered vessel. This process leads to a complete exclusion of the atherosclerotic section of the vessel from the blood stream on the one hand, and coating of the neo-lumen with a functional endothelial cell layer on the other hand (Figure 8).

The *BioStent* concept is a platform technology, i.e. by applying different cell types, a number of alternative therapies can be addressed, including (1) endobronchial stenting using respiratory epithelium, (2) oesophageal stenting using mucosa cells, or (3) urological stenting using urothelial cells, etc.

## 5. Summary

In our opinion, fibrin gel combines a number of important properties of an ideal scaffold for cardiovascular tissue engineering:

- Fibrin gel is a naturally-occurring scaffold that can be isolated as an autologous substrate from blood of the patient in question (Heselhaus, 2011);
- Starting with a cell suspension in fibrinogen solution, fibrin gel scaffolds offer immediate high cell seeding efficiency and homogenous cell distribution by gelation entrapment, with a minimal loss of cells during the seeding procedure; furthermore, there is no time-consuming cell ingrowth from the scaffold surface to the deeper parts of the scaffold (Jockenhoevel et al., 2001a);
- Polymerisation as well as degradation of the fibrin gel is controllable and can be adapted to tissue development through the use of the protease inhibitors, such as aprotinin and tranexamic acid (Cholewinski et al., 2009);
- Local, covalent immobilisation of different growth factors is possible, while PRP gels can be developed to enhance the content and delivery of growth factors (Wirz et al., 2011);
- Production of complex 3-D structures such as heart valve conduits or vascular grafts with complex side branches is possible through the use of an injection moulding technique (Flanagan et al., 2007; Jockenhoevel et al., 2001b);
- Textile-reinforced fibrin-based grafts can be implanted in the arterial circulation and function for at least 6 months *in vivo* (Koch et al., 2010);
- Fibrin-based tissue engineering can be merged with self-expanding stents to create a platform technology for cardiovascular, and other, diseases.

These properties highlight the significant potential for creation of functional, autologous implantable cardiovascular prostheses in future using tissues derived from the patient.

## 6. Acknowledgements

The authors would like to thank the Fördergemeinschaft Deutsche Kinderherzzentren e.V. for financially supporting the heart valve project and the DFG (German Research Foundation) for financially supporting the BioStent project. The multi-centre research project of the vascular graft was financially supported by the European Union Sixth Framework Program (Project STREP 013633, BioSys). The authors would also like to acknowledge their colleagues at the National University of Ireland, Galway, the Tampere University of Technology, Finland, the University Hospital Zurich, Switzerland and the colleagues of the RWTH Aachen University, Germany, for their excellent collaboration on this project. The authors also extend their thanks to the Scientific Workshop, Helmholtz Institute Aachen, for customised bioreactors and moulds, the Department of Pathology, University Hospital Aachen, for their valuable assistance with transmission- and scanning electron microscopy, respectively, and the Department of Laboratory Animal Research, University Hospital Aachen, for taking excellent care of all the animals used in these studies.

## 7. References


Bader, A., Schilling, T., Teebken, O.E., Brandes, G., Herden, T., Steinhoff, G. & Haverich, A. (1998). Tissue engineering of heart valves - human endothelial cell seeding of detergent acellularized porcine valves. *European Journal of Cardiothoracic Surgery,* Vol.14, No.3, (September 1998), pp. 279-284, ISSN 1010-7940.

Borzini, P., Mazzucco, I. (2007). Platelet rich plasma (PRP) and platelet derivatives for topical therapy. What is true from the biological point of view? *ISBT Science Series,* Vol.2, pp. 272-281, ISSN 1751-2824.

Cejna, M., Thurnher, S., Illiasch, H., Horvath, W., Waldenberger, P., Hornik, K. & Lammer, J. (2001). PTA versus Palmaz stent placement in femoropopliteal artery obstructions: a multicenter prospective randomized study. *Journal of Vascular and Interventional Radiology,* Vol.12, No.1, (January 2001), pp. 23-31, ISSN 1051-0443.

Chevallay, B. & Herbage, D. (2000). Collagen-based biomaterials as 3D scaffold for cell cultures: applications for tissue engineering and gene therapy. *Medical and Biological Engineering and Computing,* Vol.38, No.2, (March 2000), pp. 211-218, ISSN 0140-0118.

Cholewinski, E., Dietrich, M., Flanagan, T.C., Schmitz-Rode, T. & Jockenhoevel, S. (2009). Tranexamic acid - an alternative to aprotinin in fibrin-based cardiovascular tissue engineering. *Tissue Engineering Part A,* Vol.15, No.11, (November 2009), pp. 3645-3653, ISSN 1937-3341.

Cianciolo, G., Stefoni, S., Zanchelli, F., Ianelli, S., Coli, L., Borgnino, L.C., De Sanctis, L.B., Stefoni, V., De Pascalis, A., Isola, E. & La Hanna, G. (1999). PDGF-AB release during and after haemodialysis procedure. *Nephrology, Dialysis, Transplantation,* Vol.14, No.10, (October 1999), pp. 2413-2419, ISSN 1460-2385.

Claesson-Welsh, L. (1996). Mechanism of action of platelet-derived growth factor. *International Journal of Biochemistry and Cell Biology,* Vol.28, No.4, (April 1996), pp. 373-385, ISSN 1357-2725.

Epstein, G.H., Weisman, R.A., Zwillenberg, S. & Schreiber, A.D. (1986). A new autologous fibrinogen-based adhesive for otologic surgery. *Annals of Otology, Rhinology and Laryngology,* Vol. 95, No.1 (January 1986), pp. 40-45, ISSN 0096-8056.

Flanagan, T.C., Wilkins, B., Black, A., Jockenhoevel, S., Smith, T.J. & Pandit, A.S. (2006). A collagen-glycosaminoglycan model for heart valve tissue engineering applications. *Biomaterials,* Vol.27, No.10, (April 2006), pp. 2233-2246, ISSN 0142-9612.

Flanagan, T.C., Cornelissen, C., Koch, S., Tschoeke, B., Sachweh, J.S., Schmitz-Rode, T. & Jockenhoevel, S. (2007). The in vitro development of autologous fibrin-based tissue-engineered heart valves through optimised dynamic conditioning. *Biomaterials,* Vol.28, No.23, (August 2007), pp. 3388-3397, ISSN 0142-9612.

Flanagan, T.C., Sachweh, J.S., Frese, J., Schnoring, H., Gronloh, N., Koch, S., Tolba, R.H., Schmitz-Rode, T. & Jockenhoevel, S. (2009). In vivo remodelling and structural characterization of fibrin-based tissue-engineered heart valves in the adult sheep model. *Tissue Engineering Part A,* Vol.15, No.10, (October 2009), pp. 2965-2976, ISSN 1937-3341.

Freed, L.E., Vunjak-Novakovic, G., Biron, R.J., Eagles, D.B., Lesnoy, D.C., Barlow, S.K. & Langer, R. (1994). Biodegradable polymer scaffolds for tissue engineering. *Biotechnology (NY),* Vol.12, No.7, (July 1994), pp. 689-693, ISSN 0733-222X.

Grande, D.A., Halberstadt, C., Naughton, G., Schwartz, R. & Manji, R. (1997). Evaluation of matrix scaffolds for tissue engineering of articular cartilage grafts. *Journal of*

*Biomedical Materials Research,* Vol.34, No.2, (February 1997), pp. 211-220, ISSN 1549-3296.

Grenacher, L., Saam, T., Geier, A., Muller-Hulsbeck, S., Cejna, M., Kauffmann, G.W. & Richter G.M. (2004). [PTA versus Palmaz stent placement in femoropopliteal artery stenoses: results of a multicenter prospective randomized study (REFSA)]. [Article in German]. *Rofo,* Vol.176, No.9, (September 2004), pp. 1302-1310, ISSN 0340-1618.

Heselhaus, J. (2011). Kardiovaskuläres Tissue Engineering auf der Basis einer Fibringel-Matrix - Optimierung der Matrix hinsichtlich Isolation, Synthese und mechanischer Festigkeit. Dissertation, Medical Faculty, RWTH Aachen University, Germany.

Ikari, Y., Yee, K.O. & Schwartz, S.M. (2000). Role of alpha5beta1 and alphavbeta3 integrins on smooth muscle cell spreading and migration in fibrin gels. *Thrombosis and Haemostasis,* Vol.84, No.4, (October 2000), pp. 701-705, ISSN 0340-6245.

Jockenhoevel, S., Zund, G., Hoerstrup, S.P., Chalabi, K., Sachweh, J.S., Demircan, L., Messmer, B.J. & Turina, M. (2001a). Fibrin gel - advantages of a new scaffold in cardiovascular tissue engineering. *European Journal of Cardiothoracic Surgery,* Vol.19, No.4, (April 2001), pp. 424-430, ISSN 1010-7940.

Jockenhoevel, S., Chalabi, K., Sachweh, J.S., Groesdonk, H.V., Demircan, L., Grossmann, M., Zund, G. & Messmer, B.J. (2001b). Tissue engineering: complete autologous valve conduit - a new moulding technique. *Thoracic and Cardiovascular Surgeon,* Vol.49, No.5, (October 2001), pp. 287-290, ISSN 0171-6245.

Kjaergard, H.K., Weis-Fogh, U.S., Sorenson, H., Thiis, J. & Rygg, I. (1992). Autologous fibrin glue - preparation and clinical use in thoracic surgery. *European Journal of Cardiothoracic Surgery,* Vol.6, No.1, pp. 52-54, ISSN 1010-7940.

Koch, S., Flanagan, T.C., Sachweh, J.S., Tanios, F., Schnoering, H., Deichmann, T., Ella, V., Kellomaki, M., Gronloh, N., Gries, T., Tolba, R., Schmitz-Rode, T. & Jockenhoevel, S. (2010). Fibrin-polylactide-based tissue-engineered vascular graft in the arterial circulation. *Biomaterials,* Vol.31, No.17, (June 2010), pp. 4731-4739, ISSN 0142-9612.

Marx, R.E. (2004). Platelet-rich plasma: evidence to support its use. *Journal of Oral and Maxillofacial Surgery,* Vol.62, No.4, (April 2004), pp. 489-496, ISSN 0278-2391.

Meyer, M. (2004). [Molecular biology of haemostasis: fibrinogen, factor XIII]. [Article in German]. *Hamostaseologie,* Vol.24, No.2, (May 2004), pp. 108-115, ISSN 0720-9355.

Muszbek, L., Bagoly, Z., Bereczky, Z. & Katona, E. (2008). The involvement of blood coagulation factor XIII in fibrinolysis and thrombosis. *Cardiovascular and Hematological Agents in Medicinal Chemistry,* Vol.6, No.3, (July 2008), pp. 190-205, ISSN 1871-5257.

Nomura, H., Naito, M., Iguchi, A., Thompson, W.D. & Smith, E.B. (1999). Fibrin gel induces the migration of smooth muscle cells from rabbit aortic explants. *Thrombosis and Haemostasis,* Vol.82, No.4, (October 1999), pp. 1347-1352, ISSN 0340-6245.

Pakala, R., Liang, C.T. & Benedict, C.R. (2001). A peptide analogue of thrombin receptor-activating peptide inhibits thrombin and thrombin-receptor-activating peptide-induced vascular smooth muscle cell proliferation. *Journal of Cardiovascular Pharmacology,* Vol.37, No.5, (May 2001), pp. 619-629, ISSN 1533-4023.

Ross, R., Raines, E.W. & Bowen-Pope, D.F. (1986). The biology of platelet-derived growth factor. *Cell,* Vol.46, No.2, (July 1986), pp. 155-169, ISSN 0092-8674.

Schillinger, M., Sabeti, S., Loewe, C., Dick, P., Amighi, J., Mlekusch, W., Schlager, O., Cejna, M., Lammer, J. & Minar, E. (2006). Balloon angioplasty versus implantation of nitinol stents in the superficial femoral artery. *New England Journal of Medicine,* Vol.354, No.18, (May 2006), pp. 1879-1888, ISSN 0028-4793.

Tschoeke, B., Flanagan, T.C., Cornelissen, A., Koch, S., Roehl, A., Sriharwoko, M., Sachweh, J.S., Gries, T., Schmitz-Rode, T. & Jockenhoevel, S. (2008). Development of a composite degradable/nondegradable tissue-engineered vascular graft. *Artificial Organs,* Vol.32, No.10, (October 2008), pp. 800-809, ISSN 1525-1594.

Tschoeke, B., Flanagan, T.C., Koch, S., Harwoko, M.S., Deichmann, T., Ella, V., Sachweh, J.S., Kellomaki, M., Gries, T., Schmitz-Rode, T. & Jockenhoevel, S. (2009). Tissue-engineered small-caliber vascular graft based on a novel biodegradable composite fibrin-polylactide scaffold. *Tissue Engineering Part A,* Vol.15, No.8, (August 2009), pp. 1909-1918, ISSN 1937-3341.

Weis-Fogh, U.S. (1988). Fibrinogen prepared from small blood samples for autologous use in a tissue adhesive system. *European Surgical Research,* Vol.20, No.5-6, pp. 381-389, ISSN 0014-312X.

Wirz, S., Dietrich, M., Flanagan, T.C., Bokermann, G., Wagner, W., Schmitz-Rode, T. & Jockenhoevel, S. (2011). Influence of PDGF-AB on tissue development in autologous platelet-rich plasma gels. *Tissue Engineering Part A,* 2011 Apr 25. [Epub ahead of print], ISSN 1937-3341.

Witte, L.D., Kaplan, K.L., Nossel, H.L., Lages, B.A., Weiss, H.J. & Goodman, D.S. (1978). Studies of the release from human platelets of the growth factor for cultured human arterial smooth muscle cells. *Circulation Research,* Vol.42, No.3, (March 1978), pp. 402-409, ISSN 0009-7330.

Wolf, G. (1983). [Concentrated autologous tissue glue]. [Article in German]. *Archives of Otorhinolaryngology,* Vol.237, No.3, (April 1983), pp. 279-283, ISSN 0302-9530.

Ye, Q., Zund, G., Benedikt, P., Jockenhoevel, S., Hoerstrup, S.P., Sakyama, S., Hubbell, J.A. & Turina, M. (2000). Fibrin gel as a three dimensional matrix in cardiovascular tissue engineering. *European Journal of Cardiothoracic Surgery,* Vol. 17, No.5, (May 2000), pp. 587-591, ISSN 1010-7940.

# 4

# Rapid Prototyping of Engineered Heart Tissues through Miniaturization and Phenotype-Automation

Tetsuro Wakatsuki
*Department of Physiology, Biotechnology & Bioengineering Center, Medical College of Wisconsin, Milwaukee, Wisconsin, InvivoSciences, LLC, McFarland Wisconsin, U.S.A.*

## 1. Introduction

Organ transplantation is a lifesaving procedure, yet the demand for transplants far exceeds the available supply. Each year, nearly seven thousand patients in the United States die while waiting for an organ transplant [1]. Developing heart muscles *in vitro* for transplantation [3] [4] is one approach that could ultimately overcome this critical shortage. For complex tissues such as a heart muscle, the final product has to mimic physiological functions as well as the regulatory system for regenerating the recipient's tissue functions. Successful development of implantable and functional engineered tissues is currently hindered by the time-consuming and cost-intensive processes through which developers iteratively work to optimize tissue-fabrication protocols, needing to establish proper culture conditions and the ideal quantities of materials used. To overcome these burdens, it is critical to establish a prototyping process by which the developers can explore many parameters rapidly using minimal resources. To achieve this goal, 1) the prototypes of engineered tissues have to be small enough to save materials, including cells, extracellular matrices and growth factors that are necessary for the *in vitro* heart tissue development, and 2) the assay system to evaluate the heart tissue function has to be automated for rapid high-throughput analysis of prototyping results. In addition, a stable source of cells must be carefully selected to maintain the reproducibility of tissue production while scaling up the manufacturing capability.

## 2. Miniaturizing biological samples and products

Size of engineered tissue (ET) varies based on research needs. To systematically screen and analyze a large number of parameters that determine tissue function, the size of ETs has to be defined based on the required throughputs and phenotying assays for the tissue function analysis. Fully-grown cardiac muscle cells in three dimensional matrices are at least 100 um in length along their longitudinal axis [5]. Therefore, engineered heart tissues (EHTs) should be at least several times larger than the size of single cells in order to form multi-cellular functional cardiac muscles. In addition, the statistical significance of data improves significantly when using multi-cellular tissues rather than single cells [6] because the tissue samples contain from tens of thousands to over a million cells. The functional assay data collected using tissues represents an average function of many cells embedded in the tissues. By applying the photolithographical techniques, smaller, micro-scale (50-200 um) engineered tissues can be fabricated [7] to mimic intricate shapes of native tissues. However, in these micro-scale tissues, it is yet to be determined if there is an improvement of the statistical significance of functional data over that from single cells. At present, ETs in millimeter scale have been used for the functional assays with relatively good data reproducibility [8] [9]. By using ETs, a statistically significant data can be obtained using only 4-8 samples [9] instead of collecting ranging 60-600 data points per well in cell-based assays [10].

Unlike the rapid prototyping of parts with synthetic materials, the prototyping of engineered tissues with live cells has to consider the behavior of tissues after fabrication in a given culture condition. Cells are a vital element of developing functional organs and tissues. While consuming nutrients and receiving extracellular stimuli, the cells synthesize, secrete, and degrade proteins, divide, and remodel tissues. Therefore, developers of engineered tissue have to take these aspects into account for the design and development of engineered tissue samples and products. Behaviors of cells have been studied extensively to understand the molecular mechanisms by which biological functions of cell are regulated [11]. Because the cell behavior should underlie function of tissues and organs, the vast amount of information that has been accumulated through studies of cells should be used to understand tissue and organ function. However, some recent findings indicate that biology of cells cultured in two dimension (2D) is not entirely the same as that of cells cultured in a three dimensional (3D) environment, which is more similar to the *in vivo* environment in which cells reside.

## 3. 2D vs 3D cell based analysis

Even though cells cultured on a 2D surface have been assumed to closely resemble those in tissues and organs, in some aspects they are quite different from those found *in vivo. In vivo,* most cells are found embedded within a 3D extracellular matrix (ECM), and their shape and cell-ECM adhesion are different from those in 2D culture [12]. In most 2D cell culture, cells are cultured on infinitely stiff solids. However, in 3D culture, the ECM in which native cells reside can be deformed, stiffened, softened, or degraded to change the mechanics of the tissue. Stem cell linage specification has been observed to be directed by matrix stiffness [13], which strongly suggests that mechanical signaling will play a pivotal role in developing physiologically functional and implantable tissues. A difference in mechanical signal transduction between a 2D and a 3D environment was also suggested [14]. For instance, fibroblasts cultured in 3D have intrinsically lower expression levels of the small GPTase Rac than those in 2D cultured cells, which switches the migratory pattern of cells from random to persistent [15]. Reduction of the activity of focal adhesion kinase (FAK) and ERK1/2 in 3D culture, as compared to 2D cultures, has also been reported [12]. There is significant concern over the biological differences observed between 2D cell-culture systems, which have been used to study almost all signal transduction pathways to date [16], and the arguably more physiological 3D cell-culture environment.

When testing mechanical signaling, the tissues are subjected to mechanical deformations including pushing and stretching. An application of cyclic mechanical stretch to cardiac myocytes and fibroblasts up-regulates various signal transduction pathways through Rho/ROCK activation [17, 18]. Activation of extracellular signal-regulated kinases (ERKs) and up-regulation of an immediate early gene family of transcription factors, and c-fos [19] are well-characterized early responses of cells to the stretching. Up-regulated cellular processes including protein synthesis [20] and myofibril organization [21] are also apparent. The Rho/ROCK signaling pathway also regulates a hypertrophic response in cultured cardiac myocytes after application of soluble factors such as endothelin-1 treatment [22]. Therefore, an intricate balance orchestrated by soluble stimuli, mechanical deformation, and extracellular stiffness dictates the tissue development.

The tissue can function normally under homeostasis, at which the mechanical balance between extra- and intra-cellular activities is balanced. The homeostasis can be shifted by aging or other long term processes but is a quasi-steady state that is required for a tissue's physiological function. For example, physiological wound healing restores the homeostasis in the granulation tissue - fibrous connective tissue of healing wounds - by active fibroblasts contracting open wounds, secreting factors and hormones, and degrading ECMs to reconstruct the damaged tissues [23]. Pathological wound healing, such as keloids and scars in skin [24], fails to restore the homeostasis. A schematic description of potential network system is in shown in Figure 1. Intracellular molecular sensors, such as the src family kinase substrate p130Cas [25], detect the stiffness of extracellular environment to initiate its down-stream signaling events including $Ca^{2+}$ release, kinase activity, and gene regulation. Expressions of various genes are up-regulated through translocation of transcription factors from cytoplasm into nucleus. Furthermore the gene expression can be coupled to cell contraction. For instance, during the formation of actin stress-fibers in contracting fibroblasts, the balance of actin pools is shifted from a high prevalence of the monomer state to increased polymer levels. This balance shift frees up myocardin-related transcription factors (MRTF)-A, which are bound to monomer actins, and the unbound MRTF-As translocate into nucleus to promote gene expression. Therefore, MRTF-A's downstream up-regulation of wound healing gene-family including collagen type I a1 is tightly coupled to the state of actin polymerization (F- and G- actin ratio) in fibroblasts [26].

In parallel with changing the mechanical environment, the soluble factors, such as TGF $\beta_1$, binding to their specific receptors initiate the similar downstream signaling events [27, 28]. Depending on the signaling pathways, the initial signal strength can be amplified. One of recent observations indicates that signal strength is amplified in three-tiered kinase module of Raf-MEK(mitogen-activated protein kinase or extracellular signal-regulated kinase kinase)-ERK (extracellular signal-regulated kinase) by increasing concentrations of kinases involving in downstream (e.g., 1:3:6 = raf-1:MEK:ERK) in COS cells [29]. The soluble and mechanical stimuli result in the development of contractility by actin-myosin interactions, which is one of end-points of the signaling pathways. We hypothesize that the activities of actin-myosin interactions is a part of feedback mechanism by which cells sense the strength of mechanical stimuli and stiffness of extracellular space [30]. Integrin-linked kinase and ECM protein, tenascin-C, have been demonstrated to be a part of the mechanical sensory systems in zebrafish heart [31], skeletal muscles [32], and fibroblasts [33].

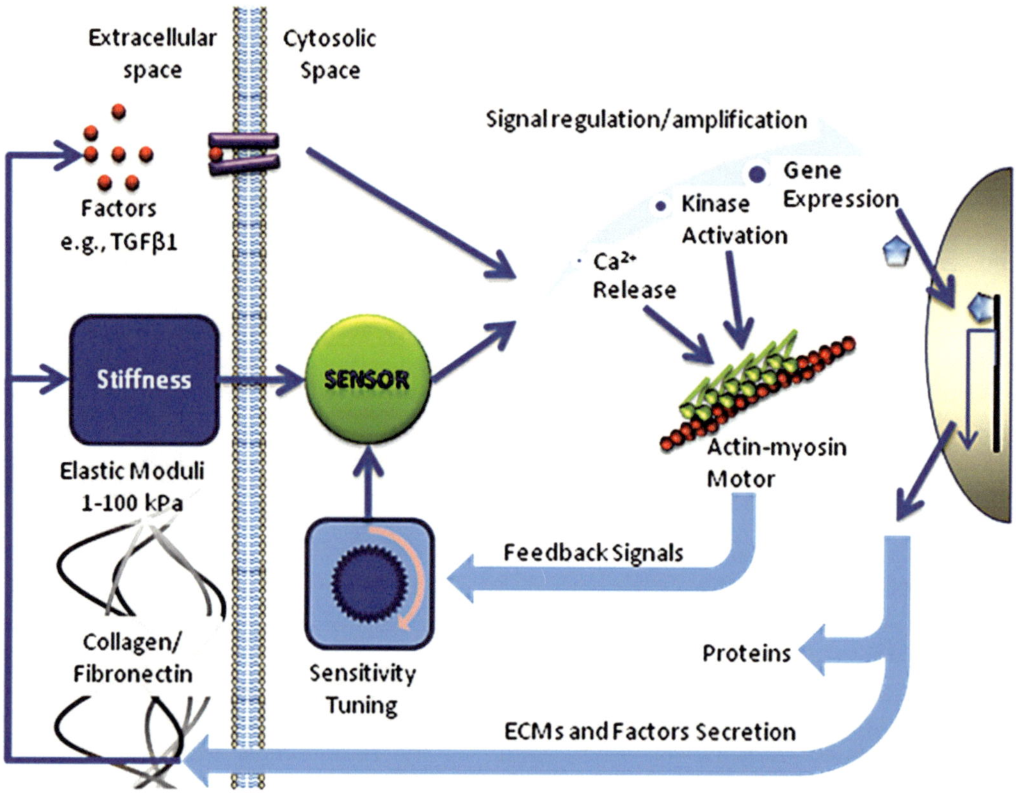


Fig. 1. Regulation of Mechanical Tissue Homeostasis Establishment

As demonstrated elegantly for p130Cas [25], a simple analogy of the biological sensory mechanism is represented as a mechanical strain gauge that measures applied forces by detecting small deformations of proteins. Depending on an elastic strength of the gauge (i.e., protein), its sensitivity can be varied. In general the sensitivity becomes higher by reducing the elasticity of the gauge material since it can deform more with a same applied force. We and others [18] assume that the molecular strain gauges are coupled to the actin-myosin forming stress fibers whose elasticity or stiffness is modulated by activity of myosin. The receptor activation initiated by binding of various soluble factors triggers and modulates myosin activity. Therefore, the sensitivity of the mechanical sensor can be modulated. Eventually the mechanical properties of cell and extracellular matrix reach a steady state to find homeostasis. However, its specific mechanism by which sensors can modulate their sensitivity and determination of the homeostasis are yet to be elucidated. Nevertheless, the concept of the establishment of homeostasis between the cells and ECM comprising "tissue" should be considered carefully while designing and developing the functional tissues *in vitro*. The mechanical properties of engineered tissues have to match with donor tissues, especially for those carrying critical mechanical functions including the cardiovascular system. As described, varying stiffness of the extracellular environment takes part in dictating stem cell fates [34]. Rapid high throughput tissue phenotyping can be used to test various soluble and mechanical stimuli to yield an optimal tissue with desired functions.

## 4. Cardiac tissue phenotyping and their potential to assay-automation

The assay system by which developers can evaluate the functional improvement that results from optimizing the tissue fabrication and culture conditions will have to accurately observe the physiological functions of cardiac tissues. The contractility of cardiac tissues is the most fundamental functional readout that can indicate the functional improvement by varying the tissue fabrication conditions. The regulatory system of cardiac contractility should also be reconstituted in the engineered tissues, which will be evaluated through rapid prototyping. As listed in Table 1, EHT-based contractility assessment is ideal for assessing cardiac functions in high-throughput, as compared to the other techniques that have traditionally been used to assess the function of myocytes, myocardium, and isolated hearts. Each method has its own advantages and disadvantages. The assay results using isolated adult cells and papillary muscles will continue to serve as references for the expected muscle cell and tissue contractility and its regulation. However, they can not report the tissue functions. Cardiac contractility assays using isolated Langendorff preparation has been gradually replaced by hemodynamic measurements using pressure-volume conductance catheters [35] as well as echo cardiography. Nevertheless, the whole heart functional assay can be used to assess functional improvement that results from the tissue transplantation.

| | Engineered Heart Tissues | Isolated Adult Cells | Papillary Muscles | Langendorff Preps |
|---|---|---|---|---|
| Sample viability period | Weeks to months | ~6 hrs | ~6-8 hrs | ~8 hrs |
| Contractility measurements with preload. | Yes (various) | No * | Yes (various) | Yes (various, PV loop) |
| Optical assessment of live cells | Yes (easy) | Yes (easy) | Yes (not easy) | No (generally) |
| High throughput testing | Yes | Yes | No | No |
| Sample damage by isolation | No | Yes | Yes | Yes |
| Comments | Reconstituted system | High data variance | | Widely used standards. |

* Isolated adult cells can be stretched by carbon fiber techniques [2]. High throughput measurements and range of stretching are limited.

Table 1. Comparison of methods for measuring cardiac contractility *in vitro*

Once the EHTs start contracting, they maintain contractility for at least 2-3 weeks. Assessment of acute and chronic changes in contractility and cell physiology can thus be monitored for long periods using EHTs. Even though EHTs are artificially reconstituted organoids, their histological structure and functional properties are similar to those in native myocardium [5, 36]. Each EHT is fabricated by pouring a gel solution containing predefined concentrations of cells, matrix proteins, and growth factors into a precisely machined well [5]. Since 1993, similar engineered tissues have been used as a model system to study signal transduction pathways that regulate contractility of fibroblasts [37], smooth muscle cells, skeletal muscle cells [8]. We were the first to introduce the use of tissue models to determine effective drug doses [38] and have recently extended the application of this technology to include phenotypic screening [39]. We demonstrated the phenotypic difference in tissue mechanics between wildtype cells and mutant cells with truncated $\alpha_1$ integrin [40]. We also extended the applications of engineered tissue models by creating cardiac tissue constructs [41]. Recently, we introduced the concept of growing mini-tissues in 96-well plates for high-throughput drug screening [42]. Initially engineered heart tissues were fabricated using a tissue fabrication mold that cast hydrogel into a ring [5]. To miniaturize the tissues, we developed tissue fabrication wells similar to those used in 96-well format. The center-to-center distance of 96-well plates is 9 millimeters. We developed 8mm x 8mm square wells to fabricate the engineered tissues. We demonstrated the automation of contractile measurements of EHTs to improve the productivity of the proposed studies. The generation of EHTs using pluripotent stem cells will open up the new pathways to develop engineered tissues using human cells. However, there are several issues that need to be solved before benefitting highly promising technology.

## 5. Multi cell types require functional cardiac patches using iPSCs

Human iPSCs, which are generated by genetic reprogramming of somatic cells to an infinitely self-renewing pluripotent state [43, 44], offer an excellent source for robust generation of large numbers of human cardiac myocytes (CMs), endothelial cells (ECs) and fibroblasts (FBs), which are all needed for producing a vital EHT *in vitro* [45]. Protocols to generate somatic cells from hPSCs typically attempt to recapitulate cardiac development in an accelerated manner by temporal application of differentiation cues[45]. For example directed differentiation protocols that achieve high yield and purity of CMs from human pluripotent stem cells apply basic FGF, TGFβ superfamily ligands, VEGF, and Wnt agonists and antagonists at precise differentiation stages [46, 47]. Recently CMs generated from human embryonic stem cells (hESCs) have been injected into rodent cardiac muscle; these CMs engrafted, survived, and resulted in functional improvement in myocardial infarction models [47];[48]. HESC-derived CMs and ECs assembled into spontaneously contracting patches and, when implanted on a rodent pericardium, survived and anastomosed [45]. This study clearly indicates that multiple cell types are necessary in generating functional cardiac patches.

Because of the requirement of multi cell types to fabricate functional cardiac patches, the high throughput rapid prototyping system can contribute to identifying the optimal combination of cell types to fabricate the best cardiac patches. Not only varying the number of CMs, ECs, and FBs to initiate the fabrication, the response to each cell type to various hormones, growth factors, and mechanical environment will have to be screened again to find the best combinations. Therefore, a rapid prototyping system that can screen various test conditions will be critically needed to establish functional tissues with desired phenotypes.

## 6. Summary

Systematically optimizing a tissue-fabrication protocol will require an iterative process of changing the parameters of fabrication and tissue culture conditions incrementally after measuring the resulting functional improvement. To test a large number of these parameters to the best outcome, many tissues have to be fabricated and their function has to be evaluated in high-throughput. To achieve this prototyping cost-effectively, the samples have to be miniaturized. Most importantly, establishing the tissue homeostasis in the engineered tissue has to be achieved either before implantation or by taking account how tissue homeostasis will be established with the acceptor's existing tissue. *In vivo* tissue organ phenotyping with transplanted tissues will be as important as availability of rapid prototyping of engineered tissues.

## 7. Acknowledgements

We thank Mr. David Glaubke for his contribution to the careful reading of manuscript and editorial comments.

## 8. Sources of funding

This research was supported in part by NIH GM087784 grant

## 9. References

[1] Roberts, M.S., Improving the Supply of Donor Organs. *JAMA: The Journal of the American Medical Association*, 2010. 304(23): p. 2643-2644.

[2] Yasuda, S.-I., et al., A novel method to study contraction characteristics of a single cardiac myocyte using carbon fibers. *Am J Physiol Heart Circ Physiol*, 2001. 281(3): p. H1442-1446.

[3] Zimmermann, W.-H., et al., Heart muscle engineering: An update on cardiac muscle replacement therapy. Cardiovascular *Research*, 2006. 71(3): p. 419-429.

[4] Zimmermann, W.-H. and R. Cesnjevar, Cardiac Tissue Engineering: Implications for Pediatric Heart Surgery. *Pediatric Cardiology*, 2009. 30(5): p. 716-723.

[5] Asnes, C.F., et al., Reconstitution of the Frank-Starling mechanism in engineered heart tissues. Biophys J, 2006. 91(5): p. 1800-10.

[6] Wakatsuki, T., K.W. Lieder, and A. Annac, Engineered Tissue Models: Innovative Tools for Early-Stage, Information-Dense, High-Throughput Screening for Drug Discovery. *American Biotechnology Laboratory*, 2006(Nov/Dec).

[7] Nelson, C.M., J.L. Inman, and M.J. Bissell, Three-dimensional lithographically defined organotypic tissue arrays for quantitative analysis of morphogenesis and neoplastic progression. *Nat. Protocols*, 2008. 3(4): p. 674-678.

[8] Vandenburgh, H., High-Content Drug Screening with Engineered Musculoskeletal Tissues. *Tissue Engineering* Part B: Reviews, 2010. 16(1): p. 55-64.

[9] Marquez, J.P., et al., High-Throughput Measurements of Hydrogel Tissue Construct Mechanics. *Tissue Eng* Part C Methods, 2009.

[10] Morelock, M.M., et al., Statistics of Assay Validation in High Throughput Cell Imaging of Nuclear Factor κB Nuclear Translocation. *ASSAY and Drug Development Technologies*, 2005. 3(5): p. 483-499.

[11] Hartwell, L.H., et al., From molecular to modular cell biology. *Nature*.

[12] Cukierman, E., et al., Taking cell-matrix adhesions to the third dimension. *Science*, 2001. 294(5547): p. 1708-12.

[13] Engler, A.J., et al., Extracellular matrix elasticity directs stem cell differentiation. *J Musculoskelet Neuronal Interact*, 2007. 7(4): p. 335.

[14] Pedersen, J.A. and M.A. Swartz, Mechanobiology in the third dimension. *Ann Biomed Eng*, 2005. 33(11): p. 1469-90.

[15] Pankov, R., et al., A Rac switch regulates random versus directionally persistent cell migration. *J Cell Biol*, 2005. 170(5): p. 793-802.

[16] Green, J.A. and K.M. Yamada, Three-dimensional microenvironments modulate fibroblast signaling responses. *Adv Drug Deliv Rev*, 2007. 59(13): p. 1293-8.

[17] Torsoni, A.S., et al., RhoA/ROCK signaling is critical to FAK activation by cyclic stretch in cardiac myocytes. *American Journal of Physiology - Heart and Circulatory Physiology*, 2005. 289(4): p. H1488-H1496.

[18] Sarasa-Renedo, A., V. Tunç-Civelek, and M. Chiquet, Role of RhoA/ROCK-dependent actin contractility in the induction of tenascin-C by cyclic tensile strain. *Experimental Cell Research*, 2006. 312(8): p. 1361-1370.

[19] Ueyama, T., et al., Activated RhoA stimulates c-fos gene expression in myocardial cells. *Circ Res*, 1997. 81(5): p. 672-8.

[20] Aikawa, R., et al., Rho family small G proteins play critical roles in mechanical stress-induced hypertrophic responses in cardiac myocytes. *Circ Res*, 1999. 84(4): p. 458-66.

[21] Hoshijima, M., et al., The low molecular weight GTPase Rho regulates myofibril formation and organization in neonatal rat ventricular myocytes. Involvement of Rho kinase. *J Biol Chem*, 1998. 273(13): p. 7725-30.

[22] Yanazume, T., et al., Rho/ROCK pathway contributes to the activation of extracellular signal-regulated kinase/GATA-4 during myocardial cell hypertrophy. *J Biol Chem*, 2002. 277(10): p. 8618-25.

[23] Tomasek, J.J., et al., Myofibroblasts and mechano-regulation of connective tissue remodelling. *Nat Rev Mol Cell Biol*, 2002. 3(5): p. 349-363.

[24] Slemp, A.E. and R.E. Kirschner, Keloids and scars: a review of keloids and scars, their pathogenesis, risk factors, and management. *Current Opinion in Pediatrics*, 2006. 18(4): p. 396-402 10.1097/01.mop.0000236389.41462.ef.

[25] Sawada, Y., et al., Force Sensing by Mechanical Extension of the Src Family Kinase Substrate p130Cas. *Cell*, 2006. 127(5): p. 1015-1026.

[26] Small, E.M., et al., Myocardin-Related Transcription Factor-A Controls Myofibroblast Activation and Fibrosis in Response to Myocardial Infarction. *Circ Res*, 2010. 107(2): p. 294-304.

[27] Shi, Y. and J. Massagué, Mechanisms of TGF-[beta] Signaling from Cell Membrane to the Nucleus. *Cell*, 2003. 113(6): p. 685-700.

[28] Du, J., et al., TRPM7-Mediated Ca2+ Signals Confer Fibrogenesis in Human Atrial Fibrillation. *Circ Res,* 2010. 106(5): p. 992-1003.
[29] Sturm, O.E., et al., The Mammalian MAPK/ERK Pathway Exhibits Properties of a Negative Feedback Amplifier. *Sci. Signal.,* 2010. 3(153): p. ra90-.
[30] Wakatsuki, T. and E.L. Elson, Reciprocal interactions between cells and extracellular matrix during remodeling of tissue constructs. *Biophysical Chemistry,* 2003. 100(1-3): p. 593-605.
[31] Bendig, G., et al., Integrin-linked kinase, a novel component of the cardiac mechanical stretch sensor, controls contractility in the zebrafish heart. *Genes & Development,* 2006. 20(17): p. 2361-2372.
[32] KJÆR, M., Role of Extracellular Matrix in Adaptation of Tendon and Skeletal Muscle to Mechanical Loading. *Physiological Reviews,* 2004. 84(2): p. 649-698.
[33] Maier, S., et al., Tenascin-C induction by cyclic strain requires integrin-linked kinase. Biochimica et Biophysica Acta (BBA) - *Molecular Cell Research,* 2008. 1783(6): p. 1150-1162.
[34] Russo, R., et al., Lack of increased expression of cell surface markers for circulating fibrocyte progenitors in limited scleroderma. *Clin Rheumatol,* 2007. 26(7): p. 1136-41.
[35] Pacher, P., et al., Measurement of cardiac function using pressure-volume conductance catheter technique in mice and rats. *Nat. Protocols,* 2008. 3(9): p. 1422-1434.
[36] Eschenhagen, T. and W.H. Zimmermann, Engineering myocardial tissue. *Circ Res,* 2005. 97(12): p. 1220-31.
[37] Kolodney, M.S. and E.L. Elson, Correlation of myosin light chain phosphorylation with isometric contraction of fibroblasts. *Journal of Biological Chemistry,* 1993. 268(32): p. 23850-5.
[38] Wakatsuki, T., et al., Effects of cytochalasin D and latrunculin B on mechanical properties of cells. *J Cell Sci,* 2001. 114(Pt 5): p. 1025-36.
[39] Wakatsuki, T., J.A. Fee, and E.L. Elson, Phenotypic screening for pharmaceuticals using tissue constructs. *Curr Pharm Biotechnol,* 2004. 5(2): p. 181-9.
[40] Zutter, M.M., et al., Collagen receptor control of epithelial morphogenesis and cell cycle progression. *Am J Pathol,* 1999. 155(3): p. 927-40.
[41] Eschenhagen, T., et al., Three-dimensional reconstitution of embryonic cardiomyocytes in a collagen matrix: a new heart muscle model system. *Faseb J,* 1997. 11(8): p. 683-94.
[42] Flanagan, N., Tissue Models Boost Drug Discovery Efforts; Eliminating Toxic and Ineffective Compounds at an Early Stage. *Genetic Engineering News,* 2005. 25(20): p. 1
[43] Takahashi, K., et al., Induction of pluripotent stem cells from adult human fibroblasts by defined factors. *Cell,* 2007. 131(5): p. 861-72.
[44] Yu, J., et al., Human induced pluripotent stem cells free of vector and transgene sequences. *Science,* 2009. 324(5928): p. 797-801.
[45] Stevens, K.R., et al., Physiological function and transplantation of scaffold-free and vascularized human cardiac muscle tissue. *Proceedings of the National Academy of Sciences,* 2009. 106(39): p. 16568-16573.
[46] Yang, P.T., et al., Wnt signaling requires retromer-dependent recycling of MIG-14/Wntless in Wnt-producing cells. *Dev Cell,* 2008. 14(1): p. 140-7.

[47] Laflamme, M.A., et al., Cardiomyocytes derived from human embryonic stem cells in pro-survival factors enhance function of infarcted rat hearts. *Nat Biotechnol*, 2007. 25(9): p. 1015-24.

[48] Caspi, O., et al., Tissue engineering of vascularized cardiac muscle from human embryonic stem cells. *Circ Res*, 2007. 100(2): p. 263-72.

# Part 2

## Skeletal Muscle

5

# Tissue Engineering of Skeletal Muscle

Klumpp Dorothee, Horch Raymund E. and Beier Justus P.
*Department of Plastic and Hand Surgery, University Hospital of Erlangen-Nürnberg*
*Germany*

## 1. Introduction

### 1.1 Has Churchill been right?

*"Fifty years hence, we shall escape the absurdity of growing a whole chicken in order to eat the breast or wing, by growing these parts separately under a suitable medium"* (Churchill 1932)

In 1932, Sir Winston Churchill predicted that it will be possible in future to grow and engineer muscle tissue *in vitro*. As a confirmation of his prediction, Vandenburgh *et al.* observed contracting muscle tissue engineered *in vitro* for the first time in 1988 (Vandenburgh, H.H. *et al.* 1988). Only one year later - in 1989 - the group showed that mechanical stimulation of embrionic myoblasts *in vitro* facilitates longitudinal growth of engineered skeletal muscle tissue (Vandenburgh, H.H. & Karlisch 1989). This rapid development raised high expectations for future clinical applications of tissue engineering (TE) of skeletal muscle. Indeed, engineered muscle tissue could be used in a wide range of clinical situations.

A frequent clinical application of skeletal muscle tissue is the microsurgical transfer of myocutaneous free flaps for the coverage of soft tissue defects. As one major disadvantage, the use of free flaps is inevitably linked with a certain morbidity at the donor site including the loss of functional muscle tissue. In this situation, engineered muscle tissue could help to reduce the donor site morbidity. Above all, the advantage of muscle TE lies in the generation of functioning muscle tissue to replace certain muscles after damage or denervation (Klumpp *et al.* 2010). For example, the treatment of facial nerve palsy is momentarily limited to the transfer of autologous muscle tissue innervated by another nerve (the trigeminus nerve, e.g.) or free transfer of distant muscle tissue (Terzis & Konofaos 2008). Though multiple techniques and modifications exist, the results yielded in those clinical situations are moderate (Kumar & Hassan 2002, Terzis & Noah 1997). Furthermore, Kim *et al* demonstrated that myoblast transplantation is a promising method for the reconstruction after partial glossectomy (Kim, J. *et al.* 2003). Herein, TE of skeletal muscle for the replacement of functional muscle tissue could offer an individual alternative.

However, a clinical application of skeletal muscle TE has not been realized to date due to certain obstacles which will be discussed in the following. Though, *in vitro* engineered tissue of skeletal muscle could already play an important role for the clinical treatment of inborn muscle diseases as well as muscle injuries. Once again, Vandenburgh and co-workers engineered dystrophic muscle tissue using it as drug screening platform for Duchenne muscular dystrophy (DMD) treatment (Vandenburgh, H. *et al.* 2009). Thus, a wide range of possible drugs can be analyzed without using time-consuming and costly *in vivo* models (Vandenburgh, H.). Vandenburgh's study demonstrates an economic approach for drug screening in general and orphan drugs in particular. Beside the relatively rare musculoskeletal disorders, engineered skeletal muscle also enables the analysis of more frequent diseases. Kaji *et al* introduced an *in vitro* model of electrically stimulated and contracting muscle tissue to analyze the insulin- and exercise-dependant glucose uptake which plays a role in insulin resistance of type 2 diabetics (Kaji *et al.*). Thus, skeletal muscle TE already plays a role for clinical treatments, though a direct bench-to-bedside approach has yet to become reality.

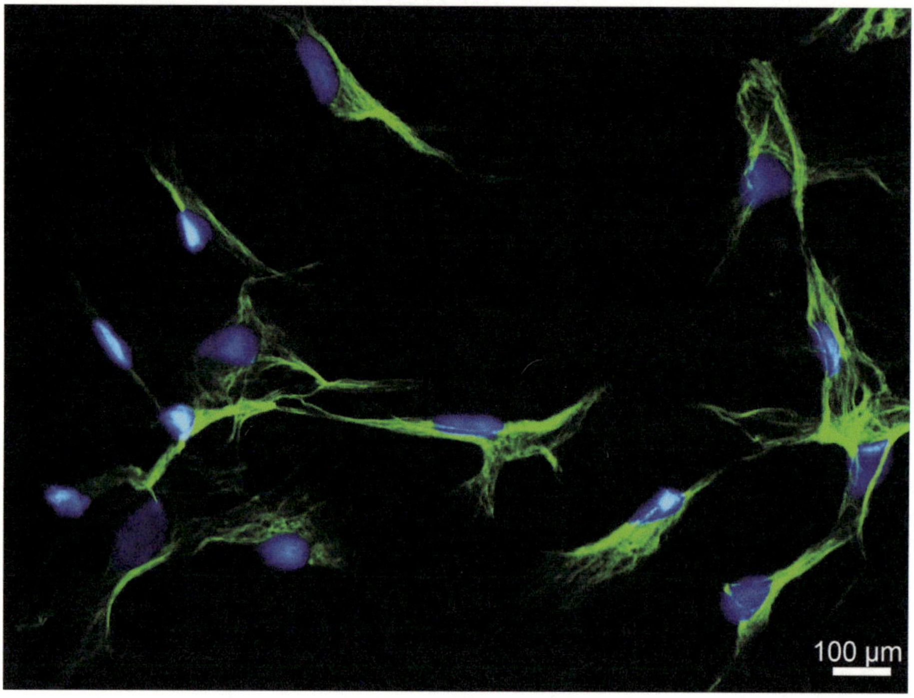


Fig. 1. Skeletal muscle precursor cells (myoblasts) *in vitro*. Immunofluorescent staining for desmin (green) an intermediate filament specifically expressed in myogenic differentiation prior to the formation of myotubes and fusion into muscle fibers. Nuclei are stained non-specifically with DAPI (Diamidine-phenylindole-dihydrochloride; blue). Magnification 400x

## 2. Finding the matrix for muscle TE

### 2.1 Materials

A variety of materials has been analyzed and reviewed regarding their suitability for skeletal muscle TE. On one side, natural materials like collagen I in the first place are preferred due to their biocompatibility and their close resemblance to the natural extracellular matrix (ECM). On the other side, synthetic materials generally show greater stability as well as cost-saving and easy handling.

Regarding collagen I - the main component of the natural ECM of mature skeletal muscle tissue - its advantage lies in high elasticity which is a pre-requisite for muscle contraction. Thus, its mechanical properties *in vivo* meet the demands for new skeletal muscle tissue perfectly. Furthermore, bovine as well as avian collagen show very low immunogenicity *in vitro* and can be safely used *in vivo* (Peng *et al.* 2010). Therefore, collagen sponges and gels have been studied *in vitro* (Madaghiele *et al.* 2008) and *in vivo* (Kroehne *et al.* 2008). However, the rapid shrinkage of hydrogels *in vitro* as well as the low stability of collagen I are important disadvantages (Beier *et al.* 2009) and limit its use in long-term experiments. Another material frequently used for tissue engineering *in vitro* and *in vivo* is fibrin. The stability of fibrin gel is dependent on the concentration of fibrinogen and thrombin and can be augmented by addition of aprotinin which inhibits fibrinolysis (Meinhart *et al.* 1999). In addition, fibrin is known to accelerate vessel ingrowth into the matrix *in vivo* due to its binding sites for vascular endothelial growth factors (VEGF), fibroblast growth factor (FGF-2) and the cytokine interleukin-1 (IL-1) (Mosesson 2005). Still, fibrin gels show a definite loss of stability after 4 weeks *in vivo* (Arkudas *et al.* 2009). As an exception to the relative instability of natural polymers, silkworm fibroin, the structure protein of silk, shows an astonishing tensile strength of 100 -300 MPa and stability (Zhao *et al.* 2003). An *in vivo* stability of over 1 year has been reported for 17% fibroin concentration but, the cell-toxic HFIP (hexafluoro-iso-propanol) as organic solvent is necessary. Even in all-aqueous dissolution of silk fibroin with concentrations of 6 to 10% fibroin, the *in vivo* stability has been found to range between 2 and 6 months (Wang, Y. *et al.* 2008). Thus, silk fibroin provides an adequate stability for tissue engineering *in vivo*. Additionally, silk is in clinical use as suture material for a long time. However, hypersensitivity and adverse reactions have been reported and put the biocompatibility of silk into question (Soong & Kenyon 1984). The chemical and immunogenic properties of silk of *Bombyx mori* silkworms have been studied intensively since and sericin, the glue protein of silk, has been identified subsequently as the immunogenic agent (Panilaitis *et al.* 2003). The use of sericin-depleted fibroin scaffolds clearly increased the biocompatibility of silk matrices and Meinel *et al* could show that the biocompatibility of pure fibroin is comparable to collagen I and even superior to poly(d,l-lactic-co-glycolic acid) (PLGA) (Meinel *et al.* 2005). Thus, silk fibroin derived from silkworms has been used extensively as sponge-like scaffold for tissue engineering *in vitro* (Mandal & Kundu 2009) and *in vivo* (MacIntosh *et al.* 2008, Unger *et al.*). But as a drawback, the high stability and tensile strength of silk fibroin comes along with low elasticity and hydrophilicity that lead to poor cell attachment *in vitro*. In addition, the low elasticity limits the use of silk fibroin scaffolds for TE of skeletal muscle.

Biodegradable synthetic polymers have also been widely used for muscle TE. Their advantages lie in easy handling and very good stability *in vitro* and *in vivo*. For example poly(l-lactic acid) (PLLA) and the more lipophilic co-polymer PLGA have been used in different orthopaedic applications due to their non-toxic properties and long-term stability in vivo (PLLA: 24 months, PLGA: approx. 6 months) (Gunatillake & Adhikari 2003). Though, inflammatory responses (Bostman 1992) and cell toxic effects *in vitro* (Ignatius & Claes 1996) have been reported due to the acidic degradation products of PLLA and PGA. One of the most frequently used synthetic polymer in TE research is poly(ε-caprolactone) (PCL). This biodegradable synthetic polymer shows a slow degradation rate resulting in a long-term stability of approximately 1 year *in vivo* (Bolgen *et al.* 2005). Furthermore, PCL is highly biocompatible and therefore suitable for *in vivo* applications (Cao *et al.* 2009). PCL has been used as films (Sarkar *et al.* 2008) and more frequently as electrospun fibers. However, PCL is also highly hydrophobic and therefore shows poor cell attachment *in vitro* (Zhang, H. & Hollister 2009). Hence, the hydrophobicity of PCL has to be attenuated before cell seeding through plasma treatment (Martins *et al.* 2009) or by coating the scaffold or blending with other materials like collagen to enhance cell attachment (Schnell *et al.* 2007, Zhang, Y.Z. *et al.* 2005). Among the variety of biodegradable synthetic polymers materials like poly(aniline)

(PANi) (Borriello *et al.*) and oxidized polypyrrole (Gomez & Schmidt 2007) stand out due to their electrical conductivity. Though their mechanical properties are similar to PCL including poor cell attachment, electrical conductivity is an interesting feature that qualifies those materials for muscle TE, especially (Li *et al.* 2006).

Taking the properties of the most frequently used biopolymers and synthetic polymers into account, the complexity of mature skeletal muscle tissue asks for a combination of different complementary materials to engineer a matrix that meets the special demands of skeletal muscle TE. Composite scaffolds containing both, synthetic polymers for stability as well as biopolymers for enhanced cell attachment and elasticity, are therefore preferred in muscle TE research. E.g. PCL has been combined with collagen (Choi *et al.* 2008), gelatine (Kim, M.S. *et al.*), PLLA (Engelhardt *et al.*) and other materials. Also the combination of silk fibroin with collagen leads to suitable mechanical properties with good cell attachment *in vitro* (Wang, G. *et al.*, Zhou *et al.*).

## 2.2 Matrices

Comparable to the wide variety of materials, the methods of processing different forms of matrices are equally numerous. Thereby, mechanical properties of a certain matrix, degradation rates and cell attachment depend on the scaffold's architecture. Whereas single-cell-layers can be easily cultured on two-dimensional scaffolds (films, micropatterned scaffolds), the architecture of three-dimensional scaffolds is more complex. To ensure cell survival and proliferation *in vitro*, a three-dimensional matrix should enable diffusion of oxygen, nutrients and metabolites as well as the migration of cells inside the scaffold. Otherwise, cells will only proliferate at the periphery but not in the scaffold's centre (Ishaug-Riley *et al.* 1998). Therefore, the most important features of matrices for three-dimensional TE are high porosity (ideally approximately 90% (Freed *et al.* 1994)), adequate pore-size (Lee *et al* have shown that a range of 50 - 200 μm pore-size are sufficient for smooth muscle cells (Lee *et al.* 2008)) and high interconnectivity of the pores (van Tienen *et al.* 2002) to enable cell migration inside the matrix.

Concerning the pore size, hydrogels usually show freely diffusion of nutrients and oxygen. Cells incorporated in hydrogels can migrate through the scaffold by degrading the gel but at the same time the stability of the gel decreases continuously. Furthermore, the architecture of hydrogels randomly spread pores (Fig. 2). However, the natural extracellular matrix within functional skeletal muscle tissue is highly orientated.

The parallel alignment of ECM and skeletal muscle tissue is the pre-requisite for effective muscle contraction and force-generation along a longitudinal axis. Curtis and Wilkinson first described the "cell guidance theory" by demonstrating that microgrooved matrices with a parallel micropattern provoke parallel aligned cell growth along the pattern of the scaffold (Curtis & Wilkinson 1997). This phenomenon is also present in myoblast culture (Choi *et al.* 2008, Huang *et al.* 2006, Huber *et al.* 2007) facilitating the generation of aligned myotubes (Gingras *et al.* 2009). Therefore, several techniques for aligned scaffold architecture have been developed including selective laser sintering or three-dimensional printing reviewed by Karande *et al* (Karande *et al.* 2004). Another method to gain spatially orientated pores in sponge-like matrices is unidirectional freeze-drying of materials like collagen (Madaghiele *et al.* 2008) or silk fibroin (Mandal & Kundu 2009, 2009). Hydrogels are gradually frozen leading to controlled formation of ice crystals which result in controlled porosity after evaporation of the aqueous part of the hydrogel. Additionally, Schoof and co-workers demonstrated that also the pore-size can be controlled by variation of the freezing-temperature (Schoof *et al.* 2001). Though, there are certain disadvantages linked to the freeze-drying method. First of all, secondary surface modifications of the sponges like coating procedures to enhance cell attachment are difficult and sparsely controllable. Secondly, the alignment of the pores is only spatial whereas the architecture of the matrix surrounding the pores usually remains at random.

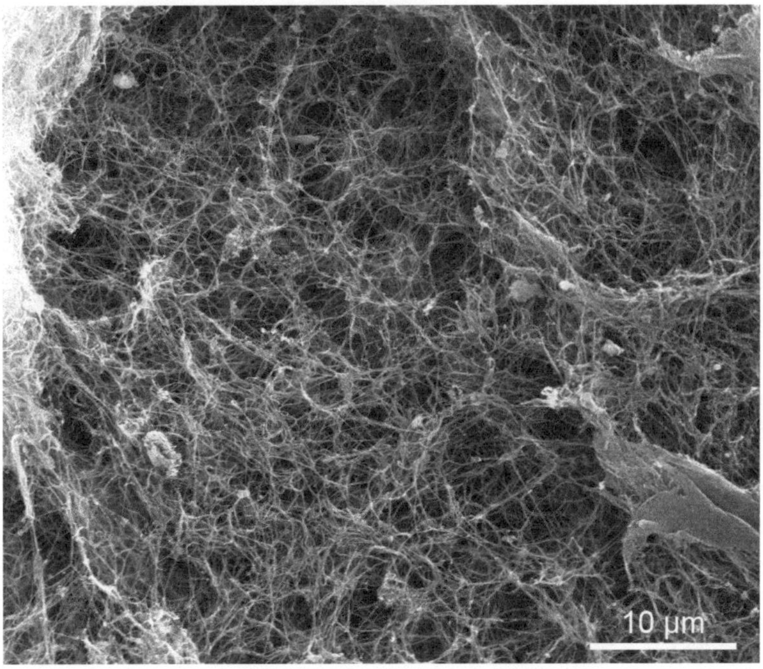


Fig. 2. Scanning Electron Microscopy (SEM) of a fibrin-collagen blend hydrogel. The aqueous part of the hydrogel is evaporated after critical point drying and fibrillar structures of fibrin and collagen is left behind. The random architecture of the hydrogel is clearly visible. 5000x magnification

On the contrary, electrospinning as an alternative method offers strict alignment of the resulting scaffolds (Ayres *et al.* 2006). The technique of electrospinning results in fibers formed by electrical voltage (Boudriot *et al.* 2006). Though, the process depends on multiple parameters, e.g. concentration and viscosity of the spinning solution, the voltage applied during the spinning process or flow rate of the spinning solution. The mechanical and chemical properties of electrospun matrices can be adjusted to the demands of the respective tissue by varying these parameters. Thus, a variety of synthetic and biopolymers can be electrospun at the micro- or nanoscale (Sell *et al.* 2009). Nanofibrous matrices electrospun from ECM proteins such as collagen I or hyaluronic acid mimic the natural ECM exactly and therefore ensure excellent cell attachment, cell viability and differentiation (Barnes *et al.* 2007). As mentioned before, biopolymers often lack the suitable stability for *in vivo* application, whereas the hydrophibicity of synthetic materials prevents rapid cell attachment. Again, the special demands of skeletal muscle TE can be met by combination of synthetic polymers and biopolymers. Different polymers can therefore be combined primarily by spinning polymer-blend solutions, core-shell spinning or co-spinning of different polymer solutions. Methods for secondary surface modification are coating

(Riboldi *et al.* 2005) or plasma treatment (Martins *et al.* 2009) of the matrix after the spinning procedure. Blending different polymers, e.g. PCL and collagen (Fig. 3), is a very simple method to generate composite matrices that combine the properties of both polymers depending on the ratio (Schnell *et al.* 2007).

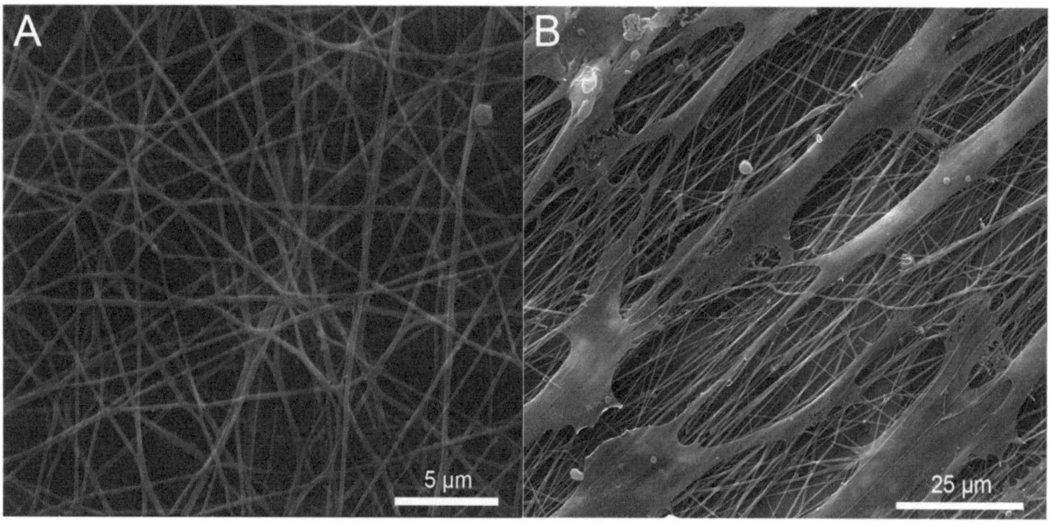


Fig. 3. Scanning Electron Microscopy (SEM) of electrospun PCL-collagen blend nanofibers. A: Randomly spun nanofibers. Magnification 10000x. B: Muscle precursor cells cultured on electrospun nanofibers with parallel alignment. The cell growth along the fibers' direction is clearly visible. Magnification 2500x

The more complex core-shell-spinning technique uses two separate polymers which are electrospun co-axially with the second polymer surrounding the first polymer at the core. Zhang *et al* used PCL as core fiber with a shell of collagen. They proved the core-shell-nanofibers to be superior to collagen-coated PCL fibers regarding cell attachment *in vitro* (Zhang, Y.Z. *et al.* 2005). Jiang and co-workers have introduced electrospun core-shell fibers as drug delivery system (DDS) (Jiang *et al.* 2005). Thereafter, the emerging field of nanofibers and nanoparticles as DDS has found its way into TE research (Sill & von Recum 2008). Controlled release of different drugs e.g. growth factors (Sahoo *et al.*) or angiogenic factors (Yang *et al.*) upgrades nanofiber matrices into "smart" matrices (Moroni *et al.* 2008).
However, electrospinning of aligned nanofibrous matrices is linked with poor control of the pore size as the main disadvantage. The generation of electrospun three-dimensional matrices, especially, results in densely packed scaffolds (fig. 3) that hinder cell infiltration (Baker & Mauck 2007, Telemeco *et al.* 2005). Therefore, co-spinning of water-soluble sacrificial fibers such as poly-(ethylene-oxide) (PEO) has been shown to overcome this problem (Baker *et al.* 2008). The sacrificial PEO fibers are interspersed inside the three-dimensional matrix and dissolve easily in water and alcohol during sterilization procedure before cell seeding. The resulting interspaces between the residual fibers then enable cells to migrate through the matrix (Baker *et al.* 2008). Though the control of pore size and interspaces in orientated nanofiber matrices is still challenging, the electrospinning technique holds great potential for TE and regenerative medicine and therefore pretends to be the most promising matrix for skeletal muscle TE at the moment.

## 3. Cell source

### 3.1 The satellite cell

Satellite cells form the major source for muscle regeneration *in vivo* after injury (Snow 1977). First described by Mauro in 1961 by electron microscopy, the term "satellite cell" was initially used for resident cells beneath the basal lamina of mature skeletal muscle fibers (Mauro 1961). Meanwhile, the "genetical footprint" of this cell population is well-known and satellite cells are specifically identified by expression of the transcription factor Paired-box 7 (Pax7) (Seale *et al.* 2000). Furthermore, satellite cells express MyoD (also known as Myf5; Fig. 4), M-cadherin, c-Met, syndecan-3 and 4 (Cornelison *et al.* 2001) and CD 34 (Beauchamp *et al.* 2000). In the past, it has been a point of discussion whether satellite cells are stem cells or myogenic progenitor cells (Zammit *et al.* 2006). Kuang *et al* proved that a small sub-population, i.e. 10% of satellite cells, shows stemness properties and repopulate the satellite cell niche *in vivo* (Kuang *et al.* 2007). These true stem cells are positive for Pax7 but negative for MyoD, whereas the majority of satellite cells are also positive for MyoD. The myogenic transcription factor MyoD marks the commitment of activated satellite cells to the myogenic line (Weintraub *et al.* 1991).

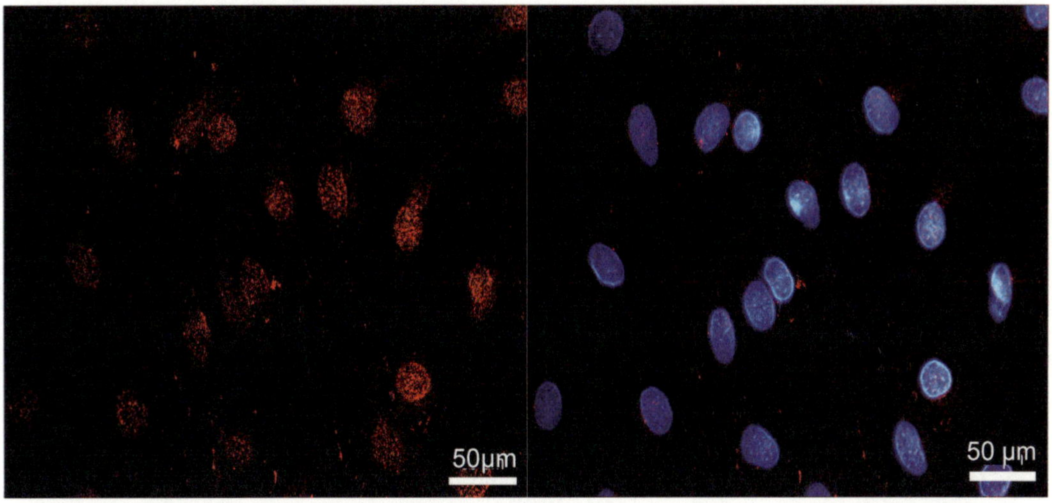


Fig. 4. Muscle precursor cells (MyoD positive cells) *in vitro*. Left side: Immnofluorescent staining for MyoD which is mainly located at the nuclei. Right side: Merge of MyoD staining and DAPI-counterstain. Magnification 400x

This myogenic imprinting renders the satellite cell to be a safe cell source for *in vivo* as well as clinical application without risking dedifferentiation and tumorigenesis. Therefore, satellite cells are the preferred cell source for clinically orientated muscle TE research (Otto *et al.* 2009). Kuang and his group have demonstrated that the $Pax7^{+}/MyoD^{+}$ cell population is renewed by the $Pax7^{+}/MyoD^{-}$ cells through asymmetric self-renewal (Kuang *et al.* 2007). MyoD-positive cells in turn regenerate injured muscle tissue by differentiation into new muscle fibers. Thus, even large muscle tissue defects can be regenerated by a relatively small cell population *in vivo* (Collins *et al.* 2005, Le Grand & Rudnicki 2007). Unfortunately, this astonishing potential of self-renewal and myogenic differentiation of satellite cells *in vivo* is usually lost when satellite cells are isolated and cultured *in vitro* (Yaffe 1968). Boonen and his group proposed the loss of the satellite cell niche at the basal lamina *in vivo* (Boonen & Post 2008). This highly specific niche preserves the stem cell status of quiescent satellite cells (Blau *et al.* 2001). Isolated satellite cells increase their expression of MyoD and differentiate after loosing contact with the basal lamina and thus loose their proliferative potential. Therefore, the generation of suitable numbers of satellite cells for muscle TE by *in vitro* culture and expansion is still challenging. Recently, Gilbert *et al* have demonstrated that the satellite cell function depends on substrate elasticity (Gilbert *et al.*). The group found the best proliferative potential of isolated satellite cells when cultured on PEG hydrogels cross-linked with laminin with an elasticity of 12 kPa which equals the elasticity of muscle tissue *in vivo*. After implantation of the cultured cells in a muscle injury model *in vivo* the engraftment rate was even comparable to freshly isolated and directly implanted satellite cells. Hence, Gilbert and co-workers showed that the satellite cell niche can be mimicked *in vitro* and thus the proliferative potential of cultured satellite cells can be preserved.

Recently, the existence of yet another cell population in adult skeletal muscle tissue has been proved: The telocyte was described by Popescu *et al* in cardiac muscle tissue first (Popescu & Faussone-Pellegrini). The typical shape of telocytes with their prolongations ("telopodes") situated in the vicinity of nerves, vessels and cardiomyocyte progenitors suggests a role in intercellular signalling as regulators in myocardial regeneration and as "nursing cells" for cardiac progenitors (Gherghiceanu & Popescu). In a recent study, Popescu and his group have identified telocytes also in skeletal muscle tissue (Popescu 2011). Beside their typical prolongations, telocytes are known to express c-kit and caveolin-1, but are Pax7 negative and thus differ from the satellite cell population. Additionally, telocytes secrete VEGF (Suciu *et al.*).

### 3.2 Stem cells

Stem cells of different origin offer a unique proliferation potential as the main advantage. To date, adult stem cells play the most important role in TE research, though other sources exist (embryonic or induced pluripotent stem cells (iPSC) e.g.) (Klumpp *et al.*). Since engineering of three-dimensional tissue of skeletal muscle asks for a large quantity of muscle cells, adult stem cells are a suitable cell source in TE research and regenerative medicine (Barile *et al.* 2009, Mollmann *et al.* 2009, Roche *et al.* 2009). Therefore, mesenchymal stromal cells (MSC) are a feasible alternative cell source for skeletal muscle TE due to their high proliferation rates *in vitro* and their low imunogenicity *in vivo* (Chen, L. *et al.* 2009) that even enables allogenic transplantation of MSCs (García-Castro J 2008, Rossignol *et al.* 2009). MSCs can be derived from different tissues, e.g. from bone marrow (BMSC) or adipose tissue derived (ADSC) (Deans & Elisseeff 2009). BMSCs are well-known and have been widely used for cytotherapy in regenerative medicine (Brazelton *et al.* 2003). However, in case of skeletal muscle TE, ADSCs should be preferred due to higher potential for myogenic differentiation as well as higher proliferation rate compared to BMSCs (Kern *et al.* 2006, Zhu *et al.* 2008). Still, the experiences of *in vivo* studies revealed a poor incorporation rate of transplanted MSCs into myofibers (Gussoni *et al.* 1997), ranging between 5 and 10% of the transplanted MSCs in DMD patients (Brazelton *et al.* 2003, Gussoni *et al.* 1997). Low incorporation rates are the main obstacle for cytotherapie in clinical settings. Though, Satija and co-workers proposed paracrine effects of transplanted MSCs *in vivo* as an important therapeutic effect (Satija *et al.* 2009). Therefore, transplanted MSCs secrete different cytokines resulting in anti-inflammatory, angiogenic and anti-apoptotic effects (Meirelles Lda & Nardi 2009, Sze *et al.* 2007) and thus facilitate local endogenous tissue repair (Nesselmann *et al.* 2008). Estrada and his group

explained the angiogenic effect of MSC through their secretion of Cyr61 (Estrada *et al.* 2009) a key factor for angiogensis and tissue repair *in vivo* (Mo *et al.* 2002). Estrada *et al* demonstrated that the sole addition of MSC secretome stimulates angiogenesis *in vitro* and *in vivo*.
Though their poor incorporation into myofibers and - compared to satellite cells - less effective myogenic differentiation *in vivo*, the paracrine effects of MSCs could augment the cell viability and myogenic differentiation of co-transplanted satellite cells.

## 4. Cell survival *in vivo* / vascularization

Whereas the generation of two-dimensional skeletal muscle tissue *in vitro* has been demonstrated by several groups before (Dennis *et al.* 2001, Strohman *et al.* 1990), engineering three-dimensional muscle tissue exceeding the size of 1 mm *in vitro* is still a challenge. Since common in vitro cultures of muscle precursor cells depend on diffusion solely, the thickness of generated tissue is limited to 500 µm to prevent apoptosis of cells in the central region of the construct (Kannan *et al.* 2005). Herein, Freed and co-workers proved the superiority of dynamic flow culturing due to enhanced diffusion capacity compared to static culture conditions (Freed *et al.* 1994). Still, the *in vitro* generation of relevant tissue sizes asks for an adequate vascularization. Levenberg and his group proved that vascularization of skeletal or cardiac muscle in vitro is possible and enhances the transport of nutrients and metabolites (Lesman *et al.* 2010, Levenberg *et al.* 2005). In their study they co-cultured muscle precursor cells with embryonic fibroblasts and endothelial cells seeded into a 3D polymer scaffold. When implanted *in vivo*, the *in vitro* generated vessels connected to vessels of the host and the tissue showed less apoptosis (Levenberg *et al.* 2005). However, even this approach does not meet the demands of a clinical setting, since the muscle tissue, engineered *in vitro*, requires an axial vascularization to enable the transplantation *in vivo* including a microsurgical anastomosis to the recipient site. In most *in vivo* experiments, matrices and muscle precursor cells are implanted subcutaneously leading to random vessel ingrowth from the constructs' periphery. In contrast to subcutaneous *in vivo* models, O. O. Erol and M. Spira introduced the arterio-venous (AV) loop model of the rat in 1980 (Erol & Sira 1980). For this *in vivo* model an AV-loop is created microsurgically between the saphenous artery and vein (Fig. 5) which can be implanted into various matrices (Polykandriotis *et al.* 2008).
Thus, vascularization in general as well as number and pattern of vessel ingrowth of different matrices can be analyzed (Arkudas *et al.* 2010, Polykandriotis *et al.* 2009) and the vascularized matrix offers a platform for tissue engineering for skeletal (Messina *et al.* 2005) or cardiac (Morritt *et al.* 2007) muscle *in vivo*. The feasibility of the AV-loop model in large animals (Beier *et al.* 2009) has been demonstrated recently and poses another step towards a more clinical setting (Beier *et al.* 2010).
Depending on the matrix architecture, a certain period of time is necessary for vascularization of the whole construct. This pre-vascularization time plays an important role for survival rates of implanted cells *in vivo*: Thus, cells implanted after this time period show significantly lower apoptosis rates (Arkudas *et al.* 2007). Vascular growth factors such as bFGF and VEGF (Yancopoulos *et al.* 2000) are frequently used to reduce the pre-vascularization time *in vivo* (Arkudas *et al.* 2007). Therefore, different approaches have been tried for controlled drug release of VEGF. Beside the use as soluble factor or immobilized in fibrin hydrogels (Arkudas *et al.* 2007), VEGF can be bound to nanoparticles (des Rieux *et al.*) or nanofibers (Vournakis *et al.* 2008) as drug delivery systems (DDS) to improve angiogenesis *in vivo* (Zisch *et al.* 2003). Kim *et al* have demonstrated the positive therapeutic

effect of nanoparticle based VEGF release in ischemic muscle tissue (Kim, J. *et al.*). Hypoxia-regulated systems enable an even more selective delivery of VEGF to ischemic sites only, e.g. in myocardial repair (Ye *et al.*). Finally, the VEGF expression of (co-) implanted MSCs could also enhance angiogenesis *in vivo.*

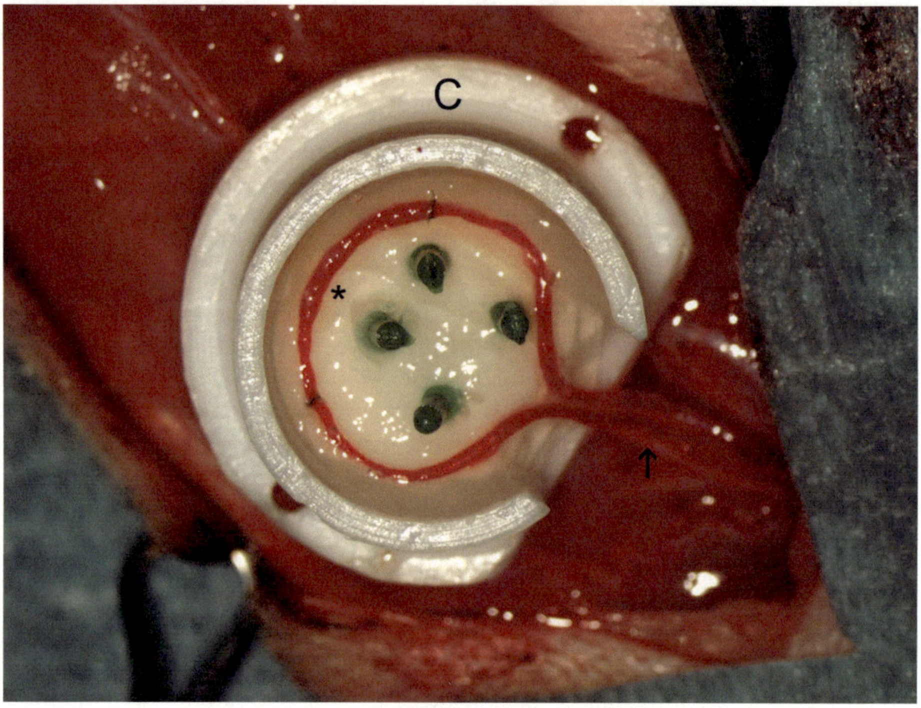


Fig. 5. Arterio-venous loop model in the rat as previously published by our group. The AV-loop (asterisk) is implanted into a Teflon chamber (C) filled with fibrin hydrogel. In this setting, the engineered tissue could be transplanted by anastomosing the pedicle (arrow) to the recipient site

Beside angiogenic growth factors, other factors such as insulin-like growth factor-1 (IGF-1) have been shown to increase survival rates of implanted cells *in vivo* (Wang, F. *et al.* 2009) and to improve myocardial regeneration (Davis *et al.* 2006, Padin-Iruegas *et al.* 2009). Thus, various methods exist to further ameliorate cell survival *in vivo* but still have to be analyzed in detail for their benefit for skeletal muscle TE.

## 5. Myogenic differentiation

### 5.1 Molecular factors for myogenic differentiation

Beside the improvement of cell survival rates, the myogenic differentiation of implanted muscle precursor cells into functional skeletal muscle tissue *in vivo* is another point that has to be addressed by future research. To date, many different molecular factors have been identified which support myogenic differentiation, e.g. akirin-1, muscle specific microRNAs and insulin-like growth factor (IGF-1).

The well-known factor IGF-1 has been demonstrated to increase the proliferation as well as myogenic differentiation of myoblasts *in vitro* by Allen and Boxhorn (Allen & Boxhorn

1989). Later on, this effect was affirmed in different *in vivo* experiments with overexpression of IGF-1 leading to muscle hypertrophy (Adams & McCue 1998) and improving muscle regeneration after trauma (Menetrey *et al.* 2000, Sato *et al.* 2003). Whereas many growth factors increase proliferation rates or differentiation of myoblasts only, IGF-1 enhances both, proliferation as well as myogenic differentiation of muscle precursor cells (Ten Broek *et al.*). Furthermore, IGF-1 overexpression improves survival rates of implanted cells *in vivo* (Wang, F. *et al.* 2009). These properties render IGF-1 as one of the most potential growth factors for myogenesis and skeletal muscle TE. Beside the mitogenic potential of IFG-1 and its positive influence on myogenesis, Haider *et al* have also shown that IGF-1 can mobilize stem cells and increase engraftment of implanted MSCs *in vivo* (Haider *et al.* 2008). The group explained this effect of IGF-1 through its activation of stromal cell derived factor (SDF)-1α and its receptor CXCR4. SDF-1α plays a crucial role in skeletal muscle regeneration and is therefore overexpressed after muscle injury as well as in dystrophic muscle to attract muscle precursor cells which express CXCR4 (Perez *et al.* 2009). Therefore, overexpression of SDF-1α through transfected MSCs (Haider *et al.* 2008) or via controlled drug release (Grefte *et al.*) improves regeneration of skeletal as well as cardiac muscle *in vivo*.

Another factor which influences early myogenic differentiation positively is akirin-1 (also known as Mighty) (Salerno *et al.* 2009). In skeletal muscle tissue, akirin-1 is known to activate quiescent satellite cells and thus promote proliferation of muscle precursor cells. Furthermore, akirin-1 consecutively induces the expression of IGF-2 and hence also increases myogenic differentiation indirectly (Marshall *et al.* 2008). Therefore, akirin-1 combines the activation of quiescent satellite cells with the promyogenic effect of downstream growth factors such as IGFs. Though promising for skeletal muscle TE, the mechanism and molecular pathways of akirin-1 still have to be analysed in detail in future. In addition, the administration of growth factors has to be critically analyzed concerning their risk of tumorigenicity *in vivo*.

Recently, a novel class of regulating factors of myogenesis has been analyzed for their promyogenic potential: Small non-coding RNAs, called microRNA (miRNA) which consist of approximately 20-22 nucleotides (Callis *et al.* 2008). Herein, certain microRNAs (miR-1, miR-133 and miR-206) have been demonstrated as muscle specific. Whereas miR-1 and miR-133 are also expressed in cardiac muscle, miR-206 is specifically expressed in skeletal muscle tissue and up-regulated in patients with muscular dystrophy (Eisenberg *et al.* 2009). Furthermore, the muscle-specific miRNAs differ in their effect on muscle precursor cells. MiR-133 increases proliferation of muscle precursor cells but also inhibits myogenic differentiation (Chen, J.F. *et al.* 2006). On the contrary, miR-1 and miR-206 have been shown to induce myogenic differentiation (Chen, J.F. *et al.* 2006, Kim, H.K. *et al.* 2006). In a recent study, Nakasa et al demonstrated that local injection of miR-1, miR-133 and miR-206 improves muscle regeneration and prevent fibrosis following muscle injury *in vivo* (Nakasa *et al.*). However, further studies are still necessary to analyze the promyogenic potential of muscle-specific microRNAs *in vitro* and *in vivo*.

### 5.2 Electrical stimulation and neurotization

Despite great efforts in the past and various molecular factors which regulate and enhance myogenesis, engineering of mature skeletal muscle tissue still remains a big challenge. Though contracting myotubes - which mark the differentiation and fusion of myoblasts in myogenesis - have been generated by various groups, the generation of adult muscle fibers depends on neural or electrical stimulation (Wilson & Harris 1993). The influence of electrical stimulation on further myogenic differentiation has been analyzed *in vitro*

(Donnelly *et al.*, Stern-Straeter *et al.* 2005) and *in vivo* (Fujita *et al.* 2007). For clinical applications, devices for electrical stimulation have to be implantable and suitable for long-term stimulation (Jarvis & Salmons 2001, Lanmuller *et al.* 2005). The *in vivo* experiments of Dennis and co-workers using implantable stimulation devices in the rat demonstrated that muscle mass as well as the maximum force of a denervated muscle can be maintained by electrical stimulation (Dennis *et al.* 2003). Thus, the physiologic stimulation via motoric innervation can be simulated to support further myogenic differentiation.

Liao *et al* showed that the combination of electrical stimulation with aligned micropatterned matrices even increases the positive effect on myogenic differentiation (Liao *et al.* 2008). Again, electrospun nanofibrous matrices offer the possibility to combine cell guidance through aligned matrix architecture with electrical stimulation via conductive nanofibers. Ghasemi-Mobarakeh and colleagues used electrospun PANi/PCL/gelatine-blend fibers as matrix for neural cell cultivation. Applying electrical stimulation to the matrices, they demonstrated enhanced cell proliferation and neurite outgrowth (Ghasemi-Mobarakeh *et al.* 2009). The use of PANi/gelatine-blend fibers offers an acceptable cell attachment *in vitro* and can be used for cultivation and electrical stimulation of muscle cells *in vitro* (Li *et al.* 2006). Finally, the influence of electrical stimulation on C2C12 murine myoblasts was analyzed by Jun and his group *in vitro*. In their experiments they found that electrical stimulation enhances myogenic differentiation via upregulation of myogenin which specifically marks early myogenesis (Jun *et al.* 2009).

However, engineering functional skeletal muscle tissue *in vivo* not only asks for highly differentiated and organised muscle fibers but also the formation of neuromuscular junctions and neurite ingrowth between the muscle fibers is necessary (Fig. 6).

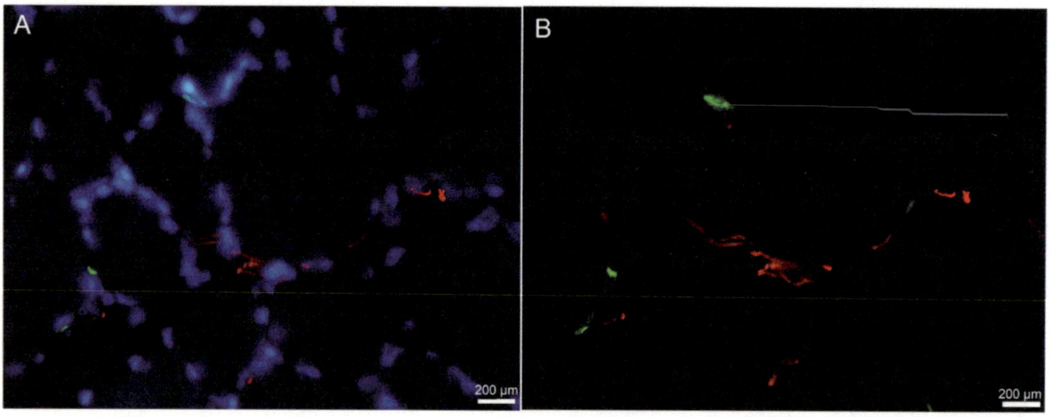


Fig. 6. Immunofluorescent staining of a cross-section of adult skeletal muscle. A: The nuclei are stained with DAPI (blue) and mark the outline of the muscle fibers. The acetylcholine receptors (specifically stained with α-bungarotoxin; green) are reached by nerve terminals (stained for specific neurofilament (NF-100), red) forming the neuromuscular junction (B: DAPI-filter excluded). Magnification 400x

The main component of the neuromuscular junction - the acetylcholine receptor (AChR) - is initially expressed in developing myofibers also in absence of neural cells, i.e. independent from motoric innervation (Witzemann 2006). The AChR clusters at this stage of myogenesis are located at the central regions of myofibers and this phenomenon known as "prepatterning", marks the development of mature myofibers in myogenesis. However, the accumulation of AChR at synaptic sites and further development into functional neuromuscular junctions depends on the specific neuronal factor agrin. Thus, the molecular signalling between developing myofibers and motor neurons is necessary for the generation of functional neuromuscular junctions (Brockhausen *et al.* 2008). The motoric innervation even defines further maturation of the developing muscle tissue into slow- or fast-twitching muscle fibers (Nehrer-Tairych *et al.* 2000). Dhawan *et al* proved that motoric neurotization of implanted muscle precursor cells *in vivo* leads to the formation of neuromuscular junctions (Dhawan *et al.* 2007). In a comprehensive study, they showed nerve-induced contractions of the *in vivo* engineered skeletal muscle tissue after explantation and analysis *in vitro*. Therefore, a successful approach for skeletal muscle TE *in vivo* will necessarily include a motor nerve for neurotization of implanted muscle precursor cells. But to date, *in vivo* models combining motoric neurotization with a pre-vascularized matrix are still rare. Recently, a new AV-loop model in the rat including motoric neurotization has been developed by our group (unpublished data).

## 6. Conclusions

As a conclusion, the main challenges in skeletal muscle TE are therefore: (1) engineering a suitable matrix for muscle TE including a clinical application, (2) improving further myogenic differentiation *in vivo* and (3) enabling the transplantation of functional skeletal muscle tissue to the recipient site including microsurgical anastomosis of an adequate vasculature as well as motoric neurotization of the engineered muscle tissue. Despite these obstacles, the achievements of the recent years demonstrate an encouraging progress of skeletal muscle TE research. Therefore, Churchill's statement concerning skeletal muscle TE *in vitro* may still come true in the future.

## 7. References


Adams, G. R. & McCue, S. A., (1998), Localized infusion of IGF-I results in skeletal muscle hypertrophy in rats. *J Appl Physiol,* 84. 5, p. 1716-22.

Allen, R. E. & Boxhorn, L. K., (1989), Regulation of skeletal muscle satellite cell proliferation and differentiation by transforming growth factor-beta, insulin-like growth factor I, and fibroblast growth factor. *J Cell Physiol,* 138. 2, p. 311-5.

Arkudas, A. et al., (2007), Axial prevascularization of porous matrices using an arteriovenous loop promotes survival and differentiation of transplanted autologous osteoblasts. *Tissue Eng,* 13. 7, p. 1549-60.

Arkudas, A. et al., (2010), Automatic Quantitative Micro-Computed Tomography Evaluation of Angiogenesis in an Axially Vascularized Tissue-Engineered Bone Construct. *Tissue Eng Part C Methods,* .

Arkudas, A. et al., (2009), Dose-finding study of fibrin gel-immobilized vascular endothelial growth factor 165 and basic fibroblast growth factor in the arteriovenous loop rat model. *Tissue Eng Part A,* 15. 9, p. 2501-11.

Arkudas, A. et al., (2007), Fibrin gel-immobilized VEGF and bFGF efficiently stimulate angiogenesis in the AV loop model. *Mol Med,* 13. 9-10, p. 480-7.

Ayres, C. et al., (2006), Modulation of anisotropy in electrospun tissue-engineering scaffolds: Analysis of fiber alignment by the fast Fourier transform. *Biomaterials,* 27. 32, p. 5524-34.

Baker, B. M. et al., (2008), The potential to improve cell infiltration in composite fiber-aligned electrospun scaffolds by the selective removal of sacrificial fibers. *Biomaterials,* 29. 15, p. 2348-58.

Baker, B. M. & Mauck, R. L., (2007), The effect of nanofiber alignment on the maturation of engineered meniscus constructs. *Biomaterials,* 28. 11, p. 1967-77.

Barile, L. et al., (2009), Bone marrow-derived cells can acquire cardiac stem cells properties in damaged heart. *J Cell Mol Med,* .

Barnes, C. P. et al., (2007), Nanofiber technology: designing the next generation of tissue engineering scaffolds. *Adv Drug Deliv Rev,* 59. 14, p. 1413-33.

Beauchamp, J. R. et al., (2000), Expression of CD34 and Myf5 defines the majority of quiescent adult skeletal muscle satellite cells. *J Cell Biol,* 151. 6, p. 1221-34.

Beier, J. P. et al., (2009), De novo generation of axially vascularized tissue in a large animal model. *Microsurgery,* 29. 1, p. 42-51.

Beier, J. P. et al., (2010), Axial vascularization of a large volume calcium phosphate ceramic bone substitute in the sheep AV loop model. *J Tissue Eng Regen Med,* 4. 3, p. 216-23.

Beier, J. P. et al., (2009), Collagen matrices from sponge to nano: new perspectives for tissue engineering of skeletal muscle. *BMC Biotechnol,* 9. p. 34.

Blau, H. M., Brazelton, T. R. & Weimann, J. M., (2001), The evolving concept of a stem cell: entity or function? *Cell,* 105. 7, p. 829-41.

Bolgen, N. et al., (2005), In vitro and in vivo degradation of non-woven materials made of poly(epsilon-caprolactone) nanofibers prepared by electrospinning under different conditions. *J Biomater Sci Polym Ed,* 16. 12, p. 1537-55.

Boonen, K. J. & Post, M. J., (2008), The muscle stem cell niche: regulation of satellite cells during regeneration. *Tissue Eng Part B Rev,* 14. 4, p. 419-31.

Borriello, A. et al., Optimizing PANi doped electroactive substrates as patches for the regeneration of cardiac muscle. *J Mater Sci Mater Med,* .

Bostman, O. M., (1992), Intense granulomatous inflammatory lesions associated with absorbable internal fixation devices made of polyglycolide in ankle fractures. *Clin Orthop Relat Res,* 278, p. 193-9.

Boudriot, U. et al., (2006), Electrospinning approaches toward scaffold engineering--a brief overview. *Artif Organs,* 30. 10, p. 785-92.

Brazelton, T. R., Nystrom, M. & Blau, H. M., (2003), Significant differences among skeletal muscles in the incorporation of bone marrow-derived cells. *Dev Biol,* 262. 1, p. 64-74.

Brockhausen, J. et al., (2008), Neural agrin increases postsynaptic ACh receptor packing by elevating rapsyn protein at the mouse neuromuscular synapse. *Dev Neurobiol,* 68. 9, p. 1153-69.

Callis, T. E. et al., (2008), Muscling through the microRNA world. *Exp Biol Med (Maywood),* 233. 2, p. 131-8.

Cao, H. et al., (2009), The topographical effect of electrospun nanofibrous scaffolds on the in vivo and in vitro foreign body reaction. *J Biomed Mater Res A,* .

Chen, J. F. et al., (2006), The role of microRNA-1 and microRNA-133 in skeletal muscle proliferation and differentiation. *Nat Genet,* 38. 2, p. 228-33.

Chen, L. et al., (2009), Analysis of allogenicity of mesenchymal stem cells in engraftment and wound healing in mice. *PLoS One,* 4. 9, p. e7119.

Choi, J. S. et al., (2008), The influence of electrospun aligned poly(epsilon-caprolactone)/collagen nanofiber meshes on the formation of self-aligned skeletal muscle myotubes. *Biomaterials,* 29. 19, p. 2899-906.

Churchill, W. S., (1932), Thoughts and Adventures First English Edition .

Collins, C. A. et al., (2005), Stem cell function, self-renewal, and behavioral heterogeneity of cells from the adult muscle satellite cell niche. *Cell,* 122. 2, p. 289-301.

Cornelison, D. D. et al., (2001), Syndecan-3 and syndecan-4 specifically mark skeletal muscle satellite cells and are implicated in satellite cell maintenance and muscle regeneration. *Dev Biol,* 239. 1, p. 79-94.

Curtis, A. & Wilkinson, C., (1997), Topographical control of cells. *Biomaterials,* 18. 24, p. 1573-83.

Davis, M. E. et al., (2006), Local myocardial insulin-like growth factor 1 (IGF-1) delivery with biotinylated peptide nanofibers improves cell therapy for myocardial infarction. *Proc Natl Acad Sci U S A,* 103. 21, p. 8155-60.

Deans, T. L. & Elisseeff, J. H., (2009), Stem cells in musculoskeletal engineered tissue. *Curr Opin Biotechnol,* 20. 5, p. 537-44.

Dennis, R. G., Dow, D. E. & Faulkner, J. A., (2003), An implantable device for stimulation of denervated muscles in rats. *Med Eng Phys,* 25. 3, p. 239-53.

Dennis, R. G. et al., (2001), Excitability and contractility of skeletal muscle engineered from primary cultures and cell lines. *Am J Physiol Cell Physiol,* 280. 2, p. C288-95.

des Rieux, A. et al., 3D systems delivering VEGF to promote angiogenesis for tissue engineering. *J Control Release,* 150. 3, p. 272-8.

Dhawan, V. et al., (2007), Neurotization improves contractile forces of tissue-engineered skeletal muscle. *Tissue Eng,* 13. 11, p. 2813-21.

Donnelly, K. et al., A novel bioreactor for stimulating skeletal muscle in vitro. *Tissue Eng Part C Methods,* 16. 4, p. 711-8.

Eisenberg, I., Alexander, M. S. & Kunkel, L. M., (2009), miRNAS in normal and diseased skeletal muscle. *J Cell Mol Med,* 13. 1, p. 2-11.

Engelhardt, E. M. et al., A collagen-poly(lactic acid-co-varepsilon-caprolactone) hybrid scaffold for bladder tissue regeneration. *Biomaterials,* 32. 16, p. 3969-76.

Erol, O. O. & Sira, M., (1980), New capillary bed formation with a surgically constructed arteriovenous fistula. *Plast Reconstr Surg,* 66. 1, p. 109-15.

Estrada, R. et al., (2009), Secretome from mesenchymal stem cells induces angiogenesis via Cyr61. *J Cell Physiol,* 219. 3, p. 563-71.

Freed, L. E. et al., (1994), Kinetics of chondrocyte growth in cell-polymer implants. *Biotechnol Bioeng,* 43. 7, p. 597-604.

Freed, L. E. et al., (1994), Biodegradable polymer scaffolds for tissue engineering. *Biotechnology (N Y),* 12. 7, p. 689-93.

Fujita, H., Nedachi, T. & Kanzaki, M., (2007), Accelerated de novo sarcomere assembly by electric pulse stimulation in C2C12 myotubes. *Exp Cell Res,* 313. 9, p. 1853-65.

García-Castro J, T. C., Madrenas J, Pérez-Simón JA, Rodriguez R, Menendez P., (2008), Mesenchymal stem cells and their use as cell replacement therapy and disease modelling tool. *J Cell Mol Med.* , 12. (6B), p. 2552-65.

Ghasemi-Mobarakeh, L. et al., (2009), Electrical stimulation of nerve cells using conductive nanofibrous scaffolds for nerve tissue engineering. *Tissue Eng Part A,* 15. 11, p. 3605-19.

Gherghiceanu, M. & Popescu, L. M., Cardiomyocyte precursors and telocytes in epicardial stem cell niche: electron microscope images. *J Cell Mol Med,* 14. 4, p. 871-7.

Gilbert, P. M. et al., Substrate elasticity regulates skeletal muscle stem cell self-renewal in culture. *Science,* 329. 5995, p. 1078-81.

Gingras, J. et al., (2009), Controlling the orientation and synaptic differentiation of myotubes with micropatterned substrates. *Biophys J,* 97. 10, p. 2771-9.

Gomez, N. & Schmidt, C. E., (2007), Nerve growth factor-immobilized polypyrrole: bioactive electrically conducting polymer for enhanced neurite extension. *J Biomed Mater Res A,* 81. 1, p. 135-49.

Grefte, S. et al., Skeletal muscle fibrosis: the effect of stromal-derived factor-1α-loaded collagen scaffolds. *Regen Med,* 5. 5, p. 737-47.

Gunatillake, P. A. & Adhikari, R., (2003), Biodegradable synthetic polymers for tissue engineering. *Eur Cell Mater,* 5. p. 1-16; discussion 16.

Gussoni, E., Blau, H. M. & Kunkel, L. M., (1997), The fate of individual myoblasts after transplantation into muscles of DMD patients. *Nat Med,* 3. 9, p. 970-7.

Haider, H. et al., (2008), IGF-1-overexpressing mesenchymal stem cells accelerate bone marrow stem cell mobilization via paracrine activation of SDF-1alpha/CXCR4 signaling to promote myocardial repair. *Circ Res,* 103. 11, p. 1300-8.

Huang, N. F. et al., (2006), Myotube assembly on nanofibrous and micropatterned polymers. *Nano Lett,* 6. 3, p. 537-42.

Huber, A., Pickett, A. & Shakesheff, K. M., (2007), Reconstruction of spatially orientated myotubes in vitro using electrospun, parallel microfibre arrays. *Eur Cell Mater,* 14. p. 56-63.

Ignatius, A. A. & Claes, L. E., (1996), In vitro biocompatibility of bioresorbable polymers: poly(L, DL-lactide) and poly(L-lactide-co-glycolide). *Biomaterials,* 17. 8, p. 831-9.

Ishaug-Riley, S. L. et al., (1998), Three-dimensional culture of rat calvarial osteoblasts in porous biodegradable polymers. *Biomaterials,* 19. 15, p. 1405-12.

Jarvis, J. C. & Salmons, S., (2001), The application and technology of implantable neuromuscular stimulators: an introduction and overview. *Med Eng Phys,* 23. 1, p. 3-7.

Jiang, H. et al., (2005), A facile technique to prepare biodegradable coaxial electrospun nanofibers for controlled release of bioactive agents. *J Control Release,* 108. 2-3, p. 237-43.

Jun, I., Jeong, S. & Shin, H., (2009), The stimulation of myoblast differentiation by electrically conductive sub-micron fibers. *Biomaterials,* 30. 11, p. 2038-47.

Kaji, H. et al., Electrically induced contraction of C2C12 myotubes cultured on a porous membrane-based substrate with muscle tissue-like stiffness. *Biomaterials,* 31. 27, p. 6981-6.

Kannan, R. Y. et al., (2005), The roles of tissue engineering and vascularisation in the development of micro-vascular networks: a review. *Biomaterials,* 26. 14, p. 1857-75.

Karande, T. S., Ong, J. L. & Agrawal, C. M., (2004), Diffusion in musculoskeletal tissue engineering scaffolds: design issues related to porosity, permeability, architecture, and nutrient mixing. *Ann Biomed Eng,* 32. 12, p. 1728-43.

Kern, S. et al., (2006), Comparative analysis of mesenchymal stem cells from bone marrow, umbilical cord blood, or adipose tissue. *Stem Cells,* 24. 5, p. 1294-301.

Kim, H. K. et al., (2006), Muscle-specific microRNA miR-206 promotes muscle differentiation. *J Cell Biol,* 174. 5, p. 677-87.

Kim, J. et al., Targeted delivery of nanoparticles to ischemic muscle for imaging and therapeutic angiogenesis. *Nano Lett,* 11. 2, p. 694-700.

Kim, J. et al., (2003), Muscle tissue engineering for partial glossectomy defects. *Arch Facial Plast Surg,* 5. 5, p. 403-7.

Kim, M. S. et al., The development of genipin-crosslinked poly(caprolactone) (PCL)/gelatin nanofibers for tissue engineering applications. *Macromol Biosci,* 10. 1, p. 91-100.

Klumpp, D. et al., (2010), Skeletal Muscle Tissue Engineering - Current Concepts and Future Perspectives. *Handchir Mikrochir Plast Chir,* .

Klumpp, D. et al., Engineering Skeletal Muscle Tissue-New Perspectives in vitro and in vivo. *J Cell Mol Med,* .

Kroehne, V. et al., (2008), Use of a novel collagen matrix with oriented pore structure for muscle cell differentiation in cell culture and in grafts. *J Cell Mol Med,* .

Kuang, S. et al., (2007), Asymmetric self-renewal and commitment of satellite stem cells in muscle. *Cell,* 129. 5, p. 999-1010.

Kumar, P. A. & Hassan, K. M., (2002), Cross-face nerve graft with free-muscle transfer for reanimation of the paralyzed face: a comparative study of the single-stage and two-stage procedures. *Plast Reconstr Surg,* 109. 2, p. 451-62; discussion 463-4.

Lanmuller, H. et al., (2005), Implantable device for long-term electrical stimulation of denervated muscles in rabbits. *Med Biol Eng Comput,* 43. 4, p. 535-40.

Le Grand, F. & Rudnicki, M. A., (2007), Skeletal muscle satellite cells and adult myogenesis. *Curr Opin Cell Biol,* 19. 6, p. 628-33.

Lee, M., Wu, B. M. & Dunn, J. C., (2008), Effect of scaffold architecture and pore size on smooth muscle cell growth. *J Biomed Mater Res A,* 87. 4, p. 1010-6.

Lesman, A., Gepstein, L. & Levenberg, S., (2010), Vascularization shaping the heart. *Ann N Y Acad Sci,* 1188. p. 46-51.

Levenberg, S. et al., (2005), Engineering vascularized skeletal muscle tissue. *Nat Biotechnol,* 23. 7, p. 879-84.

Li, M. et al., (2006), Electrospinning polyaniline-contained gelatin nanofibers for tissue engineering applications. *Biomaterials,* 27. 13, p. 2705-15.

Liao, I. C. et al., (2008), Effect of Electromechanical Stimulation on the Maturation of Myotubes on Aligned Electrospun Fibers. *Cell Mol Bioeng,* 1. 2-3, p. 133-145.

MacIntosh, A. C. et al., (2008), Skeletal tissue engineering using silk biomaterials. *J Tissue Eng Regen Med,* 2. 2-3, p. 71-80.

Madaghiele, M. et al., (2008), Collagen-based matrices with axially oriented pores. *J Biomed Mater Res A,* 85. 3, p. 757-67.

Mandal, B. B. & Kundu, S. C., (2009), Cell proliferation and migration in silk fibroin 3D scaffolds. *Biomaterials,* 30. 15, p. 2956-65.

Mandal, B. B. & Kundu, S. C., (2009), Osteogenic and adipogenic differentiation of rat bone marrow cells on non-mulberry and mulberry silk gland fibroin 3D scaffolds. *Biomaterials,* 30. 28, p. 5019-30.

Marshall, A. et al., (2008), Mighty is a novel promyogenic factor in skeletal myogenesis. *Exp Cell Res,* 314. 5, p. 1013-29.

Martins, A. et al., (2009), Surface modification of electrospun polycaprolactone nanofiber meshes by plasma treatment to enhance biological performance. *Small,* 5. 10, p. 1195-206.

Mauro, A., (1961), Satellite cell of skeletal muscle fibers. *J Biophys Biochem Cytol,* 9. p. 493-5.

Meinel, L. et al., (2005), The inflammatory responses to silk films in vitro and in vivo. *Biomaterials,* 26. 2, p. 147-55.

Meinhart, J., Fussenegger, M. & Hobling, W., (1999), Stabilization of fibrin-chondrocyte constructs for cartilage reconstruction. *Ann Plast Surg,* 42. 6, p. 673-8.

Meirelles Lda, S. & Nardi, N. B., (2009), Methodology, biology and clinical applications of mesenchymal stem cells. *Front Biosci,* 14. p. 4281-98.

Menetrey, J. et al., (2000), Growth factors improve muscle healing in vivo. *J Bone Joint Surg Br,* 82. 1, p. 131-7.
Messina, A. et al., (2005), Generation of a vascularized organoid using skeletal muscle as the inductive source. *FASEB J,* 19. 11, p. 1570-2.
Mo, F. E. et al., (2002), CYR61 (CCN1) is essential for placental development and vascular integrity. *Mol Cell Biol,* 22. 24, p. 8709-20.
Mollmann, H. et al., (2009), Stem cell-mediated natural tissue engineering. *J Cell Mol Med,* .
Moroni, L., de Wijn, J. R. & van Blitterswijk, C. A., (2008), Integrating novel technologies to fabricate smart scaffolds. *J Biomater Sci Polym Ed,* 19. 5, p. 543-72.
Morritt, A. N. et al., (2007), Cardiac tissue engineering in an in vivo vascularized chamber. *Circulation,* 115. 3, p. 353-60.
Mosesson, M. W., (2005), Fibrinogen and fibrin structure and functions. *J Thromb Haemost,* 3. 8, p. 1894-904.
Nakasa, T. et al., Acceleration of muscle regeneration by local injection of muscle-specific microRNAs in rat skeletal muscle injury model. *J Cell Mol Med,* 14. 10, p. 2495-505.
Nehrer-Tairych, G. V. et al., (2000), The influence of the donor nerve on the function and morphology of a mimic muscle after cross innervation: an experimental study in rabbits. *Br J Plast Surg,* 53. 8, p. 669-75.
Nesselmann, C. et al., (2008), Mesenchymal stem cells and cardiac repair. *J Cell Mol Med,* 12. 5B, p. 1795-810.
Otto, A., Collins-Hooper, H. & Patel, K., (2009), The origin, molecular regulation and therapeutic potential of myogenic stem cell populations. *J Anat,* 215. 5, p. 477-97.
Padin-Iruegas, M. E. et al., (2009), Cardiac progenitor cells and biotinylated insulin-like growth factor-1 nanofibers improve endogenous and exogenous myocardial regeneration after infarction. *Circulation,* 120. 10, p. 876-87.
Panilaitis, B. et al., (2003), Macrophage responses to silk. *Biomaterials,* 24. 18, p. 3079-85.
Peng, Y. Y. et al., (2010), Evaluation of the immunogenicity and cell compatibility of avian collagen for biomedical applications. *J Biomed Mater Res A,* 93. 4, p. 1235-44.
Perez, A. L. et al., (2009), CXCR4 enhances engraftment of muscle progenitor cells. *Muscle Nerve,* 40. 4, p. 562-72.
Polykandriotis, E. et al., (2009), Regression and persistence: remodelling in a tissue engineered axial vascular assembly. *J Cell Mol Med,* 13. 10, p. 4166-75.
Polykandriotis, E. et al., (2008), The venous graft as an effector of early angiogenesis in a fibrin matrix. *Microvasc Res,* 75. 1, p. 25-33.
Popescu, L. M., (2011), Identification of telocytes in skeletal muscle interstitium: implication for muscle regeneration. *J Cell Mol Med,* 15. .
Popescu, L. M. & Faussone-Pellegrini, M. S., TELOCYTES - A case of serendipity: the winding way from Interstitial Cells of Cajal (ICC), via Interstitial Cajal-Like Cells (ICLC) to Telocytes. *J Cell Mol Med,* .
Riboldi, S. A. et al., (2005), Electrospun degradable polyesterurethane membranes: potential scaffolds for skeletal muscle tissue engineering. *Biomaterials,* 26. 22, p. 4606-15.
Roche, R., Festy, F. & Fritel, X., (2009), Stem cells for stress urinary incontinence: the adipose promise. *J Cell Mol Med,* .
Rossignol, J. et al., (2009), Mesenchymal stem cells induce a weak immune response in the rat striatum after allo or xenotransplantation. *J Cell Mol Med,* .
Sahoo, S. et al., Growth factor delivery through electrospun nanofibers in scaffolds for tissue engineering applications. *J Biomed Mater Res A,* 93. 4, p. 1539-50.

Salerno, M. S. et al., (2009), Akirin1 (Mighty), a novel promyogenic factor regulates muscle regeneration and cell chemotaxis. *Exp Cell Res,* 315. 12, p. 2012-21.
Sarkar, S. et al., (2008), Fabrication of a layered microstructured polycaprolactone construct for 3-D tissue engineering. *J Biomater Sci Polym Ed,* 19. 10, p. 1347-62.
Satija, N. K. et al., (2009), Mesenchymal stem cell-based therapy: a new paradigm in regenerative medicine. *J Cell Mol Med,* 13. 11-12, p. 4385-402.
Sato, K. et al., (2003), Improvement of muscle healing through enhancement of muscle regeneration and prevention of fibrosis. *Muscle Nerve,* 28. 3, p. 365-72.
Schnell, E. et al., (2007), Guidance of glial cell migration and axonal growth on electrospun nanofibers of poly-epsilon-caprolactone and a collagen/poly-epsilon-caprolactone blend. *Biomaterials,* 28. 19, p. 3012-25.
Schoof, H. et al., (2001), Control of pore structure and size in freeze-dried collagen sponges. *J Biomed Mater Res,* 58. 4, p. 352-7.
Seale, P. et al., (2000), Pax7 is required for the specification of myogenic satellite cells. *Cell,* 102. 6, p. 777-86.
Sell, S. A. et al., (2009), Electrospinning of collagen/biopolymers for regenerative medicine and cardiovascular tissue engineering. *Adv Drug Deliv Rev,* 61. 12, p. 1007-19.
Sill, T. J. & von Recum, H. A., (2008), Electrospinning: applications in drug delivery and tissue engineering. *Biomaterials,* 29. 13, p. 1989-2006.
Snow, M. H., (1977), Myogenic cell formation in regenerating rat skeletal muscle injured by mincing. I. A fine structural study. *Anat Rec,* 188. 2, p. 181-99.
Soong, H. K. & Kenyon, K. R., (1984), Adverse reactions to virgin silk sutures in cataract surgery. *Ophthalmology,* 91. 5, p. 479-83.
Stern-Straeter, J. et al., (2005), Impact of electrical stimulation on three-dimensional myoblast cultures - a real-time RT-PCR study. *J Cell Mol Med,* 9. 4, p. 883-92.
Strohman, R. C. et al., (1990), Myogenesis and histogenesis of skeletal muscle on flexible membranes in vitro. *In Vitro Cell Dev Biol,* 26. 2, p. 201-8.
Suciu, L. et al., Telocytes in human term placenta: morphology and phenotype. *Cells Tissues Organs,* 192. 5, p. 325-39.
Sze, S. K. et al., (2007), Elucidating the secretion proteome of human embryonic stem cell-derived mesenchymal stem cells. *Mol Cell Proteomics,* 6. 10, p. 1680-9.
Telemeco, T. A. et al., (2005), Regulation of cellular infiltration into tissue engineering scaffolds composed of submicron diameter fibrils produced by electrospinning. *Acta Biomater,* 1. 4, p. 377-85.
Ten Broek, R. W., Grefte, S. & Von den Hoff, J. W., Regulatory factors and cell populations involved in skeletal muscle regeneration. *J Cell Physiol,* 224. 1, p. 7-16.
Terzis, J. K. & Konofaos, P., (2008), Nerve transfers in facial palsy. *Facial Plast Surg,* 24. 2, p. 177-93.
Terzis, J. K. & Noah, M. E., (1997), Analysis of 100 cases of free-muscle transplantation for facial paralysis. *Plast Reconstr Surg,* 99. 7, p. 1905-21.
Unger, R. E. et al., The rapid anastomosis between prevascularized networks on silk fibroin scaffolds generated in vitro with cocultures of human microvascular endothelial and osteoblast cells and the host vasculature. *Biomaterials,* 31. 27, p. 6959-67.
van Tienen, T. G. et al., (2002), Tissue ingrowth and degradation of two biodegradable porous polymers with different porosities and pore sizes. *Biomaterials,* 23. 8, p. 1731-8.
Vandenburgh, H., High-content drug screening with engineered musculoskeletal tissues. *Tissue Eng Part B Rev,* 16. 1, p. 55-64.

Vandenburgh, H. et al., (2009), Automated drug screening with contractile muscle tissue engineered from dystrophic myoblasts. *FASEB J*, 23. 10, p. 3325-34.

Vandenburgh, H. H. & Karlisch, P., (1989), Longitudinal growth of skeletal myotubes in vitro in a new horizontal mechanical cell stimulator. *In Vitro Cell Dev Biol*, 25. 7, p. 607-16.

Vandenburgh, H. H., Karlisch, P. & Farr, L., (1988), Maintenance of highly contractile tissue-cultured avian skeletal myotubes in collagen gel. *In Vitro Cell Dev Biol*, 24. 3, p. 166-74.

Vournakis, J. N. et al., (2008), Poly-N-acetyl glucosamine nanofibers regulate endothelial cell movement and angiogenesis: dependency on integrin activation of Ets1. *J Vasc Res*, 45. 3, p. 222-32.

Wang, F. et al., (2009), Fabrication and characterization of prosurvival growth factor releasing, anisotropic scaffolds for enhanced mesenchymal stem cell survival/growth and orientation. *Biomacromolecules*, 10. 9, p. 2609-18.

Wang, G. et al., Electrospun PLGA-silk fibroin-collagen nanofibrous scaffolds for nerve tissue engineering. *In Vitro Cell Dev Biol Anim*, 47. 3, p. 234-40.

Wang, Y. et al., (2008), In vivo degradation of three-dimensional silk fibroin scaffolds. *Biomaterials*, 29. 24-25, p. 3415-28.

Weintraub, H. et al., (1991), The myoD gene family: nodal point during specification of the muscle cell lineage. *Science*, 251. 4995, p. 761-6.

Wilson, S. J. & Harris, A. J., (1993), Formation of myotubes in aneural rat muscles. *Dev Biol*, 156. 2, p. 509-18.

Witzemann, V., (2006), Development of the neuromuscular junction. *Cell Tissue Res*, 326. 2, p. 263-71.

Yaffe, D., (1968), Retention of differentiation potentialities during prolonged cultivation of myogenic cells. *Proc Natl Acad Sci U S A*, 61. 2, p. 477-83.

Yancopoulos, G. D. et al., (2000), Vascular-specific growth factors and blood vessel formation. *Nature*, 407. 6801, p. 242-8.

Yang, F. et al., Genetic engineering of human stem cells for enhanced angiogenesis using biodegradable polymeric nanoparticles. *Proc Natl Acad Sci U S A*, 107. 8, p. 3317-22.

Ye, L. et al., Nanoparticle based delivery of hypoxia-regulated VEGF transgene system combined with myoblast engraftment for myocardial repair. *Biomaterials*, 32. 9, p. 2424-31.

Zammit, P. S., Partridge, T. A. & Yablonka-Reuveni, Z., (2006), The skeletal muscle satellite cell: the stem cell that came in from the cold. *J Histochem Cytochem*, 54. 11, p. 1177-91.

Zhang, H. & Hollister, S., (2009), Comparison of bone marrow stromal cell behaviors on poly(caprolactone) with or without surface modification: studies on cell adhesion, survival and proliferation. *J Biomater Sci Polym Ed*, 20. 14, p. 1975-93.

Zhang, Y. Z. et al., (2005), Characterization of the surface biocompatibility of the electrospun PCL-collagen nanofibers using fibroblasts. *Biomacromolecules*, 6. 5, p. 2583-9.

Zhao, C. et al., (2003), Structural characterization and artificial fiber formation of Bombyx mori silk fibroin in hexafluoro-iso-propanol solvent system. *Biopolymers*, 69. 2, p. 253-9.

Zhou, J. et al., Electrospinning of silk fibroin and collagen for vascular tissue engineering. *Int J Biol Macromol*, 47. 4, p. 514-9.

Zhu, Y. et al., (2008), Adipose-derived stem cell: a better stem cell than BMSC. *Cell Biochem Funct*, 26. 6, p. 664-75.

Zisch, A. H., Lutolf, M. P. & Hubbell, J. A., (2003), Biopolymeric delivery matrices for angiogenic growth factors. *Cardiovasc Pathol*, 12. 6, p. 295-310.

6

# Skeletal Muscle Tissue Engineering Using Biological Scaffolds for Repair of Abdominal Wall Defects in a Rabbit Model

Zuki Abu Bakar[1], Ayele Taddese Tsedeke[2],
Noorjahan Banu Mohamed Alitheen[3] and Noordin Mohamed Mustapha[1]
[1]*Faculty of Veterinary Medicine, Universiti Putra Malaysia, Malaysia*
[2]*Faculty of Veterinary Medicine, University of Gondar,*
[3]*Faculty of Biotechnology and Biomolecular Science, Universiti Putra Malaysia*
[1,3]*Malaysia*
[2]*Ethiopia*

## 1. Introduction

The repair of large soft tissue defects, especially abdominal wall defects, is still a challenge for surgeons and continues to be a significant problem for patients (Gangwar et al., 2006; Lai et al., 2003). Free muscle transfer from local or distant sites is commonly employed for the surgical repair of muscle-tissue defects, but this practice is frequently associated with significant donor-site morbidity (Wei et al., 1995). A potential alternative includes the *in vitro* development of a functional three-dimensional muscle for transplantation or the construction of implantable biological biomaterials to direct myogenesis at the target site. The ideal biomaterial for abdominal wall repair should possess adequate strength, no hypersensitivity reactions, and biocompatibility to facilitate tissue ingrowths, which may help long-term maintenance of mechanical strength (Lai et al., 2003). In the reconstruction of a new tissue, two components are usually very important: the cells and the matrix (scaffolds) where they are seeded.

Tissue engineering is an interdisciplinary field which applies the principles and methods of engineering and the life sciences towards the fundamental understanding of structural and functional relationships in normal and pathological tissue and the development of biological substitutes to restore, maintain or improve function (Skalak & Fox, 1988). The creation of skeletal muscle tissue using tissue engineering methods holds promise for the treatment of a variety of muscle diseases, including skeletal myopathies such as muscular dystrophy or spinal muscular atrophy, traumatic injury and aggressive tumor ablation (Guettier-Sigrist et al., 1998; Law et al., 1993). Tissues that are engineered using the patient's own cells, or immunologically inactive allogenic or xenogenic cells have the potential to overcome current problems of replacing lost tissue function and offer new therapeutic options for diseases where currently no options are available. Moreover, this technology can play a vital role in the future management of paediatrics patients (Saxena et al., 1999a).

In general "Tissue engineering" refers to the science of creating living tissue to replace, repair or augment diseased tissue. The engineered tissue may be created *in vitro* and subsequently implanted into the patient or the tissue may be created entirely *in vivo*. Regardless of the technique, tissue engineering requires at least three components: a growth-inducing stimulus (induction), responsive cells (production), and a scaffold (biomaterials) to support tissue formation (Bronzino, 2006).

Biomaterials are any material used to make devices to replace a part or a function of the body in a safe, reliable, economic and physiologically acceptable manner (Hench & Erthridge, 1982). The use of biomaterial for repair of abdominal wall defects is gaining increasing recognition and the use of biomaterials to achieve a tension–free repair has resulted in a significant reduction in post-operative pain, length of recovery period and the number of recurrence (Amid, 1997).

Currently there is an increasing demand for cheap and ideal biomaterials which can be used in reconstructive surgery for repair of traumatic wounds suffer during war, traffic accidental and natural disaster and in the restore of the functions of diseased tissues or organs. Biomaterials are either synthetic (prosthesis) such as ceramic, polymeric and composite or biologic (bioprosthesis) such as heart valve, skin and other types of tissue graft (Black, 1992). The ideal biomaterials for abdominal wall repair should possess adequate strength, no hypersensitivity reactions and biocompatibility to facilitate tissue ingrowths, which may help long term maintenance of mechanical strength (Lai et al., 2003).

Recently, new biodegradable biomaterials developed from biological materials mainly of collagen in nature have been tested for repair of body wall instead of the non-biodegradable synthetic materials. Bovine pericardium, human cadaveric fascia lata, human dura mater and collagen-based materials derived from porcine small intestine submucosa have been investigated for reconstruction of abdominal wall defects (Ueno et al., 2004; Saaverda et al., 2001; Santillan et al. 1995; Rodgers et al., 1981). However in most research, it is indicated that these collagen based biomaterials are failed to be replaced by skeletal muscle tissue or regeneration of muscle tissue is not observed as whole therefore optimal muscle recovery or regeneration may require the use of novel technology like tissue engineering. Skeletal muscle comprises approximately 48% of the body mass and is responsible for voluntary control and active movement of the body. Application of tissue engineering techniques and successful fabrication of skeletal muscle mass holds now a promising future for the restoration of 3-dimentional contour as well as the loss of function for the affected part of the body. In order to generate skeletal muscle tissue, myoblasts which are skeletal muscle tissue precursors, have been employed (Saxena, 2005).

One of the strategies for muscle tissue engineering involves the harvesting of satellite cells, their expansion *in vitro*, and their subsequent autologous implantation *in vivo* into the sites requiring repair or replacement. One of the main obstacles in the formation of new muscle tissue is the lack of an adequate support for expanded satellite cells. To overcome this obstacle, many researcher groups are trying to develop adequate synthetic and biological delivery systems for implanted cells (Conconi et al., 2005). Currently myoblast transplantations have been predominantly performed by injection of myoblast cell suspensions into mature skeletal muscle. These single cells have been shown to fuse with the host myofibers (Wernig et al., 2000). Saxena et al. (1999b) were the first to implant successfully *in vitro* cultured myoblasts into a non-muscular environment. Their group used a polyglycolic acid (PGA) mesh as a scaffold for skeletal muscle cells (Saxena et al., 2001;

Saxena et al., 1999b). Myoblasts have also been seeded onto polyglycolic acid porous polymers with successful generation of vascularized new skeletal muscle *in vivo* (Saxena and Willital, 2000). Synthetic materials, such as Dacron and Polytetrafluorethylene, have been used to repair congenital muscles defects, e.g. Onfalocele and gastrochisis (Bauer et al., 1999; Calzolari et al., 1995; Meddings *et al.*, 1993). However, all of these materials do not allow cell growth and do not follow host development. Evidence has been provided that biological materials can support *in vivo* and *in vitro* cell adhesion and proliferation.

Bovine pericardium has been used as source of natural biomaterials for a wide range of clinical applications (Jose et al., 2001; Won et al., 2000; Marques et al., 1995). However, few clinical data are available in current literature about grafting of bovine tunica vaginalis parietalis for surgical use, although up to 10x7cm or larger collagen rich sheet of tunica vaginalis parietalis can be obtained from a testis of adult cattle. Naturally derived materials, including glutaraldehyde tanned bovine pericardium (James et al., 1991), small intestine submucosa (Clarke et al., 1996; Prevel et al., 1995) and also lyophilized and glycerolized bovine pericardium and tunica vaghinalis parietalis (Hafeez, 2005), have been tried in animal models. These biomaterials are less susceptible to infection and cause less foreign body response (Badylak et al., 1998; Hiles et al., 1995). Thus, the utilization of non-edible bovine offal's of collagenous nature for the development of cheap and safe surgical patches for clinical use will be of economical importance in developing countries. However, fail to recover muscle tissue and also lack of strength over time is a concern for clinical application in which adequate tensile properties are necessary. Thus, for this reason, it is important to understand not only the biological response to degradable biomaterials, but also the expected mechanical properties of implant and replacement of tissue over time. These new collagen based biomaterials has to be improved its morphological and biomechanical properties just by seeding it with myoblast cells and must be evaluated first in animals' model before being approved for test in human.

In this study, biological collagen-based biomaterials were employed for reconstruction of abdominal wall defects in a rabbit model. These are bovine parietal pericardium and bovine tunica vaginalis parietalis which were collected from abattoir and processed by freeze-drying preservation methods and sterilized using gamma irradiations system. In recent years, this preservation method in combination with sterilization using gamma irradiation has been reported in its good output (Zuki et al., 2007; Hafeez, 2005; Hafeez et al., 2005a, 2005b).

A reason for future use of collagen-based biomaterials seems functioning as temporary scaffolds which during resorption generate new permanent tissue. However, very fast resorption rates were reported for degradable materials of synthetic or natural origin (Smith et al., 1989; Tyrell et al., 1989). Moreover, fail to recover muscle tissue and also lack of strength over time is a concern for clinical application in which adequate tensile properties are necessary.

## 2. Morphological evaluation of the myoblast seeded scaffolds *in vitro*

The parietal pericardial sacs and tunica vaginalis parietalis were collected from healthy adult cattle immediately after slaughter, placed in polyethylene plastic bags containing cold normal saline and transferred into an icebox for transportation. The BP and BTV sacs were cleaned, trimmed into rectangular sheets and washed thoroughly under running tap water and rinsed in serial changes of sterile saline (Figure 1). The BP and BTV sheets were then transferred aseptically into a sterile bottle containing 150ml of 0.05% sodium hypochlorite and shaken slowly for 10 minutes. The sodium hypochlorite were then washed off by shaking the tissue sheets in three changes of sterile normal saline for 20 minutes each change, then kept overnight at 4°C. The BP and BTV sheets were freeze-dried and sterilized by gamma irradiation at 25 KGy (Cobalt 60, JS8900, IR-174; CDM MDS NORDION, Ontario, Canada) according to the MINT (Malaysian Institute for Nuclear Technology) tissue bank work instruction Manual (1998) (Figure 2). Samples were taken from both BP and BTV grafts for histological and SEM examinations. The sections were stained with Haematoxylin and Eosin as described by Wilson and Gamble, (2002) for general histology and Masson's trichrome method as described by Jones, (2002) for demonstration of muscles and collagenous tissue.

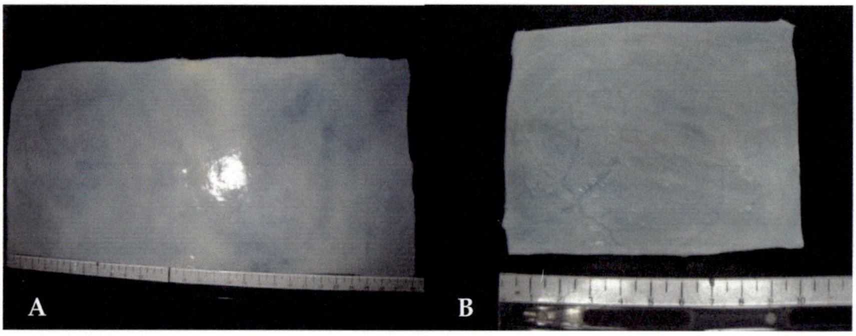


Fig. 1. Photographs A and B show BP and BTV after cleaning, respectively

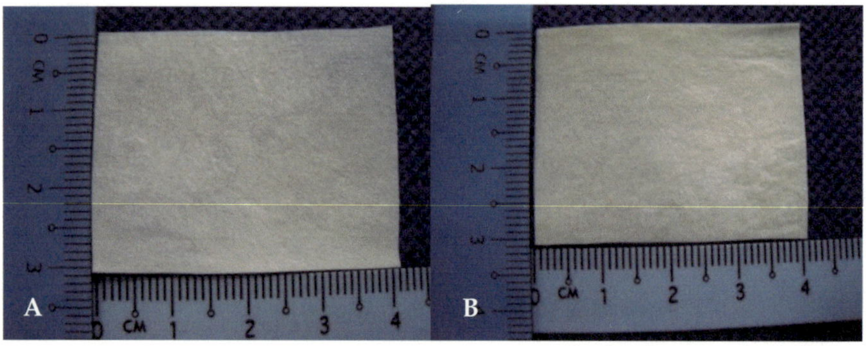


Fig. 2. Photographs A and B show 3 x 4 $cm^2$ scaffolds derived from BP and BTV, respectively. Note both have similar appearance macroscopically

Biomaterials provide mechanical stability to the construct in the short term and serve as a template for the three-dimensional organization for the developing tissue (Hutmacher, 2001). Therefore a critical step in skeletal muscle-tissue engineering is the identification of the optimal biomaterial scaffold, able to promote normal differentiation and maturation of myoblasts into myotubes and myofibers. At present, several kinds of natural scaffolds, such as porcine small intestinal submucosa (Badylak et al., 2002), acellular dermal matrix (Chung et al., 2003) and collagen (Lai et al., 2003) have been employed to repair abdominal wall defects in experimental animal models. Acellular matrices, obtained by detergent-enzymatic method do not elicit rejection responses (Roeder et al., 1999) and could be employed as promising tissue substitutes (Parnigotto et al., 2000a, 2000b; Sutherland et al., 1996).

### 2.1 Light and scanning electron microscopic evaluations of the scaffolds

Bovine parietal pericardium and bovine tunica vaginalis parietalis were used to obtain a biological scaffold possessing morphological and mechanical properties resembling those of the native tissue. Bovine pericardium originated scaffolds seems have large pores and less interconnectivity with completely acellular (Figure 3A) and is collagenous-based in nature with no cellular and vascular element (Figure 4A), and this might be as a result of slight modification of preparation methods used in the previously reported methods by Hafeez, (2005) in conjunctions with light density of collagen as compared with BTV. However, BTV originated scaffolds possessing large pores and less interconnectivity with scanty cell remnant within dense collagen bundles (Figures 3B and 4B), and this perhaps contributed by high density of the structural fiber components. A similar result has been reported by Hafeez, (2005). The light and electron microscopic examinations revealed that nearly all cellular components were removed without ultrastructural evidence of damage to fibrous components. Even though similar acellular results were obtained by Chang et al. (2002) in their *in vivo* evaluation of cellular and acellular bovine pericardia fixed with a naturally occurring crosslinking agent (genipin). But, they had used a special cell extracting methods which was not used in this study.

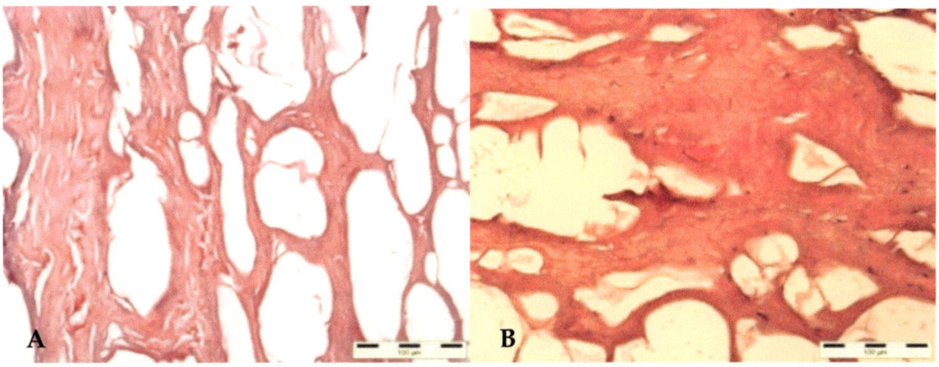


Fig. 3. Microphotographs of the collagen based of BP (A) and BTV (B) scaffolds show the pore size appears larger with less interconnectivity. Note also the scanty remnant cellular elements within a dense connective tissue matrix in B. H&E, Bar = 100μ

Scanning electron microscopic examinations revealed that the serosal/inner surface of both type of scaffolds (BP and BTV scaffolds) had irregular polygonal shape structure (Figures 5A and 6A). However, the outer layers were rough with irregular appearance of connective tissue (Figures 5B and 6B). The finding of this study is in accordance to the result reported by Hafeez (2005). They had reported that the surface of freeze-dried BP and BTV differ from fresh BP and BTV in many aspects such as the lost of the serosal layer and the exposure of underlying collagen bundles and separation of individual collagen bundles, and fibers, which appeared to be wavy and well defined. Adequate porosity with interconnected pores is required not only to achieve sufficient cell seeding density within the scaffold, but also to facilitate cell proliferation and differentiation by allowing the transport of nutrients and oxygen into and out of the scaffold (Goddard & Hotchkiss, 2007). This study has also demonstrated that the processed collagen-based biomaterials such as BP and BTV are sufficiently porous in nature (Figure 7). A similar result had been reported on bovine parietal pericardium (Chang et al., 2005).

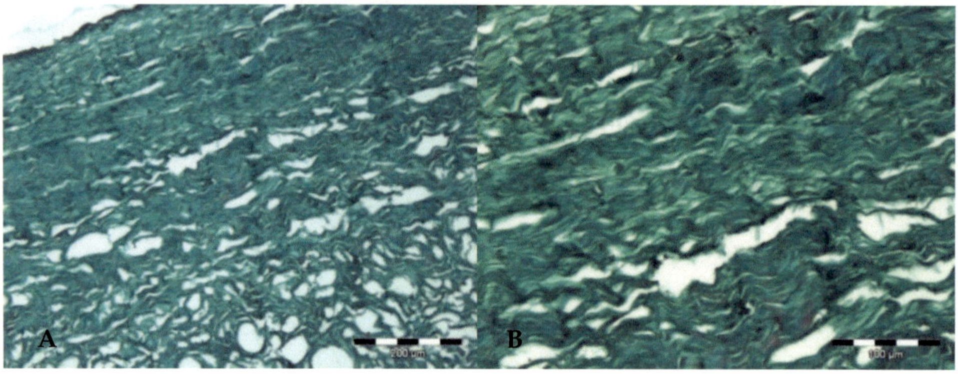


Fig. 4. Microphotographs of the masson's trichrome stained BP (A) and BTV (B) scaffolds show the scaffold is collagenous-based in nature with no cellular and vascular element in A, and dense collagenous tissue with scanty cellular elements in B. Bar = 100µ

Based on the light and scanning electron microscopic analysis the prepared pre-implanted BP and BTV are fibro-collagenous in nature with no or very few cellular remnants which make this biomaterials weak antigenicity apart from its excellent biocompatibility and biodegradability. This is a fascinating result to address the need of non-immunogenic and non-prosthetic biomaterials that could guide perhaps the regeneration of normal tissue. This study has also revealed that processed BP and BTV are porous in nature that could achieve to accommodate sufficient cell seeding density and facilitate cell proliferation and differentiation by allowing the transport of nutrient and oxygen into and out of the scaffolds (Goddard & Hotchkiss, 2007).

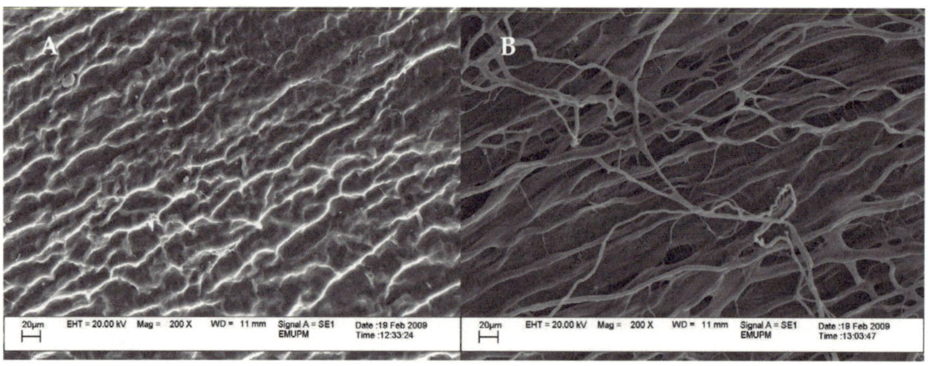


Fig. 5. SEM electronmicrographs of the A) serosal/interior and B) epipericardial surfaces of BP scaffold showing polygonal shape structure and irregular fibrous surface, respectively

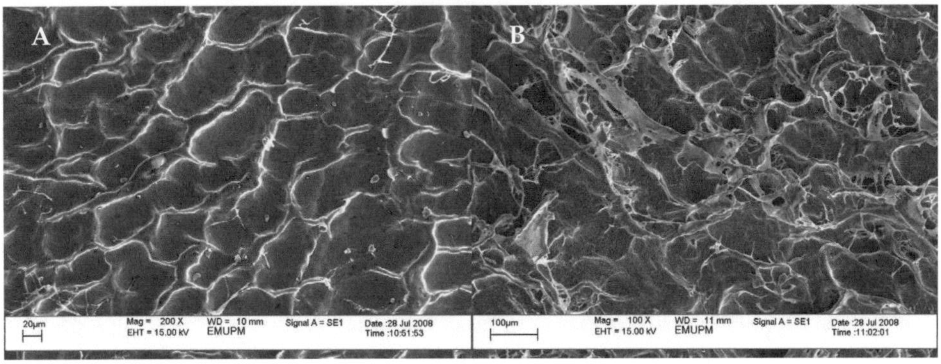


Fig. 6. SEM electronmicrograph of the A) serosal and B) scrotal surfaces of BTV scaffold showing polygonal shape structures and irregular surface, respectively

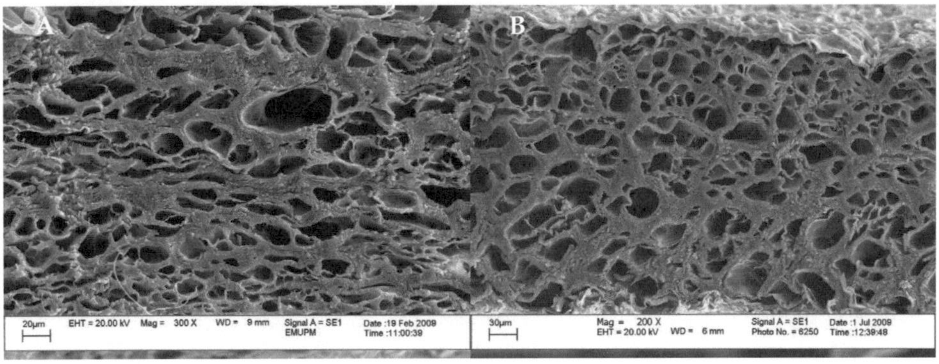


Fig. 7. SEM electronmicrographs of the cross sectioned scaffolds of BP (A) and BTV (B) showing good porosity

### 2.2 Skeletal myoblast harvesting and isolation analysis

Skeletal muscle tissue engineering depends on the unique regenerative properties of satellite cells and the ability to direct intrinsic cell programs associated with proliferation and differentiation. The development of muscle stem cells and genetically engineered myoblasts for transplantation has become in the past few years a very attractive and challenging method for treatment of patients with muscle diseases (Wu et al., 2003; Huard et al., 2003; Haider et al., 2003). However, the repair of extensive muscular defects or diseased regions may require voluminous tissue grafts seeded with large amounts of myoblasts.

Skeletal muscle tissues were obtained from the hind limbs muscle (Soleus muscles) of 5-day-old rabbit using a primary cell culture. Cell purity is assessed by desmin antibodies to prove the purity of the cultured myoblast using immunocytochemistry analysis and also quantified by flow cytometric analysis. This identification method has been also employed by many researchers despite their slight variation in techniques (Guarita-Souza et al., 2006; Ott et al., 2004; Lai et al., 2003; Winokur et al., 2003; Rando & Blau, 1994). The prepared 3 x 4 $cm^2$ scaffolds were seeded with myoblast at a density of $1.0 \times 10^7$cells. Cell morphology and growth in the scaffolds were examined using SEM.

During the initial stages of culture, the myoblast showed a rounded morphology. After the first 24 hours, the cells attached to the culture surfaces and also spindle-shaped cells started to sprout out from isolated muscle fibres and then proliferated. Therefore, this work allowed us to obtain muscle satellite cell-derived cultures, containing myoblasts, expressing the transcription factors involved in the skeletal muscle-cell differentiation program (Chen & Goldhamer, 2003). Moreover, cultured myoblasts are endowed with a high proliferation rate, so that only three weeks is needed to reach the optimal cell number for graft implantation. Hence, this myoblast isolation technique appears to give a better result than those previously used to isolate satellite cells by Lai et al. (2003), Marzaro et al. (2002) and Van Wachem et al. (1999), but it appears to give a similar results reported by Conconi et al. (2005).

Moreover the culture technique that has been used allows the preferential growth of myoblasts over fibroblasts, and the yield of myogenic cells from the initial primary culture is very high and hence the percentage of myogenic cells increases with time in culture. Thus, from a mixed culture of myoblasts and fibroblasts, a nearly pure culture of myoblasts were achieved within three weeks and therefore the purity of the myoblast was confirmed using desmin immunocytochemistry which is a common preferable method employed by many researchers to produce a better result (Guarita-Souza et al., 2006; Conconi et al., 2005; Kamelger et al., 2004). Myoblast percentages were determined using flow cytometric analysis. According to desmin immunocytochemistry and flow cytometric analysis more than 97% of the isolated skeletal myoblast cells have got myogenic phenotype (Figures 8). However, Lai et al. (2003) had reported that approximately 80% of isolated skeletal muscle cells had a myogenic phenotype in desmin immunocytochemistry analysis which perhaps indicates the superiority of the present technique to isolate myoblast.

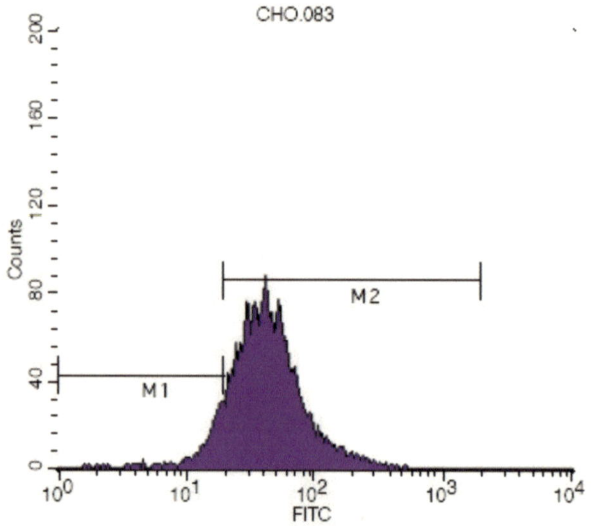


Fig. 8. Flow cytometric histograms depicted FITC-conjugated desmin positive myoblast. Note that the percentage of pure myoblast is 97% whereas non myoblast cells are 3% in flow cytometry analysis. M1= Proportion of desmin negative cells, M2= Proportion of desmin positive cells

*In vitro* test revealed the ability of myoblast to form myotube in static culture on the surface of petri-dish which was not even a tissue culture flask. However, there were only few cells on the surface with haphazard arrangement which is perhaps related to the adhesion properties of the cell towards the petri-dish as showed in figure 9. This result is in agreement with Yan et al. (2006) who reported that if the myoblast cells are plated on plain plastic tissue culture or on non-aligned collagen gels, the satellite cells differentiate into skeletal myotube but the arrangement is haphazard.

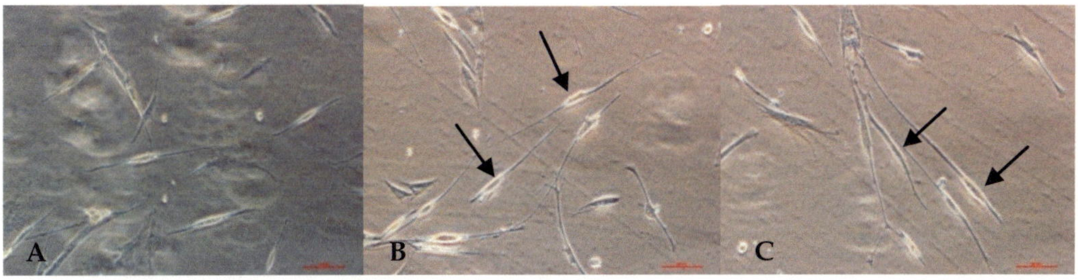


Fig. 9. The phase-contrast microphotographs showing A) the spindle-shaped myoblast at 24 hrs post-seeding, B) long spindle-shaped myoblast at 72 hrs post-seeding with some multinucleated myotube (arrows) and C) long spindle-shaped myoblast and myotube at 120 hrs post-seeding with some multinucleated myotube (arrows). Bar = 100µ

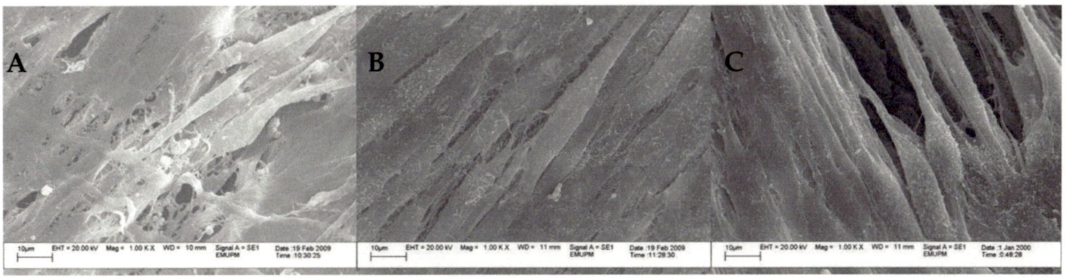


Fig. 10. SEM electronmicrographs of the myoblast seeded bovine pericardium scaffold at A) 24 hrs, B) 72 hrs and C) 120 hrs post-seeding. Note the myoblast covering almost the entire surface of the scaffolds (A), the unidirectional pattern of myotube (B) and the fusion of myotubes to form myofibres (C)

The SEM examinations revealed that both type of BP and BTV scaffolds were able to support myoblast growth and differentiation, which were evidenced by few myoblast began to cover the whole surface of the scaffolds and fused into myotube within 24 hrs (Figure 10A). At the 3rd and 5th day post-seeding, the myoblast continues to fuse and form a series of uniformly arrayed myotube (Figure 10B), whereas at 5th day post-seeding densely packed myotube with morphology reflecting myofibers were also observed (Figure 10C), and these results found to be almost consistent with result reported by Yan et al. (2006), where their study focused on tissue engineering of skeletal muscle using aligned collagen gel coated tissue flask. A similar result had been also reported by Conconi et al. (2005) on the homologous muscle acellular matrix seeded with autologous myoblasts as a tissue-engineering approach to abdominal wall-defect repair. A critical step in skeletal muscle-tissue engineering is the identification of the optimal biomaterial scaffold, able to promote normal differentiation and maturation of myoblasts into myotubes and myofibers (Conconi et al., 2005). In this work, biologically originated collagen-based biomaterials were used to obtain a scaffold possessing morphological and mechanical properties resembling those of the native tissue, and *in vitro* findings confirmed that this matrix is able to support myoblast growth and differentiation.

## 3. Post-implantation evaluations of myoblast seeded scaffolds

In this study we used a total of 36 rabbit which were randomly divided into four experimental groups (Group I, II, III and IV), comprising of equal number of animals. Group I and II served as the treatment groups, while the other groups acted as control. Three rabbits from each group were randomly selected and scarified using an intra-cardial injection of sodium pentobarbital (CEVA, Sante animale, France) at a rate of 100mg/kg at 7th, 14th and 30th day post implantation as described in Table 1.

| | | Scarifying date | | | |
|---|---|---|---|---|---|
| | | 7th Day | 14thDay | 30th Day | No of rabbits |
| Treatment groups | Group I | 3 | 3 | 3 | 9 |
| | Group II | 3 | 3 | 3 | 9 |
| Control Groups | Group III | 3 | 3 | 3 | 9 |
| | Group IV | 3 | 3 | 3 | 9 |
| | | 12 | 12 | 12 | 36 |

Group I -Myoblast seeded BP scaffolds
Group II -Myoblast seeded BTV scaffolds
Group III -Non-Seeded BP scaffolds
Group IV -Non-Seeded BTV scaffolds

Table 1. Experimental Design

### 3.1 Post-operative care and follow up analysis

The animals tolerate well to the surgical procedure and none of the rabbits died during surgical process with no post-implantation mortality which may be due to satisfactory anaesthetic technique (Table 2). Dullness, depression and partial anorexia in the immediate postoperative period was attributed to surgical trauma and inflammation at the site of reconstruction. A similar result was reported by Gangwar et al. (2006) using acellular dermal graft for repair of abdominal wall defects in rabbits. Neither the treatment groups nor its control groups in both types of scaffolds have showed any wound complication and infection as shown in figure 11, since, the complication rate after repair of large tissue defects strongly depends on applied reconstructive material. Wound infections, bowel fistulae, and repair failures can occur when synthetic materials were used for defect closure. However, naturally derived materials are less susceptible to infection (Drewa et al., 2005; Schlatter et al., 2003; Dolgin et al., 2000; Minkes, et al., 2000).

| Groups | No. of Rabbit | No. of mortality (%) | No. of Adhesion (%) | No. of Seroma (%) |
|---|---|---|---|---|
| Group I | 9 | 0(0) | 1(11.11) | 0(0) |
| Group II | 9 | 0(0) | 0(0) | 0(0) |
| Group III | 9 | 0(0) | 3(33.33) | 0(0) |
| Group IV | 9 | 0(0) | 2(22.22) | 1(11.11) |

Table 2. Post-operative complications

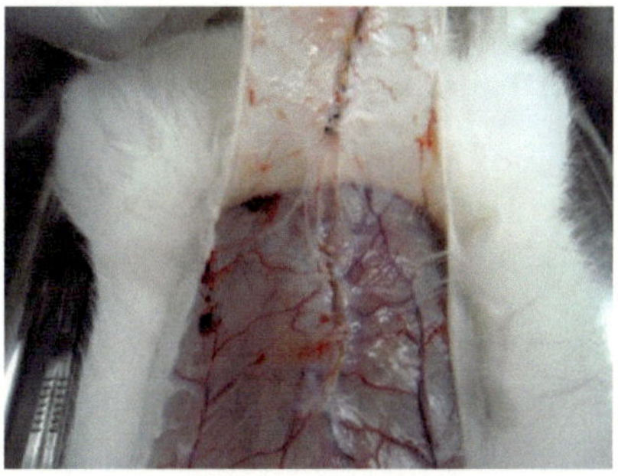

Fig. 11. Subcutaneous surface of the implanted scaffolds is covered by new/old fascia originated from surrounding fascia with overwhelming blood vessels in Group I on day 7 post-implantation

### 3.2 Macroscopic analysis

The vascular change at reconstructive site is a part of normal body response to injury. It is an attempt to increase resorption and removal of clot and debris from the wound site and finally helping in the laying down of fibrous tissue (Silver, 1982). Therefore, the present results revealed the increased vascularity at the reconstructive site in all groups of rabbit in both type of scaffolds as the post-operative day advances. Although variable degree of angiogenesis were a common feature of the treatment and its control groups in both BP and BTV scaffolds, and hence overwhelming neo-angiogenesis were recorded in the treatment groups of both type of scaffolds which might be attributed by seeding of myoblast on the scaffolds.

The study has showed that deposition of loose fibrous connective tissue and white connective tissue was observed in all the groups of rabbit regardless of the scaffolds type. However, it was on groups I and II complete neo-peritoniazation began with the appearance of glistening membrane covering the inner surface of the graft at day 7$^{th}$ of post-implantation (Figure 12) which is indeed earlier than the study conducted by Singh et al. (2008) in acellular biomaterials of porcine origin for the reconstruction of abdominal wall defects in rabbits, where complete peritoniazation the graft occurred at 21$^{st}$ post-operative day, and their finding relatively similar with the control groups of our study where early peritoniazation on the implanted scaffolds was not observed in the control groups in which myoblast was not seeded. Therefore myoblast seeding might be a good justification for early peritoniazation.

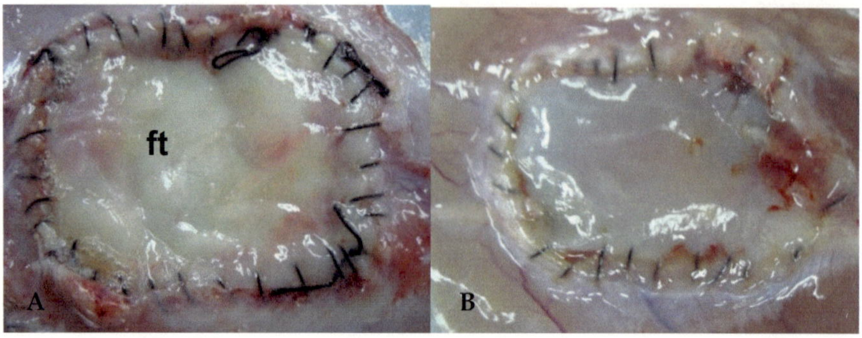


Fig. 12. Peritoneal surface of the implanted graft in Group I (A) and group II (B) on day 7 post implantation showing the surface is completely covered by newly formed white fibrous tissue including the fatty tissue (ft) and neo-peritoneum that appears smooth and shinny

Naturally derived materials, including glutaraldehyde tanned BP (James *et al.,* 1991), small intestine submucosa (Clarke et al., 1996; Prevel et al., 1995) and also lyophilized and glycerolized BP and BTV (Hafeez, 2005), have been tried in animal models. These biomaterials are less susceptible to infection and cause less foreign body response (Badylak et al., 1998; Hiles et al., 1995). However, fail to recover muscle tissue and also lack of strength over time is a concern for clinical application. Macroscopically, at day 30$^{th}$ of post-implantation, our study revealed that the control groups III and IV have showed thinning and fascial weakness, which was evidenced by pouching and distension appearance of the abdominal wall as shown in Figure 13.

Lai et al. (2003) had proven that scaffolds covered with cells had better mechanical properties than acellular/non-seeded scaffolds in body wall repair using small intestinal submucosa seeded with cells. Fauza et al. (2001) had also reported acellular collagen-based matrix alone may be insufficient scaffold for abdominal wall reconstruction. However, in our study none of the treatment groups, in both types of scaffolds have showed pouching and distension appearance of the abdominal wall, rather various degree of tissue regeneration was noticed, which was evidenced by graft opacity, and overwhelming blood vessels, fatty tissues, fibrous tissue and neo-peritoneum was observed in myoblast seeded groups of both type of scaffolds at 30 days of post-implantation (Figure 14).

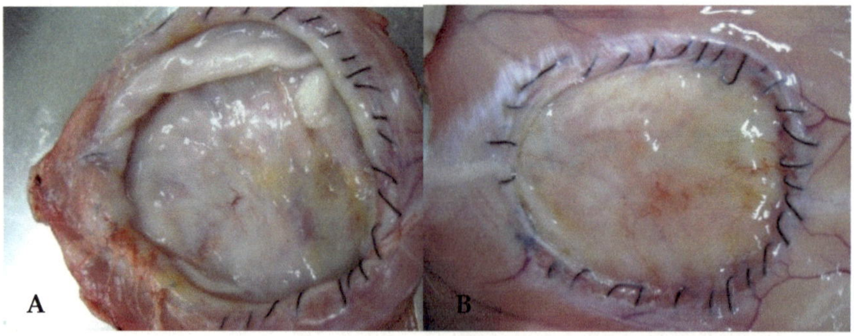


Fig. 13. Peritoneal surface of the implanted graft in group III (A) and group IV (B) on day 30 post-implantation showing the implanted graft over-stretched and lead to pouching appearances with very few blood vessels

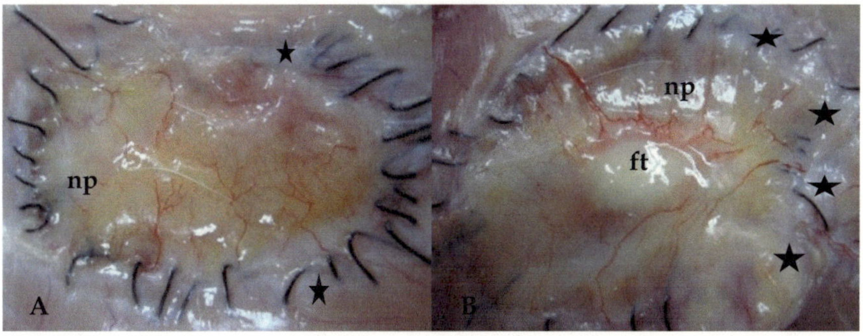


Fig. 14. Peritoneal surface of the implanted graft in group I (A) and group II (B) on day 30 post-implantation showing the neo-peritoneum (np) covered all over the surface with numerous blood vessels and white fibrous tissue at host-graft junction (stars), fatty tissue (ft)

In this study, the wound strength of repair was not evaluated. However, the firm integration was observed between the implant and host tissue produced by infiltration of fibro-collagenous tissue which was indeed further strengthens by infiltration of skeletal muscle in the treatment groups. Therefore myoblast seeded scaffolds seems to have superiority over non- seeded scaffolds in cell infiltrations as well as mechanical performances. Similarly Lai et al. (2003) have stated that the success of a tissue-engineered composite might be determined by both the strength of the scaffold and the angiogenesis to support the new cells.

The adhesion formation depends on material surface geometry and affects the correctly organized neo-peritoneum regeneration. In previous *in vitro* study in which mesothelial cells were seeded onto various biomaterials, it has been established that mesothelialization occurs early when the prosthesis is of laminar type. In contrast, when the biomaterial has the structure of a reticular mesh mesothelial deposition takes place in an irregular manner, with cells settling on the prosthetic filaments, achieving an uneven cover, therefore they speculated that early stage mesothelial deposition after implant is probably conditioned by the structural design of prosthesis (Bello'n et al., 2003).

The origin of adhesions between the organs and the parietal peritoneum has yet to be determined. These formations also appear after biomaterial implantation to repair abdominal wall defect, mainly when macroporous biomaterials, such as polypropylene mesh, are utilized. This fact has been observed by several authors in animal models, such as the Sprague-Dawley rat and the New Zealand white rabbit (Bello'n et al., 1996). Moreover, in human it is well known that intra-peritoneal positioning of conventional parietal mesh provides efficient reconstruction, but it causes visceral adhesion formation in 80–100% of the cases (Balique et al., 2005). In our study, all the adhesion found between the implanted graft and the visceral organ (caecum) was mild. 33% and 22.22% of adhesion was found in control group III and IV respectively. Absence of adhesion in treatment group II was noticed. The early formation of a mesothelium covering the myoblast seeded graft probably explains the lack of adhesion formation observed following implantation. Besides, it is likely that the delay in mesothelialization associated with non-myoblast seeded groups (control) implant gives rise to the frequent adhesions that occur at the biomaterials-visceral peritoneum interface. Ironically, a single case out of nine (11.11%) rabbits from treatment Group I have showed mild adhesion with caceum which is likely to be affected by the individual variation factors rather than by the type of the implanted grafts (Tables 3 and 4, and Figure 15).

| Adhesion degree | Date of euthanizing | | | | | |
|---|---|---|---|---|---|---|
| | Day 7th | | Day 14th | | Day 30th | |
| | S | C | S | C | S | C |
| 0 | 3 | 2 | 3 | 2 | 2 | 2 |
| 1 | 0 | 1* | 0 | 1* | 1* | 1* |
| 2 | 0 | 0 | 0 | 0 | 0 | 0 |
| 3 | 0 | 0 | 0 | 0 | 0 | 0 |

Based on Mann-Whitney Test, P= 0.270, Since P > 0.05, there is no significant difference in adhesion formation between treatment and control groups

S - Myoblast seeded BP (Group I)

C - Non seeded BP (group III)

1* - Showing minimal/ minor adhesion observed between the implanted and the underlying visceral organ

Table 3. Scoring of adhesion formation at 7th, 14th, and 30th days of post-implantation for bovine pericardium scaffolds

| Adhesion degree | Date of euthanizing | | | | | |
|---|---|---|---|---|---|---|
| | Day 7th | | Day 14th | | Day 30th | |
| | S | C | S | C | S | C |
| 0 | 3 | 2 | 3 | 2 | 3 | 3 |
| 1 | 0 | 1* | 0 | 1* | 0 | 0 |
| 2 | 0 | 0 | 0 | 0 | 0 | 0 |
| 3 | 0 | 0 | 0 | 0 | 0 | 0 |

Based on Mann-Whitney Test, P= 0.145, Since P > 0.05, there is no significant difference in adhesion formation between treatment and control groups.

S - Myoblast seeded BTV (Group II)

C - Non seeded BTV (Group IV)

1* - Showing minimal/ minor adhesion observed between the implant and the underlying visceral organ

Table 4. Scoring of adhesion formation at 7th, 14th, and 30th days of post-implantation for bovine tunica vaginalis scaffolds

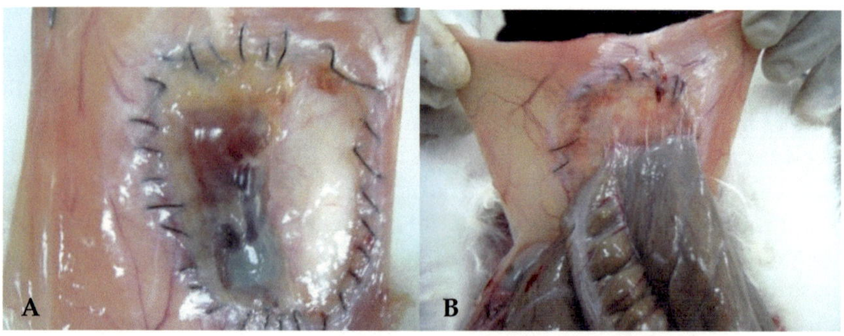


Fig. 15. Peritoneal surface of A) group IV showing seroma formed between neo-peritoneum and the implanted scaffolds at day 7 post-implantation and B) minor adhesion between the implant and caecum at day 14 post-implantation

### 3.3 Light microscopic analysis

The attack of body immune system on implant can cause failure of the biomaterials to serve as the tissue replacement. Biocompatibility involves the acceptance of biomaterial by the surrounding tissue and by the body as whole (Park & Lakes, 2007). Histopathologically, the present study revealed that the inflammatory response in all treatment and also controls groups of animals were significantly high during the first week (7th days) of post-implantation (Figure 16). It was indeed an immediate response initiated by surgical trauma when the abdominal well defects were created which perhaps correlated with the inflammatory phase of defects. A similar observations was reported by many researchers (Singh et al., 2008; Zuki et al., 2007; Gangwar et al., 2006; Gamba et al., 2002; Tung et al., 2002).

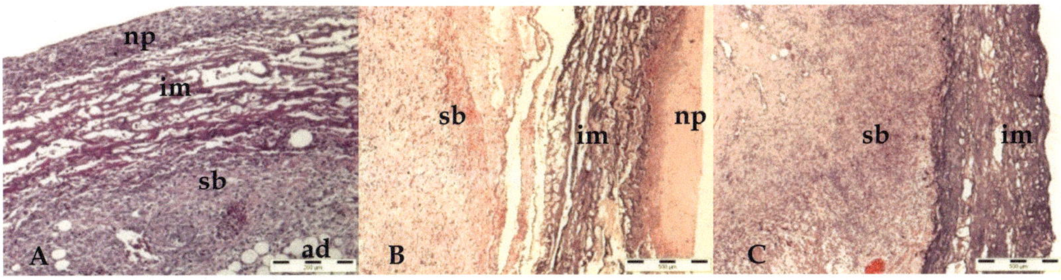


Fig. 16. Microphotographs of group I (A), group II (B) and group IV (C) at 7th day of post-implantation showing the inflammatory cells surrounding the implanted scaffold (im). Note also the presence of neo-peritoneum (np), subcutaneous tissue (sb) and adiposity (ad) in A and B but neo-peritoneum was not detected in C. H&E, Bar = 500μ

Moreover, the inflammatory process is a normal response to surgical injury and the presence of implant as foreign objects. In facts inflammation is necessary as the transitional linking stage between damaged tissue and repair of damaged tissue (Dumitriu, 1994). However, our study showed that the inflammatory process gradually decreased in every advancing period in all treatment and controls groups of both BP and BTV originated scaffolds. These findings suggest that the processing method to produce fibro-collagenous scaffolds is better to removes antigenic proteins and maintain graft integrity. It might also be associated with early degradation and resorption of the implant. As Hafeez (2005) had reported, the cervices and pores created by freeze drying and the fragmentation caused by gamma rays for sterilization in lyophilized grafts enhances inflammatory cells infiltration into the implant and leads to early degradation and resorption. Apart from its processing methods, collagen bundles of the implant in treatment group II and control group IV were not completely resorbed at 30th day of post-implantation, ironically the implant of treatment group I and its control group III has showed better resorption rate. This slight difference might be as a result of higher density of fibro- collagenous nature of BTV originated scaffolds than the BP originated scaffolds.

Previous study that had been reported by Hafeez (2005) indicated that lyophilised BP and BTV graft were started resorption at week three post-implantation in a rat model. However, the present results showed earlier resorption which can be explained by species differences as showed in figure 17. Despite the unknown exact mechanism, several factors, including the animal species, age and site of implantation may play an important role in the rate of degradation of implant (Vialle-Preles et al., 1993).

This investigation showed that in both control groups (III and IV) the implanted grafts were overwhelmed by fibroblasts, mesenchymal cells and neovascularisation. These mesenchymal and fibroblasts were frequently observed throughout the biomaterials. In fact the collagen fibers and ground substance (matrix) were synthesised and deposited by newly migrating fibroblast as it was showed by the gradual spread of newly formed tissue in the implanted graft. The newly formed collagens in the first week were delicate, immature and unorganized. Later on the 2nd and 4th weeks post-implantation, the amount of collagen fibers increased and infiltrated throughout the implant in well defined and organized form. However, there was not even a single skeletal muscle fiber formation in the implant of control groups. This finding seems consistence with previous reports of biological biomaterials mediated wound repair (Singh et al., 2008; Zuki et al., 2007; Gangwar et al., 2006; Gamba et al., 2002; Tung et al., 2002).

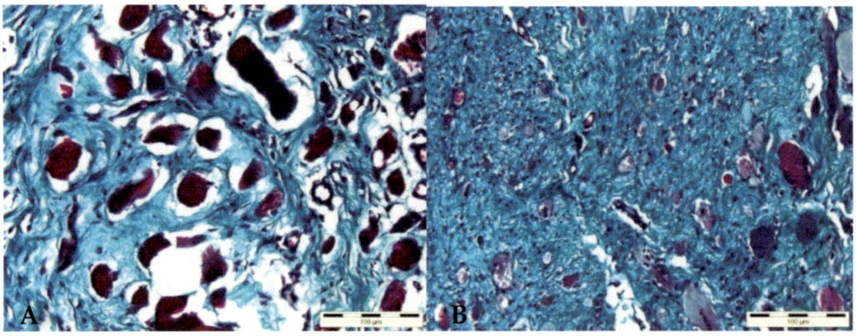


Fig. 17. Microphotographs of group I (A) and group II (B) at 14th day of post-implantation show the graft is completely replaced by collagen fiber (green) and small delicate muscle fibers (red colour). Masson trichrome stain, Bar = 500μ

In contrast to the above cases, in both the treatment groups (I and II), skeletal muscle tissue regeneration were clearly observed. The myoblast were originated from the seeded part, not from the host tissue. At 14th day of post- implantation, the newly formed muscles were young muscle fibers (Figure 17) but eventually at the 4th week post-implantation, they became matured and became well-defined muscle fibers (Figure 18). Since the seeded myoblast were labelled with fluorescence dye and therefore the newly formed muscles fibers were further confirmed by fluorescence microscopy (Figure 19).

The importance of rapid mesotheliazation of biomaterials stems from the fact that adhesion formation is inversely related to the number of mesothelial cells on the peritoneal surface (Law & Ellis, 1988). Therefore, myoblast seeding on the scaffolds has proved well organized bridging of peritoneal lining (mesotheliazation) across the wound with better vascularisation and this could be contribute to the preventions of intra-abdominal adhesion between the biomaterials and visceral organs.

The formation of foreign body giant cells is a feature of chronic inflammation which occurs in the presence of microorganism or non-phagocytosable materials for long period of time. The foreign body reaction appeared as macrophage and foreign giant cells (Anderson, 1998). In our study absence of foreign body giant cells were noted in both the treatment and control groups of the two types of scaffolds. This could suggest (indicates) the biocompatibility of the scaffolds with the host tissue. Our study has also revealed that both type of myoblast seeded scaffolds showed similar appearance histopathologically, meaning that apart from newly formed delicate collagens, newly formed young muscles were observed which eventually became matured and became well-defined muscle fibers. However, on both type of non-myoblast seeded scaffolds (control groups) only delicate, immature and unorganized collagens fibers were observed within the first week which is later on become organized and mature collagens fibers without a single skeletal muscle fibers development.

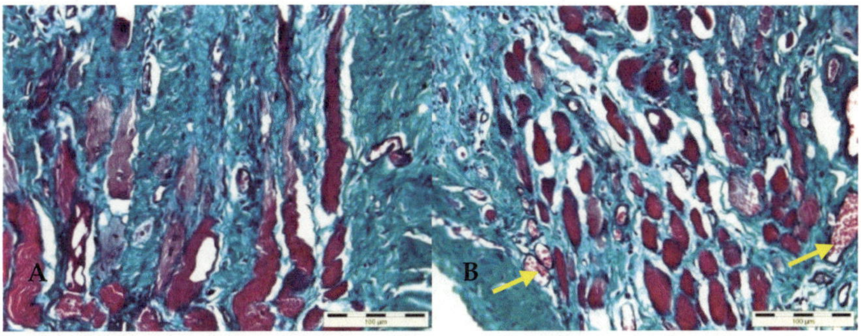


Fig. 18. Microphotographs of Group I (A) and group II (B) at 30th day of post-implantation show the graft was completely replaced by collagen fiber (Green) and well developed muscle fibers (Red) and blood vessels (arrows). Masson trichrome stain, Bar = 100μ

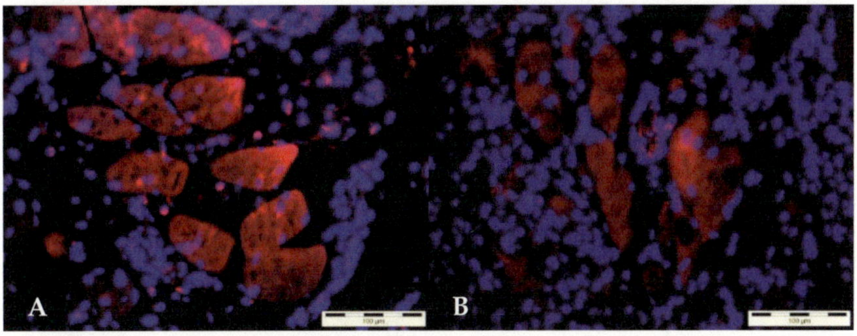


Fig. 19. Double-fluorescence microphotographs of the graph of Group I (A) and group II (B) at day 30th of post-implantation demonstrate the PKH26-labeled myofibers and its DAPI-stained nuclei. Note the well developed myotube/myofibers. Bar = 100μ

### 3.4 Scanning Electron Microscopic analysis

The SEM examinations at 7th day of post-implantation in both I and II revealed infiltration of spherical shape of inflammatory cells with scattered polygonal and spindle shapes of cells on the peritoneal surface indicating the beginning of mesotheliazation as depicted on Figure 20. On day 14th of post-implantation, the treatment groups (I and II) of both type of scaffolds, demonstrated the well organized mesotheliazation with confluent polygonal shapes of cells covering all over the peritoneal surfaces (Figure 21). By day 30th of post-implantation, the treatment groups (I and II) of both type of scaffolds, revealed the well organized mesotheliazation (neo-peritoniazation) where the polygonal shapes of cells covering all over the peritoneal surfaces with numerous cilia like structure (Figure 22).

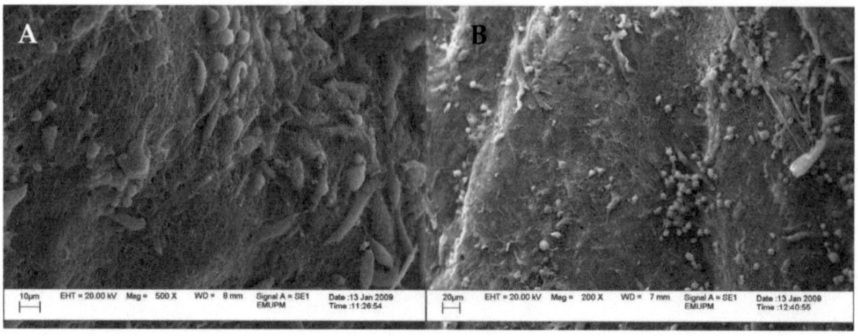


Fig. 20. Scanning electron micrographs of the peritoneal surface of group I (A) and group II (B) on 7th days of post-implantation show the infiltrated spherical shape of inflammatory cells (Yellow arrows) with scattered polygonal and spindle shapes of cells (Red arrows)

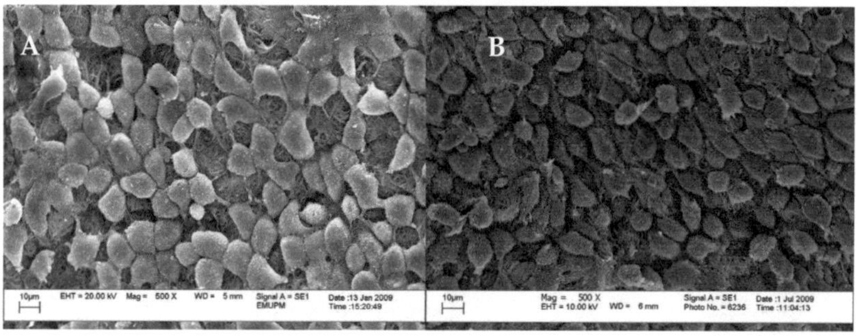


Fig. 21. Scanning electron micrographs of the peritoneal surface of Group I (A) and group II (B) on 14th day of post-implantation show the well organized mesotheliazation of the surface with polygonal shape of mesothelial cells

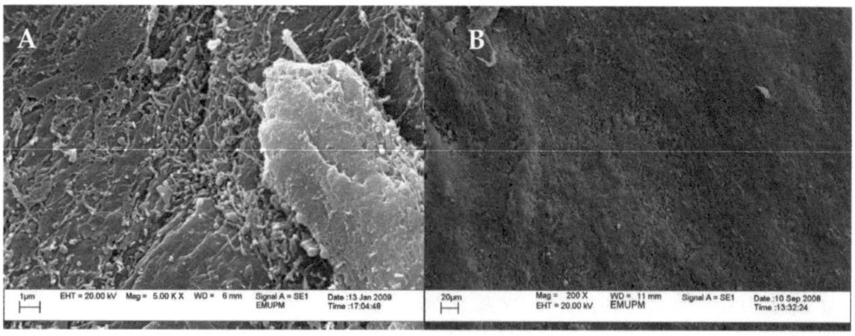


Fig. 22. Scanning electron micrographs of the peritoneal surface of Group I (A) and group II (A) on 30th day of post-implantation show the smooth appearance of the surface with numerous cilia in A

The SEM analysis of the present study has showed a clear understanding by the speed at which mesotheliazation of the myoblast seeded scaffolds in both types of scaffolds on the peritoneal surfaces. These undoubtedly give rise an optimal interface between the graft and visceral organs in avoiding the formation of adhesion and also the appearance of complications related to adhesion because as Chew et al. (2000) had reported in their study that firm adhesion may become integrated within the biomaterials which can provoke long term complications such as intestinal fistula in some cases. Moreover, the ultrastructure findings in this study are also indicates reduction of inflammatory cells as the day advances in all the treatment and control groups of both types of scaffolds.

Ultrastructural analysis has further confirmed the profound differences between treatment and its control groups in mesotheliazation/neo-peritoniazation of the peritoneal surfaces. Treatment groups/myoblast seeded scaffolds in both type of scaffolds has showed a well-organized neo-peritoneum generation which might be induced by seeding of myoblast and it could be a good explanation for absence of adhesion. In contrast, on both type of non-myoblast seeded scaffolds (control groups) uneven peritoneum with an irregular mesothelium, which is prone to develop adhesion as reported by Losi et al. (2007).

### 3.5 Fluorescence microscopic analysis

Fluorescence microscopic analysis clearly demonstrates that myoblast seeded biomaterials can be successfully transplanted into artificially created full layer of abdominal wall defects except the skin. The prerequisites for using PKH26 for labelling of cells has been demonstrated in the previous study by many researchers that the incorporation of PKH dye does not physically weaken the membrane of the cells, does not interfere with the functions of the cells and appeared to be ideal for long-term tracking of cells. Animal studies revealed no toxicity of the PKH dye and no immune response appears to generate against labelled cells (Read et al., 1991; Slezak and Horan, 1989a, 1989b; Melnicoff et al., 1988). Most importantly, Johnsson et al. (1997) had reported in their '*ex vivo* PKH26-labelling of lymphocytes for studies of cell migration *in vivo*' study of this labelling dye was found to be retained within the labelled cells and not transferred to other cells.

As expected, the seeded myoblast onto two type of naturally originated collagen based scaffolds materials did not migrate into surrounding tissue rather it remained within the defects. Therefore, cell growth and fusion within the defects suggests that this technique might be suitable for filling volume defects in ectopic locations outside skeletal muscle.

Engineering of multinucleated myotube which later on differentiated into mature striated muscle fibres and its existence for up to the end of the study period showing the superiority of this study to the previous case reported by Van Wachem et al. (1999) where failure of muscle regeneration noted. Lai et al. (2003) has also reported that scaffold covered with skeletal muscle cells did not differ from fibroblast construct in terms of hernia rate. However, Kamelger et al. (2004) have reported the successful regeneration of skeletal muscles in their comparative study of three different biomaterials in the engineering of skeletal muscle using a rat animal model. The fluorescence microscopy have clearly indicated that the similarity of the current results with their study.

In the present study, in both type of myoblast seeded scaffolds the fluorescence microscopy has clearly indicated regeneration of multinucleated skeletal muscle with well developed vascularisation all over the implant and this finding seems consistence with the results reported by Kamelger et al. (2004) and Conconi et al. (2005). However, no PKH26 labelled myoblast was observed in both type of non-myoblast seeded scaffolds, and not even a single skeletal muscle tissue has been detected in the control groups (III and IV), meaning that all the skeletal muscle obtained in the treatment groups were originated from seeding rather from host tissue which indeed confirmed by its PKH26 fluorescence dye.

Van Wachem et al. (1999) reported one of the reasons for the failure of muscle regeneration may be absence of vascularisation. However, this study has showed that early well vesicularization (neo-angiogenesis) of the implant microscopically and also macroscopically which perhaps may have an impact on the successful regeneration of skeletal muscles at large.

## 4. Conclusion

In general, lyophilized and gamma sterilized BP and BTV scaffolds have showed a tremendous potential for *in vitro* cultivation of skeletal muscle, it renders great success when used as substrate for filling of wound bed or for the delivery of cells. Myoblast harvested from primary culture are able to proliferate and form myotube *in vitro* on both type of collagens based biomaterials. This study has also showed that well vascularised biomaterials-myoblast-construct can be successfully implanted for reconstruction of abdominal wall defects which profoundly results in regeneration of skeletal muscle tissue. Myoblast seeded scaffolds did not provoke a significant inflammatory response compared to the non-seeded collagen based scaffold. These engineered myoblast-constructs have better cell infiltration and mechanical performances that ultimately avoid adhesion between the graft and visceral organs than the non-myoblast seeded construct. These finding suggests seeding of myoblast to these scaffolds may be a viable alternative to engineer skeletal muscle tissue for body replacement and remodeling.

## 5. Acknowledgement

The study was supported by e-Science research grant (project no. 06-01-04-SF0203) provided by Malaysian Government.

## 6. References

Amid, K. P. (1997). Classification of biomaterials and their related complications in abdominal wall hernia surgery. *Hernia*, vol.1, pp. 15-21.

Anderson, J. M. (1988). Inflammatory response to implants. *American Society Artificial Internal Organs*, vol.34, pp. 101-107.

Badylak, S.; Kokini, K.; Tullius, B.; Simmons-Byrd, A. & Morff, R. (2002). Morphologic study of small intestinal submucosa as a body wall repair device. *Journal of Surgical Research*, vol.103, pp. 190-202.

Badylak, S. F.; Kropp, B. & McPherson, T. (1998). SIS: A rapidly resorbable bioscaffold for augmentation cystoplasty in a dog model. *Tissue Engineering*, vol.4, pp. 397-387.

Balique, J. G.; Benchetrit, S.; Bouillot, J. L.; Flament, J. B.; Gouillat , C.; Jarsaillon, P.; Lepère, M.; Mantion, G.; Arnaud, J. P.; Magne, E. & Brunetti, F. (2005). Intraperitoneal treatment of incisional and umbilical hernias using an innovative composite mesh: four-year results of a prospective multicenter clinical trial. *Hernia*, vol.9, pp. 68-74.

Bauer, J. J.; Harris, M. T.; Kreel, I. & Gelernt, I. M. (1999). Twelve-year experience with expanded polytetrafluoroethylene in the repair of abdominal wall defects. *Mount Sinai Journal of Medicine*, vol.66, pp. 20-25.

Bello'n, J. M.; Contreras, L. A.; Buja'n , J. & Jurado, F. (1996). Effect of phosphatidylcholine on the process of peritoneal adhesion following implantation of a polypropylene mesh prosthesis. *Biomaterials,* vol.17, pp. 1369-1372.

Bello'n, J. M., Garcia-Honduvilla, N., Lopez, R., Corrales, C., Jurado, F. and Buja'n, J. (2003). In vitro mesothelialization of prosthetic materials designed for the repair of abdominal wall defects. *Journal of Material Science: Material Medicine,* vol.14, pp. 359-64.

Black, J. (1992). *Biological performance of materials,* 2nd ed. Marcel Dekker, New York.

Bronzino, J. D. (2006). *The Biomedical Engineering Handbook,* 3rd ed.CRC Taylor & Francis Group, LLC. U.S. A.

Calzolari, E.; Bianchi, F.; Dolk, H. & Milan, M. (1995). Omphalocele and gastroschisis in Europe a survey of 3 million births 1980–1990. *American Journal of Medical Genetics,* vol.58, pp. 187-194.

Chang, Y.; Chen, S-C.; Wei, H-J.; Wu, T-J.; Liang, H-C.; Lai, P-H.; Yang, H-H. & Sung, H-W. (2005). Tissue regeneration observed in a porous acellular BP used to repair a myocardial defect in the right ventricle of a rat model. *Journal of Thoracic and Cardiovascular Surgery,* vol.130, pp. 705.e1-e10.

Chang, Y.; Tsai, C-C.; Liang, H-C. & Sung, H-W. (2002). In vivo evaluation of cellular and acellular bovine pericardia fixed with a naturally occurring crosslinking agent (genipin). *Biomaterials,* vol.23, pp. 2447-2457.

Chen, J. C. & Goldhamer, D. J. (2003). Skeletal muscle stem cells. *Reproductive Biology* and *Endocrinology,* vol.1, pp. 101-107.

Chew, D. K.; Choi, L. H. & Rogers, A. M. (2000). Enterocutaneous fistula 14 years after prosthetic mesh repair of a ventral incisional hernia: a life-long risk. *Surgery,* vol.127, pp. 352-353.

Chung, S.; Hazen, A.; Levine, J. P.; Baux, G.; Olivier, W. A.; Yee, H. T.; Margiotta, M. S.; Karp, N. S. & Gurtner, G. C. (2003). Vascularized acellular dermal matrix island flaps for the repair of abdominal muscle defects. *Plastic and Reconstructive Surgery,* vol.111, pp. 225-232.

Clarke, K. L.; Lantz, G. C. & Salisbury, S. K. (1996). Intestine submucosa and polypropylene mesh for abdominal wall repair in dogs. *Journal of Surgical Research,* vol.60, pp. 107-114.

Conconi, M. T.; Coppi, P. D.; Bellini, S.; Zara, G.; Sabatti, M.; Marzaro, M.; Zanon, G. F.; Gamba, P. G.; Parnigotto, P. P. & Nussdorfe, G. G. (2005). Homologous muscle acellular matrix seeded with autologous myoblasts as a tissue-engineering approach to abdominal wall-defect repair. *Biomaterials,* vol.16, pp. 2567-2574.

Dolgin, S. E.; Midulla, P. & Shlasko, E. (2000). Unsatisfactory experience with the "minimal intervention management" for gastroschisis. *Journal of Pediatric Surgery,* vol.35, pp. 1437-1439.

Drewa, T.; Galazka, P.; Prokurat, A.; Wolski, Z.; Sir, J.; Wysocka, K. & Czajkowski, R. (2005). Abdominal wall repair using a biodegradable scaffold seeded with cells. *Journal of Pediatric Surgery,* vol.40, pp. 317-321.

Dumitriu, S. (1994). *Polymeric biomaterials.* Marcel Dekker, Inc., pp. 99-108.

Fauza, D. O.; Marler, J. J.; Koka, R.; Forse, R. A.; Mayer, J. E. & Vacanti, J. P. (2001). Fetal tissue engineering: diaphragmatic replacement. *Journal of Pediatrics Surgery,* vol.36, pp. 146-151.

Gamba, P. G.; Conconi, M. T.; Piccolo, R. L.; Zara, G.; Spinazzi, R. & Parnigotto, P. P. (2002). Experimental abdominal wall defect repaired with acellular matrix. *Journal of Pediatric Surgery,* vol.18, pp. 327-331.

Gangwar, A. K.; Sharma, A. K.; Kumar, N.; Kumar, N.; Maiti, S. K.; Gupta, O. P.; Goswami, T. K. & Singh, R.(2006). Acellular dermal graft for repair of abdominal wall defects in rabbits. *Journal of the South African Veterinary Association,* vol.77, pp. 79-85.

Goddard, J. M. & Hotchkiss, J. H. (2007). Polymer surface modification for the attachment of bioactive compounds. *Progress in Polymer Science,* vol.32, pp. 698-725.

Guarita-Souza, L. C.; Carvalho, K. A.; Woitowicz, V.; Rebelatto, C.; Senegaglia, A.; Hansen, P.; Miyague, N.; Francisco, J. C.; Olandoski, M.; Faria-Neto, J. R. & Brofman, P. (2006). Simultaneous autologous transplantation of cocultured mesenchymal stem cells and skeletal myoblasts improves ventricular function in a murine model of Chagas disease. *Circulation,* vol.114, pp. I-120-I-124.

Guettier-Sigrist, S.; Coupin, G.; Braun, S.; Warter, J. M. & Poindron, P. (1998). Muscle could be the therapeutic target in SMA treatment. *Journal of Neuroscience Research,* vol.53, pp. 663-669.

Hafeez, Y. M. (2005). Morphological and Biophysical Properties of Bovine Pericardium and Bovine Tunica Vaginalis Xenografts in a Rat Model, PhD. Thesis, University Putra Malaysia.

Hafeez, Y. M.; Zuki, A. B. Z.; Loqman, M. Y.; Noordin, M. M. & Norimah, Y. (2005b). Comparative evaluations of the processed Bovine Tunica Vaginalis implant in a rat model. *Anatomical Science International,* vol.80, pp. 181-188.

Hafeez, Y. M.; Zuki, A. B. Z.; Norimah, Y.; Asnah, H.; Loqman, M. Y.; Noordin, M. M. & Ainul-Yuzairi, M. Y. (2005a). Effect of freeze-drying and gamma irradiation on biomechanical properties of Bovine Pericardium. *Cell and Tissue Banking,* vol.6, pp. 85-89.

Haider, H. K.; Lei, Y.; Shujia, J. & Sim, E. K. (2003).Cellular myocardial reconstruction using human myoblasts. *Journal of the American College of Cardiology,* vol.42, pp. 589.

Hench, L. L. & Erthridge, E. C. (1982). Biomaterials: an Interfacial Approach, 1st ed. New York: Academic Press.

Hiles, M. C.; Badylak, S. F. & Lantz, G. C. (1995). Mechanical properties os xenogeneic small intestinal submucosa when used as an aortic graft in the dog. *Journal of Biomedical Materials Research,* vol.29, pp. 883-891.

Huard, J.; Cao, B. & Qu-Petersen, Z. (2003).Muscle-derived stem cells potential for muscle regeneration. *Birth Defects Research Part C,* vol.69, pp. 230-237.

Hutmacher, D. W. (2001). Scaffold design and fabrication technologies for engineering tissues-state of the art and future perspectives. *Journal Biomaterial Science,* vol.12, pp. 107-124.

James, N. L.; Poole-Warren, L. A.; Schindhlem, B. K.; Mitchell, R. M.; Mitchell, R. E. & Howlett, C. R. (1991). Comparative evaluation of treated BP as a xenograft for hernia repair. *Biomaterial,* vol.12, pp. 801-809.

Johnsson, C.; Festin, R.; Tufveson, G. & Totterman, T. H. (1997). *Ex vivo* pkh26-labelling of lymphocytes for studies of cell migration *In vivo. Scandinavian Journal of Immunology,* vol.45, pp. 511-514.

Jones, M. L. (2002). *Connective tissue and stains.* In: Theory and practice of histological techniques, ed. D. J. Bancroft, and M. Gamble, pp.139-162. 5th ed. Churchill Livingstne, New York.

Jose, M.; Garcia, P.; Eduardo, J. H.; Antonio, C.; Isabel, M. A. & Auroa, R. P. (2001). Ostrich pericardium, a biomaterial for construction of valve leaflets. *Biomaterials,* vol.22, pp. 2731-2740.

Kamelger, F. S.; Marksteiner, R.; Margreiter, E.; Klima, G.; Wechselberger, G.; Hering, S. & Piza, H. (2004). A comparative study of three different biomaterials in the engineering of skeletal muscle using a rat animal model. *Biomaterials,* vol.25, pp. 1649-1655.

Lai, J. Y.; Chang, P. Y. & Lin, J. N. (2003). Body wall repair using small intestinal submucosa seeded with cells. *Journal of Pediatrics Surgery,* vol.38, pp. 1752-1755.

Law, P. K.; Goodwin, T. G.; Fang, Q.; Deering, M. B.; Duggirala, V.; Larkin, C.; Florendo, J. A.; Kirby, D. S.; Li, H. J. & Chen M. (1993). Cell transplantation as an experimental treatment for Duchenne muscular dystrophy. *Cell Transplantation,* vol.2, pp. 485-505.

Losi , P.; Munao, A.; Spiller , D.; Briganti, E.; Martinelli, I.; Scoccianti, M. & Soldani, G. (2007). Evaluation of a new composite prosthesis for the repair of abdominal wall defects. *Journal of Material Science: Material Medicine,* vol.18, pp. 1939-1944.

Marques, A.; Lopes, A.; Yojo, L.; Brenda, E.; Tulio, M.; Amarante, P. M. & Torlont, H. (1995). A retrospective study of the use of BP, dura mater, and polypropylene mesh as reinforcement materials in abdominal and thoracic wall reconstruction. *Current Therapeutic Research,* vol.56, pp. 492-497.

Marzaro, M.; Conconi, M. T.; Perin, L., Giuliani, S.; Gamba, P.; De Coppi, P.; Perrino, G. P.; Parnigotto, P. P. & Nussdorfer, G. G. (2002). Autologous satellite cell seeding improves *in vivo* biocompatibility of homologous muscle acellular matrix implants. *International Journal* of *Molecular Medicine,* vol.10, pp. 177-182.

Meddings, R. N.; Carachi, R.; Gorham, S. & French, D. A. (1993). A new bioprosthesis in large abdominal wall defects. *Journal of Pediatrics Surgery,* vol.28, pp. 660-663.

Melnicoff, M. J.; Morahan, P. S.; Jensen, B. D.; Breslin, E. W. & Horan, P. K. (1988). In vivo labeling of resident peritoneal macrophages. *Journal* of *Leukocyte Biology, vol.43, pp. 387-397.*

Minkes, R. K.; Langer, J. C.; Mazziotti, M. V.; Skinner, M. A. & Foglia, R. P. (2000). Routine insertion of a silastic spring-loaded silo for infants with gastroschisis. *Journal of Pediatric Surgery,* vol.35, pp. 843-846.

MINT Tissue Bank Work Instruction manual (1998). Freeze-drying of tissues and bioburden analysis of tissue products. MTB/WI/016 and MTB/WI/014.

Ott, H. C.; Berjukow, S.; Marksteiner, R.; Margreiter, E.; Bock, G.; Laufer, G. & Hering, S. (2004). On the fate of skeletal myoblasts in a cardiac environment: down-regulation of voltage-gated ion channels. *Journal of Physiology,* vol.558, no.3, pp. 793-805.

Park, J. B. & Lakes, R. S. (2007). *Biomaterials:* An In introduction. 3rd ed. Springer Science+Business Media, LLC. pp. 1-16.

Parnigotto, P. P.; Gamba, P. G.; Conconi, M. T. & Midrio, P. (2000b). Experimental defect in rabbit uretra repaired with acellular aortic matrix. *Urological Research,* vol.28, pp. 46-51.

Parnigotto, P. P.; Marzaro, M.; Artusi, T.; Perrino, G. & Conconi, M. T. (2000a). Short bowel sindrome experimental approach to increase intestinal surface in rats by gastric homologous acellular matrix. *Journal Paediatric Surgery,* vol.35, pp. 1304-1308.

Prevel, C. D.; Eppley, B. L. & Summerlin, D. J. (1995). Small intestinal submucosa: Use in repair of rodent abdominal wall defects. *Annals of plastic surgery,* vol.35, pp. 374-380.

Rando, T. A. & Blau, H. M. (1994). Primary Mouse Myoblast Purification, Characterization, and Transplantation for Cell-mediated Gene Therapy. *Journal of Cell Biology,* vol.125, pp. 1275-1287.

Read, E. J.; Cardine, L. L. & Yu, M. Y. (1991). Flowcytometric detection of human red cells labelled with a fluorescent membrane label: potential appealication to *in vivo* survival studies. *Transfusion,* vol.31, pp. 502-508.

Rodgers, B. M.; Maher, J. M. & Talber, N. (1981). The use of preserved human dura for closure of abdominal and diaphragmatic defects. *Annual of surgery,* vol.193, pp. 606-611.

Roeder, R.; Wolfe, J.; Lianakis, N.; Hinson, T.; Geddes, L. A. & Obermiller, J. (1999). Compliance, elastic modulus, and burst pressure of small-intestine submucosa (SIS), small-diameter vascular grafts. *Journal of Biomedical Materials Research,* vol.47, pp. 65-70.

Saaverda, S.; Pelaaez, M. D.; Alvarez Zapico, J. A.; Gutierrez, S. C. & Fernandez, J. (2001). Fascia lata transplant from cadavric donor in the reconstruction of abdominal wall defects in children. *Cirugia Pediatrica,* vol.14, pp. 28-30.

Santillan, D. P.; Jasso, V. R.; Sotres-Vega, A.; Olmos, R.; Arreola, J. L.; Garc-ia, D.; Vanda, B. & Gaxiola, M (1995). Repair of thoraco-abdominal wall defects in dogs using a BP bioprosthesis. *Revista de investigation Clinica, Organo Del Hospital De Enfermedades De La Nutricion,* vol.47, pp. 439-446.

Saxena, A. K. (2005). Tissue engineering: Present concepts and strategies. *Journal of Indian Association of Pediatric Surgeons,* vol.10, pp. 14-19.

Saxena, A. K. & Willital, G. H. (2000). Skeletal muscle tissue-engineering. *International Medical Journal of Experimental & Clinical Research,* vol.6, pp. 18.

Saxena, A. K.; Marler, J.; Benvenuto, M.; Willital, G. H. & Vacanti, J. P. (1999b). Skeletal muscle tissue engineering using isolated myoblasts on synthetic biodegradable polymers: preliminary studies. *Tissue Engineering,* vol.5, pp. 525-532.

Saxena, A. K.; Willital, G. H. & Vacanti, J. P. (2001). Vascularized three-dimensional skeletal muscle tissue-engineering. *Bio-Medical Materials and Engineering,* vol.11, pp. 275-281.

Saxena, A. K.; φcker, H. & Willital, G. H. (1999a). Present status of tissue engineering for surgical indications in children. *116th Congress of the German Association for Surgery,* Munich, Germany.

Schlatter, M.; Norris, K.; Uitvlugt, N.; DeCou, J. & Connors, R. (2003). Improved outcomes in the treatment of gastroschisis using a preformed silo and delayed repair approach. *Journal of Pediatric Surgery,* vol.38, pp. 459-464.

Silver, I. A. (1982). Basic physiology of wound healing in horse. *Equine Veterinary Journal,* vol.14, pp. 7-15.

Singh, J.; Kumar, N.; Sharma, A. K.; Maiti, S. K.; Goswami, T. K. & Sharma, A. K. (2008). Acellular Biomaterials of Porcine Origin for the Reconstruction of Abdominal Wall Defects in Rabbits. *Trends in Biomaterials & Artificial Organs,* vol.22, pp. 0-0.

Skalak, R. & Fox, C. F. (1988). Tissue Engineering: Proceedings of a workshop held at Granlibakken, Lake Tahoe, CA, New York, NY: Liss. pp. 26-29.

Slezak, S. E. & Horan, P. K. (1989a). Cell-mediated cytotoxicity. A highly sensitive and informative flow cytometric assay. *Journal of Immunological Methods*, vol.117, pp. 205-214.

Slezak, S. E. & Horan, P. K. (1989b). Fluorescent *in vivo* tracking of hematopoietic cells. Part I. Technical consideration. *Blood*, vol.74, pp. 2172-2177.

Smith, S.; Gantt, N.; Rowe, M. I. & Lloyd, D. A. (1989). Dura versus goretex as an abdominal wall prosthesis in an open and closed infected model. *Journal of Pediatric Surgery*, vol.24, pp. 519-521.

Sutherland, R. S.; Baskin, L. S.; Hayward, S.W. & Cunha, G.R. (1996). Regeneration of bladder urothelium, smooth muscle, blood vessels and nerves into an acellular tissue matrix. *Journal* of *Urology*, vol.156, pp. 571–577.

Tung, W. S.; Zainol, J.; Pilly, A. G.; Yusof, N. & Yusof, L. M. (2002). Processed BTV as a Biomaterial for the Repair of Large Abdominal Wall Defects in Surgical Treatment. *The Science*, vol.2, pp. 7-11.

Tyrell, J.; Silberman, H.; Chandrasoma, P.; Niland, J. & Shull, J. (1989). Absorbable versus permanent mesh in abdominal operations. *Surgery, Gynecology and Obstetrics*, vol.168, pp. 227-232.

Ueno, T.; Pickett, L. C.; La Fuente, S. G.; Lawson, C. & Pappas, T. N. (2004). Clinical application of porcine small intestine submucosa in the management of infected or potentially contaminated abdominal defects. *Journal of Gastrointestinal Surgery*, vol.8, pp. 109-112.

Van Wachem, P. B.; Brouwer, L. A. & van Luyn, M. J. A. (1999). Absence of muscle regeneration after implantation of a collagen matrix seeded with myoblasts. *Biomaterials*, vol.20, pp. 419-426.

Vialle-Preles, M. J.; Hartmann, D. J.; Franc, S. & Herbage, D. (1993). Imminuhistochemistry study of the biological fate of a subcutaneous bovine collagen implant in rat. *Histochemistry*, vol.91, pp. 177-184.

Wei, C. Y.; Chuang, D. C.; Chen, H. C.; Lin, C. H.; Wong, S. S. & Wei, F. C. (1995). The versatility of free rectus femoris muscle flap an alternative flap. *Microsurgery*, vol.16, pp. 698-703.

Wernig, A.; Zweyer, M. & Irintchev, A. (2000). Function of skeletal muscle tissue formed after myoblast transplantation into irradiated mouse muscles. *Journal of Physiology*, vol.522, pp. 333-345.

Wilson, L. & Gamble, M. (2002). *The hematoxylins and eosin*. In:Theory and practice of histological techniques, ed. D. J. Bancroft, and M. Gamble, pp.125-138. 5th ed. Churchill Livingstne, New York.

Winokur, S. T.; Barrett, K.; Martin, J. H.; Forrester, J. R.; Simon, M.; Tawil, R.; Chung, S-A.; Masny, P. S & Figlewicz, D. A. (2003). Facioscapulohumeral muscular dystrophy (FSHD) myoblasts demonstrate increased susceptibility to oxidative stress. *Neuromuscular Disorders*, vol.13, pp. 322-333.

Won, K. L.; Ki, D. P.; Dong, K. H.; Hwal, S.; Jong-Chul, P. & Young, H. K. (2000). Heparinized BP as a novel cardiovascular bioprosthesis. *Biomaterials*, vol.2, pp. 2323-2330.

Wu, D.; Razzano, P. & Grande, D. A. (2003). Gene therapy and tissue engineering in repair of the musculoskeletal system. *Journal of Cellular Biochemistry,* vol.88, pp. 467-481.

Yan, W.; Fotadar, U.; George, S.; Yost, M.; Price, R. & Terracio, L. (2006). Tissue engineering of skeletal muscle. *Microscopy and Microanalysis,* vol.11, pp. 1254-1255.

Zuki, A. B. Z.; Hafeez, Y. M.; Loqman, M. Y.; Noordin, M. M. & Norimah, Y. (2007). Effect of Preservation Methods on the Performance of Bovine Pericardium Grafts in a Rat Model. *Anatomia Histologia Embryologia,* vol.36, pp. 349-356.

7

# Skeletal and Adipose Tissue Engineering with Adipose-Derived Stromal Cells

Jeong S Hyun, Emily R Nelson, Daniel Montoro,
Benjamin Levi and Michael T. Longaker
*Stanford University*
*United States of America*

## 1. Introduction

There are wide ranges of pathologies leading to critical adipose and skeletal tissue loss including trauma, cancer resection and congenital anomalies. With half of the adult population in the United States estimated to be affected by a chronic disease according to the World Health Organization, the biomedical burden for replacement tissue continues to rise. By 2030, the number of adults over the age of 65 is expected to double to over 70 million in the United States (Institute of Medicine, 2008). The use of autologous tissue grafts and allografts for multiple organ systems have been successfully employed clinically, however many inherent disadvantages to these strategies exist including graft failure, infection, donor site morbidity etc. (Becker et al. 2011). While synthetic materials are not limited by availability, they are prone to infection, rejection, and breakdown over time (Keefe, 2009). Due to these inherent limitations, true regenerative medicine remains the ultimate goal for tissue replacement.

In 2001, researchers isolated a new population of adult multipotent cells in lipoaspirate (Zuk, et al. 2001, 2002). Since the discovery of adipose-derived stromal cells, a tremendous amount of work has been done in characterizing this population of cells and defining their capacity for multipotent differentiation. Adipose-derived stromal cells have potential advantages over other types of stem cells such as mesenchymal stem cells in that they are widely available and easily harvested through a simple liposuction procedure without altering their viability. Adipose-derived stromal cells can differentiate into mesodermal cell lines and offer a broad range of possibilities for application within the field of tissue engineering. These cells have already been put through several clinical trials in the treatment of a diverse array of pathologies including ST-elevated myocardial infarctions, Crohn's fistulas, and spinal cord injuries (Clinical Trials.gov, 2011).

Bone regeneration is an area of great interest in the field of tissue engineering. Over sixty million Americans are expected to be diagnosed with osteoporosis or low bone mass by the year 2020 (Samelson & Hannon, 2006). This can generate critical pathology as 1 in 2 women and 1 in 4 men over the age of fifty experiences an osteoporotic fracture in their lifetime (National Osteoporosis Foundation, 2011). With limited availability of autologous bone grafts, a bioengineered tissue replacement would be an ideal clinical alternative. Adipose-derived stromal cells were shown early on to have the capability for osteogenic differentiation *in vitro* (Gimble & Gullak, 2003) and the ability differentiate into osteogenic tissue *in vivo* (Cowan & Longaker, 2004, Cowan & Longaker, 2005). With adipose-derived stromal cells osteogenic capability defined, research has focused on identifying specific methods and pathways that enhance the osteogenic capability of adipose-derived stromal cells. In this chapter, we will focus on the major pathways involved in osteogenic differentiation and their modulation to allow for greater osteogenic capability for skeletal tissue engineering.

A critical aspect in skeletal tissue engineering is the ability for the implanted cells to function within the three-dimensional structure of the surrounding skeleton. The implanted cells must have the proper biomechanical characteristics of bone to allow it to have the form and function necessary for successful skeletal regenerative tissue. The development of scaffolds with osteoinductive properties has allowed for cells to be placed within a three-dimensional structural environment mimicking the skeletal system. These scaffolds provide an environment or niche for cell differentiation and integration within the surrounding tissue and promote the proper healing of the skeletal defect. In this chapter, we will discuss several of the major synthetic scaffolds being used in skeletal tissue engineering and how certain scaffold properties can enhance osteogenic healing within the skeletal defect.

Another potential clinical application of adipose-derived stromal cells with tissue engineering is their use for soft tissue replacement. Of particular importance is the use of adipose-derived stromal cells for reconstruction in post-mastectomy patients. According to the National Cancer Institute, 12.15% of woman will develop breast cancer in their lifetime. The safety of adipose-derived stromal cells for reconstruction was shown within the setting of breast cancer once active disease was eliminated (Zimmerlan et al. 2011). In 2007, Cytori Therapeutics, Inc. received FDA approval for Celution® System apparatus which automates sorting of lipoaspirate to isolate adipose-derived stromal cells for the application of breast reconstruction. This device has already been used in Europe and Japan in the clinical setting and the application to breast reconstruction could be an avenue for a widespread clinical use of adipose-derived stromal cells meeting FDA approval.

It is important for both clinicians and scientists to understand the current research with adipose-derived stromal cells given the great potential these cells have for future therapeutics. Therefore, we will cover the current research with adipose-derived stromal cells and their potential for osteogenic and adipogenic differentiation along with some of the major relevant pathways leading to their differentiation. In addition, we will cover different scaffolds that have been used to place these cells within the three-dimensional network of the surrounding tissue. Finally, we will discuss possible directions for future research.

## 2. Adipose-derived stromal cells and osteogenesis

The generation of skeletal tissue remains an elusive clinical goal of regenerative medicine. The relative paucity of available donor sites for autogenous bone grafts limits their use in clinical application. The need for bone replacement is increasing with an aging population, leading to an increase in the number of patients diagnosed with osteoporosis with subsequent fractures. In addition, the treatment of non-union fractures across all ages and populations remains a high biomedical burden. Since the isolation and characterization of adipose-derived stromal cells, significant work has been done defining protocols for inducing osteogenic differentiation within this population of multipotent cells.

### 2.1 *In vitro* methods of human and mouse adipose-derived stromal cell harvest

Adipose derived stromal cells can be harvested following a liposuction procedure or an adipose tissue resection. While surgeons initially believed ultrasonic techniques would have a negative effect on adipose-derived stromal cell viability, it has been shown that they retain their osteogenic capacity following ultrasonic assisted liposuction (Panetta, et al, 2009). The other major difference between harvesting adipose-derived stromal cells after a liposuction procedure rather than an adipose tissue resection is the use of tumescent or saline fluid that is injected prior to liposuction procedures. This injection decreases surgical bleeding and makes the procedure technically easier to perform. However, the additional fluid injected during lipoaspiration dilutes the adipose-derived stromal cells, and thus the total number of adipose-derived stromal cells able to be isolated in liposuction tissue. Therefore, adipose-derived stromal cells harvest yields following a lipoaspiration procedure are often less than following an adipose tissue resection.

Once the lipoaspirate is collected, the stromal cells within the heterogeneous population of cells within the lipoaspirate must be isolated. The adipose tissue settles into two layers consisting of the supernatant, which contains suctioned adipocytes along with stroma and the bottom layer of liposuction aspirate fluid containing larger pieces of lipoaspirate (Levi & Longaker, 2011). Adipose-derived stromal cells can be isolated from either layer, but there is a higher yield of stromal cells from the adipocyte supernatant (Yoshimura, et al. 2006). From there, adipose-derived stromal cells are isolated though a series of digestion, washing, straining, and neutralizing steps (Yu, et al. 2011) through readily available protocols. Once the progenitor cells are isolated, they can be plated for *in vitro* expansion and manipulation in culture or direct injection into a defect site.

Adipose-derived stromal cells have been shown to be similar in profile to bone marrow and umbilical cord blood derived stem cells (Kern, et al. 2006). Adipose-derived stromal cells were first characterized by their ability to differentiate into certain mesenchymal cell lineages. Using flow cytometry analysis, researchers have attempted to define these stromal cells by specific cell surface markers. However, generating a defined list of concrete markers has been a challenge. This is likely due to variability in patient profile and number of *in vitro* passages along with variability in stages of differentiation. Despite these difficulties, the International Society for Cellular Therapy released a position paper in 2006 stating that the minimal criteria for being classified as a mesenchymal stromal cell according to cell surface markers to include the expression of CD105 (Endoglin), CD73, and CD90 while lacking expression of hematopoietic markers CD45, CD34, CD14, CD11b, CD79α, CD19, and HLA-DR (Dominici M, et al. 2006).

### 2.2 Osteogenic *in vitro* differentiation protocols

The cocktails used for osteogenic differentiation do not vary as much as those for adipogenic differentiation. In general, all osteogenic media protocols utilize β-glycerol phosphate and ascorbic acid, albeit in slightly different concentrations (**Table 1**). Interestingly, mouse adipose-derived stromal cells have been shown to be less osteogenic than human adipose-derived stromal cells when using these two components alone. Studies have shown that mouse adipose-derived stromal cells require an additional osteogenic stimulus, such as retinoic acid (Wan, et al. 2006). Moreover, growth factors such as fibroblast growth factor (FGF)-2 inhibit the osteogenic differentiation of mouse adipose-derived stromal cells, while human adipose-derived stromal cells osteogenic differentiation proceeds relatively unabated in the presence or absence of FGF-2 (Quarto & Longaker 2006; Quarto & Longaker 2008). Interestingly, preliminary data from our group has demonstrated that TGF-β acts to inhibit both osteogenesis and adipogenesis in mouse adipose-derived stromal cells. Instead, it seems that TGF-β acts to drive mouse adipose-derived stromal cells toward a chondrogenic fate (James, et al, 2009; Xu, et al. 2007). TGF-β has also been shown to inhibit osteogenic human adipose-derived stromal cell differentiation in large doses. Additionally, while retinoic acid can be utilized to augment mouse adipose-derived stromal cells differentiation, it has little effect on the ability of human adipose-derived stromal cells to undergo osteogenic differentiation (Levi, et al, 2010a; Wan, et al, 2007a) Human adipose-derived stromal cells osteogenic differentiation can, however, be enhanced by supplementation with several cytokines, such as Insulin Like Growth Factor (IGF) (Levi & Longaker 2010a), Platelet Derived Growth Factor Alpha (PDGF-α)(Levi & Longaker 2010a), Sonic Hedgehog (SHH) (James, et al. 2010), or Bone Morphogenetic Protein-2 (Knippenberg M, et al. 2006).

| Cytokine | Stimulatory or Inhibitory | Target Osteogenic Concentration |
|---|---|---|
| Retinoic Acid | Stimulatory (in mouse ASCs) | 1-10 uM |
| Sonic Hedgehog | Stimulatory | 100-750 ng/ml |
| BMP-2 | Stimulatory | 50-200 ng/ml |
| BMP-4 | Stimulatory | 10-50 ng/ml |
| IGF-1 | Stimulatory | 25-50 ng/ml |
| PDGF-α | Stimulatory | 10-20 ng/ml |
| Noggin | Inhibitory | 100-400 ng/ml |
| BMPR-IB/ALK-6/Fc Chimera | Inhibitory | 0.5-2.0 ug/ml |

Table 1. Stimulatory and Inhibitory Cytokines for ASCs for use in Osteogenic Differentiation Media

For osteogenic differentiation, cells should be plated in a 6-well plate (80,000-100,000 cells per well), a 12-well plate, (35,000-45,000 cells per well), or a 24-well plate (15,000-25,000 cells per well). After attachment, adipose-derived stromal cells can be cultured in osteogenic differentiation medium.

Early and late osteogenic differentiation has been defined to occur at specific time points, and these time points vary between species. While mouse adipose-derived stromal cells undergo early osteogenesis after 7 days in culture and late osteogenesis after 14 days in culture. In contrast, human adipose-derived stromal cells begin to differentiate much earlier than mouse adipose-derived stromal cells and show evidence of early and late osteogenesis following 3 and 7 days in culture, respectively. RNA analysis of adipose-derived stromal cells for specific osteogenic gene markers is a commonly used method to assess osteogenic differentiation in a quantitative manner. Specific gene markers for early osteogenic differentiation include *Alkaline Phosphotase (ALP), Runt Related Protein-2 (RUNX-2),* and *Collagen Ia1 (COL1A1).* Gene markers for intermediate and late osteogenesis include *Osteopontin (OPN)* and *Osteocalcin (OCN),* respectively. Alkaline phosphatase staining and quantification is another method used to assess early osteogenic differentiation of adipose-derived stromal cells (day 3 for human adipose-derived stromal cells, day 7 for mouse adipose-derived stromal cells). Late osteogenic activity can be assessed by Alizarin red or Von Kossa staining, which are assays for extracellular mineralization (day 7 in human adipose-derived stromal cells, day 14 in mouse adipose-derived stromal cells).

### 2.3 Molecular pathways of differentiation

There are several key molecular pathways that lead to the differentiation of a multipotent cell into a mature osteoblast in the process of osteogenesis. Runx-2 is considered the principal osteogenic switch and has a key role in the commitment of stem cells into an osteoblastic lineage. Runx-2 null mice show a complete deficiency in skeletal formation (Komori and Kishimoto, 1998). Runx-2 is transcriptionally upregulated by the bone morphogenetic proteins (BMP) through transduction via Smad 1 and 5 (Chen et al, 2004). Retinoic acid has also been shown to increase Runx-2 expression and stimulate mouse adipose-derived stromal cell osteogenic differentiation by increasing the expression of BMPR-IB protein (Wan, Longaker 2006). While Transforming Growth Factor-Betas (TGF-βs) have been shown to be increased during bone development and growth, their effect on precursor cell differentiation into osteoblasts depends on cell type, stage of differentiation, and levels of expression (Linkart 1996, Komori 2002).

The Wnt signaling pathway is a group of proteins involved in a wide array of developmental and cellular processes. In the canonical Wnt pathway, Wnts bind to cell-surface receptors of the Frizzled family and their co-receptors, low-density lipoprotein receptor related protein (LRP), leading to inactivation of the axin-glycogen synthase kinase 3β complex which stabilizes Beta-catenin. Beta-catenin translocates into the nucleus to promote Wnt gene expression. LRP5 knockout mice show a phenotype with low bone mass (Gong et al, 2001) while overexpression of the same gene is associated with high bone mass (Boyden et al, 2002). Wnt3A has been shown to have differing results on osteogenic differentiation depending on dosage and level of baseline differentiation along with age of the cells. (Quarto, Longaker 2010). Low dose treatment with Wnt3a enhanced bony regeneration in a mouse critical calvarial defect treated with differing doses of Wnt3a in juvenile mice, while higher doses decreased repair (Quarto, Longaker 2010). Conversely, in adult mice, increasing doses of Wnt3a enhanced bone regeneration.

In the non-canonical Wnt pathway, Wnts bind to the receptors Frizzled and Dishevelled but the non-canonical pathway does not mediate its signal through the glycogen synthase 3β kinase complex or the Beta-catenin complex, but rather through G-proteins and GTPases. While the most studied aspect of the non-canonical Wnt pathway is their regulation in cell polarity, a growing body of evidence suggests that this pathway is important in bone development. Wnt5a has been shown to increase osteoblast formation from human bone mesenchymal stem cells *in vitro* (Baksh & Tuan, 2007). In human adipose-derived stromal cells, the addition of Wnt5a *in vitro* increased the expression of Runx2 and osteocalcin and mineralized nodule formation (Santos, et al, 2010).

### 2.4 Growth factors and hormones that promote osteogenesis

Growth factors are signaling molecules that have a wide effect on cellular processes. The majority of growth factors are secreted as proteins or hormones and act in an autocrine, paracrine, or endocrine manner to affect cellular proliferation and differentiation by regulating gene expression, protein synthesis, and cell signaling. Through the modulation of specific growth factors, osteogenesis can be enhanced with the result of improved skeletal healing.

Bone morphogenetic proteins are a group of secreted cytokines that belong to the Transforming Growth Factor-β superfamily. Throughout history, skeletal tissue was known for its ability to regenerate and efforts were undertaken to try to isolate an intrinsic property secreted by these cells that led to their regenerative properties. BMPs were first discovered and named in 1965 (Urist, 1965) and the first BMP gene was isolated and cloned by Wozney in 1988 (Wozney et al, 1988). In 2002, the FDA approved the use of recombinant BMP-2 to be used on a three-dimensional scaffold for anterior lumbar fusion (McKay et al, 2007). Today, over twenty different genes belong to the BMP family mainly involved in the regulation of bone and cartilage formation (Bessa C et al. 2008). The BMP proteins signal **(Fig. 1)** through BMP receptor Type I and II, which leads to the phosphorylation of receptor-regulated SMAD 1,5, and 8, which combine with the common-mediator SMAD 4. This complex then migrates into the nucleus to affect gene expression.

A large body of research has validated the importance of BMP-2's role in bone formation and healing. BMP-2 has been shown to induce chondrocyte and osteoblast precursor formation in mesenchymal stem cells (Wall et al. 1994) and specifically shown to induce osteogenic differentiation in adipose-derived stromal cells (Dragoo et al. 2003) *in vitro*. Our group (Cowan, Longaker 2005) showed the use of BMP-2's effect combined with adipose derived stromal cells *in vivo* on bony healing by culturing adipose-derived stromal cells with rhBMP-2 for 4 weeks prior to implantation onto a PLGA scaffold into a critical-sized calvarial defect in a mouse. With the treatment of BMP-2 *ex-vivo*, cells treated with recombinant human BMP-2 showed accelerated healing over unstimulated scaffolds with complete healing in as little as two to four weeks as demonstrated by histology and microCT. Interestingly, prolonged exposure to recombinant human BMP-2 was also shown to increase osteoclast activity and increase bone turnover and resorption. Therefore, to avoid increased bone turnover, our group used direct *in vivo* injection of recombinant human BMP-2 without the prolonged *ex-vivo* treatment (Levi, Longaker, et al 2010b) into a critical calvarial defect with and without adipose-derived stromal cells. Cells treated with recombinant human BMP injected subcutaneously into the skeletal defect on post-op days 1-3 showed increased healing at 8 weeks over unstimulated scaffold alone and untreated cells. The increase in osteogenesis was observed with this method of BMP-2 augmentation, but without the stimulation of osteoclast-induced bone turnover and resportion. The current scope of research is concentrating on how to deliver BMP-2 into the skeletal wound microenvironment to promote osteogenesis and bony healing while minimizing side effects such as ectopic bone formation (Zara, Siu, Soo et al. 2011).

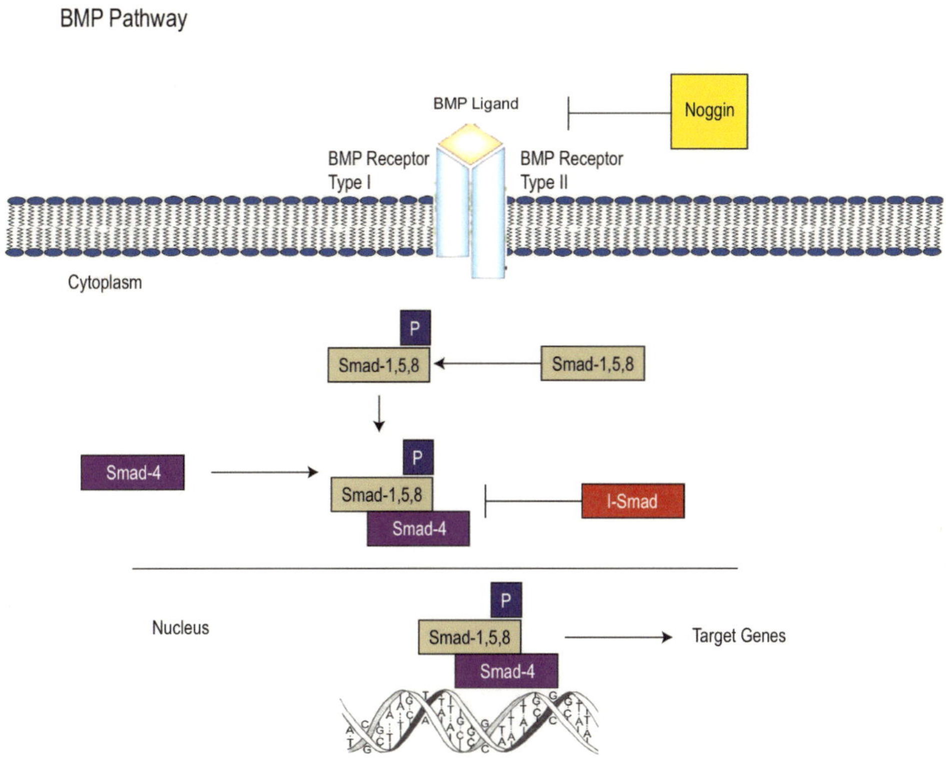


Fig. 1. Bone Morphogenetic Protein Pathway

BMP ligand binds to BMP-R1 or 2, which phosphorylates SMAD 1,5, and 8. Co-smad 4 complexes with Smad 1,5, and 8 and translocates into the nucleus and regulates gene expression

In addition to BMP-2, BMP-7 (Osteogenic Protein-1) has also been shown to be effective in the augmentation of skeletal healing. In a monkey model, recombinant human BMP-7 was injected into a scaffold at the skeletal injury site and demonstrated healing in an ulnar/tibial segmental defect model. (Cook, et al 1995). A randomized clinical trial compared the use of BMP-7 versus autogenous bone graft in the treatment of tibial non-unions (Friedlaender, 2004) and found that using BMP-7 was equivalent in clinical outcomes without the morbidity of using an autogenous bone graft. Specifically with adipose-derived stromal cells, researchers using an adenovirus vector *ex vivo* to induce BMP-7 expression showed increased bone formation both *in vitro* and *in vivo* using rat adipose-derived stromal cells (Yang, et al. 2005) and in human adipose-derived stromal cells (Kang Y, et al. 2007).

The fibroblast growth factor family has over twenty members and four receptors involved in diverse cellular processes from angiogenesis to wound healing. Similar to the BMP family, fibroblast growth factors have been shown to have a significant effect on osteogenesis (Ornitz and Marie, 2002). In particular, FGF-2 is expressed in osteoblastic cells and has been shown to have important functions in the regulation of bone and cartilage formation. FGF-2 knockout mice show decreased bone mass and bone formation compared to control mice (Montero et al. 2000). Initial work on the effect of FGF-2 on adipose-derived stromal cells showed that FGF-2 expression was critical in maintaining clonogenicity and differentiation potential (Zaragosi et al. 2006) along with the maintenance of adipose-derived stromal cells proliferation *in vitro* (Quarto and Longaker, 2006). Evaluation of FGF expression during osteogenic differentiation of adipose-derived stromal cells showed that FGF-2 is down-regulated during osteogenic differentiation (Quarto and Longaker, 2008) while FGF-18 is upregulated. This was explored further by examining different isoforms of FGF-2. The high molecular weight protein form of FGF-2 increased expression during osteogenesis while the low-molecular weight form decreased, suggesting a regulation at the translational level in the balance between maintaining multipotency and lineage commitment. In order to explore the effects of FGF-2 *in vivo,* we engineered feeder cells capable of secreting the FGF-2 protein when driven by the presence of a synthetic ligand, Shield-1, which stabilizes a protein-destabilizing domain fused to the FGF-2 gene (Kwan, Longaker, 2011). Feeder cells and adipose-derived stromal cells were both seeded on a scaffold and placed into a mouse critical-sized calvarial defect. Shield-1 was delivered intraperitoneally and the mice were followed up to 20 weeks. There was significantly improved healing with FGF-2 versus the non-injected group of control adipose-derived stromal cells and scaffold alone in the defect. A possible explanation for the improved osteogenesis and calvarial defect healing *in vivo* may be linked to the increased proliferation of osteoprogenitor cells due to FGF-2 expression.

Insulin-like growth factors are proteins with sequence homology similar to insulin under the control of growth hormone. There are two types, IGF-1 and IGF-2, which play a wide variety of cellular roles from the regulation of proliferation to apoptosis. Both IGF-1 and IGF-2 have been shown to be associated with increase in bone formation and density (Adami et al. 2010, Chen et al, 2010) while their reduction has been shown to be associated with decreased bone density (Bennett, 1984). In order to specifically identify genes upregulated during osteogenesis, our group isolated RNA from human adipose-derived stromal cells during i*n-vitro* osteogenesis and performed a microarray analysis (Lee, Longaker, 2010). Gene expression of IGF-1 was found to be elevated 3.5 fold from baseline levels at day three of osteogenic differentiation with the microarray. IGF-1's ability to increase osteogenesis was then tested by augmenting osteogenic differentiation media with IGF-1 *in vitro* (Levi, Longaker 2010a). Alkaline phosphatase activity was elevated with the addition of IGF-1 and there was increased matrix mineralization evident with the treated group stained with alizarin red. Quantitative real-time polymerase chain reaction showed elevated levels of Runx2 and Osteocalcin when adipose-derived stromal cells were treated with IGF-1.

The effect of estrogen on bone formation is well known. Decreasing levels of estrogen are one of the main culprits in the pathogenesis of osteoporosis with estrogen withdrawal causing an increase in bone remodeling (Seeman, 2003). While men do not have a comparable deceleration of hormone production, circulating levels of free estrogen also declines with men with aging leading to comparable long term loss of trabecular bone (Seeman, 2004). The addition of estrogen through 17-$\beta$ estradiol increased levels of osteogenesis in human bone marrow mesenchymal stem cells through modulation of estrogen receptor $\alpha$ and $\beta$ (Hong, et al 2006). Human adipose-derived stromal cells augmented with 17-β estradiol *in vitro* showed increased evidence of osteogenesis through increased expression of alkaline phosphatase and osteocalcin as compared to control cells without estrogen augmentation (Hong et al, 2007).

The effect of androgens on bone formation is less clear. It is recognized that patients with androgen deficiency have lower bone mass and higher bone turnover compared to patients with normal levels of androgen (Hofbauer & Khosia, 1999). The majority of *in vitro* studies indicate that testosterone and dihydrotestosterone have proliferative effects on osteoblast progenitors, but there is no consensus as to whether androgens promote or inhibit osteogenic differentiation (Vanderschueren, et al, 2004). Androgen replacement therapy *in vivo* has shown to be effective in increasing bone mass in hydogonadal patients, the effect is mitigated in patients with more modest decreases in testosterone (Vanderschueren, et al, 2004). The full effects of estrogen and androgen and differences in sex hormones still must be worked out as well as the effect of sex of the donor adipose-derived stromal cells in differences on osteogenic differentiation.

### 2.5 Gene therapy

Another strategy for improving osteogenic differentiation is gene therapy to either express or knockdown specific proteins for the desired effect altering cellular function. Strategies in genetic therapy include *in vivo* direct gene delivery using a vector for transfection or *ex vivo* techniques, which involves treating the cell of interest with modification of the genome outside of the host prior to implantation.

Gene therapy requires a vector to transfer the genetic material into the target cell. Vectors can be divided into viral and non-viral and also whether the strategy integrates the vector into the host-genome or is based on a transient effect. Viral vectors have the advantage of being efficient in their ability to infect host cells. Common viral vectors are adenovirus, lentivirus, and retrovirus. While the integrating viral vectors are efficient and express the DNA of question for the life of the cell, there is no way to completely control the site of integration leading to concern regarding mutations and oncogenic activity. Viral vectors can also elicit a host immune response causing rejection. Non-integrating strategies such as liposome based particles and electroporation transiently transfect the cells and have a decreased immune response, but are not as efficient and have variable expression.

Zhang et al. (2006) used an adenovirus vector to overexpress Runx-2 in mouse adipose-derived stromal cells to increase osteogenesis *in vitro* and *in vivo* by implanting seeded scaffolds on the back of nude mice. Several strategies have used BMP-2 gene therapy to promote bone healing and osteogenic differentiation. A segmental defect created in New Zealand white rabbits was treated with adenoviral vectors expressing BMP-2 resulting in increased healing in the treatment group (Baltzar, 2000). Lee SJ et al (2010) created a bicistronic vector for co-delivery of both Runx-2 and BMP-2 into human adipose-derived stromal cells. Using microporation for their transfection strategy, they showed increases in osteogenesis *in vitro* and implanted transfected human adipose-derived stromal cells on a PLGA scaffold subcutaneously *in vivo* showing increase in ectopic bone formation compared to control.

Osterix is a zinc finger transcription factor that is expressed in osteoblasts whose role is critical during osteoblast differentiation. Nakashima et al. (2002) first identified osterix through a screen of expressed genes during osteogenic differentiation. Through homologous recombination, they produced osterix-null mice, which showed no bone formation. Osterix appears to be downstream of Runx-2 as osterix-null mice had normal Runx-2 expression and normal cartilage formation. However, Runx-2 negative mice show impaired chondrogenesis and decreased osterix expression. Overexpression of osterix has been shown to increase osteogenic differentiation in embryonic stem cells by increased in levels of Osteocalcin and Runx-2 (Tai, et al. 2004) and in adipose-derived stromal cells (Wu L, et al 2007). Lee SJ et al. (2011) used electroporation to co-transfect Runx-2 with osterix into adipose-derived stromal cells. Transfected cells with osterix alone and osterix with Runx-2 both showed significant increase in osteogenesis over control cells.

Msx-2 (Hox-8) is a homeodomain transcription factor that functions as a transcriptional regulator of the osteocalcin promoter (Towler DA, et al 1994). Msx-2 has been found to be vital in craniofacial development, as mutations within the homeodomain of Msx-2 have been shown to lead to craniosynostosis, the premature fusion of calvarial sutures (Jabs, et al. 1993; Liu et al. 1995). Msx2 null mice have deficient endochondral bone formation, defects in skull ossification and a persistent calvarial foramen (Satokata, et al, 2000). Using *CMV-Msx2* transgenic mice, Cheng et al, (2008) showed that overexpression of Msx2 leads mesenchymal progenitor cells into an osteogenic lineage enhancing osteoblast formation along with trabecular bone formation by activation of the canonical Wnt pathway.

In looking for specific genetic targets to augment adipose-derived stromal cells for osteogenic differentiation, our group focused on the noggin protein, which is an inhibitor of BMP signaling. We evaluated the effect of noggin suppression on osteoblast differentiation by using siRNA constructs against the noggin protein in primary osteoblasts (Wan, Longaker et al, 2007b). Noggin suppression was confirmed using western blot and QT-PCR. Evaluation of BMP signaling *in vitro* with cells transfected with Noggin siRNA constructs showed an increase in BMP signaling with Smad1/5 along with evidence of osteogenic differentiation of preosteoblasts as compared to control. Noggin knockdown also increased skeletal healing *in vivo* when osteoblasts transfected with Noggin siRNA were seeded on a PLGA scaffold as demonstrated by microCT and histomorphometric analysis.

## 3. Adipogenesis and adipose-derived stromal cells

The adipogenic differentiation of multipotent stromal cells is of interest to many specialties of medicine. The growing body of research in obesity and adipose biology has opened the field of adipogenesis to numerous specialties in clinical medicine and science. Stem cell scientists, bone biologists, endocrinologists, and tissue engineers all have a vested interest in the study of adipogenesis. From a clinical standpoint, surgeons are faced with challenging reconstructive cases in patients afflicted with soft tissue deficiencies. For example, burn patients often have soft tissue atrophy and severe skin contractures and would greatly benefit from soft tissue augmentation. Similarly, the widespread use of HAART (Highly-active retroviral therapy) medications for HIV has left many HIV patients with facial lipodystrophy, which can be disfiguring and socially troublesome. In the pediatric population, patients with congenital malformations would greatly benefit from new soft tissue engineering techniques. For example, Parry-Romberg disease causes progressive hemifacial soft tissue atrophy in the face. Outcomes of these patients would greatly improve with the use of a tissue engineering approach to reconstruct their inadequate adipose compartment.

### 3.1 Adipogenic *in vitro* differentiation protocols

The adipogenic differentiation protocols of primary mesenchymal stromal cells vary as reported in the literature. For the purposes of this chapter, two of the most commonly studied mesenchymal stem cells populations were examined: bone marrow mesenchymal stem cells and adipose-derived stromal cells. Generally, a three component cocktail is used to supplement media for bone marrow mesenchymal stem cell adipogenic induction, including Insulin, Dexamethasone, and IBMX (3-isobutyl-1-methylxanthine). There are several differences between mouse and human adipogenic differentiation protocols. For example, Insulin was not used in the majority of publications in human bone marrow mesenchymal stem cell adipogenic differentiation protocols, while insulin is present in the majority of murine adipogenic differentiation protocols. Additionally, human adipogenic protocols used an increase in dexamethasone along with the addition of Indomethacin in comparison to mouse bone marrow mesenchymal stem cell protocols.

Adipose-derived stromal cells are multipotent mesenchymal cells and thus can be differentiated into adipocytes. With regards to adipose-derived stromal cell adipogenic differentiation, there are a wide variety of reports of varied differentiation cocktails with many published articles in the last year (Romo-Yanex, et al 2011; Ghosh, et al, 2010; James, et al. 2010; Valorani, et al, 2010, Lee JE, et al. 2010). Generally, a three or four component cocktail is used for adipose-derived stromal cell adipogenic differentiation, including Indomethacin, Insulin, Dexamethasone and IBMX. Similar to bone marrow mesenchymal stem cell adipogenesis protocols, there are a wide variety of recommended cocktails that become even more heterogeneous when comparing protocols for mouse adipose-derived stromal and human adipose-derived stromal cell adipogenesis. For example, though insulin concentrations in the adipose differentiation media of both mouse adipose-derived stromal cell and human adipose-derived stromal cell are relatively similar, an approximate 10-fold increase in dexamethasone concentration exists in mouse adipose-derived stromal cell adipogenesis protocols when compared to human adipose-derived stromal cell protocols

### 3.2 Molecular pathways of differentiation

At the molecular level, several key genes are expressed during the differentiation of pre-adipocytes to mature adipocytes including lipoprotein lipase (LPL), peroxisome proliferator-activated receptor (PPARγ), and enhancer binding proteins (EbP) (Kronenberg & Williams, 2008). In addition, several Hox genes have been recently identified that are highly expressed during the adipogenic differentiation of human adipose-derived stromal cell (Cowherd, et al. 1997; Cantile, et al. 2003). Other genes such as TAZ have been shown to activate the Runx-2 transcription factor and stimulate osteogenesis while inhibiting adipogenesis (Hong, et al. 2005).

PPARγ is one of the most widely studied genes involved in adipogenesis, and has been translated for clinical use in the form of the agonist rosiglitazone, as well as the similar agents (troglitazone, etc). These PPARγ agonists have been used in the treatment of diabetes as they have been shown to speed up the differentiation process of pre-adipocytes or adipoprogenitor cells *in vitro*. Rosiglitazone functions by binding to PPARγ, thus 'sensitizing' adipocytes to insulin (Mayerson, et al 2002). One of the major side effects of this class of pharmaceuticals is bone mineral density reduction, which is thought to be due to the diversion of mesenchymal stem cells to adipogenesis rather than osteogenesis *in vivo* (Bodmer, et al, 2009; Rosen, et al. 2006). To the full extent that rosiglitazone is able to induce mesenchymal stem cell adipogenesis as a single agent is unknown. However, it is believed that this class of drugs is capable of enhancing the differentiation of pre-adipocytes to adipocytes *in vivo* (Ninomiya, et al 2010). Thus, PPARγ agonists may be useful in adipocyte induction cocktails for adipose-derived stromal cells. Furthermore, from a clinical standpoint, targeting the PPARγ pathway may improve the adipogenic potential of transplanted stem cells further improving soft tissue engineering outcomes.

One of the surgical limitations of fat transfer is the high rate of adipocyte resorption over time. It is thought that adipocyte resorption during fat transfer is due to the inadequate blood supply and thus, surgeons have set out to improve circulation to transplanted adipocytes. Besides alterations in technique, recently, Yoshimura *et al* has reported enhanced adipocyte viability and vascularization when seeding adipocytes with human adipose-derived stromal cells from the same patient at the time of surgery (Suga et al 2010, Yoshimura, et al. 2008, Yoshimura, et al. 2009, Matsumoto, et al. 2006). This technique offers the use of an autologous scaffold seeded with autologous multipotent stromal cells. Human adipose-derived stromal cells are a known vasculogenic cell type that have been shown to secrete vasculogenic cytokines, thus working in an autocrine and paracrine fashion to allow for improved survival of adipocytes and enabling possible adipogenic differentiation. Therefore, potential benefits of using human adipose-derived stromal cells to differentiate into adipocytes, include their ability to proliferate after transplantation, their release of angiogenic cytokines, and their ability to differentiate into endothelial cells and undergo neovascularization (Planat-Barnard et al., 2004; Miranville, et al. 2004, Moon, et al. 2006)

## 4. Scaffold technology and adipose-derived stromal cells

Almost as important as the proper engineering of multipotent cells into the desired lineage is their placement within the three-dimensional context of the target organ. This is especially true with skeletal and soft tissue engineering as their structure is critical to their function. Recent advancements in the design and quality of scaffold technology is critical for the advancement of successful *in vivo* application of multipotent cells. With the goal of engineering functional tissue, biomimetic scaffolds are designed to provide 3-dimensional structural support for the engineered tissue, uniform pore size and structure for a matrical distribution of cells, and combinations of substrates and growth factors to promote viability and differentiation along the appropriate cell lineage. With technologies enabling the reliable production of mineralized biopolymers to nanospun fibers, scaffolds engineered from these materials promote healing *in vivo,* (Lew, et al. 1997; Osathanon, et al. 2009) and prove a powerful delivery system for adipose-derived stromal cells in the generation of functional tissue replacement.

The growing incidence of craniofacial bone defects arising from congenital malformation, surgical resection and trauma necessitates the development for skeletal regeneration of bony defects to restore functional movement and protection to the head and its critical structures. Regenerated bone exists in a dynamic environment and must provide adequate structural support and tensile strength. One of the important models for skeletal regeneration in the craniofacial context is the critical size calvarial defect, in which a bony deficit is created in an animal model that is of a size that will not heal within the lifetime of an animal. Thus, any healing observed is due to the experimental contribution. Because of the availability of experimental animals and analytical tools, the laboratory mouse is a common model for bone regeneration in calvarial critical-sized defects, which are between 3-4mm in diameter (Mooney & Siegel, 2005).

Critical to the engineering of an ideal scaffold for bone regeneration are the biocompatibility, osteoconductive, and osteoinductive properties of the materials. A biocompatible scaffold should be manufactured under sterile conditions, have an even and consistent pore size and distribution, and be non-toxic to the host tissues as it breaks down over time. Osteoconductive properties refer to the characteristics of the scaffold framework and its impact on the resulting structure of newly formed bone, while osteoinduction is the promotion and induction of host or donor cells toward mature osseous formation. A variety of inorganic polymers and synthetic nanofibers are being used as the basis for manufacturing scaffolds and vary in their capability for osteoinduction, rate of resorption, and moldability.

Polylactide (PLA) and poly lactic-co-glycolic acid (PLGA) polymers have been shown to be safe in human tissue (Frazza & Schmitt, 1971) and are resilient to fracture but lack significant osteoconduction. Calcium phosphate and calcium sulfate salts, on the other hand, tend to be brittle and difficult to mold but contain natural bone elements of crystal salts found in bone matrix and tend to be more osteoconductive scaffolds than their polymer counterparts. Of particular interest in this group is hydroxyapatite, a mineral salt composing a large proportion of naturally-occurring bone matrix with strong osteoconductivity (Chang, et al 2000).

In order to combine the tensile strength and load-bearing capabilities of a polymer scaffold with the osteoconduction and cellular substrate of mineral salts, Cowan et al employed a strategy combining osteogenic mouse adipose-derived stromal cells seeded on a PLGA scaffold coated with hydroxyapatite for bony regeneration of a 4mm critical-sized defect in mouse. (Cowan, et al, 2004) The combined osteogenic potential of adipose-derived stromal cells and osteoconduction of apatite-coated PLGA promoted complete healing within 12 weeks with major contribution from the transplanted adipose-derived stromal cells as shown by chromosomal analysis. Levi et al have recapitulated these results with apatite-coated PLGA scaffolds in mouse calvarial defect using human adipose-derived stromal cells in immunodeficient nude mice, demonstrating the potential of human cellular therapy delivered on hybrid polymer mineral scaffold for translational development. (Levi, et al. 2010b)

Although significant bone regeneration has been achieved using osteoconductive scaffolds and adipose-derived stromal cells without the use of additional growth factors (Cowan, et al 2004; Levi, et al. 2010b), the stimulation of well-characterized osteogenic pathways may enhance efficient new bone deposition in conjunction with biomimetic scaffolds and cell therapy. Bone morphogenetic proteins (BMPs) along with valproic acid and other factors have demonstrated abilities to accelerate bone formation *in vivo* (Cowan, et al. 2005) and may be candidates for delivery in combination with ASCs and apatite-coated PLGA scaffolds for increased efficiency and reliability of osteogenesis.

Similar to current clinical treatment modalities for bone defects, soft tissue deficits due to trauma, tumor, and congenital etiologies are treated with surgical grafts from autologous, allogenic, or alloplastic sources. The successes of vascularized microvascular flap transfer, the gold standard modality, are often accompanied with limitations in source material and patient morbidity. Because adipose tissue has numerous functions, including energy regulation, secretion of important paracrine and endocrine factors, and mechanical protection, it is important to develop new strategies for replacement of functional soft tissue deficits. Of major consideration for clinical translation is the regeneration of adipose tissue following mastectomy, requiring adipose stroma and a biomimetic matrix for the replacement of breast tissue.

Cellular transplants for adipogenic regeneration hinge on scaffolds that support cell adhesion and the differentiation and uptake of lipids while providing appropriate structural support. Porosity is a critical factor in the scaffold structure to allow the development of angiogenic support necessary for functional tissue formation. A number of synthetic and biological scaffolds have been studied as delivery systems of adipocyte precursor cells for effective adipose tissue generation with varying degrees of success and limitations. Inorganic synthetics such as PLGA and polyglycolic acid (PGA) offer moldable, highly regular structures and non-toxic breakdown. Gel-based delivery systems promote cell survival but lack the organizational regularity of inorganic scaffolds. Natural proteins found in connective tissue such as collagen can be engineered into scaffold design with homogeneous structure and porosity, and mimic biological matrix found commonly in adipose tissue.

Matrigel™ is a gel-like substance derived from animal connective tissue proteins and growth factors, and is readily available commercially. Despite the translational barrier of its sarcoma derivation, Kawaguchi et al and others have reported successful angiogenesis and adipogenesis following injection of Matrigel™ in mouse fat pad (Kawaguchi, et al. 1998). The mixture of components in matrigel including collagen, laminin, and growth factors demonstrated sufficiency for the generation of new vascularized adipose tissue, but because of the stochastic organization of proteins within the gel, the distribution of cells in the *de novo* tissue reflect a scattered and uneven pattern of growth. Adipogenic tissue regeneration with eventual clinical translation in mind should promote predictable growth of new soft tissue. An ideal scaffold would define the structure of the newly formed tissue to be populated by transplanted donor cells and stimulated endogenous pre-adipocytes and allow for angiogenic growth and support into newly forming tissue.

One study performed a direct *in vivo* comparison of three disparate scaffold technologies for the delivery of mouse ASCs as an assay for adipogenic ability using PGA, collagen, and hyaluronic acid gel (Itoi, et al. 2010). PGA represents the biodegradable synthetic group of scaffolds, extracellular matrix-derived collagen from tendon is a natural product that is prepped into porous sponges, and hyaluronic acid (HA) is a derivative of extracellular matrix of soft tissue involved in cell proliferation. ASCs were harvested from GFP mice to allow visualization of transplanted cell proliferation and differentiation, grown in adipogenic supplemented media, and seeded into the respective scaffolds and transplanted into athymic mice. In all groups tested, fluorescent cells were present at four weeks and a slight increase in angiogenesis was seen. In contrast, Oil Red O staining for triglycerides was positive only in PGA and collagen groups and negative in HA. At 8 weeks, the most robust GFP positive adipose-like tissue was detected in the collagen scaffold construct indicating that collagen served as the best adipogenic scaffold in this comparison.

Promising data from Davidenko et al (2010) combines the naturally-derived extracellular matrix components collagen and hyaluronic acid into a hybrid cross-linked scaffold designed to provide an array of matrix protein components with the structural advantage of collagen. Thus, this design allows for precise engineering of scaffold structure and porosity, even distribution of transplanted adipogenic cells, and physiochemical safety. Preadipocyte cells were shown to proliferate and differentiate readily on the scaffold during *in vitro* adipogenic induction and gene analysis showed increased expression of Adipsin following differentiation.

## 5. Future directions

Given the abundant availability of lipoaspirate combined with a relatively simple procedure to harvest, adipose-derived stromal cells have a potential to be critically important in the generation of tissue for clinical use in a wide range of pathologies. While there is great promise for the use of these cells, significant hurdles remain before they become available for wide spread clinical application. The ultimate translational goal is the ability to harvest the lipoaspirate and be able to process and implant the cells into the area of need in the same patient during a single surgical procedure. Research in several key areas must happen prior to this scenario being realized.

Currently, the process for sorting and isolating stem cells from the lipoaspirate is a manual process. The total number of cells typically obtained is around 300,000 cells per milliliter of lipoaspirate (Levi & Longaker, 2011). While this yield is higher then other adult-stem cell populations, the process can still be time consuming and labor intensive. For widespread clinical use to be possible, the ability to have an automated processing unit isolate the stromal cells in the lipoaspirate would be critical to decreasing the time and cost of this procedure. Cytori's Celution® System was developed to automate the sorting process and takes around 1 hour to sort 250 ml of lipoaspirate. Using Flow Cytometry, the cells sorted using the Celution® system were found to be similar to cells isolated on a manual sort with the ability to differentiate into osteogenic and adipogenic cells (Lin, et al. 2008). As more research is done characterizing specific populations within adipose-derived stromal cells, the technology should improve to better match and sort for specific populations of cells more likely to differentiate into the desired cell lineage.

The importance of scaffolds and their ability to deliver cells into the proper niche environment has been stated. As the technology improves with scaffold development, they have become much more than inert materials where cells are placed, but rather can provide a highly inductive environment for cells to be "coached" towards a certain lineage. The ability to augment scaffolds with various growth factors and cytokines has been shown to improve healing within specific wound environments (Brown, et al. 2011). These scaffolds can have the advantage of release kinetics allowing for a controlled delivery of the growth factor of interest into the wound environment (Kwan, et al, 2011). The future of scaffold technology is moving towards the ability to induce genetic therapy within its three-dimensional structure. The delivery of BMP-2 and BMP-7 through vectors within the scaffold structure into host cells has already proven to be effective (Nie, et al. 2007; Zhang, et al. 2011).

## 6. Conclusion

With the large clinical need for replacement tissue, adipose-derived stromal cells are easily obtained in large numbers and offer a viable option for future therapeutic application given their ability to differentiate into multiple cell lines. Through continued research into pathways of differentiation, the ability to maximize the yield of these multipotent cells will continue to improve. With the increased effectiveness of technology for the sorting and the application of adipose-derived stromal cells, they could soon become an important clinical tool for the treatment of a wide range of clinical pathologies.

## 7. References

Adami S, Zivelonghi A,Viaplana O, et al. Insulin-like growth factor-1 is associated with bone formation markers, PTH and bone mineral density in healthy premenopausal women. *Bone.* 2010 Jan; 46(1):244-7

Baksh D, Tuan RS. Canonical and non-canonical Wnts differentially affect the development potential of primary isolate of human bone marrow mesenchymal stem cells. *J Cell Physiol.* 2007 Sep;212(3):817-26

Baltzar AW, et al. Genetic enhancement of fracture repair: healing of an experimental segmental defect by adenoviral transfer of the BMP-2 gene. *Gene Ther* 2000 May; 7(9):734-9

Becker ST, Warnke PH, Behrens E, Wiltfang J. Morbidity after iliac crest bone harvesting over an anterior versus posterior approach. *J Oral Maxillofac Surg* 2011 Jan;69(1):48-53

Bennett AE, Wahner HW, Riggs BL, Hintz RL. Insulin-like growth factors I and II: Aging and bone density in women. *J Clin Endocrinol Metab.* 1984 Oct; 59(4):701-4

Bessa PC, Casal M Reis RL. Bone Morphogenetic proteins in tissue engineering: the road from laboratory to the clinic, part I (Basic concepts) *J Tissue Eng Regen Med* 2008 Jan;2(1):1-13

Bodmer, M., Meier, C., Kraenzlin, M. E., et al. Risk of fractures with glitazones: a critical review of the evidence to date. 2009 *Drug Saf* 32: 539-547

Boyden, LM, et al. High bone density due to a mutation in LDL-receptor-related protein 5. *N Engl J Med* 2002 May 16; 346(20), 1513–152

Brown KV, Li B, Guda T. et al. Improving bone formation in a rat femur segmental defect by controlling BMP-2 release. *Tissue Eng Part A* 2011 Feb 21

Cantile, M., Procino, A., D'Armiento, M., et al. HOX gene network is involved in the transcriptional regulation of in vivo human adipogenesis. 2003 *J Cell Physiol* 194: 225-236

Chang BS, Lee CK, Park KW, et al. Osteoconduction at porous hydroxyapatite with various pore configurations. *Biomaterials.* 2000 Jun;21(12):1291-8

Chen D, Zhao M, Mundy GR. Bone morphogenetic proteins. *Growth Factors.* 2004 Dec; 22(4)233-41

Chen L, Jiang W, Huang J et al. Insulin-like growth factor 2 (IGF-2) potentiates BMP-9-induced osteogenic differentiation and bone formation. *J Bone Miner Res.* 2010 Nov;25(11) 2447-59

Cheng SL, Shao JS, Towler DA, et al. Msx2 exerts bone anabolism via canonical Wnt signaling. *J Biol Chem.I* 2008 Jul 18;283(29):20505-22

ClinicalTrials.gov. National Institute for Health 2000-2011. Available at: http://www.clinicaltrials.gov/. Accessed March 18, 2011

Cook SD, Wolfe MN, Reuger DC. Effect of recombinant human osteogenic protein -1 on healing of segmental defects in non-human primates. *J Bone Joint Surg Am* 1995 May;77(5):734-50

Cowan CM, Shi YY, Longaker MT et al. Adipose derived adult stromal cells heal critical-size mouse calvarial defects. *Nat Biotechnology* 2004 May22;(5): 560-7

Cowan CM, Aalami OO, Longaker MT et al. Bone morphogenetic protein 2 and retinoic acid accelerate in vivo bone formation, osteoclast recruitment, and bone turnover. *Tissue Eng.* 2005 Mar-Apr;11 (3-4):645-58

Cowherd, R. M., Lyle, R. E., Miller, C. P., et al. Developmental profile of homeobox gene expression during 3T3-L1 adipogenesis. 1997 *Biochem Biophys Res Commun* 237: 470-475

Davidenko N, J.J. Campbell, E.S. Thian, C.J. Watson, R.E. Cameron. Collagen-hyaluronic acid scaffolds for adipose tissue engineering. *Acta Biomaterialia* 2010 6: 3597-3968

Dominici M, Le Blanc K, Mueller I, et al. Minimal criteria for defining multipotent mesenchymal stroma cells. The International Society for Cellular Therapy position statement. *Cytotherapy.* 2006;8(4):315-7

Dragoo JL, Choi JY, Lieberman JR, et al. Bone induction by BMP-2 transduced stem cells derived from human fat. *J Orthop Res* 2003 Jul;21(4):622-9

Friedlaender GE. Osteogenic protein-1 in treatment of tibial nonunions: current status. *Surg Technol Int.* 2004;13:249-52

Frazza, E.J. & Schmitt, E.E. A new absorbable suture. *J. Biomed. Mater.* 1971 Res. 5, 43-58

Ghosh, S., Dean, A., Walter, M., et al. Cell density-dependent transcriptional activation of endocrine-related genes in human adipose tissue-derived stem cells. 2010 *Exp Cell Res* 316: 2087-2098

Gimble J, Gullak F. Adipose-derived adult stem cells: isolation, characterization, and differentiation potential. *Cytotherapy* 2003; 5(5):362-9

Gong et al. LDL receptor-related protein 5 (LRP5) affects bone accrual and eye development. *Cell* 2001 Nov 16:107(4), 513–523

Hofbauer LC, Khosia S. Androgens effects of bone metabolism: Recent progress and controversies. *Eur J Endocrinol.* 1999 Apr;140(4):271-86

Hong, J. H., Hwang, E. S., McManus, M. T., et al. TAZ, a transcriptional modulator of mesenchymal stem cell differentiation. 2005 *Science* 309: 1074-1078

Hong L, Colpan A, Peptan IA. Modulations of 17-beta estradiol on osteogenic and adipogenic diffentiations of human mesenchymal stem cells. *Tissue Eng.* 2006 Oct;12(10):2747-53

Hong L, Colpan A, Evans CA et al. 17-Beta estradiol enhances osteogenic and adipogenic differentiation of human adipose-derived stromal cells. *Tissue Eng.* 2007 Jun;13(6):1197-203

Itoi Y, Takatori M, Hyakusoku H, Mizuno H. Comparison of readily available scaffolds for adipose tissue engineering using adipose-derived stem cells. *Journal of Plastic, Reconstructive, and Aesthetic Surgery* 2010: 63: 868-864

Jabs EW, Muller U, Mulliken JB et al. A mutation in the homeodomain of human MSX2 gene in a family affected with autosomal dominant craniosysostosis. *Cell.* 1993 Nov 5;73(3):443-50

James, A. W., Xu, Y., Lee, J. K., et al. Differential effects of TGF-beta1 and TGF-beta3 on chondrogenesis in posterofrontal cranial suture-derived mesenchymal cells in vitro. 2009 *Plast Reconstr Surg* 123: 31-43

James, A. W., Leucht, P., Levi, B., et al. Sonic Hedgehog influences the balance of osteogenesis and adipogenesis in mouse adipose-derived stromal cells. 2010 *Tissue Eng Part A* 16: 2605-2616

Institute of Medicine *Retooling for an aging America: Building the health care workforce.* Washington, D.C: National Academics Press, Inc.; 2008

Kang Y, Laio WM, Lei L, et al. *In vitro* and in *vivo* induction of bone formation based on adeno-associated virus-mediated BMP-7 gene therapy using human adipose-derived mesenchymal stem cells. *Acta Pharmacol Sin. 2007 Jun;28(6):839-49*

Kawaguchi N, Toriyama E, Nicodemau-Lena E, Innou K, Torii S, Kitagawa Y. De novo adipogenesis in mice at the site of injection of basement membrane and basic fibroblast growth factor. *Proc Natl Acad Sci USA* 1998: 95:1062-1066

Keefe MS, Keefe MA. An evaluation of the effectiveness of different techniques for intraoperative antibiotics into alloplastic implants for use in facial reconstruction. *Arch Facial Plastic Surg.* 2009 Jul-Aug;11(4):246-51

Kern S, Eichler H, Stoeve J, et al. Comparative analysis of mesenchymal stem cells from bone marrow, umbilical cord blood, or adipose tissue *Stem Cells* 2006 May; 24(5):1294-301

Knippenberg M, Helder MN, Zandieh DB. Osteogenesis versus chondrogenesis by BMP-2 and BMP-7 in adipose stem cells. *Biochem Biophys Res Commununⁿ.* 2006 Apr 14;342(3):902-8

Komori T, Kishimoto T. Cbfa1 in bone development. *Curr Opin Genet Dev* 1998 8:494–499

Komori T. Runx2, A multifactorial transcription factor in skeletal development *J Cell Biochem.* 2002;87(1):1-8

Kronenberg, H., Williams, R. H. *Williams textbook of endocrinology,* 11th Ed. Philadelphia: Saunders/Elsevier, 2008

Kwan MD, Sellmyer MA, Longaker MT. et al. Chemical control of fgf-2 release for promoting calvarial healing with adipose stem cells. *J Biol Chem.* 2011 Apr 1;286(13):11307-13

Lee J, Gepta D, Longaker MT. Elucidating mechanisms of osteogenesis in human adipose-derived stromal cells via microarray analysis. *J Craniofac Surg.* 2010 Jul;21(4): 1136-41

Lee, J. E., Kim, I., Kim, M. Adipogenic differentiation of human adipose tissue-derived stem cells obtained from cryopreserved adipose aspirates. 2010 *Dermatol Surg* 36: 1078-1083

Lee SJ, Kang SW, Lee SH et al. Enhancement of bone regeneration by gene delivery of BMP2/Runx2 bicistronic vector into adipose-derived stromal cells. *Biomaterials* 2010 Jul;31(21):5625-9

Lee SJ, Lee JM, Im GI. Eletroporation-mediated transfer of Runx2 and Osterix genes to enhance osteogenesis of adipose stem cells. *Biomaterials* 2011 Jan;32(3):760-8

Levi, B., James, A. W., Wan, D., et al. Regulation of human adipose-derived stromal cell osteogenic differentiation by insulin-like growth factor-1 and platelet-derived growth factor-alpha. *Plast Reconstr Surg* 2010a Mar 8

Levi B, James AW, Longaker MT, et al. Human adipose derived stromal cells heal critical size mouse calvarial defects. *PLoS One.* 2010b Jun 17; 5(6)

Levi B, Longaker MT. Adipose-derived stromal cells for skeletal regenerative medicine. *Stem Cells* 2011 Feb 8

Lew D, Farrell B, Bardach J, Keller J: Repair of craniofacial defects with hydroxyapatite cement. *J Oral Maxillofac Surg* 1997 55: 1441-1451

Lin K, Matsubara Y, Masuda Y, et al. Characterization of adipose tissue-derived cells isolated with the Celution system. *Cytotherapy* 2008; 10(4)$:417-26

Linkhart Ta et al. Growth factors for bone growth and repair: IGF, TGF- β, and BMP *Bone* 1996 Jul; 19(1 Suppl):1S-12S

Liu YH, Kundu R, Maxon RE Jr, et al. Premature suture closure and ectopic cranial bone in mice expressing Msx2 transgenes in the developing skull. *Proc Natl Acad Sci USA* 1995 Jun 20;92(13):6137-41

Locke M, Windsor J, Dunbar PR. Human adipose-derived stem cells: isolation, characterization, and application in surgery. *ANZ J Surg 2009* Apr;79(4)4:235-44

Matsumoto, D., Sato, K., Gonda, K., et al. Cell-assisted lipotransfer: supportive use of human adipose-derived cells for soft tissue augmentation with lipoinjection. 2006 *Tissue Eng* 12: 3375-3382

Mayerson AB, Hundal RS, Peterson KF, et al. The effects of rosiglitizone on insulin sensitivity, lipolysis, and hepatic and skeletal muscle triglyceride content in patients with type 2 diabetes. *Diabetes*. 2002 Mar;51(3):797-802

McKay WF, Peckham SM, Badura JM. A comprehensive clinical review of recombinant human bone morphogenetic protein 2 (INFUSE Bone Graft). *Int Ortop*. 2007 Dec; 31(6):729-34

Miranville, A., Heeschen, C., Sengenes, C., et al. Improvement of postnatal neovascularization by human adipose tissue-derived stem cells. 2004 *Circulation* 110: 349-355

Moon, M. H., Kim, S. Y., Kim, Y. J., et al. Human adipose tissue-derived mesenchymal stem cells improve postnatal neovascularization in a mouse model of hindlimb ischemia. *Cell Physiol Biochem* 2006 17: 279-290

Mooney MP, Siegel MI: Animal models for bony tissue engineering, in Wnek G, Bowlin G (eds): *Encyclopedia of Biomaterials and Biomedical Engineering*. New York: Marcel Dekker, 2005, pp 1-19

Montero, A., et al., Disruption of the fibroblast growth factor-2 gene results in decreased bone mass and bone formation. J. Clin. Invest. *2000,* 105, 1085–109

Nakashima K, Zhou X, de Crombrugghe B, et al. The novel zinc finger-containing transcription osterix is required for osteoblast differentiation and bone formation. *Cell* 2002 Jan 11;108(1):17-29

National Osteoporosis Foundation. http://www.nof.org/. Accessed March 18, 2011

Nie H, Wang CH. Fabrication and characterization of PLGA/HAP composite scaffolds for delivery of BMP-2 plasmid DNA. *J Control Reelease.* 2007 Jul 16;1120(1-2):111-21

Ninomiya, Y., Sugahara-Yamashita, Y., Nakachi, Y., et al. Development of a rapid culture method to induce adipocyte differentiation of human bone marrow-derived mesenchymal stem cells. 2010 *Biochem Biophys Res Commun* 394: 303-308,

Ornitz, D.M., Marie, P.J., FGF signaling pathways in endochondral and intramembranous bone development and human genetic disease. 2002 *Genes Dev*. 16, 1446–1465

Osathanon T, Giachelli CM, Somerman MJ: Immobilization of alkaline phosphatase on microporous nanofibrous fibrin scaffolds for bone tissue engineering. *Biomaterials* 2009 30:4513-4521

Panetta, N. J., Gupta, D. M., Kwan, M. D., et al. Tissue harvest by means of suction-assisted or third-generation ultrasound-assisted lipoaspiration has no effect on osteogenic potential of human adipose-derived stromal cells. 2009 *Plast Reconstr Surg* 124: 65-73

Planat-Benard, V., Silvestre, J. S., Cousin, B., et al. Plasticity of human adipose lineage cells toward endothelial cells: physiological and therapeutic perspectives. 2004 *Circulation* 109: 656-663

Quarto N, Longaker MT. FGF-2 inhibits osteogenesis in mouse adipose tissue-derived stromal cells and sustains their proliferative and osteogenic potential state. *Tissue Eng.* 2006 Jun; 12(6):1405-18

Quarto, N., Wan, D. C., Longaker, M. T. Molecular mechanisms of FGF-2 inhibitory activity in the osteogenic context of mouse adipose-derived stem cells 2008 (mASCs). *Bone* 42: 1040-1052

Quarto N, Behr B, Longaker MT. Opposite spectrum of activity of canonical Wnt signaling in the osteogenic context of undifferentiated and differentiated mesenchymal cells: implications for tissue engineering. *Tissue Eng Part A* 2010 Oct; 16(1):3185-97

Romo-Yanez, J., Montanez, C., Salazar-Olivo, L. A. Dystrophins and DAPs are expressed in adipose tissue and are regulated by adipogenesis and extracellular matrix. 2011 *Biochem Biophys Res Commun* 404: 717-722

Rosen, C. J., Bouxsein, M. L. Mechanisms of disease: is osteoporosis the obesity of bone? *Nat Clin Pract Rheumatol* 2006 2: 35-43

Samelson EJ, Hannan MT. Epidemiology of Osteoporosis. *Curr Rheumatol Rep* 2006 Feb;8(1):76-83

Santos A, Bakker Ad, Klein-Nulend J, et al. Wnt5A induces osteogenic differentiation of human adipose stem cells via rho-associated kinase ROCK. *Cytotherapy* 2010 Nov;12(7):924-32.

Satokata I, Ma L, Maas R, et al. Msx2 deficiency in mice causes pleitropic defects in bone growth and ectodermal organ formation. *Nat Genet.* 2000 Apr;24(4):391-5

Seeman E. Invited Review: Pathogenesis of osteoporosis. *J Appl Physiol.* 2003 Nov;95(5):2142-51

Seeman E. Estrogen, androgen, and the pathogenesis of bone fragility in women and men. *Curr Osteopros Rep.* 2004 Sep;2(3):90-6

Suga, H., Eto, H., Aoi, N., et al. Adipose tissue remodeling under ischemia: death of adipocytes and activation of stem/progenitor cells. 2010 *Plast Reconstr Surg* 126: 1911-1923.

Tai G. et al. Differentiation of osteoblasts from murine embryonic stem cells by overexpression of the transcriptional factor osterix. *Tissue Eng.* 2004 Sep-Oct;(9-10):1456-6

Towler DA, Rugledge SJ. Rodan GA. Msx-2/Hox8.1: A transcriptional regulator of the rat osteocalin promotor. *Mol Endocrinol.* 1994 Nov;8(11):1484-93

Urist MR. Bone: Formation by Auto-induction. *Science* 1965 Nov 12 150(698)839-9

Valorani, M. G., Germani, A., Otto, W. R., et al. Hypoxia increases Sca-1/CD44 co-expression in murine mesenchymal stem cells and enhances their adipogenic differentiation potential. 2010 *Cell Tissue*

Vanderschueren D, Vandenput L, Ohisson C, et al. Androgens and bone. *Endocr Rev.* 2004 Jun;25(3):389-425

Wall, NA, Hogan BL. TGF-Beta related genes in development. *Curr Opin. Genet. Dev.* 1994;4, 517-522 *Res* 341: 111-120

Wan DC, Shi YY, Lonagker MT. Osteogenic differentiation of mouse adipose-derived adult stromal cells requires retinoic acid and bone morphogenetic protein receptor type IB signaling. *Proc Natl Acad Sci USA*. 2006 Aug 15; 103(33):1235-40

Wan, D. C., Siedhoff, M. T., Kwan, M. D., et al. Refining retinoic acid stimulation for osteogenic differentiation of murine adipose-derived adult stromal cells. *Tissue Eng* 2007a 13: 1623-1631

Wan DC, Pomerantz JH, Longaker MT et al. Noggin Suppression Enhances in Vitro Osteogenesis and accelerates in vivo bone formation. *J Biol Chem*. 2007b Sep 7;282(36):26450-9

World Health Organization. World Health Statistics 2009. Available at http://www.who.int/whosis/whostat/2009/en/index.html. Accessed March 17, 2011

Wozney J.M. Rosen V, Celesete A. J. Et al. Novel regulators of bone formation: Molecular clones and activities. *Science* 1988:242:1528-1534

Wu L, Wu Y, Lin Y, et al. Osteogenic differentiation of adipose derived stem cells promoted by overepxression of osterix. *Mol Cell Biochem*. 2007 Jul;301(1-2):83-92

Xu, Y., Balooch, G., Chiou, M., et al. Analysis of the material properties of early chondrogenic differentiated adipose-derived stromal cells (ASC) using an in vitro three-dimensional micromass culture system. *Biochem Biophys Res Commun* 2007 359: 311-316

Yang M, Ma QJ, Zhou CY, et al. In vitro and in vivo induction of bone formation based on ex vivo gene therapy in rat adipose-derived adult stem cells expressing BMP-7. *Cytotherapy*. 2005;7(3):273-281

Yoshimura K, Shigeura T, Matsumoto D, et al. Characterization of freshly isolated and cultured cells derived from the fatty and fluid portions of liposuction aspirates. J Cell Physiol. 2006; 208:64-7

Yoshimura, K., Sato, K., Aoi, N., et al. Cell-assisted lipotransfer for facial lipoatrophy: efficacy of clinical use of adipose-derived stem cells. 2008 *Dermatol Surg* 34: 1178-1185,

Yoshimura, K., Suga, H., Eto, H. Adipose-derived stem/progenitor cells: roles in adipose tissue remodeling and potential use for soft tissue augmentation. 2009 *Regen Med* 4: 265-273

Yu G. Floyd ZE, Wu X, et al. Isolation of human adipose-derived stem cells from lipoaspirates. *Mehods Mol Biol* 2011;702:17-27

Zara JN, Siu RK, Soo C. High doses of Bone morphogenetic protein induce abnormal bone formation and inflammation in vivo. *Tissue Eng Part A* 2011 Mar 3

Zaragosi LE, Allhaud G, Dani C. Autocrine fibroblast growth factor 2 signaling is critical for self-renewal of human multipotent adipose-derived stem cells. *Stem Cells*. 2006 Nov;24(11):2412-9

Zhang X, Yan M, Ao YF et al. Runx-2 overexpression enhances osteoblastic differentiation and mineralization in adipose-derived stem cells in vitro and in vivo. *Calcif Tissue Int*. 2006 Sep; 79(3):169-7

Zhang Y, Fan W, Nortdruft L, et al. In vitro and in vivo evaluation of adenovirus combined silk fibrion scaffolds for BMP-7 gene delivery. *Tissue Eng Part C Methods*. 2011 Mar 18.

Zimmerlin L, Donnenberg AD, Rubin JP, et al. Regenerative therapy and cancer: in vitro and in vivo studies of the interaction between adipose-derived stem cells and breast cancer cells from clinical isolates. *Tissue Eng Part A*. 2011 Jan;17(1-2):93-106

Zuk PA Zhu M, Hedrick MH, et al. Multilineage cells from human adipose tissue: implications for cell-based therapies. *Tissue Eng*. 2001 Apr;7(2):211-8

Zuk PA, Zhu M, Ashjia P, et al. Human adipose tissue is a source of multipotent stem cells. *Mol Biol Cell* 2002 Dec; 13(12):4279-95

# Part 3

## Ligaments

8

# Tissue Engineering of Ligaments

Sarah Rathbone and Sarah Cartmell
*School of Materials, Materials Science Centre, The University of Manchester*
*United Kingdom*

## 1. Introduction

The main function of a ligament is to connect one bone to another bone across a joint, keeping them aligned to prevent abnormal motions and dislocations. The typical magnitude of force a ligament may experience during day-to-day activities varies. For example the anterior cruciate ligament of the knee can be exposed to daily tensile forces ranging between 67N for ascending stairs to 630N for jogging (Vunjak-Novakovic, Altman et al. 2004), whereas large loads, exceeding 1800N, can cause rupturing. Depending upon anatomical location and the extent of vascularisation, the ligament may or may not be capable of self-healing after a rupture.

Some of the most frequently ruptured ligaments occur in the knee joint, often through sporting activities such as skiing, football and basketball and the number of injuries are increasing each year (Cooper, Lu et al. 2005). Ninety percent of knee ligament injuries involve the anterior cruciate ligament (ACL) and medial collateral ligament (MCL) (Woo, Abramowitch et al. 2006). The MCL can self-heal, but the ACL cannot due to poor vascularisation. Because of this, alternative methods such as regenerative medicine have focused heavily upon the ACL with the aim of producing a fully functional tissue in vitro.

Figures indicate that approximately 250,000 people are diagnosed with ACL injuries each year in the USA (Doroski and Brink 2007), and approximately 150,000 need to undergo surgical treatment, known as an ACL reconstruction (Cooper, Lu et al. 2005). If the rupture is not treated it can cause loss of function of the associated joint which can then lead to early development of osteoarthritis (Cooper and Bailey 2006; Gentleman, Livesay et al. 2006).

The current gold standard procedure for an ACL reconstruction is surgical autografting. This involves using part of the patients own patellar tendon, hamstring or quadriceps to replace the ruptured ACL (Beasley, Weiland et al. 2005). However, these techniques cause donor site morbidity (Goulet and Germain 1997; Van Eijk, Saris et al. 2004; Cooper and Bailey 2006; Hairfield-Stein, England et al. 2007) which is associated with pain and a recovery period for the donor tissue site (Cooper, Lu et al. 2005; Hairfield-Stein, England et al. 2007). Generally 75-90% of patients have good or excellent long term success rates from these current grafting techniques (regarding functional stability and symptomatic relief upon return to normal activities) but unfortunately a substantial number of patients exist who have unsatisfactory results which could be attributed to graft failure (Vergis and Gillquist 1995). Some of these patients continue to endure pain, suffer from loss of motion secondary to the operative procedure and continue with recurrent instability (Vergis and Gillquist 1995), while others suffer from degenerative joint disease such as arthritis or experience re-injury (Hairfield-Stein, England et al. 2007). Alternatively, allografts can be used where the donor tendon is taken from a cadaver, but the disadvantages associated with this include donor scarcity, the risk of the recipient contracting a disease from the donor, or tissue rejection (Ahmed, Collins et al. 2004; Vunjak-Novakovic, Altman et al. 2004). Prosthetic replacements (synthetic grafts) have previously been used, but these have shown to be inadequate due to wear and degeneration (Mascarenhas and MacDonald 2008). It is evident that surgical ACL reconstructions have limitations and do not always give completely satisfactory long-term results in a high proportion of patients, which consequently affects their quality of life (Vergis and Gillquist 1995; Lanza, Langer et al. 2007). Because of this dilemma, regenerative medicine could be an option, where in vitro tissue engineering of ligaments can offer a solution to the problems associated with the current surgical methods (Van Eijk, Saris et al. 2004; Hairfield-Stein, England et al. 2007). Tissue engineered ligaments could provide better performance in the long run by improved biocompatibility, integration into host tissue and the ability to remodel their own extracellular matrix (Nesic, Whiteside et al. 2006).

Tissue engineering is a method which combines knowledge from material science, engineering, molecular biology and medicine (Nesic, Whiteside et al. 2006). The basic procedure normally involves using scaffolds to act as structural supports for cell growth and maturation in-vitro, where a stimulus (chemical or mechanical) may also be applied to promote the formation of a functional tissue. This concept was originally developed to repair skin and cartilage, but is now being considered as a possible option to produce neoligament tissue. To date, many different types of material have been investigated as potentially suitable scaffolds for ligament tissue engineering, focusing upon their biocompatibility, degradability, surface properties for cell attachment and overall mechanical properties. These include polymers (such as polyurethane, polylactic acid, polyglycolic acid, polycaprolactone, polyhydroxyalkanoates and alginates), silk fibroin, glasses, hydrogels and biological materials such as de-cellularised tissues.

There has been much research into the application of chemical stimulus upon cell culture in vitro. It is well documented that specific growth and differentiation factors can trigger various cellular responses such as cell differentiation, cell division and matrix remodelling (Evans 1999), making them useful in tissue engineering to influence cell behaviour. Some of the most commonly studied growth factors include transforming growth factor beta-1 (TGF-β1), basic fibroblast growth factor (bFGF) and epidermal growth factor (EGF). Bioreactors also have applications in tissue engineering, where they can be used to optimise the cell culturing conditions; for example they can improve the mass transfer of nutrients to cells in a 3D scaffold (perfusion), improve cell seeding onto a scaffold (rotation), or provide a mechanical stimulus (in tension or compression) to influence cell behaviour. For ligament tissue engineering, the bioreactor is normally used to apply tensile straining forces within physiological ranges to promote differentiation and extracellular matrix (ECM) synthesis. From the literature, the mechanical loading regimes investigated have varied from 1-10% strain, 0.01-1Hz frequency, from ½ hour – 24 hours/day over a period ranging from 1 day to six weeks. Achieving the optimal culturing conditions for a ligament tissue engineered construct can be complex, where small changes can have large affects upon cell behaviour and their final product. This chapter will review in detail the different biomaterials, loading regimes and growth factors that have been currently investigated for this purpose.

## 2. Anatomy of the ligament

### 2.1 Structure and function of a ligament

A ligament is a capsule of connective tissue made of fibres joining one bone to another across a joint where they help to guide joint motions and prevent abnormal displacement of bones relative to each other (Einhorn, O'Keefe et al. 2007). They are very strong compared to other connective tissues, such as skin, because of the high tensile loads they need to withstand (Einhorn, O'Keefe et al. 2007). Although there are several hundred ligaments in the body, many of the examples given have focused upon the ACL because it is the most frequently injured knee ligament. Figure 1 indicates where the main knee ligaments are located around and within the knee joint.

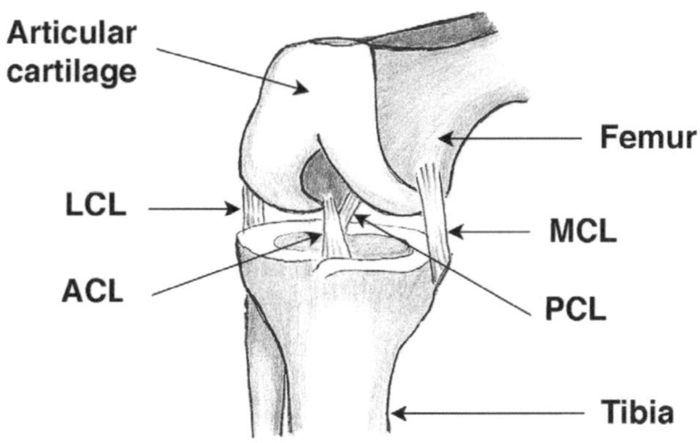


Fig. 1. The diagram illustrates where the fours main knee ligaments are located; the anterior cruciate ligament (ACL), the posterior cruciate ligament (PCL), the medial collateral ligament (MCL), and the lateral collateral ligament (LCL)

There are three main types of connective tissue within the human body, connective tissue proper (loose and dense regular connective tissue), fluid connective tissue (transports substances in blood) and supporting connective tissue (cartilage and bone). Skeletal ligament is a dense regular connective tissue, which is comprised of fibroblasts (connective tissue cells), and extracellular matrix (proteins and water making up the connective tissue). The periodical change in direction of collagen fibres gives the connective tissue a distinct undulating pattern. The fibroblasts (located within in the ECM) are responsible for producing the ECM components to maintain and repair the connective tissue. After an injury, these cells become mobile, migrating to the wounded tissue to increase the synthesis of specific proteins to aid tissue repair (Rogers 1983; Alberts, Johnson et al. 2000). The ECM is composed of two main classes of macromolecules; polysaccharide chains of glycosaminoglycans (which have adhesion functions and attract water), and fibrous proteins such as collagen, elastin and reticular fibres which give structural support to the tissue (Alberts, Johnson et al. 2000). The ground substance of the ECM is a hydrophilic water-like gel containing the polysaccharides and fibrous proteins, allowing diffusion of waste products and nutrients between the tissue cells and capillaries (Hansen, Masouros et al. 2006).

Ligament connective tissue is classed as dense regular tissue because the closely packed collagen fibres are aligned in an ordered regular, way, giving tensile strength and support to the tissue. The basic structure of collagen is very similar in all collagen types, where its formation starts with the synthesis of polypeptide chains on the ribosome which are composed mainly of glycine, hydroxylysine and hydroxyproline repeats (Carpenter and Hankenson 2004). Inside the cell, three polypeptide α-chains coil together into a right-hand twist to make a triple helix (super helix) forming the procollagen molecule which then becomes exocytosed from the cell. The procollagen molecules polymerize in the extracellular space firstly by aggregating together into a microfibrils, then aggregating into fibrils, where finally, the fibres become stabilized by covalent cross-links which form within and between the tropocollagen molecules (Alberts, Johnson et al. 2000). It is this extensive cross-linking, particularly in collagen I, which gives the collagen fibrils their stability and great tensile strength, which in turn makes the tissue very strong so that the ligament can resist deformation from stretching forces (Doroski, Brink et al. 2007). Collagen type I, III and V are all structural components of ligament fibrils (Posthemus, September et al. 2009), where type I and III provide tensile strength and type V regulates fibre assembly and diameter (September, Schwellnus et al. 2007). Collagen type X is present where the ligament integrates into the bone. Tenascin-c, another type of protein found in ligament ECM, regulates the tissues response to mechanical loading (September, Schwellnus et al. 2007). The individual collagen fibrils are randomly orientated, but as they aggregate into fibres they gain a more parallel orientation with the longitudinal axis of the ligament, giving the tissue a crimping pattern (wavy appearance) (Goulet, Germain et al. 1997).

### 2.2 Bone attachment

Ligaments attach to the bone surfaces as an aggregation of collagen fibre bundles (Ellenbecker 2000), either by direct insertion, or both direct and indirect insertion. The ACL inserts into the bone by direct insertion, the most common ligament insertion type, where its collagenous fibres attach directly to the bone tissue. The collagenous fibres blend into the fibrocartilaginous layer, interweave through the fibrocartilage zone, through the mineralized fibrocartilage zone, then enter the bone (Beasley, Weiland et al. 2005; Woo, Abramowitch et al. 2006). The calcified collagenous fibres which anchor the ligament firmly to the bone are known as Sharpey's fibres (Einhorn, O'Keefe et al. 2007). The medial collateral ligament (MCL), which is also a knee ligament, is inserted into the bone by both direct and indirect insertion, where the superficial fibres (near to the surface) merge with the periosteum (the connective tissue surrounding the bone), while other fibres penetrate the bone deeper and attach to the bone directly at acute angles (Woo, Abramowitch et al. 2006).

### 2.3 Characteristic components of ligament tissue

It is the variation in ratio between collagen types and other ECM components which gives each ligament type its diversity and characteristic mechanical behaviour (Woo, Abramowitch et al. 2006). Due to the absence of specific markers, ligaments can only be distinguished from other ligaments and other tissue types (eg tendon) by structural, molecular and mechanical properties. Although there is no single specific marker in the ligament, tenscin-c has been considered to be a marker due to its characteristically high amounts in the ligament (Doroski, Brink et al. 2007), where the presence of collagen types I and III, tenomodulin, biglycan, decorin, elastin and fibronectin are also characteristic of ligament tissue (Vunjak-Novakovic, Altman et al. 2004; Chen, Huang et al. 2008). The total amounts and specific ratios of the ECM components, ground substance and cells are characteristic properties unique for each type of connective tissue which relate to its anatomical location and function (Vunjak-Novakovic, Altman et al. 2004). Connective tissue types can be differentiated by the presence and total quantities of collagen, tenascin-c, elastin, fibronectin, decorin, biglycan, ratios of collagen types, crimping pattern and collagen fibril diameter. Table 1 lists the main components of a typical ligament as a percentage of their wet weights, whereas those in table 2 are the dry weights of the main collagen types and ratio of collagen type I : III which are unique to the ACL, as reported by various authors.

| Tissue type (wet weight) | Collagen type I (%) | Other collagens such as type III, V, VI (%) | Elastin (%) | Fibronectin and other glycoproteins (%) | Proteoglycans (%) | Water (%) | Author |
|---|---|---|---|---|---|---|---|
| Ligament (general) | 20 | 3-5 | 1-2 | 1-2 | <1 | 70 | (Einhorn, O'Keefe et al. 2007) |

Table 1. The biochemical constituents of wet ligament tissue

| Tissue type (dry weight) | Collagen type I (%) | Collagen type III (%) | Collagen type V (%) | Ratio of collagen I : III | Author |
|---|---|---|---|---|---|
| ACL | 70-80 | 8-10 | 10-12 | 9:1 | (Woo, Abramowitch et al. 2006; Doroski, Brink et al. 2007) |

Table 2. The collagen content of the ACL in dry tissue

## 3. Cell adherence

It is essential for fibroblasts to attach to a substrate and spread out to enable them to grow, proliferate, mature and produce functional tissue. In order for these cells to adhere to their substrate, ECM adhesion proteins such as fibronectin, vitronectin or collagen are required to adsorb to the substrate first, where the cells will then subsequently adhere to the adhesion proteins. Fibronectin can exist in two major forms; (1) soluble plasma fibronectin, a constituent of plasma, and (2) insoluble cellular fibronectin, a component of the ECM (Pankov and Yamada 2002). Cellular fibronectin can be expressed by different cell types including fibroblasts (Pankov and Yamada 2002), where it is found in ligaments and other connective tissues. Studies have found it be up-regulated during ligament formation in embryogenesis where it guides the migrating cells (Laurencin and Freeman 2005).

Integrin-mediated binding enables the cell to become connected to its surroundings by linking the interior of the cell to the ECM proteins (outside the cell). The contact made with the ECM can generate intercellular signals which can affect gene expression, morphology, cell survival (Johansson, Svineng et al. 1997), control cell adherence, cell migration, cytoskeletal organization (Sechler, Corbett et al. 1997) regulate growth, proliferation, differentiation, and subsequently affect development or maintenance of the ligament tissue (Alberts, Johnson et al. 2000; Vunjak-Novakovic, Altman et al. 2004). Two known intracellular pathways involved upon integrin-fibronectin binding are the Ras-MAPK (Ras-mitogen activated protein kinase) and the FAK (focal adhesion kinase) pathways.

Fibronectin mediated cell adhesion studies have been conducted by various researchers as a technique for improving cell attachment in vitro. Research has shown that cell retention of rat MSC's on fibronectin-coated surfaces was improved (Dennis and Caplan 1993), whilst other studies indicate that it increases the adhesive strength of cells, suggesting it occurred due to the increased number of bonds between fibroblasts and fibronectin-coated glass slides (Athanassiou and Deligianni 2001).

## 4. Response of fibroblasts to mechanical stimulus

### 4.1 Mechanical transduction

Mechanical forces play a major role in the formation and architecture of native tissues in vivo, but also help maintain healthy tissue (homeostasis) in adult tissue. During daily activities, human body tissues are subjected to mechanical forces of various magnitudes, depending upon the activity and posture during these movements. The ACL has been shown to withstand forces of up to 1730N in people aged 16-26 years, but much less in people aged 48-86 years, with a mean average of approximately 734N (Noyes and Grood 1976). However, the forces which can be tolerated become significantly reduced when they are perpendicular to the bone insertion sites (Einhorn, O'Keefe et al. 2007). The externally applied forces can alter the cells structure, mechanical properties, behaviour, and function (Miyazaki, Hasegawa et al. 2000) which are required for tissue homeostasis (Fulton 1984; Altman, Lu et al. 2002). Tissue homeostasis occurs through ECM remodelling (re-organization) which involves ECM degradation by apoptosis (programmed cell death) and the formation of new tissue by cell proliferation (multiplication). An equilibrium between proliferation and apoptosis is essential during ligament growth, healing, tissue homeostasis and adaptation to exercise (Chuen, Chuk et al. 2004).

The cells are believed to sense mechanical forces either through their cell surface integrin receptors or through ion channels in the cell membrane, and respond accordingly. These external forces can alter the cell structure, its mechanical properties, behaviour, and function (Miyazaki, Hasegawa et al. 2000), where, for example they may increase or decrease ECM production, or regulate proliferate and differentiate. Transducing (converting) external physical forces into cellular signals across the membrane is known as mechanical transduction (Ingber 1999). Although the mechanisms of transduction are not well understood, many investigations have been carried out upon integrin receptors and the cell cytoskeleton that support the theory that the cell senses external mechanical forces from the ECM via the integrin receptors (subsequently causing deformation and reorganisation of the cell cytoskeleton) (Pertigliano, McAllister et al. 2006) and activates specific cellular pathways.

#### 4.1.1 Cytoskeletal tension

Upon cell-substrate binding via the integrin receptors, this exerts a force upon the cell cytoskeleton, which generates an intracellular tension. In effect, this links the cell membrane to the nucleus which influences gene expression, by relaying signals from the plasma membrane to the nucleus (Matyas, Edwards et al. 1994). The cytoskeleton is composed of a network of protein filaments within the cytoplasm, which maintains cell shape, giving it structure and support, and allowing the cell to bear stress without splitting (Alberts, Johnson et al. 2000). The cytoskeleton has a number of other functions including connecting each fibroblast to other cells and to the substrate, generating tension within the cell to produce stress fibres, and also assisting the cell in locomotion (Fulton 1984). There are three proteins in the cytoskeletal network, actin, tubulin and vimentin, which assemble to form the three main structural filaments of the network system (actin filaments, microtubules, intermediate filaments respectively) (Portner, Bagel-Heyer et al. 2005). Both actin and the intermediate filaments are involved in connecting the fibroblasts' internal structure to other cells and to the ECM. When the fibroblast encounters a suitable substrate, it extends its projections (filopodia), which then attach to the substrate allowing the rest of the cell to adhere. This generates a small tension, where the cell subsequently spreads (Fulton 1984; Alberts, Johnson et al. 2000), promoting formation of stress fibres within the cell. This enables the cell to withstand the tension generated from the cell-ECM contact and make connections with the nucleus to modulate cell behaviour. The effects include activating specific genes (Fulton 1984; Altman, Lu et al. 2002) and generating key proteins (including degrading enzymes and ECM components) to remodel the ECM for promoting new tissue formation (Vunjak-Novakovic, Altman et al. 2004; Portner, Bagel-Heyer et al. 2005).

#### 4.1.2 Intracellular cell signalling

The most widely accepted mechanism for mechanical transduction is that involving intracellular signalling pathways. Once the cell binds its substrate, the integrin receptors act as mechanoreceptors, receiving then relaying mechanical signals from the ECM through the cell (as biochemical signals) to the nucleus via intracellular pathways. This can either promote gene expression, regulate growth, proliferation or differentiation, subsequently affecting development or maintenance of the connective tissue (Alberts, Johnson et al. 2000; Vunjak-Novakovic, Altman et al. 2004). The most studied intracellular pathway is the Ras-mitogen-activated protein kinase (Ras-MAPK) pathway, which is considered to be the one which acts as a general, but unspecific signal transducer, converting the signal from the applied mechanical stress and relaying it through the cell interior (Chiquet, Sarasa-Renedo et al. 2003). The Ras-MAPK pathway becomes activated once the receptor receives an extracellular mechanical signal. Briefly, the received signal causes a kinase on the receptor to become activated which then activates a GTP-binding protein (Ras protein). The Ras protein causes downstream phosphorylations by activating the first MAP kinase (Raf) in the chain to phosphorylate the next MAP kinase (Mek), which phoshporylates and activates the next MAP kinase (Erk), which subsequently activates other proteins. Eventually the signal reaches the gene regulatory proteins in the nucleus which interact with transcription factors and promoters to regulate gene expression and protein activity (Alberts, Johnson et al. 2000).

#### 4.1.3 Ion channels

Besides mechanical transduction via integrin receptors, external forces (stress) can also be conveyed into the cell through stretch-induced ion channels which open and close in response to cell membrane deformation (Matyas, Edwards et al. 1994). Ion channels are cation specific channels located in the cell plasma membrane, which allows rapid diffusion of the ions down their electro-chemical gradients across the lipid bilayer. The channels are gated, only opening briefly to allow specific cations (such as calcium, sodium or potassium) to pass through, and then close again. The channel gates open in response to several types of specific stimulus, one those being mechanical stress which operates mechanically-gated channels (Alberts, Johnson et al. 2000), subsequently affecting the cell behaviour. Cyclic stretching has been reported to stimulate $Ca^{2+}$ influx into osteoblasts, and it is thought that mechanical stretch-induced $Ca^{2+}$ signal transmission may involve the actin filaments (Wang 2006).

## 5. Injuries sustained and healing potential

Ligaments in skeletally mature people are very strong and stiff at high loads, however, there can be some variation in the mechanical properties of each ligament type depending upon the individuals' gender and age. Strength and stiffness of the ACL has been found to significantly reduce as age increases in adults (Noyes and Grood 1976) and can be significantly lower in adult females compared to their male counterparts (Chandrashekar, Mansouri et al. 2006). Ligaments reach their maximum strength when the loading forces are aligned with the ligament fibres and aligned with the direction of bone insertion, becoming three times as strong as when the force acts perpendicular to the bone insertion sites (Einhorn, O'Keefe et al. 2007). They rupture when the load (externally applied force) becomes too excessive to withstand and the collagen fibres tear apart. The ACL, for example, normally ruptures in the mid-substance (middle region) when the knee joint experiences too much force, but it can also tear at the bone insertion sites. The position of the tibia relative to the femur can increase the magnitude of the stress placed onto the knee joint. The stress becomes greatest when the tibia is fully extended and internally rotated simultaneously, increasing the tension upon specific fibre bundles which are trying to resist deformation and abnormal motion. A rupture of the mid-substance occurs when the cross-links between the collagen fibrils slip allowing the tropocollagen helix to over-stretch, allowing the tissue to tear (Laurencin and Freeman 2005).

Generally, the blood supply to ligaments is sparse when compared to other tissue types (such as the skin), affecting their healing potential (Einhorn, O'Keefe et al. 2007) which can be limited further by anatomical location, age and gender. The healing capacity of the mature ACL is very low due to its anatomical location. It is encapsulated within the knee joint (intrasynovial), being surrounded by the lubricating synovial fluid, and is poorly vascularised (Amiel, Frank et al. 1984; Ahmed, Collins et al. 2004; Cooper, Lu et al. 2005). As a result it cannot self repair (Cooper, Lu et al. 2005; Lu, Cooper et al. 2005), therefore medical treatment is necessary. There is no direct blood supply from the fibrocartilage zone of the bone to the ACL, so the ACL relies mainly upon diffusion of nutrients and waste through the joint fluid from and to the blood vessels of the surrounding synovial tissue respectively (Beasley, Weiland et al. 2005). The surrounding synovial tissue is vascularized by the medial genicular artery, and the lateral inferior genicular artery, forming a vascular plexus (network of vessels) around the knee. It is the small vessels from the plexus which supply the ligament with the essential nutrients by diffusion (Zantop, Patterson et al. 2005). It possible that a few of these small blood vessels may actually penetrate the ACL and directly supply it with nutrients (Arnoczky 1983). As a result of poor vascularization to the midsubstance, the ACL has a low healing capacity and can not self repair (Carpenter and Hankenson 2004). In contrast to the ACL, the MCL which is extrasynovial can self heal

spontaneously because it has a greater vascularisation and receives more blood (Carpenter and Hankenson 2004).
After injury, those ligaments which are well vascularized have three stages of healing; inflammation, cellular proliferation and migration, ECM repair and finally ECM remodelling (Laurencin and Freeman 2005). Generally, these stages promote fibroblast proliferation. Fibroblasts and macrophages then migrate to the injured site and granulation tissue forms (stroma). GAG, elastin and collagen are synthesised to form new ECM, and finally the ECM is remodelled, where it initially forms into a disordered tissue but later becomes more organized (Alberts, Johnson et al. 2000; Laurencin and Freeman 2005). With avascular, or poorly vascularised ligaments, this process is not carried out, or only in a limited way, which prevents them from self-healing spontaneously and surgery may be needed to repair them. With the case of the ACL, if left untreated, this could eventually cause osteoarthritis in the knee because the ACL has failed to maintain correct bone alignment and control normal motion across the knee joint (Foster, Butcher et al. 2005; Utukuri, Somayaji et al. 2006).

## 6. Surgical treatment

After rupture, the ligament is normally repaired by surgery, which can be by suturing or grafting. Based upon clinical investigations, surgical grafting has become the gold standard for ligament repair (Einhorn, O'Keefe et al. 2007). In the case of the ACL, a surgical reconstruction is performed which is the only method shown to at least partially restore function, helping to improve the patient's quality of life. This method involves implanting a graft to replace the damaged ligament.
Three main types of grafts can be used; autografts, allografts or synthetic grafts. The current gold standard procedure for reconstructing an ACL is autografting, which involves using a ligament or tendon from another part of the patient's body and using it to replace the damaged ACL. This can be a section from the patients patellar tendon (joining knee cap to tibia) or the hamstring tendon (joining calf muscle to bone in heal) (Beasley, Weiland et al. 2005). Patellar tendon is often used because its strength and mechanical properties are similar to or exceed that of normal native ACL (Fenwick, Hazleman et al. 2002). The central third of the patellar tendon is removed with a piece of knee cap (bone plug) attached to one end, and a section of the tibia attached at the other end (Beasley, Weiland et al. 2005). The damaged ACL is removed, a bone tunnel (channel) is drilled out from the femur and tibia, and the graft is threaded through and screwed into leg bones. This reconstruction operation takes approximately 2 ½ hours. Allografting involves using a ligament or tendon from a different human donor, normally a cadaver (corpse). The procedure is the same as autografting, and may give the same results, but disadvantages include donor scarcity, the risk of the recipient contracting a disease from the donor, or tissue rejection (Vunjak-Novakovic, Altman et al. 2004). Clinical outcomes in the short term can be good, with 80% success rate in restoring knee stability (Einhorn, O'Keefe et al. 2007). Unfortunately, both autografting and allografting can be unsatisfactory methods for long term performance in some patients, where they suffer from instability (Woo, Abramowitch et al. 2006) due to mechanical failure, fatigue or creeping, (gradual stretching of the tissue under constant load). This occurs due to a slight mismatch in mechanical properties between the graft and native ligament tissue, where the injury may then recur at a later date (Lu, Cooper et al. 2005). It has been suggested that even two years after the implantation the tendon graft

remains structurally and mechanically different to normal native ligament and never actually becomes "ligamentised" (Fenwick, Hazleman et al. 2002). Another disadvantage of autografting is donor site morbidity which can cause pain, swelling, local nerve damage, scarring, stiffness weakness or infection (Einhorn, O'Keefe et al. 2007). Synthetic grafts have also been used such as carbon, the Gortex prosthesis, the Stryker-Dacon ligament, the Leeds-Keio artificial ligament, LARS ligament and Kennedy ligament augmentation devices, but creeping, fatigue and limited integration between host tissue and the graft have occurred several years after implantation (Ahmed, Collins et al. 2004; Cooper, Lu et al. 2005). The advantages and disadvantages of the current grafting methods are summarized in table 3.

| | **Autograft** | **Allograft** | **Synthetic graft** |
|---|---|---|---|
| **Advantages** | No rejection. No disease transmission. No donor scarcity. | No donor site morbidity. | No donor site morbidity. No tissue disease transmission. No donor scarcity. |
| **Disadvantages** | Donor site morbidity. Patellar fracture. Quadriceps weakness. Limited bone integration. Mismatch in different tissue properties, causing mechanical failure, creeping, fatigue. Recurring injury. | Donor scarcity. Limited bone integration. Tissue rejection. Mismatch in different tissue properties, causing mechanical failure, creeping, fatigue. Recurring injury. | Limited bone integration (weak graft-host tissue interface). Mismatch in different tissue properties causing mechanical failure. Creeping (stretching & loosening). Poor long-term instability. Fatigue. Recurring injury. |

Table 3. The advantages and disadvantages of three types of graft

## 7. Tissue engineering as regenerative medicine

### 7.1 Background

Tissue engineering is a rapidly developing area in regenerative medicine which uses knowledge of biological, chemical and engineering techniques to regenerate new tissue in-vitro (Cooper, Lu et al. 2005). Some tissues in the body are capable of self repair after injury, while others are not, and tissue engineering is a relatively new technique which could offer alternative methods to restore tissues and organ functions (Quaglia 2008). The techniques enable various biophysiological parameters to be controlled in order to develop a functional tissue ready for implantation (Lanza, Langer et al. 2007). The procedure involves using a scaffold to act as a structural support for cell growth and maturation in vitro to eventually produce a functional tissue to repair or replace damaged tissue. This concept was originally developed to repair skin, cartilage and bone, but is now being considered as a possible option to produce neo-ligament tissue rather than using the traditional surgical grafting approach. Although the current methods of treatment may help to fully restore the ligaments in some patients, its long-term success in others is unsatisfactory, indicating that there is a need to find more successful, alternative methods of treatment for full restoration of ligaments. Unlike synthetic grafts which can degrade and lose strength over time, tissue engineered implants could perform better in the long term with their biocompatibility, improved integration into surrounding host tissue, and their ability to remodel the ECM as and when required to (Nesic and Whiteside 2006). Tissue engineering also has the advantage of producing an immediately functional tissue (Vunjak-Novakovic, Altman et al. 2004), but the successful incorporation of the soft tissue implant into bone could be a challenge.

For applications in ligament tissue engineering, a scaffold is required to be biocompatible, biodegradable, allow cell adherence, have sufficient surface area and volume for cell in-growth, be sufficiently strong to withstand mechanical loading forces in vitro and in vivo, and posses a similar stiffness to the native ligament tissue (if it is to be implanted before it degrades) (Christian, Jones et al. 2001 ; Cooper and Lu 2005; Probhakar, Brocchini et al. 2005; Gentleman, Livesay et al. 2006; Sahoo, Ouyang et al. 2006). These points are summarized in table 4. Often, three-dimensional (3-D) scaffolds are preferred to the two-dimensional (2-D) scaffolds because they not only allow cell in-growth, but can also retain cells in their differentiated state. From the literature, it has been reported that fibroblasts cultured in 2-D monolayers have de-differentiated (reverted back to their undifferentiated state) during cell culture (Schulze-Tanzil, Mobasheri et al. 2004), which may not be desirable for tissue engineering purposes.

| Scaffold requirements | The purpose of this feature |
|---|---|
| Biocompatible | Avoids immunorejection (a cytotoxic response could kill the cells) |
| Biodegradable | To degrade at the same rate at which neotissue forms to avoid the need for surgical removal |
| Enable cell adherence | To allow cells attachment for growth and proliferation to occur |
| Provide sufficient surface area/volume | To provide sufficient space for cell spreading and growth |
| Possess comparable strength/stiffness | To withstand cyclic mechanical loading forces with magnitudes and strains similar to those found in vivo |
| Surgical implantation | Ease of fixing/bonding to bone (bio-active) |

Table 4. The main requirements of a scaffold with respect to their application

### 7.2 Suitable cell types for ligament tissue engineering

Fibroblasts and mesenchymal stem cells (MSC's) have been considered to be the preferred cell type for seeding onto scaffolds in tissue engineering (Doroski, Brink et al. 2007). Some reports suggest that MSC's are a potentially better source for ligament tissue engineering than ligament fibroblasts due to their higher expression of collagen type I and III (Ge, Goh et al. 2005). Other reports however, feel that ACL fibroblasts are more appropriate because of the characteristic ratios of collagen types produced during tissue repair (Fu, Harner et al. 1993). In one particular study MSC's were isolated from a human ACL and the results demonstrated that both ACL-derived MSC's and bone marrow MSC's expressed marker genes for ligament fibroblasts, but mRNA expression levels for collagen I and III were higher in the ACL-derived MSC's. It was concluded from this study that ACL-derived MSC's have an increased potential to form ligament fibroblasts in comparison to bone marrow MSC's (Huang, Chen et al. 2008). Co-culturing MSC's with ligament fibroblast has been shown to successfully induce MSC differentiation into fibroblasts, where this conclusion was based upon the mRNA expression of key ligament genes (collagen I collagen III, and tenascin-c) and synthesis of these key ligament proteins (Fan, Liu et al. 2008). This feature makes them an attractive cell choice for ligament tissue engineering.

### 7.2.1 Mesenchymal stem cells (MSC's)

Mesenchymal stem cells (MSC's) are multipotent progenitor cells, meaning they can differentiate into specific cell types of various cell lineages. They are found in multiple adult tissue types including bone marrow, muscle, synovial tissue and adipose tissue (Centeno, Busse et al. 2008), where they have the potential to produce cartilage, bone, muscle, tendon, ligament or fat (Papathanasopoulos and Gaiannoudis 2008) in response to the appropriate stimuli (Lanza, Langer et al. 2007). MSC's can be encouraged to move down specific cell lineages by using media which contains hormones such as dexamethasone, hydrocortisone, or growth factors such as transforming growth factor β (TGF-β) (Papathanasopoulos and Gaiannoudis 2008), cytokines, transcription factors (Lanza, Langer et al. 2007) or using purely mechanical stimulus (Altman, Horan et al. 2001). MSC's have been used successfully to regenerate articular cartilage in animal models and to regenerate bone in humans (Papathanasopoulos and Gaiannoudis 2008). Because they can be easily isolated and expanded (Papathanasopoulos and Gaiannoudis 2008; Yu, Chen et al. 2008), with the capacity to differentiate, this makes them desirable for tissue engineering applications (Papathanasopoulos and Gaiannoudis 2008). One of the first areas in which they were applied was in tendon and ligament tissue engineering (Lanza, Langer et al. 2007).
Another appealing feature of MSC's is their immunosuppressive and anti-inflammatory effects. They express low levels of major histocompatibility complex (MHC) class I molecules on their surface (preventing natural killer cells deleting them), and no class II MHC, allowing them to escape recognition by alloreactive T helper cells (Zhao, Liao et al. 2004; Lanza, Langer et al. 2007; Popp, Eggenhofer et al. 2008; Swart, Martens et al. 2008). However, it has been reported that MSC's infused into allogeneic MHC-mismatched mice have been rejected (Swart, Martens et al. 2008). In contrast, this was not the case when genetically modified MSC's were injected into a baboon (Zhao, Liao et al. 2004).

## 7.3 Biomaterials suitable for ligament tissue engineering

To date, many different materials in their various physical forms have been investigated as substrates for tissue engineering applications in general. These include synthetic polymers, natural polymers, glasses, silk, hydrogels, composites and many more. Only those related to ligament tissue engineering, are covered in this chapter (summarized in table 5).

Synthetic polymers such as polylactic acid, polyglycolic acid and polylactide-co-glycolide (PLA, PGA and PLAGA respectively) are approved by the USA food and drugs agency (FDA) for a variety of clinical applications (Cooper, Lu et al. 2005). These polymers degrade by hydrolysis of ester bonds (water breaks up the molecule), and the components are removed by the natural pathways of the body (Rezwan, Ghen et al. 2006), making them biocompatible. Some of these have been produced into cell scaffolds and tested for their suitability as substrates. PLAGA fibres have been used to make 3-D braided scaffolds, which consisted of 3 regions - the attachment site for the femur bone at one end, the main ligament region in the middle, and the attachment site for the tibia bone at the other end. The results indicated that the scaffold was biocompatible by the observed attachment, spreading and growth of the ACL fibroblasts initially seeded onto it (Cooper, Lu et al. 2005). PLAGA has also been produced into nano-fibres which were electrospun into a knitted PLAGA scaffold to increase the surface area for cell attachment. It significantly improved MSC attachment and proliferation, but also demonstrated that cell function had improved due to the increased mRNA expression of type I collagen and decorin (Sahoo, Ouyang et al. 2006). Braided PLA, PGA and PLAGA have also been found to enhanced rabbit ACL fibroblast attachment and support high cell numbers, being highest on PLA (Lu, Cooper et al. 2005).

Alginates are a natural linear polysaccharide copolymers extracted from brown algae belonging to the phaeophyceae (Hua and Wang 2009). They are currently used in the food, cosmetic and agricultural industries (Hua and Wang 2009). Because they are easy to process, with good biocompatibility and low toxicity, they have been studied in drug stabilization and drug delivery (Lee and Mooney 2001; Drury and Mooney 2003) and for tissue engineering purposes (Sakai, Masuhara et al. 2005), where bone marrow cells have been successfully cultured on them (Wang, Shelton et al. 2003). Other report have also confirmed there suitability by enabling cell adhesion, migration, proliferation and differentiation to take place (Zhao, Deng et al. 2003).

Polyhydroxyalkanoates (PHA's) are currently under investigation for their uses in tissue engineering. They are naturally derived biocompatible polyesters which are produced by microorganisms as carbon and energy stores in unbalanced growing conditions become (Chen and Wu 2005). They are known to be biocompatible because they are found naturally occurring in mammal blood and tissues, where their purity can be increased by removing long-chain fatty acids and the endotoxin lipopolysaccharide, preventing any adverse reactions (Zhao, Deng et al. 2003; Chen and Wu 2005). They range from hard and brittle to soft and elastomeric, but can also be blended (combined) with other types of PHA or modified at the surface to alter their mechanical properties and biocompatibility, and produced either as a film or a foam (Rezwan, Ghen et al. 2006). So far, PHA's have been used for a number of different applications including tissue regeneration, repair devices, sutures and bone marrow scaffolds (Shishataskaya, Volova et al. 2004). One report compared the mechanical and surface properties of several modified PHA's and it was concluded that hydrophilicity and a low tackiness were found to be more important than the surface roughness for cell attachment and growth. Substrate stiffness also appeared to influence cell attachment, where the stiffer, more brittle PHA's, retained a significant number of viable cells on their surfaces (Rathbone, Furrer et al. 2009). Because of their diversity in surface texture, flexibility and their biocompatibility, PHA's show potential as cell substrates in tissue engineering.

Two types of glasses that have been used for medical research are bioglass (developed in the 1969 by Larry Hench) and controlled release glass (CRG) which is a phosphate-based glass, developed in the 1970's. Bioglass does not dissolve completely in fluids, but changes chemically upon its partial degradation, and is currently regarded as the most biocompatible material for bone regeneration due to its bioactivity and osteoconductivity (Wu, Hsu et al. 2009). Unlike bioglass, CRG dissolves completely in fluids at a predetermined rate, leaving no solid residues because phosphorous pentoxide is a main component within its formulation. The metal ions in CRG are found naturally occurring within the body (Probhakar, Brocchini et al. 2005) and upon glass degradation (dissolution) the released ions become removed by the bodies own metabolic system without causing a toxic response, avoiding the need for surgical removal if implanted into the body. Because CRG has a controllable solubility in body fluids and do not need surgical removal, this makes them an ideal scaffold material for promoting neotissue formation in vivo (Ahmed, Lewis et al. 2003). Very little work has been carried out with phosphate based glasses (CRG) for soft tissue engineering, however Bitar and colleagues cultured several cell types, including tendon fibroblasts, upon glass disks with various dissolution rates, where they successfully attached, proliferated while maintaining their phenotype, and it was concluded that the glass (of specific composition) would be ideal scaffold materials for engineering of both hard and soft tissues (Bitar, Salih et al. 2004). In fibrous form, phosphate based glasses have high tensile strength, making them useful for tensile applications, but it is also possible to produce them with dimensions similar to ligament collagen fibres in vivo, particularly the diameter, which can potentially assist cell attachment and spreading. Bitar et al suggested that the fibre diameter of the phosphate based glass which they tested influenced cell attachment (Bitar, Salih et al. 2008).

Many other materials have been investigated. Silk fibroin scaffolds have supported and enhanced ligament- specific differentiation of human MSC's. The silk was cabled into 6-cord wire rope matrices, improving its elasticity. The authors suggested that the silk matrix had similar a hierarchical structure to the collagen fibres in native ACL, making the mechanical properties comparable to ACL in stiffness and strength (Wang, Kim et al. 2006). Silk fibroin has also been produced as a microporous silk sponge and incorporated into a knitted silk mesh. After seeding them with rabbit MSC's, the constructs were implanted into rabbits, where at 24 weeks the cells were well distributed throughout a regenerating ACL, and producing key ligament proteins (collagen I and III, and tenacin-c). Also a direct ligament–bone insertion was achieved resembling native ACL-bone insertion (Fan, Liu et al. 2008). Collagen hydrogels have been shown to successfully promote higher production of type I and type III collagen from the cells, where the tissue formation was improved with a ligament-like organisation (Noth, Schupp et al. 2005; Gentleman, Livesay et al. 2006). Poly desamino-tyrosine ethyl carbonate scaffolds have proved to be successful in supporting fibroblast growth while possessing the necessary strength for use as an ACL graft (Laurencin and Freeman 2005).

Composites (consisting of more than one type of material) have also been constructed and analysed. Gelatin with silk fibroin has been produced into microporous sponges around Silk cables, where Fan and colleagues co-cultured rabbit MSC's with ACL fibroblasts on the scaffold, which enabled MSC differentiation into ligament fibroblasts. They detected mRNA expression of collagen type I and III, and tenascin-c with the corresponding protein production (Fan, Liu et al. 2008).

| Material | Physical form | Affect on cells/material properties | Author |
|---|---|---|---|
| Synthetic polymers | | | |
| **PLAGA** | Braided | Improved fibroblast attachment, spreading & growth. | (Cooper, Lu et al. 2005) |
| **PLAGA** | Electrospun PLAGA nano-fibres onto knitted PLAGA scaffold. | Improved porcine MSC attachment & proliferation. Cells gave higher expression of type I collagen, decorin and biglycan genes in comparison to cells on just a knitted PLAGA scaffold. | (Sahoo, Ouyang et al. 2006) |
| **PGA coated with BioGlass** | Mesh coated with Bioglass. | Increased fibroblast proliferation (208f cell line). | (Day, Boccaccini et al. 2004) |
| **PLA, PGA, PLAGA coated with fibronectin** | Braided | Enhanced rabbit ACL fibroblast adhesion and supported high cell numbers, (highest for PLA). | (Lu, Cooper et al. 2005) |
| **Degrapol$^R$ PU** | Fibre-fleece | Supported fibroblast adhesion & proliferation. | (Milleret, Simonet et al. 2009) |
| Natural polymers | | | |
| **Collagen hydrogel** | Hydrogel + collagen fibres | Increased production of type I & III collagen fibres, giving better tissue formation, and ligament-like organization in the tissue. | (Noth, Schupp et al. 2005; Gentleman, Livesay et al. 2006) |
| **Silk fibroin** | Rope matrix | Enhanced ligament- specific differentiation of human MSC's. | (Wang, Kim et al. 2006) |
| **Silk fibroin** | Microporous silk mesh rolled around braided silk cord | MSC seeded construct was implanted into pigs. At 24 weeks MSC's differentiated into fibroblast-like cells, expressing collagen I and III, tenascin-c. | (Fan, Liu et al. 2009) |
| **Silk fibroin** | Microporous silk sponge incorporated into knitted silk mesh | Rabbit MSC seeded constructs implanted into rabbits. At 24 weeks cells were well distributed throughout the regenerating ACL, producing key ligament proteins (coll I & III, Tenascin-c), direct ligament –bone insertion with 4 zones was reconstructed resembling native ACL-bone insertion. | (Fan, Liu et al. 2008) |

| | | | |
|---|---|---|---|
| **Silk fibroin** | Silk fibroin electrospun onto knitted silk base (random/aligned) | Aligned fibres showed improved cell proliferation and collagen production compared to random orientation. | (Teh, Goh et al. 2008) |
| **Composites** | | | |
| **Gelatin + silk fibroin** | Microporous sponge around Silk cables | Co-cultured rabbit MSC + ACL fibroblasts on the scaffold allowing MSC differentiation into ligament fibroblasts (mRNA expression of Coll 1 & 3, Tenascin-c and corresponding protein production). | (Fan, Liu et al. 2008) |
| **Glass** | | | |
| **Phosphate-based glass** | Disks | Increased adhesion & proliferation of fibroblasts when CaO content was 46-48 mol%. | (Bitar, Salih et al. 2004) |
| **Phosphate-based glass** | Fibres | Increased adhesion & proliferation of fibroblasts when CaO content was 46mol%. | (Bitar, Knowles et al. 2005) |
| **Bioactive glass** | Disks | Supported rabbit MSC adherence and proliferation. | (Meseguer-Olmo, Bernabeu-Esclapez et al. 2008) |
| **Collagen fibres** | Braided/plied (cross-linked) | Implanted into goats, analysed over 11 months post implantation, concluded they were loosing strength, therefore not suitable as ACL substitute. | (Chvapil, Speer et al. 1993) |

Table 5. Some of the various different materials previously used as scaffolds (for ligament tissue engineering) their physical forms, and their suitability for cell cultures

### 7.4 Material surface modifications using fibronectin

Because fibronectin is known to function as a cell adhesion protein in vivo, it has been studied in vitro as a method for improving cell attachment. It has been used to modify the surface of various biomaterials to improve cell attachment to the surface, making it useful for tissue engineering. Reports have shown that fibronectin has improved cell retention of rat MSC's on fibronectin-coated surfaces (Dennis and Caplan 1993), increased the adhesive strength of cells, which was probably due to the increased number of bonds between fibroblasts and fibronectin-coated glass slides (Athanassiou and Deligianni 2001). Other studies concluded that braided PLLA and PLAGA polymers coated with fibronectin (10μg/ml) improved attachment of rabbit ACL cells and effected long term matrix production (Lu, Cooper et al. 2005). From the work carried out by Garcia and colleagues, their results indicated that cell adhesion strength (in osteosarcoma cells) increased on glass surfaces in a concentration-dependent fashion as fibronectin concentrations increased from

0.1-1μg/ml. Plates coated with a concentration of 20μg/ml have been shown to improve cell adhesion of human MSC's in comparison to uncoated plates (Salasznyk, Williams et al. 2004), being more affective than coatings of collagen I or IV. However, in contrast to Salasznyk's findings, Vohra and colleagues who also used a fibronectin concentration of 20μg/ml (Vohra, Hennessy et al. 2008), suggested that although the fibronectin coating improved cell attachment compared to the negative control, MSC's preferred to attach to a collagen I coating in comparison to the fibronectin and negative control.

### 7.5 Bioreactor culture of tissue engineered ligaments

A bioreactor is a vessel designed to contain cultures, where the environmental conditions can be optimised and carefully controlled to encourage certain biological and biochemical processes to take place (Martin, Wendt et al. 2004). Currently, many different types of bioreactors exist. They can be used to improve mass transfer of nutrients, waste products and oxygen through the culture medium, improve cell attachment, cell growth and proliferation. Bioreactors have been used in tissue engineering to apply mechanical forces to cell constructs (mechanical loading), and reported to promote differentiation of mesenchymal stem cells (MSC) into ligament fibroblasts (Zhang, Wang et al. 2004; Meyer, Buchter et al. 2005), induce alignment of fibroblasts with the direction of the applied force, upregulate mRNA expression of key ligament genes and produce helically organized collagen fibres (Altman, Horan et al. 2001). Mechanical conditioning can also be used to improve the structural and functional properties of a tissue once it has been engineered. Only those bioreactors related to ligament tissue engineering will be discussed here.

When forces are applied to cells, the magnitude of the applied force, the way in which it is applied (constantly or alternating), the duration of time and the direction of forces (translational or rotational) will have a specific effect upon cell behaviour. These complex forces are experienced by the native tissue under physiological conditions. A specific combination or sequence of any of these can influence the cell to give a positive or negative response. Therefore variations in mechanical loading regimens can affect and alter the gene expression, and hence protein production and regulate tissue formation (Nesic, Whiteside et al. 2006).

## 8. Response of cells to mechanical stimuli in vitro

Many in vitro studies investigating the affects of mechanical loading have been performed with bioreactors. Kaplan and colleagues used a bioreactor to apply mechanical stimulation to mesenchymal progenitor cells seeded into a collagen 3-D gel matrix. The bioreactor applied multidimensional forces concurrently - translational (2mm) + rotational ($90^{o}$) - at a frequency of 0.0167 Hz (one complete cycle of stress and relaxation per minute) constantly for 21 days, which was chosen to mimic the unique combination of forces experienced by the ligament under physiological conditions in vivo (Altman, Horan et al. 2001). Their results induced cell alignment parallel to the direction of the stretching force with an elongated cell morphology, mRNA expression of specific genes (type I and III collagen, tenascin-c and fibronectin), helically organized type I collagen fibres orientated in the direction of force, with the selective differentiation of human MSC's into a ligament cell lineage rather than towards alternative lineages. The controls (non- loaded cells), showed few of these features. They concluded that "the mechanical forces could play a role in differentiation and not just promote formation of specific tissue types from the already differentiated cells" and "the mechanical stimulation appeared to cause a selective differentiation to ligament-like cells" (Altman, Horan et al. 2001). Mechanical cyclic loading in compressed acellular collagen gels has been reported to encourage collagen fibril aggregation, promote fusion, resulting in an increase in its mechanical strength (Cheema, Chuo et al. 2007). Table 6 summarizes some of the mechanical loading conditions carried out in various studies, and the effects which they had upon the cells.

| Frequency (Hz) | Translational stretching (cyclic) | Rotational | Period of time | Type of bioreactor | Effect upon cells | Author |
|---|---|---|---|---|---|---|
| 1 | Stretched by 5% of its initial length. Cyclic, uniaxial | No | 24 hours | Custom built tensile bioreactor | Tendon fasicles showed up-regulation of collagen | (Screen, Shelton et al. 2005) |
| 0.0167 | 10% Increased by 2mm in length when stretched. Cyclic, biaxial | 25% Twisted it by 90° | 21 days | Custom built tensile/ compression bioreactor system | Human MSC's acquired spindle-shaped morphology & cell alignment, expression of collagen I & III, tenascin-c, fibronectin, helically organized collagen fibres, differentiation into ligament cell lineage. | (Altman, Horan et al. 2001) |
| | 12% Cyclic, uniaxial | No | 14 days | Custom built bioreactor | Human MSC's showed increased expression of collagen 1 & 3, Fn, elastin in loaded samples. Observed elongated fibroblasts embedded in a wavy orientated collagenous tissue with ligament-like ECM. | (Noth, Schupp et al. 2005) |

| Frequency (Hz) | Translational stretching (cyclic) | Rotational | Period of time | Type of bioreactor | Effect upon cells | Author |
|---|---|---|---|---|---|---|
| 0.5 | Flexion | No | 0.5, 2. 4 hrs | Flexible Silicone coated films | Osteoblasts (MG63) showed BMP-2, BMP-4, proto-oncogene C-Fos upregulation at 0.5hrs.<br>Alk-3 upregulation at 2 & 4 hrs | (Sakoda, Shin et al. 1999) |
| 0.167 | Flexion cyclic | No | 24 & 48 hrs continuous | Flexercell (flexible substratum) | Human ACL cells inc expression for coll I at 48hrs | (Miyaki, Ushida et al. 2001) |
| 0.5 | Stretched by 4% & 8%<br>Cyclic, uniaxial | No | 4 hrs | Microgrooved silicon dishes | Osteoblast (MG63) showed slight upregulation in gene expression of MMP-1 | (Yang, Im et al. 2005) |
| 10 cycles /min (0.167Hz) | Stretched by 10%<br>Cyclic, uniaxial | No | 24 hrs | Silicon dishes, membrane coated with collagen I | Human ACL cells inc mRNA expression in Coll I & III | (Kim, Akaike et al. 2002) |
| 1 | 5%<br>Cyclic, uniaxial | No | 1 hr/day for 15 days | - | Human ACL cells showed up regulation in mRNA expression of coll I & III, tenascin-c, fibronectin. | (Schlenker, Kreja et al. 2006) |

| Frequency (Hz) | Translational stretching (cyclic) | Rotational | Period of time | Type of bioreactor | Effect upon cells | Author |
|---|---|---|---|---|---|---|
| 0.25 | 10%<br>Cyclic, uniaxial | No | 8hrs/day for 7dasy | Custom built tensile machine | Tracheal fibroblasts showed up reg of procollagen 1 a1,TGF-β1, CTGF. | (Webb, Hitchcock et al. 2006) |
| 1 | 3% & 10%<br>Cyclic, uniaxial | No | 8 hrs & 48 hrs | Flexcell plates coated with collagen I | Human MSC's<br>Up reg of coll I & III, tenascin-c with 10% 48 hrs | (Chen, Huang et al. 2008) |
| | 2% in centre of well, 17% at periphery of well.<br>Cyclic, uniaxial | No | 24 hrs | Culture plates with flexible membrane | Rabbit ACL cells<br>Became spindle shaped, aligned perpendicular to force, inc coll synthesis in coll I only (at 17% - periphery). | (Toyoda, Matsumoto et al. 1998) |
| 1 | 0.05 (5%)<br>and<br>0.075 (7.5%)<br>strains.<br>Cyclic, uniaxial | No | 0.5, 1, 2, 4, 16, 24 hrs | Flexercell membrane coated with collagen I | Human ACL fibroblasts showed upregulation in mRNA expression for Coll I & III between 16-24 hrs (0.05 strain).<br>There was up reg of coll 1 at 0.5hrs and at 24hrs, a decrease at 4hrs (0.075 strain). | (Hsieh, Tsai et al. 2000) |
| 0.5 | 7% | No | 2 hrs | ST150 STREX | Human ACL fibroblasts showed upregulation in mRNA expression for collagen 1 (6 fold). Activated stress fibre formation by shifting distribution of integrin receptors ($\alpha_5\beta_3$) to peripheral edges of the cells. | (Tetsunaga, Furumatsu et al. 2009) |

| Frequency (Hz) | Translational stretching (cyclic) | Rotational | Period of time | Type of bioreactor | Effect upon cells | Author |
|---|---|---|---|---|---|---|
| 0.5 | 5%<br>Cyclic, uniaxia | No | Ranging 15 mins – 24 hrs | - | Human fibroblasts stretching + 10ng/ml TGF-B enhanced synthesis of collagen (mRNA express). | (Kadi, Fawzi-Grancher et al. 2006) |
| 0.5 | 4%<br>8%<br>Cyclic, uniaxial | No | 4 hrs/day for 2 days. Then rest for 24 hrs. | A modified Flexcell silicone substrate | Human ACL fibroblasts showed a higher proliferation rate at 8% and more cells were aligned perpendicular to stretch direction than 4% or the static controls. | (Park, Kim et al. 2006) |
| 0.5 | 5%<br>Cyclic, uniaxial | No | 24 hrs | Bioflex culture plates | Pig ACL fibroblasts showed upregulated collagen I mRNA expression, but not collagen 3 or biglycan. Estrogen + loading inhibited mRNA expression of all 3 genes, but estrogen alone upregulated collagen I & III. | (Lee, Liu et al. 2004) |
| 1 | 5%<br>Or<br>4 different amplitudes ranging 0.5 – 5%<br>Cyclic, uniaxial | No | 4hrs<br>OR<br>1 hr/day for 5 days/week over 9 weeks. | Custom built multi-test-station apparatus. | Bovine synoviocytes at 4hrs showed proliferation increased by 24%.<br>The 9 week tests showed collagen fibrils aligned in direction of load in the strained scaffolds, but random on unstrained controls.<br>The larger the amplitude of strain, the more cells & ECM present on scaffold. | (Raif, Seedhom et al. 2007) |

| Frequency (Hz) | Translational stretching (cyclic) | Rotational | Period of time | Type of bioreactor | Effect upon cells | Author |
|---|---|---|---|---|---|---|
| 1 | Preload with 2 & 10mN. Then 10% strain. Cyclic, uniaxial | No | 24 hrs | Custom built cell straining system | Preloading human dermal fibroblasts on collagen gels influenced their anabolic & catabolic processes. Preloading to 10mN then cyclic loading increased expression levels for latent MMP-1, MMP-9, MMP-2. Preloading to 2mN then cyclic loading increased stiffness of constructs | (Berry, Shelton et al. 2003) |
| 0.1 | 10% Cyclic, uniaxial | No | 1hr/day for 6 weeks | Custom built bioreactor for dynamic stretching | Human extensor tenocytes seeded onto PGA fibres formed tendon complex structure after in vitro loading, but became further matured by in vivo mechanical stimulation when implanted into mice, showing better collagen fibre alignment, more mature fibril structure & stronger mechanical properties. | (Wang, Liu et al. 2008) |
| 0.0125<br>0.125 | 0.06%<br>0.6%<br>3%<br>6% | No | 24 hrs | Custom built bioreactor for dynamic stretching | NIH 3T3 mouse fibroblasts strained to 0.6% showed increased spreading along contours of scaffold when loaded (for both frequencies) in comparison to static controls. No significant difference was seen between the 2 frequencies for each strain variation. The addition of TGF-1 (1, 10, 100ng/ml) had no affect on cell spreading, suggesting cell morphology & adaption may be affected purely by short-term mechanical loading. | (Puk, Miller et al. 2006) |

Table 6. The mechanical loading regimens which have been applied to tissue engineered constructs in vitro and the effects they have had upon various cell types

| Cell type | TGF concentration | Combined with other GF's | Effects | Author |
|---|---|---|---|---|
| MSC | hr TGF-β1<br>10ng/ml<br>Static culture | | 10ng/ml increased GAG expression & proliferation after 14days, whereas 0.1-1ng/ml did not. | (Chen, Tsai et al. 2005) |
| h ACL explants | TGF-β1<br>0.6ng/ml<br>Static culture | Ascorbic acid 25ug/ml | Cell number increased. Collagen production increased by 3 times. | (Meaney Murray, Rice et al. 2003) |
| h MSC | TGF-β<br>5ng/ml<br>Static culture | TGF+Insulin<br><br>Adding ascorbic acid to TGF promoted greatest ratio of collagen: total protein production | TGF alone encouraged differentiation into fibroblasts (regarding morphology & alignment). Collagen 1 & 3 express increased. Negligible expression of BSP & OSP. | (Moreau, Chen et al. 2005) |
| h MSC | TGF-β<br>0.1, 0.5, 1, 5 ng/ml<br>Static culture | | Proliferation rate increased as concentration increased (0.1-5ng/ml).<br>1ng/ml TGF + FCS increased matrix production. | (Locklin, Oreffo et al. 1998) |
| Rabbit ACL fibroblasts | TGF-β1<br>0.01, 0.1, 1ng/ml<br>Static culture | | Increased collagen & non-collagenous protein synthesis as concentration increased, highest being at 1ng/ml. The increase was mostly for collagen 1 which increased by 1.5 times. | (Marui, Niyibizi et al. 1997) |
| Sheep ACL explants | hr TGF-β1<br>10, 50, 100ng/ml<br>Static culture | | Increased proliferation (seen at 96 hrs). | (Spindler, Imro et al. 1996) |
| Human and rabbit mesenchymal progenitor cells | hr TGF-β1<br>Static culture | | Chondrogenic | (Chen, Tsai et al. 2005) |

Table 7. The effects of TGF-β1 upon fibroblasts and MSC's

## 9. Response of cells to chemical stimuli in vitro

During a cells life cycle, the cell needs to receive the appropriate signals at specific time points to instruct it to grow, proliferate, differentiate or synthesise ECM. These chemical signals are often provided by growth and differentiation factors. They bind the cell surface receptors on target cells to activate specific intracellular signalling pathways, controlling cell growth, proliferation, migration and differentiation (Quaglia 2008). The effect that a differentiation or growth factor has upon a cell can vary depending on the cell type, the stage in the cells life cycle and its environmental conditions. Such an example is transforming growth factor-β (TGF-β) which can be an inhibitor or stimulator of inflammation, or ECM remodelling by inducing mRNA expression of integrins, collagen and fibronectin (Evans 1999).

In order for tissue engineering to be successful, it is necessary to create an artificial environment for the cells, and possibly mimic the in vivo environment, to promote formation of new tissue. Such an environment can be created by using, not only a suitable scaffold or mechanical stimulus, but also a chemical stimulus by incorporating growth factors into the culture media (Tabata 2003). From published research work, many authors report using various different growth factors such as basic fibroblast growth factor (bFGF), platelet derived growth factor (PDGF), transforming growth factor-β1 (TGF- β1), epidermal growth factor (EGF) and insulin-like growth factor (IGF) to encourage cells to proliferate, differentiate or increase the production of collagen (Schmidt, Georgescu et al. 1995; Spindler, Imro et al. 1996; Scherping, Schmidt et al. 1997; Murray, Rice et al. 2003; Pertigliano, McAllister et al. 2006). In vitro cell culture studies which used bFGF, PDGF, EGF or TGF-β1 individually, have shown them to increase the proliferation of ACL fibroblasts (Schmidt, Georgescu et al. 1995; Spindler, Imro et al. 1996; Scherping, Schmidt et al. 1997; Murray, Rice et al. 2003), whereas a combination of TGF with bFGF or EGF has promoted MSC differentiation into fibroblasts (Pertigliano, McAllister et al. 2006). Growth factors have also been identified for their roles in regulating healing and repair of connective tissue after injury (Spindler, Imro et al. 1996). It is thought that the response of musculoskeletal tissues to injury or mechanical stress is modulated by growth factors such as PDGF and TGF-β (Spindler, Imro et al. 1996).

### 9.1 In vitro studies with TGF-β1

TGF-β1 has been used in various studies ranging in concentration from 0.1-15ng/ml. Specific concentrations have been shown to increase fibroblast proliferation, increase collagen production, glycosaminoglycan expression, and encourage MSC's to differentiate into fibroblasts (Spindler, Imro et al. 1996; Locklin, Oreffo et al. 1998; Meaney Murray, Rice et al. 2003; Chen, Tsai et al. 2004; Chen, Tsai et al. 2005; Moreau, Chen et al. 2005; Giannouli and Kletsas 2006; Marenzara, Wilson-Jones et al. 2006). Table 7 summarises the effects that different concentrations of TGF-β1 have had on fibroblasts and MSC's.

## 10. Characterisation of a tissue engineered ligament

When creating a tissue-engineered ligament, it is important to be aware of the characteristic components and properties found in native ligament as a comparing standard for functionality (shown in table 1). The ACL characteristics are demonstrated below in table 8.

The variation in tensile strength has been correlated to age and gender, where strength and stiffness of the ACL has been found to be lower in adult females (Chandrashekar, Mansouri et al. 2006), and can also significantly reduce as age increases, being 2-3 times higher in younger people, aged 16-26 years, compared to those aged at approximately 60 years old (Noyes and Grood 1976).

| Collagen fibre arrangement | Fibroblast distribution and orientation | Tensile strength - maximum force upon ACL at failure (N) | Maximum elongation of ACL at failure (mm) | Stiffness - Young's modulus of elasticity (MPa) | Author |
|---|---|---|---|---|---|
| Aligned in a fairly parallel orientation with the longitudinal axis of the ligament | Sparsely distributed throughout the ECM (approximately 20% if the tissue volume), aligned on collagen fibre bundles | 556-1730 | 8-12 | 9-13 | (Noyes and Grood 1976; Laurencin, Ambrosio et al. 1999; Azangwe, Mathias et al. 2001; Chandrashekar, Mansouri et al. 2006; Doroski, Brink et al. 2007) |

Table 8. Fibre organisation and mechanical properties of the ACL

## 11. Conclusions

Tissue engineering of ligaments is still in its early stages, but its prospects have great potential. Tissue engineering has the ability to overcome the limitations of autografts and allografts by generating a tissue with the correct structural and biomechanical properties for a more successful transplant, hopefully giving a better long-term mechanical performance. The benefits of using autologous cells from the patient reduces the risk of tissue rejection or transmission of infectious diseases associated with allografts, and also avoids donor site morbidity associated with allografts. Advances in research in this area continue to optimise a combination of the factors discussed in this chapter such as choice of biomaterial, cell type and stimuli for potential synergistic effects.

## 12. References

Ahmed, I., C. Collins, et al. (2004). Processing, characterisation and biocompatibility of iron-phosphate glass fibres for tissue engineering. *Biomaterials* 25(16): 3223-3232.

Ahmed, I., M. Lewis, et al. (2003). Phosphate glasses for tissue engineering: Part 2. Processing & characterisation of a ternary-based $P_2O_5$-CaO-$Na_2O$ glass fibre system. *Biomaterials* 25(3): 501-507.

Alberts, B., A. Johnson, et al. (2000). Molecular Biology of the Cell 4th Edition, Garland Science.

Altman, G., R. Horan, et al. (2001). Cell differentiation by mechanical stress. *FASEB Journal* 16: 270-272.

Altman, G., H. Lu, et al. (2002). Advanced bioreactor with controlled application of multi-dimensional strain for tissue engineering. *Journal of Biomechanical engineering* 124: 742-749.

Arnoczky, P. (1983). Anatomy of the anterior cruciate ligament. *Clinical orthopaedics and related research* 172: 19-25.

Athanassiou, G. and D. Deligianni (2001). Adhesion strength of individual human bone marrow cells to fibronectin. Integrin $\beta_1$-mediated adhesion. *Journal of Material Science: Materials in Medicine* 12: 965-970.

Azangwe, G., K. Mathias, et al. (2001). Preliminary comparison of the rupture of human and rabbit anterior cruciate ligaments. *Clinical Biomechanics* 16(10): 913-917.

Beasley, L., D. Weiland, et al. (2005). Anterior cruciate ligament reconstruction: A literature review of the anatomy, biomechanics, surgical considerations, and clinical outcomes. *Operative Techniques in Orthopaedics* 15(1): 5-19.

Berry, C., J. Shelton, et al. (2003). Influence of external uniaxial cyclic strain on oriented fibroblast-seeded collagen gels. *Tissue Engineering* 9(4): 613-624.

Bitar, M., J. Knowles, et al. (2005). Soluble phosphate glass fibres for repair of bone-ligament interface. *Journal of materials science. Materials in medicine* 16: 1131-1136.

Bitar, M., V. Salih, et al. (2008). Iron-phosphate glass fiber scaffolds for the hard-soft interface regeneration: The effect of fiber diameter and flow culture condition on cell survival and differentiation. *Journal of Biomedical Materials Research* A 87(4): 1017-1026.

Bitar, M., V. Salih, et al. (2004). Soluble phosphate glasses in vitro studies using human cells of hard and soft tissue origin. *Biomaterials* 25(12): 2283-2292.

Carpenter, J. and K. Hankenson (2004). Animal models of tendon and ligament injuries for tissue engineering applications. *Biomaterials* 25(9): 1715-1722.

Centeno, J., D. Busse, et al. (2008). Regeneration of meniscus cartilage in a knee treated with percutaneously implanted autologous mesenchymal stem cells. *Medical Hypotheses* 71: 900-908.

Chandrashekar, N., H. Mansouri, et al. (2006). Sex-based differences in the tensile properties of the human anterior cruciate ligament. *Journal of biomechanics* 39(16): 2943-2950.

Cheema, U., C.-B. Chuo, et al. (2007). Engineering functional collagen scaffolds: Cyclic loading increases material strength and fibril aggregation. *Advanced Functional Materials* 17: 2426-2431.

Chen, C., Y. Tsai, et al. (2005). Type I and II collagen regulation of chondrogenic differentiation by mesenchymal progenitor cells. *Journal of Orthopaedic Research* 23: 446-453.

Chen, G. and Q. Wu (2005). The application of polyhydroxyalkanoates as tissue engineering materials. *Biomaterials* 26(33): 6565-6578.

Chen, Y., C. Huang, et al. (2008). Effects of cyclic mechanical stretching on the mRNA expression of tendon/ligament-related and osteoblast-specific genes in human mesenchymal stem cells. *Connective tissue Research* 49(1): 7-14.

Chiquet, M., A. Sarasa-Renedo, et al. (2003). How do fibroblasts translate mechanical signals into changes in extracellular matrix production? *Matrix Biology* 22(1): 73-80.

Christian, P., I. Jones, et al. (2001 ). Monomer transfer moulding and rapid prototyping methods for fibre reinforced thermoplastics for medical applications. *Composites part A: Applied science and manufacturing* 32(7): 969-976.

Chuen, F., C. Chuk, et al. (2004). Immunohistochemical characterization of cells in adult human patellar tendons. *Journal of Histochemistry and Cytochemistry* 52(9): 1151-1157.

Chvapil, M., D. Speer, et al. (1993). Collagen fibres as a temporary scaffold for replacement of ACL in goats. *Journal of Biomedical Materials Research* 27: 313-325.

Cooper, J. and L. Bailey (2006). Evaluation of the anterior cruciate ligament, medial collateral ligament, Achilles tendon and patellar tendon as cell sources for tissue-engineered ligament. *Biomaterials* 27: 2747-2754.

Cooper, J., H. Lu, et al. (2005). Fiber-based tissue-engineered scaffold for ligament replacement: design considerations and in vitro evaluation. *Biomaterials* 26(13): 1523-1532.

Day, R., A. Boccaccini, et al. (2004). Assessment of polyglycolic acid mesh and bioactive glass for soft-tissue engineering scaffolds. *Biomaterials* 25(27): 5857-5866.

Dennis, J. and A. Caplan (1993). Porous ceramic vehicles for rat-marrow-derived (*Rattus norvegicus*) osteogenic cell delivery: Effects of pre-treatment with fibronectin or laminin. *Journal of Oral Implantology* XIX(2): 106-115.

Doroski, D. and D. Brink (2007). Techniques for biological characterization of tissue-engineered tendon and ligament. *Biomaterials* 28(2): 187-202.

Drury, J. and D. Mooney (2003). Hydrogels for tissue engineering: scaffold design variables and applications. *Biomaterials* 24: 4337-4351.

Einhorn, T., R. O'Keefe, et al. (2007). Orthopaedic basic science. Foundations of clinical practice, American Academy of Orthopaedic Surgeons (AAOS).

Ellenbecker, T. (2000). Knee ligament rehabilitation. New York, Churchill Livingstone. New York.

Evans, C. (1999). Cytokines and the role they play in the healing of ligaments and tendons. *Current opinion Sports medicine* 28(2): 71-76.

Fan, H., H. Liu, et al. (2008). Enhanced differentiation of mesenchymal stem cells co-cultured with ligament fibroblasts on gelatin/silk fibroin hybrid scaffold. *Biomaterials* 29: 1017-1027.

Fan, H., H. Liu, et al. (2009). Anterior cruciate ligament regeneration using mesenchymal stem cells and silk scaffold in large animal model. *Biomaterials* 30(28): 4967-4977.

Fan, H., H. Liu, et al. (2008). In vivo study of anterior cruciate ligament regeneration using mesenchymal stem cells and silk scaffolds. *Biomaterials* 29: 3324-3337.

Fenwick, S., B. Hazleman, et al. (2002). The vasculature and its role in the damaged and healing tendon. *Arthritis Research* 4(4): 252-260.

Foster, A., C. Butcher, et al. (2005). Changes in arthroscopic findings in the anterior cruciate ligament deficient knee prior to reconstructive surgery. *The Knee* 12(1): 33-35.

Fu, F., C. Harner, et al. (1993). Biomechanics of knee ligaments. *The Journal of Bone and Joint Surgery* 75-A: 11.

Fulton, A. (1984). The cytoskeleton-cell architecture and choreography, Chapman & Hall London.

Ge, Z., J. Goh, et al. (2005). Selection of cell source for ligament tissue engineering. *Cell transplant* 14(8): 573-583.

Gentleman, E., G. Livesay, et al. (2006). Development of ligament-like structural organization and properties in cell-seeded collagen scaffolds in vitro. *Annals of Biomedical engineering* 34(5): 726-736.

Goulet, F., L. Germain, et al. (1997). Principles of tissue engineering. The need for bioengineered tendons & ligaments RG Landes company.

Hairfield-Stein, M., C. England, et al. (2007). Development of self-assembled tissue engineered ligament from bone marrow stromal cells. *Tissue Engineering* 13(4): 703-710.

Hansen, U., S. Masouros, et al. (2006). (iii) Material properties of biological tissues related to joint surgery. *Current Orthopaedics* 20(1): 16-22.

Hsieh, A., C. Tsai, et al. (2000). Time-dependent increases in type III collagen gene expression in medial collateral ligament fibroblasts under cyclic strains. *Journal of Orthopaedic Research* 18: 220-227.

Hua, S. and A. Wang (2009). Synthesis, characterisation and swelling behaviours of sodium alginate-g-poly(acrylic acid)/sodium humate superabsorbent. *Carbohydrate Polymers* 75: 79-84.

Huang, T., Y. Chen, et al. (2008). Isolation and characterization of mesenchymal stromal cells from human anterior cruciate ligament. *Cytotherapy* 10(8): 806-814.

Ingber, D. (1999). How cells (might) sense microgravity. *Faseb* 13: S3-S15.

Johansson, S., G. Svineng, et al. (1997). Fibronectin-Integrin interactions. *Frontiers in Bioscience* 2: d126-146.

Kadi, A., S. Fawzi-Grancher, et al. (2006). Effect of cyclic stretching and TGF-beta pathway on the extra cellular matrix synthesis in tissue engineering. *Journal of Biomechanics* 39(Suppl 1): S579-580.

Kim, S.-G., T. Akaike, et al. (2002). Gene expression of type I and type III collagen by mechanical stretch in anterior cruciate ligament cells. . *Cell Structure and Function* 27: 139-144.

Lanza, R., R. Langer, et al. (2007). Principles of Bone Tissue Engineering - Tendons & Ligaments. Third edition, Elsevier Inc.

Laurencin, C., A. Ambrosio, et al. (1999). Tissue engineering: Orthopaedic applications. *Annual Review of Biomedical engineering* 1: 19-46.

Laurencin, C. and J. Freeman (2005). Ligament tissue engineering: An evolutionary materials science approach. *Biomaterials* 26: 7530-7536.

Lee, C.-Y., X. Liu, et al. (2004). The combined regulation of estrogen and cyclic tension on fibroblast biosynthesis derived from anterior cruciate ligament. *Matrix Biology* 23: 323-329.

Lee, K. and D. Mooney (2001). Hydrogels for tissue engineering. *Chemical Reviews* 101(7): 1873-1886.

Locklin, R., R. Oreffo, et al. (1998). Effects of TGF-β and bFGF on the differentiation of human bone marrow stromal fibroblasts. *Cell Biology International* 23(3): 185-194.

Lu, H., J. Cooper, et al. (2005). Anterior cruciate ligament regeneration using braided biodegradable scaffolds: in vitro optimization studies. *Biomaterials* 26(23): 4805-4816.

Martin, I., D. Wendt, et al. (2004). The role of bioreactors in tissue engineering. *Trends in Biotechnology* 22(2): 80-86.

Marui, T., C. Niyibizi, et al. (1997). Effect of growth factors on matrix synthesis by ligament fibroblasts. *Journal of Orthopaedic Research* 15(1): 18-23.

Mascarenhas, R. and P. MacDonald (2008). Anterior cruciate ligament reconstruction: A look at prosthetics - past, present and possible future. *McGill Journal of Medicine* 11(1): 29-37.

Matyas, J., P. Edwards, et al. (1994). Ligament tension affects nuclear shape in situ: An in vitro study. *Connective Tissue Research* 31(1): 45-53.

Meaney Murray, M., K. Rice, et al. (2003). The effect of selected growth factors on human anterior cruciate ligament cell interactions with a three-dimensional collagen-GAG scaffold. *Journal of Orthopaedic Research* 21(2): 238-244.

Meseguer-Olmo, L., A. Bernabeu-Esclapez, et al. (2008). In vitro behaviour of adult mesenchymal stem cells seeded on a bioactive glass ceramic in the $SiO_2$-CaO-$P_2O_5$ system. *Acta Biomaterialia* 4: 1104-1113.

Meyer, U., A. Buchter, et al. (2005). Design and performance of a bioreactor system for mechanically promoted three-dimensional tissue engineering. *British Journal of Oral and Maxillofacial surgery* 44(2): 134-140.

Milleret, V., M. Simonet, et al. (2009). Cyto- and hemocompatibility of a biodegradable 3D-scaffold material designed for medical applications. *Journal of Biomedical Materials Research* B Applied Biomaterials 91B(1): 109-121.

Miyaki, S., T. Ushida, et al. (2001). Mechanical stretch in anterior cruciate ligament derived cells regulates type I collagen and deocrin expression through extracellular signal-regulate kinase 1/2 pathway. *Materials Science and Engineering* C 17: 91-94.

Miyazaki, H., Y. Hasegawa, et al. (2000). A newly designed tensile tester for cells & its applications to fibroblasts. *Journal of Biomechanics* 33(1): 97-104.

Moreau, J., J. Chen, et al. (2005). Growth factor induced fibroblast differentiation from human bone marrow stromal cells in vitro. *Journal of Orthopaedic Research* 23: 164-174.

Murray, M., K. Rice, et al. (2003). The effect of selected growth factors on human anterior cruciate ligament cell interactions with a three-dimensional collagen-GAG scaffold. *Journal of orthopaedic research* 21(2): 238-244.

Nesic, D., R. Whiteside, et al. (2006). Cartilage tissue engineering for degenerative joint disease. *Advanced drug delivery reviews* 58(2): 300-322.

Noth, U., K. Schupp, et al. (2005). Anterior cruciate ligament constructs fabricated from human mesenchymal stem cells in a collagen type I hydrogel. *Cytotherapy* 7(5): 447-455.

Noyes, F. and E. Grood (1976). The strength of the anterior cruciate ligament in humans and Rhesus monkeys. *The Journal of bone and joint surgery* 58-A(8): 1074-1081.

Pankov, R. and K. Yamada (2002). Fibronectin at a glance. *Journal of Cell Science* 115: 3861-3863.

Papathanasopoulos, A. and P. Gaiannoudis (2008). Biological considerations of mesenchymal stem cells and endothelial progenitor cells. *Injury International Journal of the care of the injured* 39S2: S21-S32.

Park, S., I. Kim, et al. (2006). Biological responses of ligament fibroblasts and gene expression profiling on micropatterned silicone substrates subjected to mechanical stimuli. *Journal of Bioscience and Bioengineering* 102(5): 402-412.

Pertigliano, F., D. McAllister, et al. (2006). Tissue engineering for anterior cruciate ligament reconstruction: A review of current strategies. Arthroscopy: *The journal of Arthroscopic and related surgery* 22 (4) 441-451. 22(4): 441-451.

Popp, F., E. Eggenhofer, et al. (2008). Mesenchymal stem cells can induce long-term acceptance of solid organ allografts in synergy with low-dose mycophenolate. *Transplant Immunology* 20: 55-69.

Portner, R., S. Bagel-Heyer, et al. (2005). Bioreactor design for tissue engineering. *Journal of Bioscience and Bioengineering* 100(3): 235-245.

Posthemus, M., A. September, et al. (2009). The association between the COL12A1 gene and anterior cruciate ligament ruptures. *British Journal of Sports Medicine* DOI: 10.1136/bjsm.2009.060756.

Probhakar, R., S. Brocchini, et al. (2005). Effect of glass composition on the degradation properties and ion release characteristics of phosphate glass-polycaprolactone composites. *Biomaterials* 26(15): 2209-2218.

Puk, C., D. Miller, et al. (2006). The effects of short-term stimulation on fibroblast spreading in an *in vitro* 3D system. *Journal of Biomedical Materials Research* 76A: 665-673.

Quaglia, F. (2008). Bioinspired tissue engineering: The great promise of protein delivery technologies. *International Journal of Pharmaceutics.*

Raif, E., B. Seedhom, et al. (2007). Cyclic Straining of cell-seeded synthetic ligament scaffolds: Development of apparatus and methodology. *Tissue Engineering* 13(3): 629-640.

Rathbone, S., P. Furrer, et al. (2009). Biocompatibility of polyhydroxyalkanoate as a potential material for ligament and tendon scaffold material. *Journal of Biomedical Materials Research* Part A DOI: 10.1002/jbm.a.32641.

Rezwan, K., Q. Ghen, et al. (2006). Biodegradable and bioactive porous polymer/inorganic composite scaffolds for bone tissue engineering. *Biomaterials* 27(18): 3413-3431.

Rogers, W. (1983). Cells and tissues. An introduction to histology and cell biology, Academic press Inc.(London) Ltd.

Sahoo, S., H. Ouyang, et al. (2006). Characterization of a novel polymeric scaffold fro potential application in tendon/ligament tissue engineering. *Tissue Engineering* 12(1): 1-9.

Sakai, S., H. Masuhara, et al. (2005). Transition of mechanical property of porous alginate scaffold with cells during culture period. *Journal of Bioscience and Bioengineering* 100(1): 127-129.

Sakoda, S., H. Shin, et al. (1999). Mechanical stretching of human osteoblast-like cells stimulates bone morphogenic proteins and macrophage colony-stimulating factor productions. *Pathophysiology* 6: 63-69.

Salasznyk, R., W. Williams, et al. (2004). Adhesion to vitronectin and collagen I promotes osteogenic differentiation of human mesenchymal stem cells. *Journal of Biomedicine and Biotechnology* 2004(1): 24-34.

Scherping, S., C. Schmidt, et al. (1997). Effect of growth factors on the proliferation of ligament fibroblasts from skeletally mature rabbits. *Connective tissue research* 36(1): 1-8.

Schlenker, H.-J., L. Kreja, et al. (2006). Effect of mechanical strain on human tenocytes seeded on a novel polylactide scaffold for tissue engineering of ligaments. *Journal of Biomechanics* 39(Suppl 1): S222.

Schmidt, C., H. Georgescu, et al. (1995). Effect of growth-factors on the proliferation of fibroblast from the medial collateral and anterior cruciate ligaments. *Journal of orthopaedic research* 13(2): 184-190.

Schulze-Tanzil, G., A. Mobasheri, et al. (2004). Cultivation of human tenocytes in high-density culture. *Histochemistry and Cell Biology* 122: 219-228.

Screen, H., J. Shelton, et al. (2005). Cyclic tensile strain upregulates collagen synthesis in isolated tendon fascicles. *Biochemical and Biophysical Research Communications* 336: 424-429.

Sechler, J., S. Corbett, et al. (1997). Modulatory Roles for integrin activation and the synergy site of fibronectin during Matrix assembly. *Molecular biology of the cell* 8: 2563-2573.

September, A., M. Schwellnus, et al. (2007). Tendon and ligament injuries: The genetic component. *British Journal of Sports Medicine* 41(4): 241-246.

Shishataskaya, E., T. Volova, et al. (2004). Tissue response to the implantation of biodegradable polyhydroxyalkanoate sutures. *Journal of Materials Science: Materials in Medicine*. 15: 719-728.

Spindler, K., A. Imro, et al. (1996). Patellar tendon and anterior cruciate ligament have different mitogenic responses to platelet-derived growth factor and transforming growth factor beta. *Journal of orthopaedic research* 14(4): 542-546.

Swart, J., A. Martens, et al. (2008). Mesenchymal stem cells: A future for the treatment of arthritis? *Joint Bone Spine* 75: 379-382.

Tabata, Y. (2003). Tissue regeneration based on growth factor release. *Tissue Engineering* 9(Supplement 1): S5-S15.

Teh, T., J. Goh, et al. (2008). Comparative study of random and aligned nanofibrous scaffolds for tendon/ligament tissue engineering. *Journal of Biomechanics* 41 (S1)(S1): S527.

Tetsunaga, T., T. Furumatsu, et al. (2009). Mechanical stretch stimulates integrin $\alpha V\beta 3$-mediated collagen expression in human anterior cruciate ligament cells. *Journal of Biomechanics* 42: 2097-2103.

Toyoda, T., H. Matsumoto, et al. (1998). Tensile load and the metabolism of anterior cruciate ligament cells. *Clinical Orthopaedics and Related Research* 353: 247-256.

Utukuri, M., H. Somayaji, et al. (2006). Update on paediatric anterior cruciate injuries. *The knee* 13(5): 345-352.

Van Eijk, F., D. Saris, et al. (2004). Tissue engineering of ligaments: A comparison of bone marrow stromal cells, anterior cruciate ligament, and skin fibroblasts as a cell source. *Tissue Engineering* 10(5-6): 893-903.

Vergis, A. and J. Gillquist (1995). Graft failure in Intra-Articular anterior cruciate ligament reconstructions: A review of the literature. *Arthroscopy: The Journal of Arthroscopic and related surgery* 11(3): 312-321.

Vohra, S., K. Hennessy, et al. (2008). Comparison of mesenchymal stem cell and osteosarcoma cell adhesion to hydroxyapatite. *Journal of Material Science: Materials in Medicine* 19: 3567-3574.

Vunjak-Novakovic, G., G. Altman, et al. (2004). Tissue engineering of ligaments. *Annual Review of Biomedical Engineering* 6: 131-156.

Wang, B., W. Liu, et al. (2008). Engineering of extensor tendon complex by an *ex vivo* approach. *Biomaterials* 29: 2954-2961.

Wang, J. (2006). Mechanobiology of tendon. *Journal of Biomechanics* 39: 1563-1582.

Wang, L., R. Shelton, et al. (2003). Evaluation of sodium alginate for bone marrow cell tissue engineering. *Biomaterials* 24: 3475-3481.

Wang, Y., H. Kim, et al. (2006). Stem cell-based tissue engineering with silk biomaterials. *Biomaterials* 27(36): 6064-6082.

Webb, K., R. Hitchcock, et al. (2006). Cyclic strain increases fibroblast proliferation, matrix accumulation, and elastic modulus of fibroblast-seeded polyurethane constructs. *Journal of Biomechanics* 39: 1136-1144.

Woo, S., S. Abramowitch, et al. (2006). Biomechanics of knee ligaments: injury, healing and repair. *Journal of Biomechanics* 39(1): 1-20.

Wu, S., H. Hsu, et al. (2009). Preparation of porous 45S5 Bioglass-derived glass-ceramic scaffolds by using rice husk as a porogen additive. *Journal Material Science. Materials in Medicine* 20(6): 1229-1236.

Yang, G., H.-J. Im, et al. (2005). Repetitive mechanical stretching modulates IL-1β induced COX-2, MMP-1 expression, and $PGE_2$ production in human patellar tendon fibroblasts. *Gene* 363: 166-172.

Yu, B.-Y., P.-Y. Chen, et al. (2008). The behaviours of human mesenchymal stem cells on the poly(3-hydroxybutyrate-*co*-3-hydroxyhexanoate) (PHBHHx) membranes. *Desalination* 234: 204-211.

Zantop, T., W. Patterson, et al. (2005). Anatomy of the anterior cruciate ligament. *Operative techniques in Orthopaedics* 15(1): 20-28.

Zhang, L., X. Wang, et al. (2004). Cyclic stretching and co-culture with fibroblasts promote the differentiation of rat mesenchymal stem cells to ligament fibroblasts. *Tissue engineering* 15: P63.

Zhao, K., Y. Deng, et al. (2003). Polyhydroxyalkanoate scaffolds with good mechanical properties & biocompatibility. *Biomaterials* 24: 1041-1045.

Zhao, R., L. Liao, et al. (2004). Mechanisms of and perspectives on the mesenchymal stem cell in immunotherapy. *Journal of Laboratory and Clinical Medicine* 143(5): 284-291.

# 9

# Potential of Tissue-Engineered Ligament Substitutes for Ruptured ACL Replacement

Goulet F.[1,2,3], Chabaud S.[1,2,4], Simon F.[1,2,4],
Napa I.D.[1,2,4], Moulin V.[1,2,4] and Hart D.A.[5]
*[1]Centre LOEX de l'Université Laval,*
*[2]Génie tissulaire et régénération: LOEX,*
*Centre de recherche FRSQ du Centre hospitalier affilié universitaire de Québec,*
*[3]Département de réadaptation,*
*[4]Département de chirurgie, Faculté de médecine, Université Laval, Québec, QC,*
*[5]McCaig Institute for Bone and Joint Health, University of Calgary, Calgary, Alberta*
*Canada*

## 1. Introduction

Anatomical descriptions of anterior cruciate ligament (ACL) can be found in Egyptian papyrus scrolls dating back to 3000 BC. Hippocrates (460-370BC) has described knee subluxation with ACL injury. Even today, the ACL of the knee joint is still the frequent target of trauma, especially in the young and active population (Murray, 2009). Rupture of the ACL of knee affects over 175,000 patients a year worldwide (Myer et al., 2004). Such injury often results from twisting or turning the knee while landing (e.g. non-contact rupture), valgus stress (*knock-knee position*), or hyperextension directly related to contact or collision. (Wu et al., 2010). For unknown reasons, this major knee joint stabilizer fails to heal after suture repair, (Murray et al., 2007). Often, other knee structures are directly affected when partial or complete ACL injuries occur, impairing knee function (Englund et al., 2009). The tibia and femur are more likely to rub against each other, leading to cartilage defects and osteoarthritis. While a variety of rehabilitation protocols have been assessed, rehabilitation doesn't insure long-term maintenance of knee function. The clinical decision to replace a torn ACL is based on the extent of the damage caused by the injury and subsequent functioning of the knee, and the choice of ACL substitute that is grafted is made accordingly (e.g. autograft, allograft, synthetic, or potentially, a tissue engineered autologous construct). The surgery of the ACL is usually performed several weeks after the injury in order to allow the swelling and inflammation to subside, although in some countries the surgical intervention can occur within hours or days of the injury. After the surgery, physical rehabilitation is necessary to strengthen the surrounding muscles, stabilize the knee joint, prevent arthrofibrosis, and restore range of motion.

### 1.1 Function of the ACL on the knee joint

The ACL has both proprioceptive and mechanical functions. It is the first restraint to anterior translation of the tibia. The ultimate tensile load and stiffness of the human femur-

ACL-tibia complex are 2160 ± 157 N and 242 ± 28 N/mm, respectively (Dargel et al., 2007; Woo, 2006). The forces applied on the ACL during normal walking are 303N or less (Peterson & Zantop, 2007). The highest loads and strains on the ACL during daily function are quadriceps-powered extension of the knee, moving it from approximately 40 degrees of flexion to full extension (This explains why open kinetic chain exercises are avoided during early phases of rehabilitation). Secondly, the ACL forces the tibia to internally rotate during anterior tibial translation, indicating that the ACL primarily restrains internal rotational moments during anteroposterior translation, (Fukubayashi et al., 1982).

### 1.2 Compromised healing of the injured ACL

In contrast with extra-articular knee ligaments (e.g. the collateral ligaments), the healing of the intra-articular ACL in vivo is poorly understood. A complete tear of the ACL (midsubstance or at the insertion), would not leave a residual structure on which to form a scar. However, a partial tear should leave at least a partial template to form a scar. Interestingly, even if bleeding occurs after ACL injury, no fibrin-platelet plug is observed to form inside the joint, even at the injury site. One possible explanation for this observation is that circulating intra-articular plasmin breaks down the fibrin plug as fast as it can form (Murray, 2009). Recent work has shown that after trauma to the joint, the production of plasmin is upregulated via the increased secretion of urokinase plasminogen activator (Rosc et al., 2002). With the additional circulating plasmin, the fibrin network is quickly destabilized within the joint environment and no fibrin-platelet plug forms. This premature loss of the fibrin-platelet plug would have the significant clinical benefit of preventing overall joint scarring and stiffness (arthrofibrosis) and thus maintenance of joint mobility after injury. As formation of a fibrin-platelet plug is an essential first step for wound healing of musculoskeletal tissue outside the joint, the loss of this fibrin-platelet plug inside the joint may be the key mechanism behind the failure of intra-articular tissues to heal (Murray, 2009).

However, other possibilities likely also exist. In studies of partial injuries to the ACL in either rabbit (Lo et al., in preparation; Heubner et al., in preparation) or sheep models (Heard et al., 2011), the response to injury leads to an inflammation. In fact, a partial injury to one band of the ACL leads to an initial response not dissimilar to that of the MCL at 3 weeks post-injury, which then fails to mature and ultimately fails to lead to functional repair, possibly due to persistent inflammation (Lo et al., in preparation). Thus, the failure of healing (non-union) and laxity observed in the ACL after suture repair may also be attributed to the adverse influence of synovial fluid (Murray et al., 2009; Woo et al., 2000), to alterations in the cellular metabolism after injury (Amiel et al., 1989), and to intrinsic cell deficiencies (Kobayashi et al., 2000). The superior capacity of the MCL to increase its blood supply through angiogenesis and increased flow is essential for ligament healing to occur, and may be the major difference in healing potential between the ACL and MCL (Kobayashi et al, 2000). Several growth factors have been studied in tendons post-surgery. The expression of Aggrecan, Versican, Biglycan, Lumican, Decorin, TGF-β and β-FGF mRNAs have been monitored in vivo, using a rabbit model of flexor tendon injury (Berglund et al., 2006). The role and the mode of interactions between these various factors, as well as the sequential post-traumatic modulation of their expression remain to be defined. However, it is clear that whatever the mechanisms involved, the intra-articular environment is not conducive to normal ACL healing, and this also has implications for the incorporation of ACL reconstructions with either natural tissues or tissue engineered tissues.

### 1.3 Impact of torn ACL on the knee strength, stability and osteoarthritis (OA)

Injury to the anterior cruciate ligament (ACL) is regarded as critical to the physiological kinematics of the femoral-tibial joint, with its disruption eventually causing long-term functional impairment. Both the initial trauma and the pathologic motion pattern of the injured knee may result in primary degenerative lesions of the secondary stabilisers of the knee, each of which are associated with the early onset of OA (Von Porat et al., 2004). Consequently, there is a wide consensus that young and active patients may profit from reconstructing the ACL. Several factors have been identified as significantly influencing the biomechanical characteristics and the functional outcome of an ACL reconstructed knee joint. These factors are: (1) individual choice of autologous graft material using either patellar tendon-bone grafts or quadrupled hamstring tendon grafts, or another type of ACL substitute, (2) anatomical bone tunnel placement within the footprints of the native ACL, (3) adequate substitute tension after cyclic graft preconditioning, and (4) graft fixation close to the joint line using biodegradable graft fixation materials that provide an initial fixation strength exceeding those loads commonly expected during rehabilitation. Based on such factors, the literature encourages mid- to long-term clinical and functional outcome assessments after ACL reconstruction (Dargel et al., 2007).

At follow-up, patellofemoral OA is associated with higher activity level, meniscal injury, extension and flexion deficit, and ACL reconstruction. Although risk factors for posttraumatic OA are multifactorial, the primary risk factor that stood out in the study of Neuman et al. (2009), was whether a meniscectomy had been performed. Early activity modification and neuromuscular knee rehabilitation might also have been related to the low prevalence of radiographic knee OA. In patients with ACL injury willing to restrict themselves to moderate activity level to avoid re-injury, initial treatment without ACL reconstruction should be considered. (Neuman et al., 2009). Interestingly, the incidence of OA after an ACL tear appears to be similar whether or not the ACL has been reconstructed (Lohmander et al., 2007; Oiestad et al., 2010; Frobell et al., 2011), possibly due to concurrent injury to other joint structures at the time of the ACL tear. However, quality of life and a return to knee stability is gained by reconstruction, particularly in younger active patients.

## 2. Current surgical approaches to restore knee function after ACL injury

New options for ACL replacement are under development and some clinical studies are in progress (Altman et al., 2008; Murray, 2009). One of these options is tissue engineering of a tissue replacement, but the majority of reconstructions are performed with autologous or allogeneic tissues. By definition, tissue engineering involves a combination of knowledge and skills in biology and in biomechanics (Functional Tissue Engineering Conference Group, 2008). Such multidisciplinary science has opened the door to the conception and fabrication of new biocompatible materials for ACL reconstruction (Woo, 2009). However, independently from the type of ACL substitute chosen, successful reconstruction of the ACL depends on anatomic placement of a graft ligament substitute (Cole et al., 2000). Strength has been and remains a major consideration in the choice of grafts (Frank & Jackson, 1997). Accurate tunnel placement minimizes graft excursion and impingement against the roof of the intercondylar notch. This will result in maximum knee stability and motion (Fineberg et al., 2000; Steiner et al., 2008).

### 2.1 Autologous tendon for torn ACL replacement

Orthopaedic autograft reconstruction of the ligament often uses a bone-to-bone technique for optimal repair (Frank & Jackson, 1997; Paxton et al., 2009). The central part of the patellar tendon, including bone fragments from the tibia and the femur, or the hamstring tendons is often used as an ACL substitute. Bone-patellar tendon-bone (BPTB) autografts have been proclaimed as the "gold standard" in ACL reconstruction (Woo et al., 2006). Biomechanically, a 10-mm wide BPTB graft has stiffness and ultimate load values of 210 ± 65 N/mm and 1784 ± 580 N, respectively (Wilson et al., 1999), which compare well with those of the young human femur-ACL-tibia complex (242 ± 28 N/mm and 2160 ± 157 N, respectively), (Woo et al., 2006). It also shows the advantage of having bone blocks available for graft fixation in the osseous tunnels that leads to better knee stability. Unfortunately, ligament creep or laxity is observed post-surgery, adding to the morbidity associated to the partial loss of an healthy tendon, knee chronic pain, loss of motion, knee instability, quadriceps weakness and patella rupture. The mechanisms by which such grafts lose function is not known in detail, but it may involve replacement of the transplanted tissue by a scar-like matrix after neovascularization, infiltration of cells from the intraarticular environment, and a persistent inflammation as a consequence of the surgery. Thus, many orthopedic surgeons conclude that it is clinically highly justified to explore other strategies for ACL replacement.

### 2.2 Cadaver allograft ACL implantation

The use of allograft substitutes (ACL taken from cadavers) overcomes the need for autologous tissues, avoiding donor site morbidity. However, this approach has deficiencies considering the risks of disease transmission, graft rejection, delayed physiologic integration and inflammation (Murray, 2009). Allografts are relatively expensive and their availability might be limited. Nevertheless, patellar tendons taken from cadavers are still used for ruptured ACL replacement, depending on the short- and long-term needs of each patient that is treated. Cadaver allografts are readily strong and they are progressively remodelled by the cells of the host in situ post-grafting.

## 3. Essential factors needed to translate tissue engineering into effective treatment modalities

Over the last fifteen years, tissue engineered ACL substitutes have raised the interest of orthopedic surgeons for obvious reasons. The concept of producing a ligament in vitro solves the issue of morbidity associated with autologous tendon grafts. Depending on the technological approach developed to reconstruct an ACL in vitro, the availability of tissue-engineered products widens the spectrum of surgical options and provides a mid- to long-term solution to prevent knee joint instability. However, the features of a tissue engineered ACL depend on the type of biomaterials that compose its scaffold. The potential and the functionality of any reconstructed ligament must be assessed in animal models. This is the critical step to evaluate the feasibility, the viability and the factors of failure associated with all new ACL replacement strategies under development.

### 3.1 Animal models: strengths and limitations

Rabbit and goat models have been used to assess torn ACL replacement strategies (Xerogeanes et al, 1998). The advantage of the rabbit remains the wide choice of antibodies and biomarkers commercially available to analyze the various constituents of ACL substitutes post-grafting and ex-vivo. The main limitation of this model is its size, and the fact that it doesn't compare with the human knee joint. The goat is larger and the size of its knee joint is close to the human structures. This facilitates development of a pre-clinical protocol for the implantation of a new type of ACL replacement. However, significant differences are observed between the magnitude of force experienced by the goat ACL and its anteromedial and posterolateral bundles when compared with the corresponding human ACL. Nonetheless, the caprine animal model is widely used for ACL characterization in vitro and in vivo (Tischer et al., 2009; Robayo et al, 2011; Tremblay et al, 2011). The most challenging preclinical model to study ACL repair is certainly the dog knee joint. As in humans, spontaneous ACL injuries are frequently observed in large dog knee joints. The medial meniscus is commonly damaged along with the ACL. The canine model also involves several other issues, including early OA development after torn ACL injury (Murray et al., 2006). For this reason, an ACL substitute tested with success in the canine knee joint would certainly indicate serious potential for use in the human knee. As well, success in the canine model could also have veterinary implications for prevention of OA development in valuable companion, working and show animals.

### 3.2 Cell-seeded (stem cells, autologous cell sources) or acellular ACL substitutes?

There are two groups of cell-seeded ACL substitutes: tissues populated with live cells (e.g. BPTB or semitendinous autologous ACL grafts), and grafts containing dead cells (e.g. cadavers allografts). In addition to the risks of immune and/or inflammatory reaction, autologous or allogenic dead cells must be eliminated post-implantation. The cellular debris present in a cell-seeded ACL substitute may slow down or delay its colonization and impair the growth of host cells that migrate into the grafts from the bone insertion sites towards the middle ligament substance. ACL graft vascularization post-implantation might also be delayed by the presence of microstructures that are subjected to an ongoing necrosis process in the implant. ACL substitutes already populated with live autologous cells that will not be rejected by the host have the potential to be quickly regenerated in situ, following neovascularization to restore blood supply, as well as reinnervation.

Interestingly, tissue engineered ACL substitutes can be seeded with live cells before implantation to initiate matrix deposition and remodeling in the tissue. This may enhance the integration of the implant in situ post-grafting. However, the choice of the cells source remains controversial. The use of mesenchymal stem cells (MSC) is an attractive option as these cells have a long lifespan and are not fully differentiated, suggesting that they could be tolerated from one individual to another. The use of human serum instead of serum of animal source, seems to have a direct and a positive effect on the phenotype of human synovium-derived mesenchymal stem cells (MSC), (Tateishi et al., 2008). The results of this study indicated that human serum (HS) is superior for the culture of human MSCs compared with fetal bovine serum (FBS) in terms of cellular expandability, without losing chondrogenic or osteogenic differentiation capacity. Coupled with the advantage in eliminating the potential risk accompanied with the use of xeno-derived materials, pooled, well-characterized HS could be a useful reagent to promote cellular expansion for clinical synovial stem cell-based therapy. Nonetheless, the use of a "universal" human cell population for large-scale production of ACL substitutes would be ideal for obvious technical and economic reasons (Hart et al., 2005). Some issues regarding the source of autologous cells present the advantage to be non-immunogenic, but their amplification in vitro could also involve some risks of phenotypic changes.

The source of the MSC is also an issue that needs to be addressed for optimal development of a functional tissue engineered construct for replaced of a ruptured ACL. MSC have been obtained from multiple sources including bone marrow (BM), muscle, synovial membranes (SM), and more recently, from synovial fluid (SF) itself (reviewed in Peng & Huard, 2003; McGonagle & Jones, 2008; Chanda et al., 2010; Augello et al., 2010; Fong et al., 2011). Recent studies have indicated that SM-derived MSC are effective in generating tissue engineered constructs that can serve as articular cartilage repair tissues (Ando et al., 2007; Shimomura et al., 2010), but whether such MSC can also serve as effective cells to generate a replacement ACL have not been explored in detail (McGonagle & Jones, 2008). Very recent studies have shown that porcine (Ando et al., in preparation) and normal human (Krawetz et al., in preparation) SF and SM MSC exhibit very similar properties, so this source may be a preferred population in the future for ACL repair constructs. Why such MSC are present in the SF is not known, but likely they are there to facilitate repair normal repair of microdamage to intraarticular tissues such as cartilage, menisci and ligaments, so they exhibit good potential for tissue engineering purposes.

The use of fibroblasts isolated from skin instead of ACL biopsies to populate tissue-engineered ACL substitute led to comparable results in the goat knee joint (Tremblay et al., 2011). The fibroblasts isolated from both sources secrete the same types of collagen (I and III), but a skin microbiopsy is more readily harvested than a tissue sample collected under arthroscopy from a ruptured ACL in an injured knee joint. Ultimately, acellular ACL substitutes made of biodegradable and biocompatible biomaterials remain the safest option. However, such substitutes will have to be colonized by cells from the host post-implantation. Therefore, tissue-engineered acellular ACL substitutes implies a delayed remodeling of the graft (Robayo et al., 2011), which may impair its long-term stability in situ. Cells seem to play an essential role in the development of such tissue in vitro. However, to keep the best of both options, it would be important to proceed in two phases to obtain a viable and efficient ACL substitute. The first phase would be based on the use of cells to structure and strengthen the ligament scaffold in culture. When strong enough to sustain the biomechanical stress in the joint, the ligament could be lyophilized before implantation in order to kill the cells initially present in the implant and allow cell colonization by the host, being sure that the structure of the scaffold will play its role, and thus enhance integration.

### 3.3 Biomaterials: synthetic, native or both?

To facilitate host cell migration, growth, colonization and remodeling of ACL substitutes post-implantation, the scaffold of the graft must be porous and entirely biocompatible. The graft must be biodegradable and the products generated by this degradation should also be readily eliminated through physiological mechanisms. Collagen I is the major and the natural component of the native ACL matrix. Therefore, this protein is highly suitable, alone or in combination with other biocompatible constituents, to build the scaffold of a tissue-engineered ACL substitute (Goulet et al., 2004). For example, a collagen-platelet rich plasma (collagen-PRP) bridging scaffold, added in a central ACL defect, can stimulate healing of the ACL histologically and biomechanically (Murray et al., 2006; Spindler et al., 2009). Other types of tissue-engineered biomaterial scaffolds have been described for their potential as ACL replacement alternatives, as long as they can be implanted and sustain the biomechanical stress to which they are subjected in vivo (Altman et al., 2008; Sandmann and Tischer, 2010). More recently, the concept of anatomic double bundle ACL reconstruction has been developed to replace antero-medial and postero-lateral bundles of the ACL. Such approach aims at improving the biomechanical functionality of tissue-engineered grafts, even when the knee is subjected to rotatory loads (Woo et al., 2006).

### 3.4 Role of biomechanical stimulation in generating an effective tissue-engineered construct

The main advantage of including living cells in any tissue-engineered ACL substitute is the possibility to stimulate early matrix synthesis and remodelling in the reconstructed tissue in vitro. Cyclic stretching is an example of biomechanical stimuli that increase collagen synthesis by cells in the tissue, thereby strengthening the resultant scaffold made of aligned fibre bundles (Goulet et al., 2000, Kaneko et al., 2009). Thus, the ACL substitute has already reached a first step towards matrix synthesis, and the process can continue to progress in response to the in vivo strains that are applied to the graft in the post-transplantation environment (Goulet et al., 2004).

### 3.5 Placement and effective repair of the biomechanical environment

The anatomic placement of a graft ligament substitute is crucial for its proper integration and functional behavior (Cole et al., 2000). As previously mentioned, accurate tunnel placement should be performed with respect to the original insertion sites of the torn ACL, to limit graft excursion and impingement against the roof of the intercondylar notch, (Fineberg et al., 2000; Steiner et al., 2008). The ACL substitute must be able to resist an initial graft tension ranging from 44 to 88 N (Woo et al., 2006). An *in vivo* study on goats found no significant differences in knee kinematics and *in situ* forces, between high (35 N) and low (5 N) initial tension at 6 weeks after surgery (Abramowitch et al., 2003). Viscoelastic studies have revealed that the tension in the graft can decrease by as much as 50% within a short time after fixation because of its stress relaxation behaviour. Variation in outcomes may also be influenced by additional variables associated with the individual such as age, sex, genetics, obesity, muscle strength, activity, and re-injury. Therefore, a return to joint stability and functional state is the main objective of the clinical management of the knee joint following ACL injury. A tissue-engineered ACL substitute that could stabilize the joint and maintain its functional state post-grafting is certainly needed. However, if the integrity of the grafted material is to continue to function and progress in biomechanical properties, then likely the intra-articular environment will have to be manipulated to limit surgery-induced inflammation and promote anabolic conditions.

## 4. Tissue engineered ACL substitutes assessed in vivo and under development

Several tissue-engineered ACL substitutes have been proposed in the literature. Cross-linked bovine collagen fibers, chitosan-based hyaluronan hybrid polymer fibers, alginate-based chitosan hybrid polymer fibers, polyglycolic acid scaffolds, silk matrix, and other biodegradable scaffolds seeded or not with living cells (Altman et al., 2008). However, presently only ligament augmentation devices have been subjected to clinical trials. In contrast, many biosynthetic and biological ACL substitutes are being assessed in preclinical models. Such is the case with the tissue engineered ACL substitute that was developed using bone blocks, type I collagen and autologous cells (Goulet et al., 2004). The scaffold for this ACL substitute is made of bovine type I collagen that is seeded with autologous fibroblasts isolated from the ACL of the goat to be grafted. This ligament is produced entirely in vitro (Fig.1).

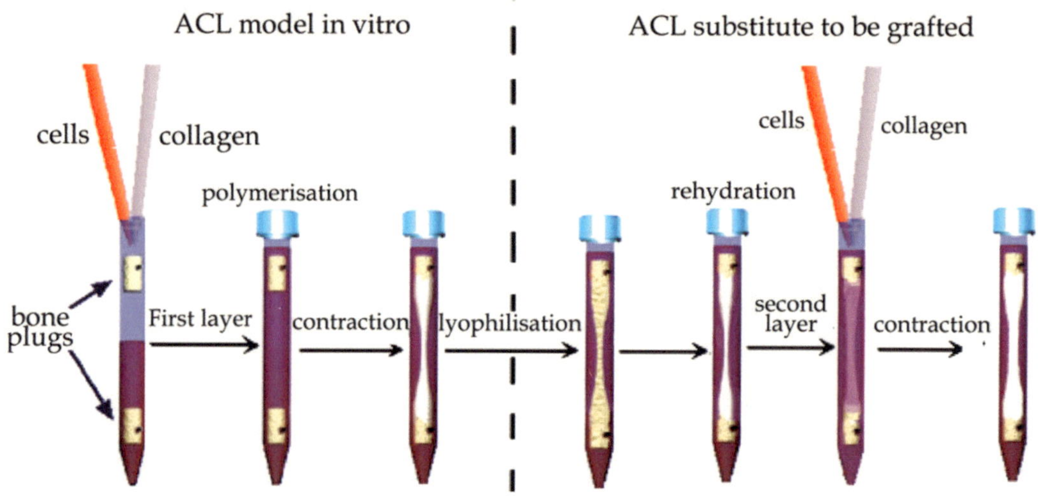


Fig. 1. Steps of production of a tissue engineered ACL substitute

A type I collagen solution (1 mg/ml), is mixed with autologous fibroblasts (0,5 x $10^6$ cells /ml) in a test tube containing two bone plugs fixed with a pin at the top and at the bottom of it. The collagen polymerizes and is contracted by the cells to obtain a ligament that can be used as an in vitro model. To produce a ligament for implantation into a knee joint, a surgical thread is placed between the bones (not shown). A lyophilization step is then added to obtain a scaffold that is stronger. Once rehydrated, a second layer of collagen including cells is subsequently added to the construct. The collagen polymerizes and the ACL substitute is ready to be grafted into a knee joint.

The ACL substitute is maintained under static tension in culture (Fig.2), with the tension applied by pulling one of the bone plugs in a programmed manner. This process also induces the alignment of the collagen fibers in the direction of the applied tension.

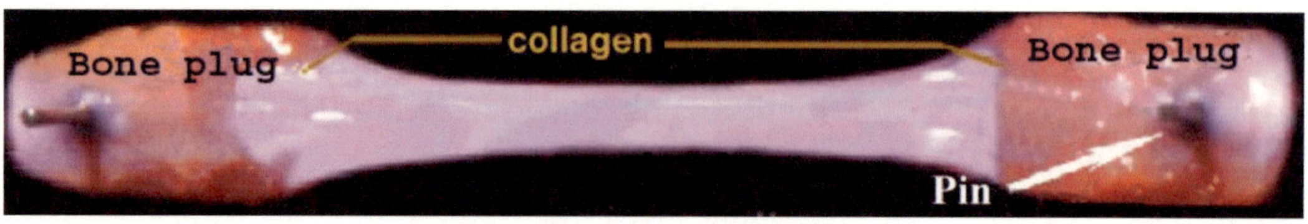


Fig. 2. Tissue engineered ACL substitute

Macroscopic view of a tissue-engineered ACL substitute maintained in culture for two months under static tension, anchored with two spongious bone plugs. Note the central portion of the tissue that was contracted by the live cells that populate its type I collagen scaffold. A surgical thread, bioresorbed within a month in vivo, reinforces the links between the bones.

The main features that orthopedic surgeons want to know regarding such ACL substitutes produced in vitro are their strength and their stiffness. Rupture assays were performed to evaluate the mechanical properties of the engineered ligament constructs before implantation (Table 1). Any natural collagen gel that is not reinforced using a thread or another type of support cannot resist a tension that is higher than 0.5 N (Fig3A). Once lyophilized, the collagen scaffold shows an ultimate strength of 2 N (Fig.3B), while the addition of a bioresorbed surgical thread will allow the structure to resist a tensile load of more than 60 N.

| Structure of the matrix | Ultimate strength (N) | Location of the rupture site |
|---|---|---|
| 1 layer of hydrated collagen (gel) | 0.2-0.5 (+ 5%) | Interface bone-ligament |
| + 1 layer of hydrated collagen lyophilized (rehydrated) | 2 (+ 5%) | Middle of the ligament |
| + 1 layer of hydrated collagen lyophilized (rehydrated) around a resorbable surgical thread: Maxon 3.0) | > 60N (grafted in goat knee joint) | Not determinated |

Averaged cross-sectional area surface: 100 mm$^2$ (+ 5%)
N= Newton

Table 1. Averaged ultimate strength values (n=12) of ex-vivo ACL grafts (n=6)

The tissue-engineered ACL was grafted into goat knee joints for periods varying from one to 13 months (Goulet et al., 2004; Robayo et al, 2011; Tremblay et al, 2011). An immobilising plaster cast was placed on the leg during the first week post-surgery to limit motion (to favour healing) and to prevent the goat from irritating the wound (Fig.4A). Following removal of the cast, the goats gradually returned to putting weight on the affected leg, walking freely thereafter (Fig.4B).

Such implant is vascularized within a month in situ post-grafting. Nerve endings, Sharpey's fibers, and fibrocartilage were observed in all of the grafts at six months post-implantation (n=3). Electron microscopic analyses demonstrated that the diameter of the collagen fibers synthesized by the cells in vivo, were comparable with native ACL fibers (Goulet et al., 2004). One of the issues that could impede the potential use of this ACL substitute for clinical application would be the difficulty to harvest a torn ACL biopsy to isolate the cells that are required. Recently, experiments conducted with tissue engineered ACLs seeded with autologous skin fibroblasts and grafted in goat knee joints for six months led to the same observations that were made with ACL cells (Fig.5). Thus, animal experimentation using skin fibroblasts has also shown promising signs of integration into the goat knee joint (Tremblay et al., 2011). Interestingly, analyses by zymography demonstrated that caprine and human fibroblasts, isolated from ACL and skin biopsies, secrete the same types of gelatinases (collagenases), suggesting that they share similar potential for collagen remodeling (data not shown).

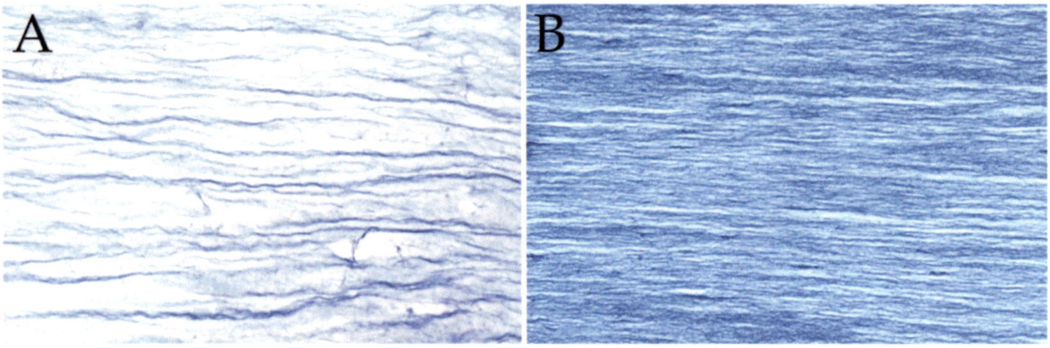


Fig. 3. Collagen scaffolds of a tissue engineered ACL subjected to static tension in culture. Photomicrographs of histological microsections of the type I collagen scaffold of tissue-engineered ACL stained with the Trichrome de Masson's method. Note the good alignment of the collagen fibers that was induced by the horizontal tension applied on the tissue for 24 hrs in culture. Shortly after collagen polymerization, the density of the fibers in the tissue-engineered scaffold is low (A). After lyophilisation and rehydration of the scaffold, the collagen fibers become closer to each other, leading to an increased matrix density and strength. (x60)

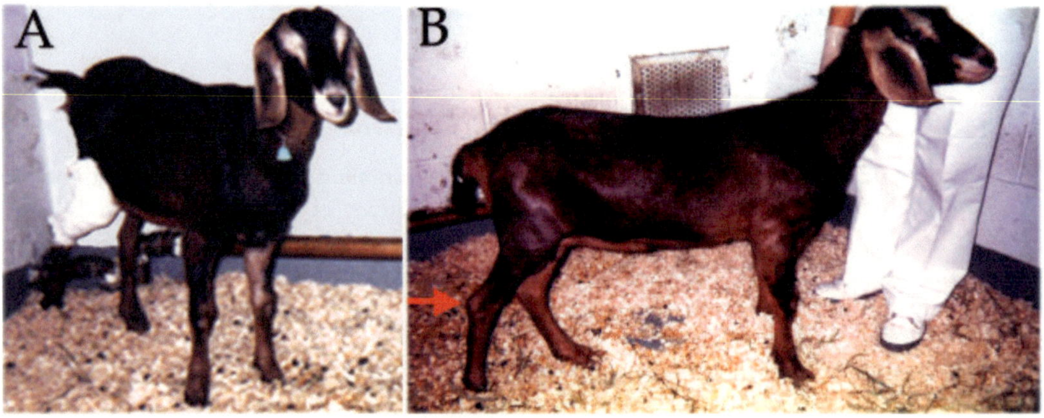


Fig. 4. A goat grafted with an autologous tissue-engineered ACL.
Photograph of a goat three days (A) and one month (B) after the implantation in the right knee joint of a tissue-engineered ACL seeded with autologous fibroblasts. The animal could jump and run only one month post-implantation

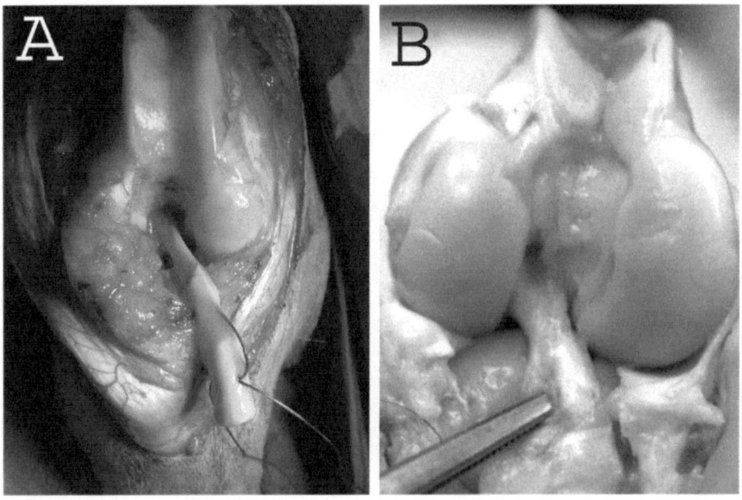


Fig. 5. Implantation of a tissue engineered ACL substitute in a goat knee joint.
Macroscopic view of a tissue-engineered ACL substitute at the time of implantation (A) and six months later (B). The articular cartilage didn't show any sign of degeneration before and after tissue-engineered ACL implantation

## 5. Conclusions

The fundamental and clinical issues associated with the management of torn ACL replacement remain those that were described by Frank and Jackson (1997) fifteen years ago. Back then, the concept of an effective tissue-engineered ligament for permanent ACL replacement was a new and exciting distant possibility. Today, such a concept is a very realistic outcome, with the evolution of nanotechnology, combined with a wider spectrum of biomaterials developed through interdisciplinary skills and know-how (Functional Tissue Engineering Conference Group, 2008). The development of tissue-engineered ACLs in vitro presents several advantages for the patients and for the orthopaedic surgeons. It avoids healthy tissue morbidity and thus, reduces the period of rehabilitation post-surgery. It allows the production of a substitute that has the proper length, diameter and scaffold to fit the joint and stabilize it. The modulation of matrix fibers in vitro by mechanical stimulation, cell-matrix interactions and structural modifications of the scaffold, are good indicators that the tissue-engineered ACL substitute will become a viable surgical option in a very near future.

However, generation of an effective engineered ACL replacement that can withstand the in vivo loading environment is only a critical first step in a process. Implantation and post-implantation events are also critical to sustain the potential of such engineered constructs, and these include correct placement of the grafts to mimic normal loading patterns, as well as controlling the intraarticular environment of minimize the catabolic influence of persistent inflammation and maximizing the anabolic activities in this compartment. Such control could contribute to the long-term effectiveness of the constructs via prevention of loss of function via alteration of their creep properties and prevention of conversion to a structure with scar-like properties. With such control, knee function can be restored and risk for OA development later in life may be abrogated.

## 6. Acknowledgment

The authors gratefully acknowledge the collaboration of Dr. Réjean Cloutier and Dr. Jean Lamontagne, orthopaedic surgeons, and the financial support of the Natural Sciences and Engineering Research Council of Canada (NSERC), Canadian Institutes of Health Research (CIHR), The Arthritis Society, the Alberta Innovates Health Solutions OA Team Grant, the Alberta Foundation for Medical Research, and the Réseau ThéCell, FRSQ.

## 7. References

Abramowitch, S.D.; Papageorgiou, C.D.; Withrow, J.D.; Gilbert, T.W. & Woo, S.L. (2003). The effect of initial graft tension on the biomechanical properties of a healing ACL replacement graft: a study in goats. *Journal of Orthopaedic Research*, Vol.21, No.4, (July 2003), pp. 708-715, ISSN 0736-0266

Altman, G.H.; Horan, R.L.; Weitzel, P. & Richmond, J.C. (2008). The use of long-term bioresorbable scaffolds for anterior cruciate ligament repair. *The Journal of the American Academy of Orthopaedic Surgeons*, Vol.16, No.4, (April), pp. 177-187, ISSN 1067-151X

Amiel, D.; Ishizue, K.K.; Harwood, F.L.; Kitabayashi, L. & Akeson, W.H. (1989). Injury of the anterior cruciate ligament: the role of collagenase in ligament degeneration. *Journal of Orthopaedic Research*, vol.7, No.4, pp. 486-493, ISSN 0736-0266

Ando, W.; Tateishi, K.; Hart, D.A.; Katakai, D.; Tanaka, Y.; Takata, K.; Hashimoto, J.; Fugie, H.; Shino, K.; Yoshikawa, H. & Nakamura, N. (2007). Cartilage repair using an in vitro generated scaffold-free tissue-engineered construct derived from porcine synovial mesenchymal stem cells. *Biomaterials,* Vol.28, No.36, (December), pp. 5462-5470, *ISSN* 0142-9612

Augello, A.; Kurth, T.B. & DeBari, C. (2010). Mesenchymal stem cells: a perspective from in vitro cultures to in vivo migration and niches. *European Cells and Materials*, Vol.20, (September), pp. 121-133, ISSN 1473-2262

Berglund, M.; Reno, C.; Hart, D.A. & Wiig, M. (2006). Patterns of mRNA expression for matrix molecules and growth factors in flexor tendon injury: Differences in the regulation between tendon and tendon sheath. *The Journal of Hand Surgery*, Vol.31A, No.8, (October), pp. 1279-1287, *ISSN* 1753-1934

Cole, J.; Brand, J.C. Jr; Caborn, D.N. & Johnson, D.L. (2000). Radiographic analysis of femoral tunnel position in anterior cruciate ligament reconstruction. *American Journal of Knee Surgery*, Vol.13, No.4, (April), pp. 218-222, *ISSN* 0899-7403

Chanda, D.; Kumar, S. & Ponnazhagan, S. (2010). Therapeutic potential of adult bone marrow-derived mesenchymal stem cells in diseases of the skeleton. *Journal of Cellular biochemistry*, Vol.111, No.2, (october), pp. 249-257, ISSN 0730-2312

Dargel, J.; Gotter, M.; Mader, K.; Pennig, D.; Koebke, J. & Schmidt–Wiethoff, R. (2007). Biomechanics of the anterior cruciate ligament and implications for surgical reconstruction. *Strategies in Trauma and Limb Reconstruction,* Vol.2, No.1, (April), pp. 1–12, *ISSN* 1828-8936

England, M.; Guermazi, A. & Lohmander, S.L. (2009). The role of the meniscus in knee osteoarthritis: a cause or a consequence. *Radiologic Clinics of North America*, Vol.47, No.4, (July), pp. 703-712, *ISSN* 0033-8389

Fineberg, M.S.; Zarins, B. & Sherman, O.H. (2000). Practical considerations in anterior cruciate ligament replacement surgery. *Arthroscopy*, Vol.16, No.7, (October), pp.715-724, ISSN 0749-8063

Fong, E.L.; Chan, C.K. & Goodman, S.B. (2011). Stem cell homing in musculoskeletal injury. *Biomaterials*, Vol.32, No.2, (January), pp. 395-409, *ISSN* 0142-9612

Frank, C.B. & Jackson, D.W. (1997). The science of reconstruction of the anterior cruciate ligament. *Journal of Bone and Joint Surgery*, Vol.79, No.10, (October), pp. 1556-1576, ISSN 00219355

Frobell, R.B.; Roos, E.M.; Roos, H.P.; Ranstam, J. & Lohmander, L.S. (2010). A randomized trial of treatment for acute anterior cruciate ligament tears. *The New England Journal of Medicine*. Vol.363, Vol.4, (July), pp. 331-342, *ISSN* 0028-4793

Functional Tissue Engineering Conference Group. (2008). Evaluation Criteria for Musculoskeletal and Craniofacial Tissue Engineering Constructs: A conference Report. *Tissue Engineering (part A)*, Vol.12, No.4, (December), pp. 2089-2104, *ISSN* 1937-3368

Fukubayashi, T.; Torzilli, P.A.; Sherman, M.F. & Warren, R.F. (1982). An in vitro biomechanical evaluation of anterior-posterior motion of the knee. Tibial displacement, rotation, and torque. *The Journal of Bone Joint Surgery (American volume)*, Vol.64, No.2, (February), pp. 258–264, ISSN 0021-9355

Goulet, F.; Rancourt, D.; Cloutier, R.; Germain, L.; Poole, A.R. & Auger, F.A. (2000). Tendons and ligaments. *In Principles of Tissue Engineering* (2nd Ed.).(R. Lanza, R. Langer and J.Vacanti, Eds). Academic Press Ltd, San Diego, pp. 711-722, ISBN 978-0-12-370615-7

Goulet, F.; Rancourt, D.; Cloutier, R.; Tremblay, J.; Bouchard, M.; Stevens, L.-M.; Labrosse, J.; Dupuis, D.; Lamontagne, J. & McKee, M. (2004). Implantation of bioengineered anterior cruciate ligament substitutes: histological, ultrastructural and biomechanical analyses. *Applied Bionics and Biomechanics*, Vol.1, No.2, pp. 115-121, ISSN 1176-2322

Hart, D.A.; Shrive, N.G. & Goulet, F. (2005). Tissue engineering of ACL replacements. *Sports Medicine and Arthroscopy Review*, Vol.13, No.3, (September), pp. 170-176, *ISSN* 1062-8592

Heard, B.J.; Achari, Y.; Chung, M.; Shrive, N.G. & Frank, C.B. (2011). Early joint tissue changes are correlated with a set of inflammatory and degradative synovial biomarkers at ACL autograft and its sham surgery in an ovine model. *Journal of Orthopaedic Research*, (March 8), [E-pub ahead of print], ISSN 0736-0266

Kaneko, D.; Sasazaki, Y.; Kikuchi, T.; Ono, T.; Nemoto, K.; Matsumoto, H. & Toyama, Y. (2009). Temporal Effects of Cyclic Stretching on Distribution and Gene Expression of Integrin and Cytoskeleton by Ligament Fibroblasts In Vitro. *Connective Tissue Research*, Vol.50, No.4, pp. 263–269, ISSN (printed): 0300-8207, ISSN (electronic) 1607-8438

Kobayashi, K.; Healey, R.M.; Sah, R.L.; Clark, J.J.; Tu, B.P.; Goomer, R.S.; Akeson, W.H.; Moriya, H. & Amiel, D. (2000). Novel method for the quantitative assessment of cell migration: a study on the motility of rabbit anterior cruciate (ACL) and medial collateral ligament (MCL) cells. *Tissue Engineering*, Vol.6, No.1, (February), pp. 29-38, *ISSN* 2152-4847

Lohmander, L.S.; Englund, P.M.; Dahl, L.L. & Roos, E.M. (2007). The Long-term Consequence of Anterior Cruciate Ligament and Meniscus Injuries: Osteoarthritis. *The American Journal of Sports Medicine*, Vol.35, No.10, (October), pp. 1756-1769, ISSN 0363-5465

McGonagle, D. & Jones, E. (2008). A potential role for synovial fluid mesenchymal stem cells in ligament regeneration. *Rheumatology*, Vol.47, No.8, (August), pp. 1114-1116, ISSN 1462-0324

Myer, G.D.; Ford, K.R. & Hewett, T.E. (2004). Rationale and clinical techniques for anterior cruciate ligament injury prevention among female athletes. *Journal of Athletic Training*, Vol.39, No.4, (December), pp. 352-364, ISSN 1062-6050

Murray, M.M. (2009). Current status and potential for primary ACL repair. *Clinics in Sports Medicine*, Vol.28, No.1, (January), pp. 51–61, ISSN 0278-5919

Murray M.M., Spindler KP, Devin C, Snyder BS, Muller J, Takahashi M, Ballard P, Nanney LB, Zurakowski D. (2006). Use of a collagen-platelet rich plasma scaffold to stimulate healing of a central defect in the canine ACL. *Journal of Orthopaedic Research*, Vol.24, No.4, (April), pp. 820-830, ISSN: 0736-0266

Murray, M.M.; Spindler, K.P.; Abreu, E.; Muller, J.A.; Nedder, A.; Kelly, M.; Frino, J.; Zurakowski, D.; Valenza, M.; Snyder, B.D. & Connolly, S.A. (2007). Collagen-platelet rich plasma hydrogel enhances primary repair of the porcine anterior cruciate ligament. *Journal of Orthopaedic Research*, Vol.25, No.1, (January), pp. 81-91, ISSN: 0736-0266

Neuman, P.; Kostogiannis, I.; Fridén, T.; Roos, H.; Dahlberg, L.E. & Englund, M. (2009). Patellofemoral osteoarthritis 15 years after anterior cruciate ligament injury - a prospective cohort study. *Osteoarthritis and Cartilage*, Vol.17, No.3, (March), pp. 284-290, ISSN 1063-4584

Oiestad, B.E.; Holm, I.; Aune, A.K.; Gunderson, R.; Myklebust, G.; Engbretsen, L.; Fosdahl, M.A. & Risberg, M.A. (2010) Knee function and prevalence of knee osteoarthritis after anterior cruciate ligament reconstruction: a prospective study with 10 to 15 years of follow-up. *The American Journal of Sports Medicine*, Vol.38, No.11, (July), pp. 2201-2210, ISSN 0363-5465

Paxton, J.Z.; Donnelly, K.; Keatch, R.P. & Baar, K. (2009). Engineering the bone-ligament interface using polyethylene glycol diacrylate incorporated with hydroxyapatite. *Tissue Engineering (Part A)*, Vol.15, No.6, (June), pp. 1201-1209, *ISSN* 1937-3368

Peng, H. & Huard, J. (2003). Stem cells in the treatment of muscle and connective tissue diseases. *Current Opinion in Pharmacology*, Vol.3, No.3, (June), pp. 329-333, ISSN 1471-4892

Petersen, W. & Zantop, T. (2007). Anatomy of the anterior cruciate ligament with regard to its two bundles. *Clinical Orthopaedics and Related Research*, Vol.454, (January), pp. 35-47, ISSN 0009-921X

Robayo, L.M.; Moulin, V.; Tremblay, P.; Cloutier, R.; Lamontagne, J.; Larkin, A.-M.; Chabaud, S.; Simon, F.; Islam, N. & Goulet, F. (2011). New ligament healing model based on tissue-engineered collagen scaffolds. *Wound Repair and Regeneration*, Vol.19, No.1, (January), pp. 38-48, ISSN 1067-1927

Rosc, D.; Powierza, W.; Zastawna, E.; Drewniak, W.; Michalski, A. & Kotschy, M. (2002). Post-traumatic plasminogenesis in intraarticular exudate in the knee joint. *Medical Science Monitor*, Vol.8, No.5, (May), pp. CR371–CR378, ISSN 1234-1010

Sandmann, G.H. & Tischer, T. (2010). Tissue engineering of the anterior cruciate ligament and meniscus using acellularized scaffolds. In *Tissue engineering* (D. Eberli, Ed.). pp. 437-458, InTech, ISBN 978-953-307-079-7

Shimomura, K.; Ando, W.; Tateishi, K.; Nansai, R.; Fujie, H.; Hart, D.A.; Kohda, H.; Kita, K.; Kanamoto, T.; Mae, T.; Nakata, K.; Shino, K.; Yoshikawa, H. & Nakamura, N. (2010). The influence of skeletal maturity on allogenic synovial mesenchymal stem cell-based repair of cartilage in a large animal model. *Biomaterials*, Vol.31, No.31, (November), pp. 8004-8011, *ISSN* 0142-9612

Spindler, K.P.; Murray, M.M.; Carey, J.L.; Zurakowski, D. & Fleming, B.C. (2009). The Use of Platelets to Affect Functional Healing of an Anterior Cruciate Ligament (ACL) Autograft in a Caprine ACL Reconstruction Model. *The Journal of Orthopaedic Research*, Vol.27, No.5, (May), 631–638, ISSN 0736-0266

Steiner, M.E.; Murray, M.M. & Rodeo, S.A. (2008). Strategies to improve anterior cruciate ligament healing and graft placement. *The American Journal of Sports Medicine*, Vol.36, No.1, (January), pp. 176-189, ISSN 0363-5465

Tateishi,K.; Ando, W., Higuchi, C.; Hart, D.A.; Hashimoto, J.; Nakata, K.; Yoshikawa, H. & Nakamura N. (2008). Comparison of human serum with fetal bovine serum for expansion and differentiation of human synovial MSC: Potential feasibility for clinical applications. *Cell Transplantation*, Vol.17, No., (5), (January), pp. 549-557, ISSN 0963-6897

Tischer, T.; Ronga, M.; Tsai, A.; Ingham, S.J.M.; Ekdahl, M.; Smolinski, P. & Fu, F.H. (2009). Biomechanics of the goat three bundle anterior cruciate ligament. *Knee Surgery, Sports Traumatology, Arthroscopy*, Vol.17, No.8, (August), pp. 935-940, ISSN 0942-2056

Tremblay, P.; Cloutier, R.; Lamontagne, J.; Belzil, A.-M.; Larkin, A.-M.; Chouinard, L.; Chabaud, S.; Laverty, S.; Lussier, B. & Goulet, F. (2011). Potential of skin fibroblasts for application to anterior cruciate ligament tissue engineering. *Cell Transplantation*, Vol.20, No.X, pp. XX-XX (In press), ISSN 0963-6897

Wilson TW, Zafuta MP, Zopitz M: (1999). A biomechanical analysis of matched bone-patellar-tendon-bone and double-looped semitendinosus and gracilis tendon grafts. *The American Journal of Sports Medicine*, Vol.27, No.2, (Mars-April), pp. 202-207, ISSN 0363-5465

von Porat, A.; Roos, E.M. & Roos, H. (2004). High prevalence of osteoarthritis 14 years after an anterior cruciate ligament tear in male soccer players: a study of radiographic and patient relevant outcomes. *Annals of the Rheumatic Diseases*, Vol.63, No.3, (March), pp. 269-273, ISSN 0003-4967

Woo, S.L.; Vogrin, T.M. & Abramowitch, S.D. (2000). Healing and repair of ligament injuries in the knee. *Journal of the American Academy of Orthopaedic Surgeons*, Vol.8, No.6, (November-December), pp. 364–372, ISSN 1067-151X

Woo, S. L.-Y.; Wu, C.; Dede, O.; Vercillo, F. & Noorani, S. (2006). Biomechanics and anterior cruciate ligament reconstruction. *Journal of Orthopaedic Surgery and Research*, Vol.1, (September 25), pp. 2, ISSN 1749-799X

Woo, S.L.-Y. (2009). Tissue engineering: use of scaffolds for ligament and tendon healing and regeneration. *Knee Surgery Sports Traumatology Arthroscopy*, Vol.17, No.6, pp. 559–560, ISSN 0942-2056

Wu, J.L.; Seon, J.K.; Gadikota, H.R.; Hosseini, A.; Sutton, K.M.; Gill, T.J. & Li, G. (2010). In situ forces in the anteromedial and posterolateral bundles of the anterior cruciate ligament under simulated functional loading conditions. *American Journal of Sports Medicine*, Vol.38, No.3, (March), pp. 558-563, ISSN 0363-5465

Xerogeanes, J.W.; Fox, R.J.; Takeda, Y.; Kim, H.S.; Ishibashi, Y.; Carlin, G.J. & Woo, S.L. (1998). A functional comparison of animal anterior cruciate ligament models to the human anterior cruciate ligament. *Annals of Biomedical Engineering*, Vol.26, No.3, (May-June), pp. 345-352, ISSN 0090-6964

# Part 4

## Cartilage

# 10

# Joint Cartilage Tissue Engineering and Pre-Clinical Safety and Efficacy Testing

Thomas G. Koch[1,2], Lorenzo Moroni[3],
Younes Leysi-Derilou[4] and Lise C. Berg[5]
[1]*Department of Biomedical Sciences, Ontario Veterinary College, University of Guelph, Guelph,*
[2]*Orthopaedic Research Lab, Aarhus University Hospital, Aarhus C,*
[3]*Tissue Regeneration Department, MIRA institute for Biomedical Technology and Technical Medicine, University of Twente, Enschede,*
[4]*Department of Biomedical Sciences, University of Guelph, Guelph, ON*
[5]*Department of Basic Animal and Veterinary Sciences, Faculty of Life Sciences, University of Copenhagen, Frederiksberg C*
[1,4]*Canada*
[2,5]*Denmark*
[3]*The Netherlands*

## 1. Introduction

### 1.1 Cartilage injuries, the triad and quartet of tissue engineering

Presently, focal cartilage injuries in humans are treated by abrasion arthroplasty with or without subchondral bone microfracture, autologous chondrocyte or mesenchymal stem cells (MSC) implantation and osteochondral plugs used in a sequential fashion pending severity and duration of the problem (Figure 1) (Williams et al., 2010; Steinert et al., 2007). Current strategies in human medicine for treatment of diffuse joint degeneration rely on replacement of the whole degenerated joint with inert implants. Excellent treatment outcome has been achieved for up to 15 years or more, but approximately 20% of treated patients require revision procedures after this time (Steadman et al., 2001). For younger patients this current state-of-the-art may translate to two or more revision surgeries during their lifetime. A biological solution to repair damaged cartilage that would provide life-long pain relief would be a major medical achievement.

The tissue engineering triad refers to the use of cells, scaffolds and cytokines to engineer tissues in vitro or in vivo. Such engineered tissues can potentially be utilized for tissue replacement strategies, for pharmacological screening of agents for therapeutic or toxic effects, or to gain insight into tissue developmental processes. Cartilage tissue, engineered using this triad of components often exhibit hyaline cartilage morphology, but the tissue has inferior mechanical properties when compared to native joint cartilage (Grad et al., 2011; Schulz and Bader, 2007). A fourth component of tissue engineering, namely mechanical stimulation has been added to the classical triad in order to better replicate the in vivo environment of joint cartilage (Grad et al., 2011; Schulz and Bader, 2007). Using this "quartet" of tissue engineering (cells, scaffolds, cytokines, and mechanical stimuli), cartilage with better mechanical properties has been produced. The mechanical properties of cartilage produced using mechanical stimulation may or may not be better than static culture depending on the cell source, timing of mechanical stimulation, method and duration of mechanical stimulation, and other variables. It is clear that each component of this tissue engineering "quartet" should be studied in detail since different cells respond differently to different cytokines, scaffold composition and topography as well as mode and timing of mechanical stimuli. This underlines the complexity of tissue engineering where each of these areas is a separate research field.

Here, we will briefly discuss the cells used for cartilage tissue engineering followed by more detailed discussion of selected topics on scaffolds, cytokines and means of mechanical stimuli. Pre-clinical animal models of cartilage repair and future perspectives follow these tissue-engineering considerations.

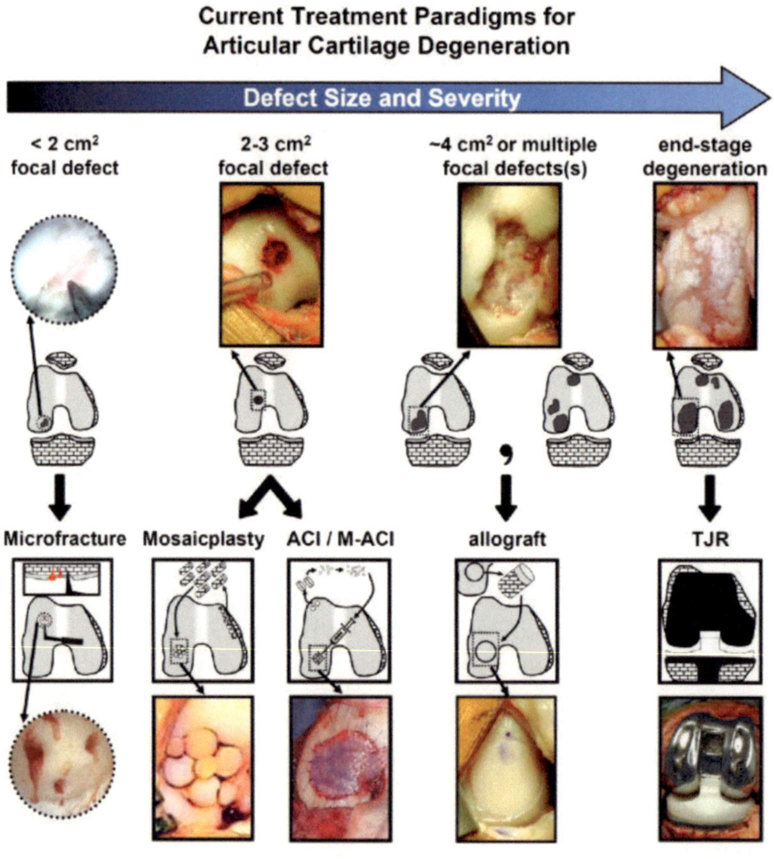


Fig. 1. Current articular cartilage treatment algorithm. ACI = autologous chondrocyte implantation, MACI = matrix-assisted autologous chondrocyte implantation, TJR = total joint replacement. Reprinted with permission from Williams et al. (2010)

## 2. Cells of cartilage tissue engineering

Articular hyaline cartilage is a very specialized tissue characterized by low cellularity, extensive extracellular matrix, lack of vascular, lymphatic, and nervous supply, and insufficient number of resident progenitor cells needed for mounting an appropriate regenerative response to injury. The result is minimal intrinsic repair capacity.

The obvious choice of cell for cartilage repair is the chondrocyte from hyaline cartilage since it is normally responsible for the production and maintenance of joint cartilage. However, a number of factors have limited the use or therapeutic success of chondrocytes. Chondrocytes only constitute about 5% of joint cartilage and insufficient chondrocyte numbers are therefore often retrieved for immediate therapeutic use, which in turn necessitate ex vivo cell expansion prior to re-implantation. Chondrocytes unfortunately tend to dedifferentiation towards the fibroblast cell lineage when expanded in culture making them less suitable for transplantation (Schulze-Tanzil, 2009). Another limitation to the use of autologous chondrocytes is the need for two surgical procedures, weeks apart to harvest and later implant the cells, which adds time, cost, anesthetic risk and risks of donor site morbidity (Schulze-Tanzil, 2009; Williams et al., 2010). Interestingly, osteoarthritic chondrocytes seem to perform equally well in vitro when compared to chondrocytes from healthy joints (Dehne et al., 2009; Stoop et al., 2007).

Stem cells or cells with chondrogenic potential from various tissue sources have been investigated for cartilage repair. Adult cells such as bone marrow multipotent stromal cells (BM-MSCs) are attractive due to the decreased cost and risks associated with collection compared to autologous chondrocytes. Concerns raised with the use of BM-MSCs include potentially a decreased yield and differentiation potential of MSCs from the bone marrow with increasing age making the technique age-dependent. This is paradoxical to the fact that clinical cases often occur in aged individuals (Wilson et al., 2010; Kasper et al., 2009). Adult cells with chondrogenic potential have been isolated from a large number of tissues and have been reviewed elsewhere (Solchaga et al., 2011; Hildner et al., 2011; O'Sullivan et al., 2011). Embryonic and recently, induced pluripotent stem cells (iPS cells), may also be valuable cell sources for cartilage repair (Hiramatsu et al., 2011; Medvedev et al., 2010; Waese and Stanford, 2011; Fecek et al., 2008). Neonatal stem cells with chondrogenic potential have been isolated from fresh umbilical cord blood of humans as well as horses (Berg et al., 2009; Koch et al., 2007; Santourlidis et al., 2011; Zhang et al., 2011). Human cord blood derived stem cells have showed increased proliferation capacity and broader differentiation potential compared to stem cells from bone marrow and adipose tissue (Zhang et al., 2011; Kogler et al., 2004). Also, cord blood MSCs may be more chondrogenic than bone marrow MSCs (Berg et al., 2009; Zhang et al., 2011).

So far a reliable cell source and method of isolating cells with high chondrogenic potential has not been reported. From a tissue-engineering point of view this is a major limitation at the moment, since it precludes development of predictable and reproducible protocols for cartilage production.

## 3. Biomaterials & scaffold fabrication technologies

Joint cartilage regeneration can be achieved by two strategies, namely cell- and scaffold-based therapies. Cell therapy has already reached the clinics in the form of autologous chondrocyte implantation (ACI) or matrix-assisted ACI (MACI). However, this regenerating technique appears to assist only partially in the repair process as it leads to the formation of mechanically inferior fibrocartilage compared to native joint cartilage. Although clinical outcome is better five years after surgery when compared to the baseline, recent comparative studies showed that ACI and MACI were not significantly better than marrow-stimulating techniques (e.g. microfractures) or reconstructive techniques (e.g.

mosaicarthroplasty) using autografts, allografts or synthetic material (Ebert et al., 2011; Vasiliadis et al., 2010). The influence of scaffold composition and surface topography on cell function and differentiation is being increasingly recognized (Rosso et al., 2005; Milner and Siedlecki, 2007; Thakar et al., 2008; Patel et al., 2010). Improvement to the MACI technique and other scaffold-based approaches may therefore be made through precise engineering of three-dimensional (3D) porous biomaterials - 3D scaffolds - that promote cell lodging, migration, and differentiation while providing mechanical support during tissue repair.

Different biomaterials are used to fabricate scaffolds, predominantly natural and synthetic polymers. In general two classes of polymers can be distinguished: (i) hydrogels formulated as cell carriers for minimally invasive surgeries; (ii) solid polymers designed for optimal mechanical stability. Hereafter, these biomaterials, their physicochemical and mechanical properties, and the correspondent fabrication technologies implemented to make 3D scaffolds will be discussed.

### 3.1 Hydrogels

A number of general properties should be possessed by hydrogels. Obviously, these materials need to elicit an appropriate host response and display satisfactory biocompatibility. If a specific hydrogel composition would be associated with prolonged inflammation, the resulting immune response toward the encapsulated cells might affect the success of the implantation. The gelation mechanism is also important. Typically hydrogels are formed by ionic or covalent cross-linking. Ionic cross-links are very dynamic and may be formed and disrupted in presence of a multivalent ionic fluid environment, which is like physiological fluids. This may hamper the control over the degradation properties of the resulting hydrogels. Conversely, covalent cross-links are more stable and confer enhanced mechanical and physical properties to the hydrogels. Yet, they are often more toxic and more difficult to break than ionic cross-links. This implies that the cross-linking yield should always be maximum and different routes for hydrogel degradation should be envisioned in the design of covalent cross-linking biodegradable hydrogels (Lee and Mooney, 2001). Alternative methods to form cross-links have been developed by exploiting the phase transition characteristics of polymers. Temperature sensitive hydrogels have been synthesized for tissue engineering applications. These polymers are subjected to a solid to gel phase transition at a specific temperature, called lower critical solution temperature, which can be tailored to be at body temperature (Fedorovich et al., 2009; Vihola et al., 2005). Alternatively, stereocomplexation of amphiphilic polymers has also been used as a strategy to form physical cross-linked hydrogels. These hydrogels display customized physical and mechanical properties depending on the polymer molecular weight and relative concentration of the amphiphilic blocks (Hiemstra et al., 2005; Hiemstra et al., 2006b). Finally, mechanical and degradation properties are important to consider when choosing a hydrogel for a specific tissue. Ideally, a hydrogel should possess a similar stiffness to that of the targeted tissue and a degradation rate matching the speed of tissue formation.

*Natural hydrogels.* Among natural hydrogels, alginate, chitosan, hyaluronic acid and its derivatives, and collagen have been widely investigated for cartilage repair. Alginate is derived from brown algae and is composed of linear block copolymers of mannuronic (M) and guluronic (G) acids. It has been widely used in drug delivery and tissue engineering applications because of its biocompatibility, relative low cost, abundance, and easy gelation in presence of divalent cations (Shapiro and Cohen, 1997; Terada et al., 2005). Ionic cross-links are very dynamic and can lead to unpredictable and uncontrollable dissolution of the gel in a physiological environment. Therefore covalent cross-links were introduced and were shown to improve the mechanical properties and the control over the degradation rate of these hydrogels. Alginate is a rather inert biomaterial. In order to enhance cell-material interactions, alginate hydrogels have been modified with different biological factors to promote cell adhesion or differentiation into specific tissue phenotypes (Hao et al., 2007; Hsiong et al., 2008). An important issue is that alginate degradation products often are larger than the threshold size for renal clearance. Although previous approaches revealed new opportunities to control the degradation of alginates (Lee et al., 2000), more efforts are needed to translate these gels into the clinics.

Chitosan is a polysaccharide forming the exoskeleton of many seashells. As it is mainly comprised of glucosamine, a component of cartilage extra-cellular matrix (ECM), chitosan is an attractive biomaterial for regenerative therapies of the skeletal system (Chenite et al., 2000; Madihally and Matthew, 1999). Its positive charge allows complexation with other negatively charged ECM proteins present in articular cartilage like glycosaminoglycans. This would induce *in vivo* sequestration of growth factors embedded in the surrounding cartilaginous ECM. Chitosan can be processed in the form of a gel exploiting its pH-dependent solubility, drawn into solid fibers, or foamed by freeze-drying (Chenite et al., 2000; Yamane et al., 2005). This processing versatility enables the fabrication of scaffolds with fine-tunable mechanical properties. Although the intrinsic batch-to-batch variability during chitosan extraction is still a major drawback, its physical, mechanical, and biological properties are promising for articular cartilage regeneration.

Hyaluronic acid is a highly hydrophilic proteoglycan present in the extra-cellular matrix (ECM) of several tissues. In cartilage, it contributes to maintain homeostasis and physical integrity thanks to its viscosity, capacity to retain water, and interactions with chondrocytes and other ECM proteins. When used as a biomaterial, hyaluronic acid can be covalently cross-linked or esterified to improve its mechanical properties and physical stability. In the latter case, hylans are formed. These biomaterials are currently used in the clinics as hydrogels or membranes to assist in autologous chondrocyte implantation (Brun et al., 1999; Campoccia et al., 1998; Hollander et al., 2006). Yet, the less-than-optimal mechanical properties and the potential presence of endotoxins and impurities are drawbacks that still need to be improved to consider these biomaterials a fully viable alternative to synthetic polymers. Furthermore, hyaluronic acid is present in high concentrations in the surrounding tissues of malignant tumors (Knudson et al., 1989). This evidence should be further studied to ensure that no risks are associated with the use of hyaluronic acids or its derivatives as biomaterials.

Collagen is one of the extra cellular matrix protein most used to form natural hydrogels. It is easy to obtain from different sources as it is the most common protein present in numerous tissues and organs. Jellification typically occurs through thermally reversible physical cross-links. These gels can be easily remodeled through production of collagenases by encapsulated cells, thus offering the opportunity to study cell-matrix interactions during tissue development (Mueller et al., 1999; Pachence, 1996; Nehrer et al., 1997). Collagen gels can also be formed in the presence of other proteins such as chondroitin sulphate or hyaluronic acid, so that the resulting semi-interpenetrating network can be functionalized to display biological moieties of interest in cartilage regeneration. Gelatin, a derivative of collagen formed by breaking its natural triple-helix structure, has also been used as a hydrogel for cartilage regeneration (Choi et al., 1999). The main drawbacks of collagen and its derivatives lye in the potential immunogenic risks due to its origin, batch variation, and poor mechanical properties. Chemical cross-linking strategies by exposure to glutaraldehyde or genipin confer enhanced physical stability and mechanical properties to the resulting hydrogels, however without reaching similar stiffness to that of articular cartilage.

*Synthetic hydrogels.* Among synthetic hydrogels, a number of derivatives based on poly(ethylene glycol) (PEG) have been developed and used for cartilage tissue engineering. This biomaterial is relatively inert, highly hydrophilic, and very versatile for the functionalization of its backbone with different biological moieties. PEG is known to have anti-fouling properties and is often used in vascular applications as a coating to prevent thrombogenesis. PEG can be synthesized either as a linear, branched, or star-shaped polymer. In its star-shaped form, different peptides or growth factors can be coupled depending on the number of arms. These biological factors can be covalently linked to the PEG chains through passive or active linkers comprised of an enzymatically sensitive peptide sequence, thus being able to release the payload depending on cellular activity (Adelow et al., 2008; Lutolf et al., 2003). PEG or its oxidized version poly ethylene oxide (PEO) can also be grafted to other polymers such as poly propylene oxide (PPO) or poly lactic acid (PLA). PEO-PPO-PEO is a tri-block copolymer, also commercially known as pluronic®, that jellifies through a temperature sensitive phase change. These copolymers have been conventionally used as drug delivery vehicles and recently also explored in tissue engineering applications for skeletal regeneration (Fedorovich et al., 2009; Park et al., 2009; Batrakova and Kabanov, 2008). Yet, PEO-PPO copolymers are not degradable limiting their potential use. PEG-PLA copolymers have been synthesized for tissue engineering applications. Gelation occurs via stereocomplexation of L- and D-lactic acid. By changing the molecular weight of the PLA and PEG blocks, it is possible to vary the mechanical properties of the resulting gels. Proteins and other biological factors can be easily incorporated during gelation in these degradable hydrogels, conferring them a high versatility and potential for clinical applications (Hiemstra et al., 2006a; Hiemstra et al., 2006c). An alternative derivative of PEG is synthesized by linking acrylate groups to the main chain. PEG-diacrylate (PEGDA) is a photopolymerizable hydrogel that cross-links in presence of an initiator under UV light. PEGDA has been extensively used in cartilage tissue engineering as an inert encapsulating system or after modification with different peptides, growth factors, or in semi-interpenetrating networks with ECM proteins (Hwang et al., 2006; Lee et al., 2006; Sharma et al., 2007). It is a reproducible hydrogel system with fine-tunable physicochemical and mechanical properties that enable tissue regeneration without the potential risks associated to natural polymers, thus making it a promising candidate for clinical use.

### 3.2 Solid polymers

Within solid polymers, poly lactic acid (PLA), poly glycolic acid (PGA) and copolymers (PLGA) have been broadly used in tissue engineering as well as for cartilage regeneration (Anderson and Langone, 1999; Babensee et al., 2000; Chu et al., 1997; Freed et al., 1993; Honda et al., 2000; Sarazin et al., 2004). These biomaterials are approved by the food and drug administration (FDA) as they activate a minimal or mild foreign body reaction, and as such are considered biocompatible. The mechanical properties and degradation rate can be tailored by varying the molecular weight and copolymer ratio. They have already been studied for drug delivery (Jang and Shea, 2003; Uhrich et al., 1999; Richardson et al., 2001; Nof and Shea, 2002; Sengupta et al., 2005) and are suitable for tissue engineering applications, as the degradation products (lactic and glycolic acids) obtained due to hydrolysis are normally present in the metabolic pathway and can be naturally eliminated by the body. However, their bulk degradation may be associated with the formation and accumulation of large amounts of degradation products in a short time frame (months vs. years) that cannot be easily discarded, resulting in local inflammation in tissues (Bostman et al., 1989) and enzymatic hydrolysis (Fu et al., 2000). Another polyester commonly used in tissue engineering is poly ε-caprolactone (PCL). This polymer is characterized by good biocompatibility and mechanical properties. It degrades at a much slower rate than PLA, PGA, and PLGA, which makes it attractive when long-term implants and controlled release applications are desired (Honda et al., 2000; Wang, 1989; Hutmacher et al., 2001; Choi and Park, 2002). It has been also shown that PCL can selectively adsorb vitronectin, a protein that is known to facilitate stem cell chondrogenic differentiation. Conversely, PLA selectively adsorb fibronectin and seems to be better suited to induce stem cell osteogenic differentiation. A different family of thermoplastic polymers that has been investigated for tissue engineering is poly ethylene oxide terephthalate-co-poly butylene terephtalate (PEOT/PBT). These polyether-ester copolymers possess good physical properties like elasticity, toughness and strength (Bezemer et al., 1999). By varying the molecular weight of the starting PEG segments and the weight ratio of PEOT and PBT blocks, it is possible to tailor physico-chemical and mechanical properties (Bezemer et al., 1999; van Dijkhuizen-Radersma et al., 2002; Deschamps et al., 2002; Olde Riekerink et al., 2003; Woodfield et al., 2004; Moroni et al., 2006b), degradation rate (Deschamps et al., 2002), and protein adsorption (Mahmood et al., 2004). PEOT/PBT copolymers have been demonstrated to be biocompatible both in vitro and in vivo for skin, cartilage, and bone regeneration (Malda et al., 2004; van Blitterswijk et al., 1993; Bakker et al., 1988; Beumer et al., 1994a; Beumer et al., 1994b). A further modulation in degradation rate and drug release profile can be achieved by substituting part or all of the terephtalate domains with succinate blocks during the copolymerization reaction (van Dijkhuizen-Radersma et al., 2003; van Dijkhuizen-Radersma et al., 2004; van Dijkhuizen-Radersma et al., 2005). PLA, PGA, PLGA copolymers, PCL, and PEOT/PBT copolymers have proven to be interesting biomaterials to fabricate 3D scaffolds. Although their properties can be customized for specific purposes, some concerns over their degradation mechanism and rate still remain.

To obviate bulk degradation, surface eroding polymers have been developed such as polyortho-esters (POEs) (Choi and Heller, 1978), polyphospho-esters (PPEs) (Wang et al., 2001a; Wang et al., 2001b), and polyanhydrides (PAs) (Leong et al., 1986). Surface erosion is a degradation mechanism (Andriano et al., 1999; Burkoth et al., 2000), which affects the stability of the scaffolds to a lesser extent and elicits a lower in vivo inflammatory response, as compared to polyesters and polyether-esters previously considered. PPEs display adequate mechanical properties also for hard tissue engineering. Although PPAs, PAs, and POEs have been used in some cases for hard tissues repair, they might be more suitable for soft tissue engineering due to their generally low mechanical properties. More recently, PLA based polymers modified with photosensible chemical groups like fumarates or acrylates have been developed for biomedical applications (Melchels et al., 2009; Melchels et al., 2006). These biomaterials offer the advantage of being processed by rapid prototyping technologies with high resolution, thus enabling the fabrication of sophisticated scaffold geometries. Yet, some issues may still arise from remnant toxicity due to the acrylic groups.

### 3.3 Scaffold fabrication technologies

A plethora of fabrication technologies have been developed and characterized to fabricate three-dimensional (3D) scaffolds for tissue engineering applications (Figure 2). Although these techniques can also be applied to hydrogels, they have been mostly used with solid polymers due to the intrinsic advantage of the former as a minimally invasive injectable biomaterial. Foams and textiles are the two predominant types of scaffolds used in tissue engineering. Foams can be fabricated by gas foaming, freeze drying, or porogen leaching (Barry et al., 2004; Sproule et al., 2004; Schoof et al., 2001; Ma and Zhang, 2001; Claase et al., 2003; Sarazin and Favis, 2003). Textile scaffolds can be produced by wet or melt spinning, creating fibers that are randomly deposited on top of each other, woven, or knitted (Cima et al., 1991; Freed et al., 1994; Niklason and Langer, 1997). Once deposited, thermal or chemical treatments can be applied to improve fiber bonding, thus enhancing structural stability and mechanical properties (Kim and Mooney, 1998a; Kim and Mooney, 1998b; Mikos et al., 1993). These methods are relatively easy to implement, but offer a limited control over mechanical properties, interconnectivity of of pore network, pore size and shape, and porosity. It has recently been shown that scaffolds mimicking the structural and architectural characteristics of the targeted tissue support enhanced tissue formation, and fabrication technologies that enable a fine control over scaffold pore network and strut size are needed. In the specific case of articular cartilage, three different zones can be distinguished: (i) deep, (ii) middle, and (iii) surface zones. In each of these regions, the alignment of collagen type II is distinct and results in the overall arch-like architecture typical of articular cartilage. Furthermore, chondrocytes behave in a different manner and are responsible for the production of different extra-cellular matrix proteins, namely different proteoglycans, depending on the zone where they are located (Klein et al., 2009; Klein et al., 2003; Schuurman et al., 2009). This has lead researchers to focus their attention on technologies that enable the fabrication of layer-by-layer scaffolds. This would enable not only a better control of different cartilage zones, but also the construction of more complex tissue such as osteochondral grafts. A simple and effective solution is to cross-link monolithic hydrogel regions containing chondrocytes of the different regions or other cells on top of each other (Elisseeff et al., 2002; Lee et al., 2008; Hwang et al., 2007).

Where a better control over cell and tissue spatial distribution is desired, rapid prototyping (RP) technologies offer an appealing solution. RP is based on computer aided design (CAD) and computer aided manufacturing (CAM) to build porous 3D scaffolds in a layer-by-layer controlled manner. An extensive number of biomaterials can be processed by these techniques in a custom-made shape (Hutmacher, 2001; Yang et al., 2002; Yeong et al., 2004), with tailored mechanical properties to the specific application considered (Hollister, 2005; Taboas et al., 2003; Lin et al., 2004). The outcomes are 3D scaffolds that possess fine-tunable porosity, pore size and shape, and have a completely interconnected pore network, which permits a better cell migration and nutrient perfusion than 3D scaffolds fabricated with more conventional techniques like foaming or spinning (Malda et al., 2004; Sachlos and Czernuszka, 2003). In addition, the fabrication of personalized scaffolds can be envisioned with the acquisition and processing of computer tomography (CT) and/or magnetic resonant imaging (MRI) anatomical data from patient datasets (Hollister, 2005; Moroni et al.,

2007a). Since RP is based on layer-by-layer processing, it is also theoretically possible to change the pore network structural and architectural characteristics in space in order to better mimic specific ECM and cell spatial arrangement. In practice, this has not been extensively explored. Moroni et al. (Moroni et al., 2006a) have been active in studying how mechanical properties can be optimized with different pore networks for the regeneration of different types of cartilage. Similarly, Woodfield et al. (2005) produced scaffolds with pores varying along the longitudinal axis in the attempt to mimic the spatial distribution of chondrocytes in articular cartilage. Oh et al. (2010) used a similar approach and fabricated scaffolds with different pore size ranges by a more conventional freeze-drying method. In these scaffolds, chondrogenic differentiation in adipose derived mesenchymal stem cells was better supported with pores in the 370-400 μm range. The resulting scaffolds showed enhanced tissue formation, while cell and ECM distribution resembled more closely that of native hyaline cartilage. By combining different materials, it was also possible to fabricate osteochondral constructs that functionally supported both bone and cartilage regeneration (Moroni et al., 2008; Sherwood et al., 2002).

From these studies it is clear that 3D scaffolds fabricated by RP offer many advantages and a broad flexibility as a model to study new strategies for tissue engineering. However, they do not possess biologically active properties to improve communication with the adhered or encapsulated cells. This can be achieved by two means: (i) implementing drug delivery vehicles into the scaffolds, (ii) improving the fabrication resolution to achieve true synthetic ECM substitutes. In the former case, biological factors have been encapsulated or covalently linked into hydrogels (Ehrbar et al., 2007; Saha et al., 2007; Kong and Mooney, 2007) or incorporated into microspheres and fibers (Martins et al., 2010; Richardson et al., 2001), which were directly used for tissue engineering applications (see Delivery Methods section). Controlled drug delivery strategies showed to improve cartilage regeneration when combined with 3D scaffolds. However, local and spatial control over the release of biological factor is still lacking and might contribute to better mimic the native tissue architecture during regeneration. In the latter case, 3D scaffolds with feature dimensions in the range of ECM have been recently fabricated by two-photon polymerization (2PP). This technology exploits *in situ* polymerization of photosensible polymers at specific wavelengths (Ovsianikov et al., 2011). With 2PP, a number of rather complex structures can be fabricated with nanometric resolution (Ovsianikov et al., 2010). These can be interesting tools to study fundamental cell-material interaction at a single cell level, but the fabrication time window might be too long when clinically relevant scaffold dimensions are needed. An alternative technique that has been extensively used to mimic the ECM of tissues is electrospinning (ESP). Here, fiber meshes are fabricated by spinning a biomaterial solution into an electric field that destabilizes the solution flow and form a continuous jet of fibers collected on a target plate. The fiber deposition architecture can be modified depending on the geometry and electric properties of the collector plate (Zhang and Chang, 2008; Zhang and Chang, 2007). ESP has been used with a large number of biomaterials (Li and Xia, 2004) and also enables the incorporation of growth factors, proteins, or cells during fabrication (Li et al., 2006a; Patel et al., 2008). The fabricated scaffolds possess physical and surface properties that have already been shown to support cartilage tissue engineering (Li et al., 2006b; Li et al., 2005). These meshes may suffer from a lack of cell penetration due to the high fiber density and small pore size. However, spinning of more biomaterials with different degradation rate can obviate this. Virtually, water-soluble biomaterials would allow a fast degradation and a better infiltration of cells into the scaffold pores.

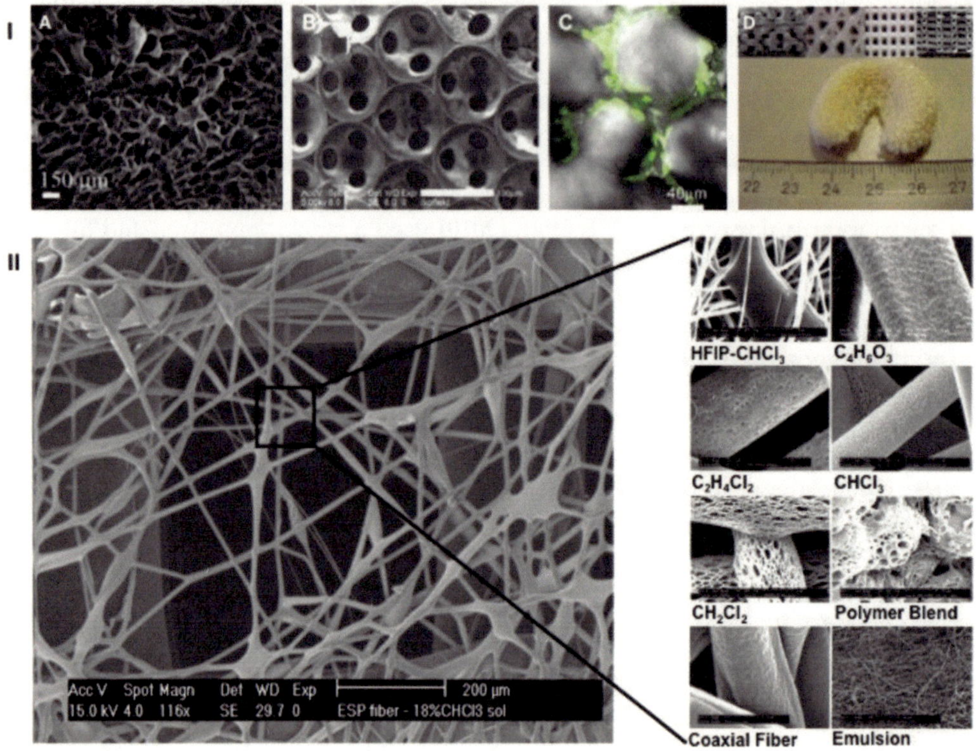


Fig. 2. Scaffold fabrication technologies. Different scaffold fabrication technologies can be used to create 3D porous biomaterials for tissue engineering applications. (I) Conventional foams can be obtained by (I-A) salt leaching or (I-B, I-C) by inverted colloidal crystal (ICC) template. In both case a porogen material is used to form a defined volume with the selected biomaterial and later removed through selective dissolution. In case of ICC, the resulting pore network is improved in terms of interconnectivity and (C) cell seeding efficiency (cells green fluorescent). (I-D) Textile scaffolds can be fabricated by rapid prototyping technologies. (I-D) Here, a meniscus shaped scaffold is shown. Insets display different fiber deposition methods, which affect the formed pore size and shapes. Not only solid polymers, but also hydrogels can be processed (no picture shown but available at Landers et al (2002) (Landers et al., 2002). (II) ECM mimicking meshes can be fabricated by electrospinning. (III-A) The electrospun fibers typically have a smooth surface, but (II-B) depending on the solvent used can also display different surface morphologies. Panel (I-A, I-B, I-C) modified by Deschamps et al (2002) and Nichols et al. (2009). Panel (I-D) modified by Moroni et al (2007). Panel (II) modified by Moroni et al. (2008)

## 4. Cytokines release

Cartilage engineering is not only a result of cells and scaffolds coming together to form a 'cartilage-like' structure. As mentioned earlier there are two relevant cell choices – chondrocytes and MSCs. Chondrocytes need to be in an environment resembling the physiological properties of cartilage to maintain their phenotype. MSCs will need to go through the process of chondrogenesis. Chondrogenesis is an intricate process where MSCs go through stages of differentiation to become fully matured chondrocytes (Figure 3). Only when the cells have committed to the chondrogenic fate will they start to lay down cartilage-type extracellular matrix (ECM). The differentiation process involves a range of stage specific molecules that include phenotypic determinants, adhesion molecules, and signalling molecules as described in great detail by Chen et al. (2009).

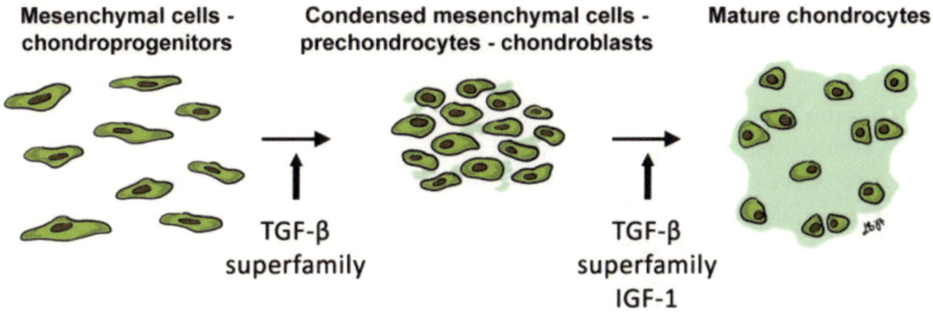


Fig. 3. Chondrogenic differentiation. Chondrogenesis is an intricate process where MSCs go through stages of differentiation to become fully matured chondrocytes. Members of the TGF-β superfamily and IGF-1 are examples of important regulators in the differentiation process. Figure by L. C. Berg

When attempting to produce tissue that resembles true cartilage both chondrocytes and MSCs will need the right molecular signals. Differentiation can be promoted by adding chondrogenic cytokines and growth factors to the construct. Some of the major players in the intricate network of chondrogenic cell signalling are well described in the literature, but other members are only known in very little detail or not at all. There is no question we still have much to learn, before we fully understand the temporal and spatial importance of the cytokines and growth factors involved in cartilage synthesis. However, studies have shown that even a simpler version with addition of only a few key molecules will strongly increase the quantity and quality of cartilage-like tissue produced.

In the following sections, we will present some of these key factors currently being used in cartilage engineering, and the delivery methods available.

### 4.1 Growth factors

#### 4.1.1 Transforming growth factor-β superfamily

The transforming growth factor-β (TGF-β) superfamily includes the TGF-β and the bone morphogenetic (BMP) proteins. The most well described family members in relation to cartilage tissue engineering are TGF-$\beta_1$, TGF-$\beta_3$, BMP-2 and BMP-7 (also known as osteogenic protein-1 (OP-1)), but other members may prove to be equally important.

TGF- $\beta_1$, TGF-$\beta_2$, and TGF-$\beta_3$ have all been shown to be chondrogenic in vitro on MSCs (Kulyk et al., 1989; Blunk et al., 2002). They are believed to be involved in the early stages of the differentiation process (Chen et al., 2009). In vitro they are known to stimulate synthesis of ECM including collagens, fibronectin (Ignotz and Massague, 1986), and proteoglycans (Chen et al., 1987), and decrease the expression of collagen type 1 in MSCs (Kurth et al., 2007). Studies on human MSCs have shown that TGF-$\beta_2$ and TGF-$\beta_3$ are more chondrogenic than TGF-$\beta_1$ (Barry et al., 2001; Chen et al., 2009).

Unfortunately, the very promising results with TGF-β from in vitro studies have not been reproduced in animal models. In mice and rabbits intra-articular administration of TGF-β has been shown to have a number of side effects including inflammation and osteophyte formation (Bakker et al., 2001; van Beuningen et al., 2000). These side effects appear to be linked to the non-cartilaginous tissues of the joint (Blaney Davidson et al., 2007a; Scharstuhl et al., 2002), and can potentially be prevented if the growth factor only comes into contact with the cartilage.

The BMPs involved in chondrogenesis include BMP-2, -3, -4, -5, -6, -7, and -9. They support chondrocyte phenotype and stimulate synthesis of proteoglycans (Li et al., 2003; Sailor et al., 1996). BMP-2, and -7 have been shown to be the more chondrogenic of the BMPs in MSCs in vitro (Majumdar et al., 2001; Sekiya et al., 2005; Fan et al., 2010), and both have shown promising results in animal models (Badlani et al., 2009; Blaney Davidson et al., 2007b). The chondrogenic effect of BMP-7 is synergistically increased when used in combination with TGF-β (Xu et al., 2006; Kim and Im, 2009) or insulin-like growth factor (IGF)-1 (Loeser et al., 2003; Chubinskaya et al., 2007a). BMP-2 is also a strong inducer of osetogenesis and is used clinically for bone fusion (Tang et al., 2011; Hamilton et al., 2011), which needs to be taken into consideration if using BMP-2 for cartilage generation. Only BMP-2 and BMP-7 are currently approved for clinical use in human patients (Haleem and Chu, 2010).

In addition to the TGF-βs and BMPs, this large family of molecules also includes activin and growth differentiation factor 5 (GDF5) that have been shown to be chondrogenic inducers in MSCs (Jiang et al., 1993; Bai et al., 2004).

#### 4.1.2 Insulin-like growth factor-1

IGF-1 is primarily synthesized in the liver regulated by growth hormone (GH). The responsiveness of cells to IGF-1 decreases with age and osteoarthritis (Boehm et al., 2007; Loeser et al., 2002). IGF-1 is capable of inducing chondrogenesis in MSCs on its own, but it is much more effective in combination with other growth factors e.g. TGF-β (Longobardi et al., 2006) and BMP-7 (Loeser et al., 2003; Chubinskaya et al., 2007b).

### 4.2 Delivery methods

Most growth factors and cytokines have short half-lives (Nimni, 1997), and their effect would therefore be limited to the time right after implantation of the construct. Subsequent supplementary treatments entail several risks depending on the choice of route of administration.

Systemic delivery presents a number of potential problems. Since the amount of active reagent actually reaching the site of interest will be a fraction of the initial dose, the administered dosage will have to be up regulated. This will increase both the cost of treatment and the risk of adverse reactions. In route to the target site the growth factors and cytokines will come into contact with a number of other tissues, where they may cause undesirable effects. Local injection into the construct provides a more controlled administration, but repeated joint injections are not desirable due to the risk of infection and added cost to the patient.

Because of these issues, a number of delivery systems have been developed and tested, where growth factors and cytokines are released into the local environment of the construct in a time and dose controlled manner. Some of these delivery systems are closely linked to the scaffold materials, and may not work well with all types of scaffolds. Other systems are linked to the cells. A requirement for the successful delivery system is that the active molecules are protected against degradation until time of release. The systems range from very basic soaking of the scaffold in growth factor suspension (Kanematsu et al., 2004) to highly sophisticated release systems, where attempts are being made to closely mimic the stage specific differentiation process or even create dual tissues by releasing different molecules at controlled time points, dosages, and locations in the scaffold (Wang et al., 2009; Suciati et al., 2006). These highly complex systems are still in their infancy.

#### 4.2.1 Direct attachment or incorporation

The simplest version of delivery system is a direct attachment of growth factor to the surface of the scaffold material by soaking the scaffold in growth factor suspension (Kanematsu et al., 2004), or incorporation of the growth factors and cytokines into biodegradable scaffolds, usually hydrogels (Yamamoto et al., 2001; Nelson et al., 2011). The factors are then released passively from the surface of the scaffold or as the scaffold degrades. Current applications of hydrogels primarily include drug delivery, since the scaffolds are too soft to play the structural role needed in cartilage constructs (Woodfield et al., 2002).

*Molecule bound.* Another way of incorporating growth factors and cytokines in scaffolds is to bind them to intermediary components in the construct. This method may prolong the effect of the growth factors by slowing down the release process. A study in rabbits using MSCs, elastic block copolymer scaffolds and TGF-$\beta_3$, showed that chondroitin sulphate-bound TGF-$\beta_3$ had a slower release profile than TGF-β directly incorporated into the scaffold (Park et al., 2010). Similar results have been achieved by binding TGF-$\beta_3$ (Park et al., 2008) or bFGF (Jeon et al., 2006) to heparin.

*Loaded structures.* The cytokines can also be delivered to the constructs in loaded structures. Microspheres (Kim et al., 2003; Elisseeff et al., 2001; Fan et al., 2007; Fan et al., 2004), liposomes (Hunziker et al., 2001), and micro sponges (Fan et al., 2010) loaded with growth factors have all been tested in studies in cartilage tissue engineering. Their advantage is a more controlled release rate, while the growth factors are kept relatively protected from their surroundings thus preserving their activity.

*Gene therapy.* A different approach to delivery of growth factors and cytokines important to chondrogenesis is to make the MSCs themselves produce the factors necessary to the cartilage construct. This can be achieved by using gene therapy techniques. The use of gene therapy in cartilage tissue engineering has been thoroughly reviewed by Steinert et al. (2008). Here, we will simply provide a short introduction to the concept. There are several different methods available ranging from a simple direct delivery of genetic material at the defect site to complex procedures implemented as part of the cartilage engineering process. No matter which method is used, the most important factors in gene therapy are related to how well the gene material is transferred to the target cell, and how efficient the now transgenic cells are at producing the desired molecules.

The gene therapy techniques fall into two categories - in vivo and ex vivo techniques. Common to them is the need for good vectors. These can be non-viral or viral. The non-viral vectors are safer but less efficient, while the viral vectors are more efficient but pose a potential safety risk especially if they are injected into the patient (Steinert et al., 2008).

In vivo delivery is the cheaper option, where vectors harbouring the gene material are introduced to the cells directly at the site of injury. This method is simple and fast, but it is difficult to control the efficacy of transfer as well as the safety of the procedure. Especially if viral vectors are used there is a risk of the vectors inserting themselves into the genetic material of host cells in the area. The in vivo method is particularly useful in tissues where it is not possible to remove cells for ex vivo transfection. Its use in studies on cartilage repair has been very limited. The ex vivo method is more time consuming, expensive, and technically challenging. The genetic material is transferred to the cells, before the cells are used in the patient. Using this method makes it possible to test transfection rate, and the risks associated with use of viral vectors are eliminated. Ex vivo gene transfer has been tested in a number of studies involving MSCs and cartilage tissue engineering. MSCs transfected with BMP-7 yielded better cartilage repair than non-treated control cells (Mason et al., 2000). Similarly, gene induced expression of TGF-β1 and BMP-2 promoted chondrogenesis in MSCs, while induced expression of IGF-1 did not (Palmer et al., 2005). A subsequent study from the same group showed that the use of combinations of those three chondrogenic genes had a strong synergistic effect on chondrogenesis (Steinert et al., 2009).

## 5. Mechanical stimuli and bioreactors

During daily activities, the cartilage is exposed to direct compression, hydrostatic pressure, or shear. It has been suggested that mechanical loading may increase extracellular matrix (ECM) synthesis during cartilage engineering (Portner et al., 2009). The hypothesis that mechanical stimulation enhances cartilage formation is based on studies of developmental biology where restriction of joint loading after birth leads to poor post-natal cartilage adaptation (Williamson et al., 2003a; Williamson et al., 2001; Williamson et al., 2003b; Mikic et al., 2004; Mikic et al., 2000). Overall, mechanical stimulation leads to increased cell expansion as well as increased extracellular matrix proteins production compared to regular static cultures (Portner et al., 2009).

A bioreactor is defined as any device in which a biological/biochemical process is performed under controlled conditions. When compared to static cultures, bioreactors offer a number of advantages such as uniformed distribution and increased mass transfer, control of pH, temperature, gas supply ($O_2$ and $CO_2$), nutrients, waste product removal, and the opportunity to incorporate mechanical stimuli. Detailed reviews on the general concepts of bioreactors for tissue engineering and the application of bioreactors for the purpose of cartilage engineering are available (Godara et al., 2008; Chen et al., 2006; Chung and Burdick, 2008; Portner et al., 2009; Haasper et al., 2008; Huang et al., 2010b; Schulz and Bader, 2007; Concaro et al., 2009; Grad et al., 2011). Selected physical outcome parameters used in cartilage engineering in defined and described in Text Box 1.

Bioreactors for cellular therapeutic use have being grouped into two categories. Bioreactors for tissue engineering and bioreactors for cell infusion therapies. However, the simpler systems used for infusion therapies have also been utilized for tissue engineering purposes due to ease and cost. Bioreactors for isolation and expansion of cells for cell infusion therapies are largely similar to or adapted from pharmaceutical monoclonal antibody production systems or industrial yeast-based methods. Spinner flasks and rotating wall vessels are examples of bioreactors for cell expansion (Concaro et al., 2009). In spinner flasks a magnetic stir bar moves the medium. The media movement provides the cell-scaffold constructs with nutrients and oxygen and facilitates waste removal by overcoming the normal diffusion limit of 100-200 μm. In this system a balance has to be struck between homogenous mass transfer including uniform pH gradient and shear gradients that can cause cell damage. Stirring at 50 rpm is a common starting point (Concaro et al., 2009). If the speed of the rotating outer wall of rotating wall vessels is calibrated to the mass of the cell carrier constructs, then a microgravity environment is created where the cells are exposed to low shear stress and high mass transfer. However, hitting this "soft spot" of equilibrium can be challenging and failure hereof may lead to shear stress and constructs colliding with the walls, which adversely affect cell function (Concaro et al., 2009). However, rotating wall vessels have been utilized to study microgravity's effect on cartilage tissue engineering both on Earth and in space (Vunjak-Novakovic et al., 2002; Marolt et al., 2006). Marked differences where noted between cartilage engineered on Earth and in space, on Mir, confirming that physical forces modulate musculoskeletal tissue such as cartilage (Vunjak-Novakovic et al., 2002). Bioreactors for cartilage engineering have largely been used to evaluate response to compression, but bioreactors applying electrical fields, ultrasound, centrifugal forces, shear forces, perfusion of 3D constructs, tension of cell layers, hydrostatic pressure, and hydrostatic pressure with perfusion have also been reported (Schulz and Bader, 2007). These bioreactors are mostly custom made, but an increasing number of commercial bioreactors are becoming available for various purposes (Yeatts and Fisher, 2011). Comparison between research groups is therefore exceedingly difficult. One major concern with many of the studies is the validation of the bioreactor prior to use and the continued calibration. Thorough evaluation and validation of the bioreactors prior to and continuously through their use by applying objective measurable parameters is critical in order to evaluate the results and conclusions. Many of the bioreactors can only stimulate the tissue and have limited, if any, possibilities of sampling and analyzing the tissue or culture medium without terminating the culture process.

Most cartilage bioreactor studies have reported work using uniaxial direct compression (Grad et al., 2011). A general starting point for direct compression studies of cells adhered to a scaffold is 10% or 15% compression at a frequency of 1Hz (Grad et al., 2011; Terraciano et al., 2007; Mouw et al., 2007; Huang et al., 2004). Recently, mechanical stimulation showed improved cartilage formation of porcine chondrocyte cartilage constructs compared to static culture, but no difference was noted between perfusion and perfusion-compression constructs (Tran et al., 2011). In the compression group, constructs where cultured with perfusion alone at a flow rate of 0.5ml/min for the first week followed by 1 Hz sinusoidal unconfined compression, 4 hours a day, 5 days a week, starting with a load of 0.5 N until 20 N by the third week (Tran et al., 2011). Perfusion was maintained in this compression group for the 4-week duration of the study and compared to a control group of perfusion only. Biochemistry revealed a higher glycosaminoglycan (GAG) content, but a lower collagen content in the bioreactor construct compared to native cartilage. The discrepancy between GAG and collagen content could be due to enzymatic collagen degradation or simply reflect immature cartilage since rabbit studies have shown that the cartilage collagen network does not mature until 3 months of age (Julkunen et al., 2009; Riesle et al., 1998). Rabbit joint cartilage at birth is homogenous with collagen fiber alignment parallel to the surface, but as the cartilage matures the collagen fibers organize into the three zones seen in adult cartilage (Julkunen et al., 2011). However, the specific cartilage organization and biochemical content differ from joint to joint and even between opposing joint surfaces of the same joint (Julkunen et al., 2011). Recently, bovine bone marrow (BM) derived MSCs exposed to long-term dynamic compression in chondrogenic culture were shown to exert improved mechanical properties compared to static culture as previously reported in studies of chondrocyte cultures (Huang et al., 2010a). Interestingly immediate mechanical stimulation of cartilage constructs appear detrimental and a more physiological approach of initial stem cell chondrogenic differentiation using TGF-β without mechanical stimulation, as in utero, followed by mechanical stimulation, as post-natal, appear more effective (Terraciano et al., 2007; Mouw et al., 2007; Thorpe et al., 2008). These observations would concur with the observations of normal rabbit joint cartilage development as described above, where the cartilage at birth is a homogenous structure that remodels post-natally in response to loading.

Hydrostatic pressure in cartilage tissue engineering has been reviewed elsewhere (Elder and Athanasiou, 2009). Hydrostatic pressure does not exert any measurable strain on the cell and the mechanistic effect are believed to be mediated through deformation of transmembrane ion transport proteins leading to increased intracellular calcium concentrations (Schulz and Bader, 2007; Kornblatt and Kornblatt, 2002; Elder and Athanasiou, 2009). The effect of hydrostatic pressure is incompletely understood, but both dynamic and static pressure has been investigated in both normal chondrocytes, osteoarthritic chondrocytes and as a mean for chondrogenic cell differentiation (Elder and Athanasiou, 2009). Physiological hydrostatic pressure of 5-10 Mpa are generally anabolic when applied in a dynamic fashion to both 2D and 3D chondrocyte layers, but if static pressure is applied anabolic effect is only noted in 3D layers (Elder and Athanasiou, 2009). Such physiological loads can also promote stem cell differentiation, chondrocyte redifferentiation, and exert chondroprotective effects and osteoarthritic chondrocytes (Elder and Athanasiou, 2009). Super-physiological pressures (20-50 Mpa) are particular detrimental to cell metabolism if applied for more than 2 hours (Elder and Athanasiou, 2009). Schulz and Bader have reviewed the mechanism of hydrostatic pressure stimulation and why prolonged loading leads to cartilage damaged (Schulz and Bader, 2007). Interestingly, compression likely acts through hydrostatic pressure as well. The reason is that negatively charged proteoglycans provide frictional resistance by binding water and preventing the water from being squeezed out of the tissue. The net effect is increased hydrostatic pressure within the tissue. However, the water does shift out of the tissue if the pressure persists. In this case, the load is increasingly carried by the collagen fibers until the fibers are orientated parallel to the load direction. Load beyond this point leads to tearing of the collagen network.

One trend in cartilage bioreactor design is towards a more tribological approach, where the science and technology of interacting surfaces in relative motion is used to develop multi-axial bioreactors that more closely mimic the movement of a natural joint (Figure 4) (Grad et al., 2011). Multiaxial compression and shearing stimulation represent more closely in vivo conditions and are associated with broader chondrogenic gene expression profiles compared to uniaxial compressive loading (Grad et al., 2011). Another trend is the ability of bioreactors to simultaneously provide mechanical stimuli and evaluate the physical properties of the developing tissue (Portner et al., 2009). The ability to monitor the construct as it develops allows for a reduction in sample size since temporal changes can be assessed on the same sample. It may even be possible to perform online temporal molecular evaluation of the cells, the conditioned culture medium and extracellular matrix composition and responses in closed bioreactor system through the application of high-resolution magic angle spinning nuclear magnetic resonance (HR-MAS NMR), matrix-assisted laser desorption and ionization time-of-flight mass spectrometry (MALDI-TOF MS), combinatory phase sensitive scanning acoustic microscopy (PSAM), and confocal laser scanning microscopy (Schulz and Bader, 2007). Such advanced bioreactors can be valuable tools in preconditioning cell-scaffolds prior to clinical research use as well as model systems for in vitro evaluation of cell response to a variety of physical as well as chemical stimuli. These systems may also have value as model systems to reduce the number of animals used for in vivo studies by allowing replacement and refinement prior to in vivo studies. However, their utility for commercialization of tissue replacement strategies may be hampered by high production costs and violation of GMP due to the often semi-sterile productions systems.

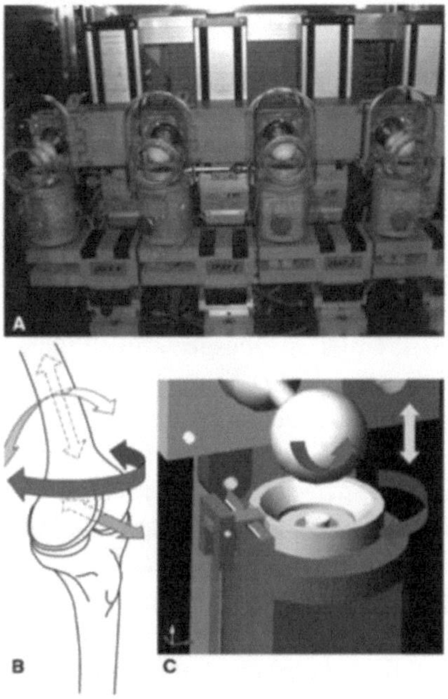


Fig. 4. Advanced bioreactor. (A) Advanced tribological bioreactor capable of multiaxial stimulation of tissue-engineered constructs. Each chamber can be individually controlled to simulate the movements of complex joints, e.g. the knee (b and C) Reprinted with permission from Grad et al. (2011)

Soft lithography produced microfluidic bioreactors have been advocated as the next generation of tissue engineering bioreactors to gain mechanistic insights due to affordability, flexibility and precision (Godara et al., 2008). Mesenchymal stem cells proliferation, motility and osteogenic differentiation in response to various culture regimes was studied using time-lapse imaging of 96 culture chambers within a single microfluidic chip (lab-on-a-chip) (Gomez-Sjoberg et al., 2007). However, one of the challenges is absorption of small hydrophobic molecules, typically hormones, by the plastic used for producing the microchips (Gomez-Sjoberg et al., 2010). Such methods may lend themselves well to studying the issue of MSC population heterogeneity, incomplete MSC chondrogenic programming, evaluation of novel chondrogenic inducing small molecules and signaling pathways involved in chondrogenic differentiation as identified elsewhere (Huang et al., 2010c). A potential limitation of such micro-bioreactors is that they mainly evaluate 2D structures and cartilage tissue tends to prefer a 3D environment.

As outlined, sophisticated bioreactors are currently available in a few labs for the generation and real-time analysis of cartilage constructs. However, the mechanical properties of the produced cartilage remain significantly inferior to that of native cartilage although histological, biochemical and gene expression assays may indicate otherwise. A major drawback of most publications to date is the omission of a relevant positive biological control sample such as native cartilage. Most studies compare the test sample to a negative control sample consisting of the empty cell scaffold/carrier. Such study designs make it difficult to determine the biological relevance/utility of the engineered tissue. This being said, it is probably unknown when an engineered tissue is "good enough" for clinical implantation. This will likely depend on the constructs used as well as the patient age, lesion location and severity, and the accompanying rehabilitation program. Mature native tissue of the tissue of interest is likely the best biological yardstick presently, but the degree of tissue maturation preferable at the time of transplantation remains undetermined.

As alluded to already, another major challenge is the development of aseptic, cost effective, bioreactor systems that will fulfill current regulatory requirements. A significant segment of the bioreactor industry is dedicated to the development of disposable systems as reviewed elsewhere (D'Aquino, 2006; Eibl and Eibl, 2009; Singh, 1999). These systems include wave-mixed, orbital shaken and stirred bioreactors (Eibl et al., 2010). These systems may provide cost-effective solutions due to savings on operative utility, cleaning and validation costs as well as reduced water and cleaning agent consumption for cleaning. Some of the challenges are real-time analysis of the process due to range limitations of disposable sensors, pre-validation that would shift the regulatory burden from the end-user to the manufacturer, incompatibilities with certain chemicals and temperatures, and limited accessories such as valves and sampling systems (Eibl et al., 2010; D'Aquino, 2006). These systems have not yet been significantly evaluated for cartilage engineering.

De novo tissue engineering has also been investigated, where the so-called cell niche is relied upon to direct transplanted cells or tissue pieces towards the appropriate tissue has also been investigated (Grad et al., 2011; Shah et al., 2010; Stevens et al., 2005). This approach is often autologous in nature and associated with minimal handling and laboratory exposure of the cells. ACI and MACI are examples of this strategy. However, the resulting repair tissue remains inferior to native cartilage. Recently, self-assembling nanofibers were shown to promote cartilage repair of full thickness chondral defects in rabbits (Shah et al., 2010). Alternatively, tissue can be made in vivo at a site distant to the ultimate repair site for later relocation to the injury site. Such de novo tissue formation was elegantly demonstrated for bone regeneration in a rabbit model (Stevens et al., 2005). Bone formation was induced by alginate injection under the tibial periosteum and later the neo-bone was removed and transplanted into an induced cortical bone defect where it promoted bone healing. Cartilage tissue was also generated using this model by adding molecules to the gel that inhibited angiogenesis and promoted chondrogenesis.

The influence of oxygen tension during cell and tissue culture has also been investigated in relation to cartilage engineering and is an area, which appears to deserve further investigation. Low oxygen levels have been reported to be a more potent promoter of chondrogenesis than dynamic compression (Meyer et al., 2010). Previous studies have shown controversial findings in relation to the effect of oxygen tension on chondrogenic differentiation of mesenchymal stem cells with some reporting increased proliferation rates and chondrogenic potency and others reporting reduced proliferation and differentiation potency (Grayson et al., 2007; Merceron et al., 2010; Krinner et al., 2009; Holzwarth et al., 2010).

## 6. Pre-clinical animal models for safety and efficacy evaluation of engineered cartilage

Cartilage defects are "quality-of-life" lesions, as opposed to life-threatening conditions, for which therapies are available today that provide palliative relief for a large number of patients. Since long-term safety of new cell-based treatment modalities have yet to be determined it is possible that adverse effects could ultimately lead to worse quality of life than the initial cartilage problem caused. Thorough evaluation of efficacy and long-term safety is therefore prudent before introducing new cell-based therapies for such lesions. In vivo studies can generally be categorized into models where the animal is used as the bioreactor directly at the injury site or at a distant site for later relocation to the injury site, as discussed above, or as models were animals are used to evaluate in vitro engineered constructs.

Members of the Orthopaedic Trauma Association (OTA) recently discussed pre-clinical animal models for cartilage and bone repair (Lansdowne, 2010; Martineau, 2010). This topic has also been reviewed by a number of authors (Pearce et al., 2007; Reinholz et al., 2004; Chu et al., 2010; Koch and Betts, 2007). The consensus with regard to cartilage defects is that the research question determines the choice of model, since there is no ideal pre-clinical animal model. General issues to consider are the lesion model (chondral, osteochondral, degenerative), most appropriate model (cost, availability, joint structure, age), cartilage similarity (thickness, structure, cell density, biochemistry, biomechanics), and the nature of the lesion (area, depth, location) (Martineau, 2010). Small animal models are rodents and rabbits. Rodent models have limited translational value, but are cost effective models for mechanistic studies of chondrogenesis, generation of proof-of-concept data and bridging in vitro and large animal studies. Rabbits are easy to handle and cost effective, but have thin cartilage with excellent endogenous repair potential as well as highly flexible joints which support a relatively low body weight (low loading of the joints). Large animal models include minipigs, goats, and horses (Martineau, 2010). Mini-pigs do not comply with rehabilitation programs, but advantages comprise joint size sufficiently large to allow arthroscopy, partial or full-thickness chondral defects, growth plate closure, poor endogenous chondral repair potential and the possibility of making 6-8 mm diameter size lesions. Goats (as well as sheep) are similar to mini-pigs in that they do not comply with rehabilitation programs, but allow arthroscopic approaches and large defects can be made in cartilage that have limited intrinsic repair potential. Goats and sheep studies are more expensive due to increased housing and handling costs. Horses can comply with rehabilitation programs, allow for defect sizes similar to human defects, exert low intrinsic repair potential and arthroscopic treatment modalities and make follow-up assessment possible. The drawback of equine studies is high cost, high loading forces, and hard bone, and it is difficult to achieve protected weight bearing in the horse.

The OTA study group on bone defect models had the following universal considerations: consider the 3R's of animal use and reduce, refine and replace before choosing the most relevant animal model; perform a pilot study if inexperienced with the model; animal models do not account for co-morbidities (e.g. obesity, diabetes etc); ensure optimal animal care before and after surgery by consulting with veterinary specialists in surgery (www.acvs.org or www.ecvs.org), anesthesia (www.acva.org or www.ecva.eu.org), and laboratory animal medicine (www.aclam.org or www.eclam.org); ideally standardize your model with a negative and a positive control although in comparative studies comparison with the current "golden standard" treatment is sufficient (Lansdowne, 2010).

The argument for increased use of domestic animal species as pre-clinical animal models is based on the fact that many human conditions have a spontaneous counter-part in animals that would allow for more physiologically relevant studies compared to induced lesions in inbreed subpopulations of small laboratory animals. Veterinary medicine today often rivals human medicine with regard to diagnostic and treatment modalities, some companion animals largely live in the same epigenetic environment as humans, and the veterinary market is of significant monetary value in itself (Koch et al., 2009; Koch and Betts, 2007). Most recently, induced pluripotent stem cells (iPS cells) have been generated from equine cells for the purpose of treating sporting horses as well as utilizing the horse as a pre-clinical animal model of comparative human disorders (Nagy et al., 2011). Stem cells and animal therapies have recently been reviewed elsewhere (Figueroa et al., 2011). The sentiment from 2004 by Fiester and colleagues that successful treatment of a spontaneous animal disorder may significantly advance stem cell research remains valid today and could be expanded to include the field of regenerative medicine (Fiester et al., 2004).

## 7. Conclusion and future perspectives

The field of cartilage tissue engineering is a very diverse research field investigating the use of bioreactors, scaffold compositions and designs in combination with a wide range of cells and cytokines. It is still an emerging research field and careful pre-clinical assessment of the potential treatment modalities is advised to avoid long-term adverse effects (Figure 5). However, these treatment modalities hold the potential of providing life-long solutions to the currently incurable problem of joint cartilage damage. Domestic animal models may provide valuable translation significance between proof-of-principle studies in small laboratory animals and expensive human pre-clinical trials.

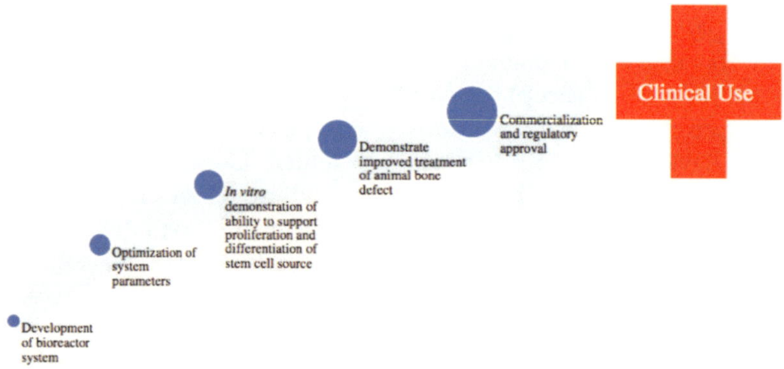


Fig. 5. Clinical roadmap for bone tissue engineering bioreactors. This roadmap is equally useful for cartilage engineering. In vivo proof-of-principle can be shown in laboratory animal models, but domestic animal models may be of more translation value prior to human clinical trials. Reprinted with permission from Yeatts and Fisher (2011)

**Text Box 1.** Selected physical outcome parameters used in cartilage engineering
Stress, strain and frequency are often referred to in tissue engineering so a brief description of these terms is warranted. Stress, often denoted as sigma ($\sigma$), is the amount of force acting over a given cross-sectional area. Stress is expressed as force per area units. The unit of stress is pascal (Pa) or newtons (N) per area in the metric system, and psi in the English system. Conversions of these different units are as follows: 1 psi = $lb_F/in^2$, 1 Pa = $1.45 \times 10^{-4}$ psi, 1 Pa = 1 $N/m^2$, 1 kPa = 1 where Newton (N) is the SI unit of force. 1 N = kg $m/s^2$, e.g. the net force required to accelerate a mass of one kilogram at a rate of one meter per second squared. Engineering strain, often denoted as $\varepsilon$, is the nominal change in length of a material ($\varepsilon = \Delta L/L_0 = (L - L_0)/L_0$). The unit of strain is often mm/mm, cm/cm, etc, or no unit at all since it is the ratio of a given measuring system. Frequency, denoted with the SI unit hertz (Hz), is the number of cycles per second, typically of a sine wave. Shear stress, $\tau$, as exerted on cells in perfusion systems with a laminar flow profile between the ingress and egress plates can be calculated as follows: $\tau = (6\mu Q)/(bh^2)$, where $\mu$ is the viscosity of the medium, $Q$ is the volumetric flow rate, $b$ is the width of flow channel, and $h$ is the distance between the two plates (Shiragami and Unno, 1994; Nagel-Heyer et al., 2005). The shear stress unit is force per area, e.g. $N/m^2$ or similar representation of force per area using psi, Pa and $in^2$ or $cm^2$. Other physical parameters often used in cartilage engineering such as Darcy's law, Reynolds number (Re), mass flow rate, flow velocity has been reviewed elsewhere (Concaro et al., 2009).

## 8. References


Adelow, C., T. Segura, J.A. Hubbell, and P. Frey. 2008. The effect of enzymatically degradable poly(ethylene glycol) hydrogels on smooth muscle cell phenotype. *Biomaterials*. 29:314-326.

Anderson, J.M., and J.J. Langone. 1999. Issues and perspectives on the biocompatibility and immunotoxicity evaluation of implanted controlled release systems. *J Control Release*. 57:107-113.

Andriano, K.P., Y. Tabata, Y. Ikada, and J. Heller. 1999. In vitro and in vivo comparison of bulk and surface hydrolysis in absorbable polymer scaffolds for tissue engineering. *J Biomed Mater Res*. 48:602-612.

Babensee, J.E., L.V. McIntire, and A.G. Mikos. 2000. Growth factor delivery for tissue engineering. *Pharm Res*. 17:497-504.

Badlani, N., Y. Oshima, R. Healey, R. Coutts, and D. Amiel. 2009. Use of bone morphogenic protein-7 as a treatment for osteoarthritis. *Clinical orthopaedics and related research*. 467:3221-3229.

Bai, X., Z. Xiao, Y. Pan, J. Hu, J. Pohl, J. Wen, and L. Li. 2004. Cartilage-derived morphogenetic protein-1 promotes the differentiation of mesenchymal stem cells into chondrocytes. *Biochemical and biophysical research communications*. 325:453-460.

Bakker, A.C., F.A. van de Loo, H.M. van Beuningen, P. Sime, P.L. van Lent, P.M. van der Kraan, C.D. Richards, and W.B. van den Berg. 2001. Overexpression of active TGF-beta-1 in the murine knee joint: evidence for synovial-layer-dependent chondro-osteophyte formation. *Osteoarthritis and cartilage / OARS, Osteoarthritis Research Society*. 9:128-136.

Bakker, D., C.A. van Blitterswijk, S.C. Hesseling, and J.J. Grote. 1988. Effect of implantation site on phagocyte/polymer interaction and fibrous capsule formation. *Biomaterials*. 9:14-23.

Barry, F., R.E. Boynton, B. Liu, and J.M. Murphy. 2001. Chondrogenic differentiation of mesenchymal stem cells from bone marrow: differentiation-dependent gene expression of matrix components. *Exp Cell Res*. 268:189-200.

Barry, J.J., H.S. Gidda, C.A. Scotchford, and S.M. Howdle. 2004. Porous methacrylate scaffolds: supercritical fluid fabrication and in vitro chondrocyte responses. *Biomaterials*. 25:3559-3568.

Batrakova, E.V., and A.V. Kabanov. 2008. Pluronic block copolymers: evolution of drug delivery concept from inert nanocarriers to biological response modifiers. *J Control Release*. 130:98-106.

Berg, L., T. Koch, T. Heerkens, K. Bessonov, P. Thomsen, and D. Betts. 2009. Chondrogenic potential of mesenchymal stromal cells derived from equine bone marrow and umbilical cord blood. *Vet Comp Orthop Traumatol*. 22:363-370.

Beumer, G.J., C.A. van Blitterswijk, and M. Ponec. 1994a. Biocompatibility of a biodegradable matrix used as a skin substitute: an in vivo evaluation. *J Biomed Mater Res*. 28:545-552.

Beumer, G.J., C.A. van Blitterswijk, and M. Ponec. 1994b. Degradative behaviour of polymeric matrices in (sub)dermal and muscle tissue of the rat: a quantitative study. *Biomaterials*. 15:551-559.

Bezemer, J.M., D.W. Grijpma, P.J. Dijkstra, C.A. van Blitterswijk, and J. Feijen. 1999. A controlled release system for proteins based on poly(ether ester) block-copolymers: polymer network characterization. *J Control Release*. 62:393-405.

Blaney Davidson, E.N., P.M. van der Kraan, and W.B. van den Berg. 2007a. TGF-beta and osteoarthritis. *Osteoarthritis and cartilage / OARS, Osteoarthritis Research Society*. 15:597-604.

Blaney Davidson, E.N., E.L. Vitters, P.L. van Lent, F.A. van de Loo, W.B. van den Berg, and P.M. van der Kraan. 2007b. Elevated extracellular matrix production and degradation upon bone morphogenetic protein-2 (BMP-2) stimulation point toward a role for BMP-2 in cartilage repair and remodeling. *Arthritis Res Ther*. 9:R102.

Blunk, T., A.L. Sieminski, K.J. Gooch, D.L. Courter, A.P. Hollander, A.M. Nahir, R. Langer, G. Vunjak-Novakovic, and L.E. Freed. 2002. Differential effects of growth factors on tissue-engineered cartilage. *Tissue Engineering*. 8:73-84.

Boehm, A.K., M. Seth, K.G. Mayr, and L.A. Fortier. 2007. Hsp90 mediates insulin-like growth factor 1 and interleukin-1beta signaling in an age-dependent manner in equine articular chondrocytes. *Arthritis and rheumatism*. 56:2335-2343.

Bostman, O., E. Hirvensalo, S. Vainionpaa, A. Makela, K. Vihtonen, P. Tormala, and P. Rokkanen. 1989. Ankle fractures treated using biodegradable internal fixation. *Clin Orthop Relat Res*:195-203.

Brun, P., R. Cortivo, B. Zavan, N. Vecchiato, and G. Abatangelo. 1999. In vitro reconstructed tissues on hyaluronan-based temporary scaffolding. *J Mater Sci Mater Med*. 10:683-688.

Burkoth, A.K., J. Burdick, and K.S. Anseth. 2000. Surface and bulk modifications to photocrosslinked polyanhydrides to control degradation behavior. *J Biomed Mater Res*. 51:352-359.

Campoccia, D., P. Doherty, M. Radice, P. Brun, G. Abatangelo, and D.F. Williams. 1998. Semisynthetic resorbable materials from hyaluronan esterification. *Biomaterials*. 19:2101-2127.

Chen, J., R.L. Horan, D. Bramono, J.E. Moreau, Y. Wang, L.R. Geuss, A.L. Collette, V. Volloch, and G.H. Altman. 2006. Monitoring mesenchymal stromal cell developmental stage to apply on-time mechanical stimulation for ligament tissue engineering. *Tissue Eng*. 12:3085-3095.

Chen, J.K., H. Hoshi, and W.L. McKeehan. 1987. Transforming growth factor type beta specifically stimulates synthesis of proteoglycan in human adult arterial smooth muscle cells. *Proceedings of the National Academy of Sciences of the United States of America*. 84:5287-5291.

Chen, W.H., M.T. Lai, A.T. Wu, C.C. Wu, J.G. Gelovani, C.T. Lin, S.C. Hung, W.T. Chiu, and W.P. Deng. 2009. In vitro stage-specific chondrogenesis of mesenchymal stem cells committed to chondrocytes. *Arthritis and rheumatism*. 60:450-459.

Chenite, A., C. Chaput, D. Wang, C. Combes, M.D. Buschmann, C.D. Hoemann, J.C. Leroux, B.L. Atkinson, F. Binette, and A. Selmani. 2000. Novel injectable neutral solutions of chitosan form biodegradable gels in situ. *Biomaterials*. 21:2155-2161.

Choi, N.S., and J. Heller. 1978. Drug Delivery devices manufactured from poly(orthoesters) and poly(orthocarbonates). *US Patent*. 4.093.709.

Choi, S.H., and T.G. Park. 2002. Synthesis and characterization of elastic PLGA/PCL/PLGA tri-block copolymers. *J Biomater Sci Polym Ed*. 13:1163-1173.

Choi, Y.S., S.R. Hong, Y.M. Lee, K.W. Song, M.H. Park, and Y.S. Nam. 1999. Studies on gelatin-containing artificial skin: II. Preparation and characterization of cross-linked gelatin-hyaluronate sponge. *J Biomed Mater Res*. 48:631-639.

Chu, C.R., J.S. Dounchis, M. Yoshioka, R.L. Sah, R.D. Coutts, and D. Amiel. 1997. Osteochondral repair using perichondrial cells. A 1-year study in rabbits. *Clin Orthop Relat Res*:220-229.

Chu, C.R., M. Szczodry, and S. Bruno. 2010. Animal models for cartilage regeneration and repair. *Tissue Eng Part B Rev*. 16:105-115.

Chubinskaya, S., A. Hakimiyan, C. Pacione, A. Yanke, L. Rappoport, T. Aigner, D.C. Rueger, and R.F. Loeser. 2007a. Synergistic effect of IGF-1 and OP-1 on matrix formation by normal and OA chondrocytes cultured in alginate beads. *Osteoarthritis and cartilage / OARS, Osteoarthritis Research Society*. 15:421-430.

Chubinskaya, S., M. Hurtig, and D.C. Rueger. 2007b. OP-1/BMP-7 in cartilage repair. *International orthopaedics*. 31:773-781.

Chung, C., and J.A. Burdick. 2008. Engineering cartilage tissue. *Adv Drug Deliv Rev*. 60:243-262.

Cima, L.G., J.P. Vacanti, C. Vacanti, D. Ingber, D. Mooney, and R. Langer. 1991. Tissue engineering by cell transplantation using degradable polymer substrates. *J Biomech Eng*. 113:143-151.

Claase, M.B., D.W. Grijpma, S.C. Mendes, J.D. De Bruijn, and J. Feijen. 2003. Porous PEOT/PBT scaffolds for bone tissue engineering: preparation, characterization, and in vitro bone marrow cell culturing. *J Biomed Mater Res A*. 64:291-300.

Concaro, S., F. Gustavson, and P. Gatenholm. 2009. Bioreactors for tissue engineering of cartilage. *Adv Biochem Eng Biotechnol*. 112:125-143.

D'Aquino, R. 2006. Bioprocessing Systems Go Disposable. *Chemical Engineering Progress*.

Dehne, T., C. Karlsson, J. Ringe, M. Sittinger, and A. Lindahl. 2009. Chondrogenic differentiation potential of osteoarthritic chondrocytes and their possible use in matrix-associated autologous chondrocyte transplantation. *Arthritis Res Ther*. 11:R133.

Deschamps, A.A., M.B. Claase, W.J. Sleijster, J.D. de Bruijn, D.W. Grijpma, and J. Feijen. 2002. Design of segmented poly(ether ester) materials and structures for the tissue engineering of bone. *J Control Release*. 78:175-186.

Ebert, J.R., W.B. Robertson, J. Woodhouse, M. Fallon, M.H. Zheng, T. Ackland, and D.J. Wood. 2011. Clinical and magnetic resonance imaging-based outcomes to 5 years after matrix-induced autologous chondrocyte implantation to address articular cartilage defects in the knee. *The American journal of sports medicine*. 39:753-763.

Ehrbar, M., S.C. Rizzi, R.G. Schoenmakers, B.S. Miguel, J.A. Hubbell, F.E. Weber, and M.P. Lutolf. 2007. Biomolecular hydrogels formed and degraded via site-specific enzymatic reactions. *Biomacromolecules*. 8:3000-3007.

Eibl, R., and D. Eibl. 2009. Application of Disposable Bag-Bioreactors in Tissue Engineering and for the Production of Therapeutic Agents. *Adv Biochem Eng Biotechnol*.

Eibl, R., S. Kaiser, R. Lombriser, and D. Eibl. 2010. Disposable bioreactors: the current state-of-the-art and recommended applications in biotechnology. *Appl Microbiol Biotechnol*. 86:41-49.

Elder, B.D., and K.A. Athanasiou. 2009. Hydrostatic pressure in articular cartilage tissue engineering: from chondrocytes to tissue regeneration. *Tissue Eng Part B Rev*. 15:43-53.

Elisseeff, J., W. McIntosh, K. Fu, B.T. Blunk, and R. Langer. 2001. Controlled-release of IGF-I and TGF-beta1 in a photopolymerizing hydrogel for cartilage tissue engineering. *Journal of orthopaedic research : official publication of the Orthopaedic Research Society*. 19:1098-1104.

Elisseeff, J.H., A. Lee, H.K. Kleinman, and Y. Yamada. 2002. Biological response of chondrocytes to hydrogels. *Ann N Y Acad Sci*. 961:118-122.

Fan, H., H. Liu, R. Zhu, X. Li, Y. Cui, Y. Hu, and Y. Yan. 2007. Comparison of chondral defects repair with in vitro and in vivo differentiated mesenchymal stem cells. *Cell transplantation*. 16:823-832.

Fan, H., H. Tao, Y. Wu, Y. Hu, Y. Yan, and Z. Luo. 2010. TGF-beta3 immobilized PLGA-gelatin/chondroitin sulfate/hyaluronic acid hybrid scaffold for cartilage regeneration. *Journal of biomedical materials research. Part A*. 95:982-992.

Fan, Z., S. Chubinskaya, D.C. Rueger, B. Bau, J. Haag, and T. Aigner. 2004. Regulation of anabolic and catabolic gene expression in normal and osteoarthritic adult human articular chondrocytes by osteogenic protein-1. *Clin Exp Rheumatol*. 22:103-106.

Fecek, C., D. Yao, A. Kacorri, A. Vasquez, S. Iqbal, H. Sheikh, D.M. Svinarich, M. Perez-Cruet, and G.R. Chaudhry. 2008. Chondrogenic derivatives of embryonic stem cells seeded into 3D polycaprolactone scaffolds generated cartilage tissue in vivo. *Tissue engineering. Part A*. 14:1403-1413.

Fedorovich, N.E., I. Swennen, J. Girones, L. Moroni, C.A. van Blitterswijk, E. Schacht, J. Alblas, and W.J. Dhert. 2009. Evaluation of Photocrosslinked Lutrol Hydrogel for Tissue Printing Applications. *Biomacromolecules*.

Fiester, A., H. Scholer, and A. Caplan. 2004. Stem cell therapies: time to talk to the animals. *Cloning Stem Cells*. 6:3-4.

Figueroa, R.J., T.G. Koch, and D.H. Betts. 2011. Stem Cells and Animal Therapies. *In* Comprehensive Biotechnology: 2nd Edition. M. Moo-Young, editor. Elsevier.

Freed, L.E., J.C. Marquis, A. Nohria, J. Emmanual, A.G. Mikos, and R. Langer. 1993. Neocartilage formation in vitro and in vivo using cells cultured on synthetic biodegradable polymers. *J Biomed Mater Res*. 27:11-23.

Freed, L.E., G. Vunjak-Novakovic, R.J. Biron, D.B. Eagles, D.C. Lesnoy, S.K. Barlow, and R. Langer. 1994. Biodegradable polymer scaffolds for tissue engineering. *Biotechnology (N Y)*. 12:689-693.

Fu, K., D.W. Pack, A.M. Klibanov, and R. Langer. 2000. Visual evidence of acidic environment within degrading poly(lactic-co-glycolic acid) (PLGA) microspheres. *Pharm Res*. 17:100-106.

Godara, P., C.D. McFarland, and R.E. Nordon. 2008. Mini Review: Design of bioreactors for mesenchymal stem cell tissue engineering. *Journal of Chemical Technilogy and Biotechnology*. 83:408-420.

Gomez-Sjoberg, R., A.A. Leyrat, B.T. Houseman, K. Shokat, and S.R. Quake. 2010. Biocompatibility and Reduced Drug Absorption of Sol-Gel-Treated Poly(dimethyl siloxane) for Microfluidic Cell Culture Applications. *Anal Chem*.

Gomez-Sjoberg, R., A.A. Leyrat, D.M. Pirone, C.S. Chen, and S.R. Quake. 2007. Versatile, fully automated, microfluidic cell culture system. *Anal Chem*. 79:8557-8563.

Grad, S., D. Eglin, M. Alini, and M.J. Stoddart. 2011. Physical Stimulation of Chondrogenic Cells In Vitro: A Review. *Clinical orthopaedics and related research*.

Grayson, W.L., F. Zhao, B. Bunnell, and T. Ma. 2007. Hypoxia enhances proliferation and tissue formation of human mesenchymal stem cells. *Biochemical and biophysical research communications*. 358:948-953.

Haasper, C., J. Zeichen, R. Meister, C. Krettek, and M. Jagodzinski. 2008. Tissue engineering of osteochondral constructs in vitro using bioreactors. *Injury*. 39 Suppl 1:S66-76.

Haleem, A.M., and C.R. Chu. 2010. Advances in Tissue Engineering Techniques for Articular Cartilage Repair. 20:76-89.

Hamilton, D.K., J.S. Smith, D.L. Reames, B.J. Williams, D.R. Chernavvsky, and C.I. Shaffrey. 2011. Safety, efficacy, and dosing of recombinant human bone morphogenetic protein-2 (rhBMP-2) for posterior cervical and cervico-thoracic instrumented fusion with a minimum two-year follow-up. *Neurosurgery*.

Hao, X., E.A. Silva, A. Mansson-Broberg, K.H. Grinnemo, A.J. Siddiqui, G. Dellgren, E. Wardell, L.A. Brodin, D.J. Mooney, and C. Sylven. 2007. Angiogenic effects of sequential release of VEGF-A165 and PDGF-BB with alginate hydrogels after myocardial infarction. *Cardiovasc Res*. 75:178-185.

Hiemstra, C., Z. Zhong, P.J. Dijkstra, and J. Feijen. 2005. PEG-PLA hydrogels by stereocomplexation for tissue engineering of cartilage. *J Control Release*. 101:332-334.

Hiemstra, C., Z. Zhong, L. Li, P.J. Dijkstra, and J. Feijen. 2006a. In-situ formation of biodegradable hydrogels by stereocomplexation of PEG-(PLLA)8 and PEG-(PDLA)8 star block copolymers. *Biomacromolecules*. 7:2790-2795.

Hiemstra, C., Z.Y. Zhong, X. Jiang, W.E. Hennink, P.J. Dijkstra, and J. Feijen. 2006b. PEG-PLLA and PEG-PDLA multiblock copolymers: synthesis and in situ hydrogel formation by stereocomplexation. *J Control Release*. 116:e17-19.

Hiemstra, C., Z.Y. Zhong, S.R. Van Tomme, W.E. Hennink, P.J. Dijkstra, and J. Feijen. 2006c. Protein release from injectable stereocomplexed hydrogels based on PEG-PDLA and PEG-PLLA star block copolymers. *J Control Release*. 116:e19-21.

Hildner, F., C. Albrecht, C. Gabriel, H. Redl, and M. van Griensven. 2011. State of the art and future perspectives of articular cartilage regeneration: a focus on adipose-derived stem cells and platelet-derived products. *Journal of tissue engineering and regenerative medicine.*

Hiramatsu, K., S. Sasagawa, H. Outani, K. Nakagawa, H. Yoshikawa, and N. Tsumaki. 2011. Generation of hyaline cartilaginous tissue from mouse adult dermal fibroblast culture by defined factors. *The Journal of clinical investigation.* 121:640-657.

Hollander, A.P., S.C. Dickinson, T.J. Sims, P. Brun, R. Cortivo, E. Kon, M. Marcacci, S. Zanasi, A. Borrione, C. De Luca, A. Pavesio, C. Soranzo, and G. Abatangelo. 2006. Maturation of tissue engineered cartilage implanted in injured and osteoarthritic human knees. *Tissue Eng.* 12:1787-1798.

Hollister, S.J. 2005. Porous scaffold design for tissue engineering. *Nat Mater.* 4:518-524.

Holzwarth, C., M. Vaegler, F. Gieseke, S.M. Pfister, R. Handgretinger, G. Kerst, and I. Muller. 2010. Low physiologic oxygen tensions reduce proliferation and differentiation of human multipotent mesenchymal stromal cells. *BMC cell biology.* 11:11.

Honda, M., T. Yada, M. Ueda, and K. Kimata. 2000. Cartilage formation by cultured chondrocytes in a new scaffold made of poly(L-lactide-epsilon-caprolactone) sponge. *J Oral Maxillofac Surg.* 58:767-775.

Hsiong, S.X., N. Huebsch, C. Fischbach, H.J. Kong, and D.J. Mooney. 2008. Integrin-adhesion ligand bond formation of preosteoblasts and stem cells in three-dimensional RGD presenting matrices. *Biomacromolecules.* 9:1843-1851.

Huang, A.H., M.J. Farrell, M. Kim, and R.L. Mauck. 2010a. Long-term dynamic loading improves the mechanical properties of chondrogenic mesenchymal stem cell-laden hydrogel. *Eur Cell Mater.* 19:72-85.

Huang, A.H., M.J. Farrell, and R.L. Mauck. 2010b. Mechanics and mechanobiology of mesenchymal stem cell-based engineered cartilage. *J Biomech.* 43:128-136.

Huang, A.H., A. Stein, and R.L. Mauck. 2010c. Evaluation of the complex transcriptional topography of mesenchymal stem cell chondrogenesis for cartilage tissue engineering. *Tissue Eng Part A.* 16:2699-2708.

Huang, C.Y., K.L. Hagar, L.E. Frost, Y. Sun, and H.S. Cheung. 2004. Effects of cyclic compressive loading on chondrogenesis of rabbit bone-marrow derived mesenchymal stem cells. *Stem Cells.* 22:313-323.

Hunziker, E.B., I.M. Driesang, and E.A. Morris. 2001. Chondrogenesis in cartilage repair is induced by members of the transforming growth factor-beta superfamily. *Clinical orthopaedics and related research*:S171-181.

Hutmacher, D.W. 2001. Scaffold design and fabrication technologies for engineering tissues--state of the art and future perspectives. *J Biomater Sci Polym Ed.* 12:107-124.

Hutmacher, D.W., T. Schantz, I. Zein, K.W. Ng, S.H. Teoh, and K.C. Tan. 2001. Mechanical properties and cell cultural response of polycaprolactone scaffolds designed and fabricated via fused deposition modeling. *J Biomed Mater Res.* 55:203-216.

Hwang, N.S., S. Varghese, H.J. Lee, P. Theprungsirikul, A. Canver, B. Sharma, and J. Elisseeff. 2007. Response of zonal chondrocytes to extracellular matrix-hydrogels. *FEBS Lett.* 581:4172-4178.

Hwang, N.S., S. Varghese, Z. Zhang, and J. Elisseeff. 2006. Chondrogenic differentiation of human embryonic stem cell-derived cells in arginine-glycine-aspartate-modified hydrogels. *Tissue Eng.* 12:2695-2706.

Ignotz, R.A., and J. Massague. 1986. Transforming growth factor-beta stimulates the expression of fibronectin and collagen and their incorporation into the extracellular matrix. *The Journal of biological chemistry.* 261:4337-4345.

Jang, J.H., and L.D. Shea. 2003. Controllable delivery of non-viral DNA from porous scaffolds. *J Control Release*. 86:157-168.

Jeon, O., S.W. Kang, H.W. Lim, J. Hyung Chung, and B.S. Kim. 2006. Long-term and zero-order release of basic fibroblast growth factor from heparin-conjugated poly(L-lactide-co-glycolide) nanospheres and fibrin gel. *Biomaterials*. 27:1598-1607.

Jiang, T.X., J.R. Yi, S.Y. Ying, and C.M. Chuong. 1993. Activin enhances chondrogenesis of limb bud cells: stimulation of precartilaginous mesenchymal condensations and expression of NCAM. *Developmental biology*. 155:545-557.

Julkunen, P., T. Harjula, J. Iivarinen, J. Marjanen, K. Seppanen, T. Narhi, J. Arokoski, M.J. Lammi, P.A. Brama, J.S. Jurvelin, and H.J. Helminen. 2009. Biomechanical, biochemical and structural correlations in immature and mature rabbit articular cartilage. *Osteoarthritis Cartilage*. 17:1628-1638.

Julkunen, P., J. Iivarinen, P.A. Brama, J. Arokoski, J.S. Jurvelin, and H.J. Helminen. 2011. Maturation of collagen fibril network structure in tibial and femoral cartilage of rabbits. *Osteoarthritis Cartilage*. 18:406-415.

Kanematsu, A., S. Yamamoto, M. Ozeki, T. Noguchi, I. Kanatani, O. Ogawa, and Y. Tabata. 2004. Collagenous matrices as release carriers of exogenous growth factors. *Biomaterials*. 25:4513-4520.

Kasper, G., L. Mao, S. Geissler, A. Draycheva, J. Trippens, J. Kuhnisch, M. Tschirschmann, K. Kaspar, C. Perka, G.N. Duda, and J. Klose. 2009. Insights into mesenchymal stem cell aging: involvement of antioxidant defense and actin cytoskeleton. *Stem Cells*. 27:1288-1297.

Kim, B.S., and D.J. Mooney. 1998a. Development of biocompatible synthetic extracellular matrices for tissue engineering. *Trends in biotechnology*. 16:224-230.

Kim, B.S., and D.J. Mooney. 1998b. Engineering smooth muscle tissue with a predefined structure. *J Biomed Mater Res*. 41:322-332.

Kim, H.J., and G.I. Im. 2009. Combination of transforming growth factor-beta2 and bone morphogenetic protein 7 enhances chondrogenesis from adipose tissue-derived mesenchymal stem cells. *Tissue engineering. Part A*. 15:1543-1551.

Kim, S.E., J.H. Park, Y.W. Cho, H. Chung, S.Y. Jeong, E.B. Lee, and I.C. Kwon. 2003. Porous chitosan scaffold containing microspheres loaded with transforming growth factor-beta1: implications for cartilage tissue engineering. *Journal of controlled release : official journal of the Controlled Release Society*. 91:365-374.

Klein, T.J., J. Malda, R.L. Sah, and D.W. Hutmacher. 2009. Tissue engineering of articular cartilage with biomimetic zones. *Tissue Eng Part B Rev*. 15:143-157.

Klein, T.J., B.L. Schumacher, T.A. Schmidt, K.W. Li, M.S. Voegtline, K. Masuda, E.J. Thonar, and R.L. Sah. 2003. Tissue engineering of stratified articular cartilage from chondrocyte subpopulations. *Osteoarthritis Cartilage*. 11:595-602.

Knudson, W., C. Biswas, X.Q. Li, R.E. Nemec, and B.P. Toole. 1989. The role and regulation of tumour-associated hyaluronan. *Ciba Found Symp*. 143:150-159; discussion 159-169, 281-155.

Koch, T.G., L.C. Berg, and D.H. Betts. 2009. Current and future regenerative medicine - principles, concepts, and therapeutic use of stem cell therapy and tissue engineering in equine medicine. *Can Vet J*. 50:155-165.

Koch, T.G., and D.H. Betts. 2007. Stem cell therapy for joint problems using the horse as a clinically relevant animal model. *Expert Opin Biol Ther*. 7:1621-1626.

Koch, T.G., T. Heerkens, P.D. Thomsen, and D.H. Betts. 2007. Isolation of mesenchymal stem cells from equine umbilical cord blood. *BMC Biotechnol*. 7:26.

Kogler, G., S. Sensken, J.A. Airey, T. Trapp, M. Muschen, N. Feldhahn, S. Liedtke, R.V. Sorg, J. Fischer, C. Rosenbaum, S. Greschat, A. Knipper, J. Bender, O. Degistirici, J. Gao, A.I. Caplan, E.J. Colletti, G. Almeida-Porada, H.W. Muller, E. Zanjani, and P. Wernet. 2004. A new human somatic stem cell from placental cord blood with intrinsic pluripotent differentiation potential. *J Exp Med*. 200:123-135.

Kong, H.J., and D.J. Mooney. 2007. Microenvironmental regulation of biomacromolecular therapies. *Nat Rev Drug Discov*. 6:455-463.

Kornblatt, J.A., and M.J. Kornblatt. 2002. The effects of osmotic and hydrostatic pressures on macromolecular systems. *Biochim Biophys Acta*. 1595:30-47.

Krinner, A., M. Zscharnack, A. Bader, D. Drasdo, and J. Galle. 2009. Impact of oxygen environment on mesenchymal stem cell expansion and chondrogenic differentiation. *Cell proliferation*. 42:471-484.

Kulyk, W.M., B.J. Rodgers, K. Greer, and R.A. Kosher. 1989. Promotion of embryonic chick limb cartilage differentiation by transforming growth factor-beta. *Developmental biology*. 135:424-430.

Kurth, T., E. Hedbom, N. Shintani, M. Sugimoto, F.H. Chen, M. Haspl, S. Martinovic, and E.B. Hunziker. 2007. Chondrogenic potential of human synovial mesenchymal stem cells in alginate. *Osteoarthritis and cartilage / OARS, Osteoarthritis Research Society*. 15:1178-1189.

Landers, R., A. Pfister, U. Hubner, H. John, R. Schmelzeisen, and R. Mullhaupt. 2002. Fabrication of soft tissue engineering scaffolds by means of rapid prototyping techniques. *Journal of materials science*. 37:3107-3116.

Lansdowne, J.L. 2010. Preclinical Models for Bone Defects. *In* Orthopaedic Trauma Association Annual Meeting and Basic Science Focus Forum Symposium. Orthopaedic Trauma Association, AO Research Institute Davos, Davos Platz 7270, Switzerland.

Lee, H.J., J.S. Lee, T. Chansakul, C. Yu, J.H. Elisseeff, and S.M. Yu. 2006. Collagen mimetic peptide-conjugated photopolymerizable PEG hydrogel. *Biomaterials*. 27:5268-5276.

Lee, K.Y., K.H. Bouhadir, and D.J. Mooney. 2000. Degradation behavior of covalently cross-linked poly(aldehyde guluronate) hydrogels. *Macromolecules*. 33:97-101.

Lee, K.Y., and D.J. Mooney. 2001. Hydrogels for tissue engineering. *Chem Rev*. 101:1869-1879.

Lee, S.J., S.H. Oh, J. Liu, S. Soker, A. Atala, and J.J. Yoo. 2008. The use of thermal treatments to enhance the mechanical properties of electrospun poly(epsilon-caprolactone) scaffolds. *Biomaterials*. 29:1422-1430.

Leong, K.W., J. Kost, E. Mathiowitz, and R. Langer. 1986. Polyanhydrides for controlled release of bioactive agents. *Biomaterials*. 7:364-371.

Li, C., C. Vepari, H.J. Jin, H.J. Kim, and D.L. Kaplan. 2006a. Electrospun silk-BMP-2 scaffolds for bone tissue engineering. *Biomaterials*. 27:3115-3124.

Li, D., and Y.N. Xia. 2004. Electrospinning of nanofibers: Reinventing the wheel? *Advanced Materials*. 16:1151-1170.

Li, J., K.S. Kim, J.S. Park, W.A. Elmer, W.C. Hutton, and S.T. Yoon. 2003. BMP-2 and CDMP-2: stimulation of chondrocyte production of proteoglycan. *Journal of orthopaedic science : official journal of the Japanese Orthopaedic Association*. 8:829-835.

Li, W.J., Y.J. Jiang, and R.S. Tuan. 2006b. Chondrocyte phenotype in engineered fibrous matrix is regulated by fiber size. *Tissue Eng*. 12:1775-1785.

Li, W.J., R. Tuli, C. Okafor, A. Derfoul, K.G. Danielson, D.J. Hall, and R.S. Tuan. 2005. A three-dimensional nanofibrous scaffold for cartilage tissue engineering using human mesenchymal stem cells. *Biomaterials*. 26:599-609.

Lin, C.Y., N. Kikuchi, and S.J. Hollister. 2004. A novel method for biomaterial scaffold internal architecture design to match bone elastic properties with desired porosity. *J Biomech*. 37:623-636.

Loeser, R.F., C.S. Carlson, M. Del Carlo, and A. Cole. 2002. Detection of nitrotyrosine in aging and osteoarthritic cartilage: Correlation of oxidative damage with the presence of interleukin-1beta and with chondrocyte resistance to insulin-like growth factor 1. *Arthritis and rheumatism*. 46:2349-2357.

Loeser, R.F., C.A. Pacione, and S. Chubinskaya. 2003. The combination of insulin-like growth factor 1 and osteogenic protein 1 promotes increased survival of and matrix synthesis by normal and osteoarthritic human articular chondrocytes. *Arthritis and rheumatism*. 48:2188-2196.

Longobardi, L., L. O'Rear, S. Aakula, B. Johnstone, K. Shimer, A. Chytil, W.A. Horton, H.L. Moses, and A. Spagnoli. 2006. Effect of IGF-I in the chondrogenesis of bone marrow mesenchymal stem cells in the presence or absence of TGF-beta signaling. *Journal of bone and mineral research : the official journal of the American Society for Bone and Mineral Research*. 21:626-636.

Lutolf, M.P., G.P. Raeber, A.H. Zisch, N. Tirelli, and J.A. Hubbell. 2003. Cell-responsive synthetic hydrogels. *Advanced Materials*. 15:888-892.

Ma, P.X., and R. Zhang. 2001. Microtubular architecture of biodegradable polymer scaffolds. *J Biomed Mater Res*. 56:469-477.

Madihally, S.V., and H.W. Matthew. 1999. Porous chitosan scaffolds for tissue engineering. *Biomaterials*. 20:1133-1142.

Mahmood, T.A., R. de Jong, J. Riesle, R. Langer, and C.A. van Blitterswijk. 2004. Adhesion-mediated signal transduction in human articular chondrocytes: the influence of biomaterial chemistry and tenascin-C. *Exp Cell Res*. 301:179-188.

Majumdar, M.K., E. Wang, and E.A. Morris. 2001. BMP-2 and BMP-9 promotes chondrogenic differentiation of human multipotential mesenchymal cells and overcomes the inhibitory effect of IL-1. *Journal of cellular physiology*. 189:275-284.

Malda, J., T.B. Woodfield, F. van der Vloodt, F.K. Kooy, D.E. Martens, J. Tramper, C.A. van Blitterswijk, and J. Riesle. 2004. The effect of PEGT/PBT scaffold architecture on oxygen gradients in tissue engineered cartilaginous constructs. *Biomaterials*. 25:5773-5780.

Marolt, D., A. Augst, L.E. Freed, C. Vepari, R. Fajardo, N. Patel, M. Gray, M. Farley, D. Kaplan, and G. Vunjak-Novakovic. 2006. Bone and cartilage tissue constructs grown using human bone marrow stromal cells, silk scaffolds and rotating bioreactors. *Biomaterials*. 27:6138-6149.

Martineau, P.A. 2010. Articular Cartilage Injury: Choosing a Pre-Clinical model in Orthopaedic Trauma. *In* Orthopaedic Trauma Association Annual Meeting and Basic Science Focus Forum Symposium. Orthopaedic Trauma Association, AO Research Institute Davos, Davos Platz 7270, Switzerland.

Martins, A., A.R. Duarte, S. Faria, A.P. Marques, R.L. Reis, and N.M. Neves. 2010. Osteogenic induction of hBMSCs by electrospun scaffolds with dexamethasone release functionality. *Biomaterials*. 31:5875-5885.

Mason, J.M., A.S. Breitbart, M. Barcia, D. Porti, R.G. Pergolizzi, and D.A. Grande. 2000. Cartilage and bone regeneration using gene-enhanced tissue engineering. *Clinical orthopaedics and related research*:S171-178.

Medvedev, S.P., E.V. Grigor'eva, A.I. Shevchenko, A.A. Malakhova, E.V. Dementyeva, A.A. Shilov, E.A. Pokushalov, A.M. Zaidman, M.A. Aleksandrova, E.Y. Plotnikov, G.T. Sukhikh, and S.M. Zakian. 2010. Human Induced Pluripotent Stem Cells Derived from Fetal Neural Stem Cells Successfully Undergo Directed Differentiation into Cartilage. *Stem cells and development*.

Melchels, F.P., J. Feijen, and D.W. Grijpma. 2009. A poly(d,l-lactide) resin for the preparation of tissue engineering scaffolds by stereolithography. *Biomaterials*.

Melchels, F.P., D.W. Grijpma, and J. Feijen. 2006. Photo-crosslinking of functionalised lactide oligomers for the fabrication of osteochondral tissue engineering scaffolds. *J Control Release*. 116:e98-100.

Merceron, C., C. Vinatier, S. Portron, M. Masson, J. Amiaud, L. Guigand, Y. Cherel, P. Weiss, and J. Guicheux. 2010. Differential effects of hypoxia on osteochondrogenic potential of human adipose-derived stem cells. *American journal of physiology. Cell physiology*. 298:C355-364.

Meyer, E.G., C.T. Buckley, S.D. Thorpe, and D.J. Kelly. 2010. Low oxygen tension is a more potent promoter of chondrogenic differentiation than dynamic compression. *J Biomech*. 43:2516-2523.

Mikic, B., A.L. Isenstein, and A. Chhabra. 2004. Mechanical modulation of cartilage structure and function during embryogenesis in the chick. *Ann Biomed Eng*. 32:18-25.

Mikic, B., T.L. Johnson, A.B. Chhabra, B.J. Schalet, M. Wong, and E.B. Hunziker. 2000. Differential effects of embryonic immobilization on the development of fibrocartilaginous skeletal elements. *J Rehabil Res Dev*. 37:127-133.

Mikos, A.G., Y. Bao, L.G. Cima, D.E. Ingber, J.P. Vacanti, and R. Langer. 1993. Preparation of poly(glycolic acid) bonded fiber structures for cell attachment and transplantation. *J Biomed Mater Res*. 27:183-189.

Milner, K.R., and C.A. Siedlecki. 2007. Fibroblast response is enhanced by poly(L-lactic acid) nanotopography edge density and proximity. *Int J Nanomedicine*. 2:201-211.

Moroni, L., M. Curti, M. Welti, S. Korom, W. Weder, J.R. De Wijn, and C.A. Van Blitterswijk. 2007a. Anatomical 3D fiber-deposited scaffolds for tissue engineering: Designing a neotrachea. *Tissue Engineering*. 13:2483-2493.

Moroni, L., J.R. de Wijn, and C.A. van Blitterswijk. 2006a. 3D fiber-deposited scaffolds for tissue engineering: influence of pores geometry and architecture on dynamic mechanical properties. *Biomaterials*. 27:974-985.

Moroni, L., and J.H. Elisseeff. 2008. Biomaterials engineered for integration. *Materials Today*. 11:44-51.

Moroni, L., D. Hamann, L. Paoluzzi, J. Pieper, J.R. de Wijn, and C.A. van Blitterswijk. 2008. Regenerating articular tissue by converging technologies. *PLoS ONE*. 3:e3032.

Moroni, L., F.M. Lambers, W. Wilson, C.C. van Donkelaar, J. de Wijn, R. Huiskesb, and C.A. van Blitterswijk. 2007b. Finite Element Analysis of Meniscal Anatomical 3D Scaffolds: Implications for Tissue Engineering. *Open Biomed Eng J*. 1:23-34.

Moroni, L., G. Poort, F. Van Keulen, J.R. de Wijn, and C.A. van Blitterswijk. 2006b. Dynamic mechanical properties of 3D fiber-deposited PEOT/PBT scaffolds: An experimental and numerical analysis. *J Biomed Mater Res A*:605-614.

Mouw, J.K., J.T. Connelly, C.G. Wilson, K.E. Michael, and M.E. Levenston. 2007. Dynamic compression regulates the expression and synthesis of chondrocyte-specific matrix molecules in bone marrow stromal cells. *Stem Cells*. 25:655-663.

Mueller, S.M., S. Shortkroff, T.O. Schneider, H.A. Breinan, I.V. Yannas, and M. Spector. 1999. Meniscus cells seeded in type I and type II collagen-GAG matrices in vitro. *Biomaterials*. 20:701-709.

Nagel-Heyer, S., C. Goepfert, F. Feyerabend, J.P. Petersen, P. Adamietz, N.M. Meenen, and R. Portner. 2005. Bioreactor cultivation of three-dimensional cartilage-carrier-constructs. *Bioprocess Biosyst Eng*. 27:273-280.

Nagy, K., H.K. Sung, P. Zhang, S. Laflamme, P. Vincent, S. Agha-Mohammadi, K. Woltjen, C. Monetti, I.P. Michael, L.C. Smith, and A. Nagy. 2011. Induced Pluripotent Stem Cell Lines Derived from Equine Fibroblasts. *Stem Cell Rev*.

Nehrer, S., H.A. Breinan, A. Ramappa, G. Young, S. Shortkroff, L.K. Louie, C.B. Sledge, I.V. Yannas, and M. Spector. 1997. Matrix collagen type and pore size influence behaviour of seeded canine chondrocytes. *Biomaterials*. 18:769-776.

Nelson, D.M., P.R. Baraniak, Z. Ma, J. Guan, N.S. Mason, and W.R. Wagner. 2011. Controlled Release of IGF-1 and HGF from a Biodegradable Polyurethane Scaffold. *Pharmaceutical research*.

Nichols, J.E., J. Cortiella, J. Lee, J.A. Niles, M. Cuddihy, S. Wang, J. Bielitzki, A. Cantu, R. Mlcak, E. Valdivia, R. Yancy, M.L. McClure, and N.A. Kotov. 2009. In vitro analog of human bone marrow from 3D scaffolds with biomimetic inverted colloidal crystal geometry. *Biomaterials*. 30:1071-1079.

Niklason, L.E., and R.S. Langer. 1997. Advances in tissue engineering of blood vessels and other tissues. *Transpl Immunol*. 5:303-306.

Nimni, M.E. 1997. Polypeptide growth factors: targeted delivery systems. *Biomaterials*. 18:1201-1225.

Nof, M., and L.D. Shea. 2002. Drug-releasing scaffolds fabricated from drug-loaded microspheres. *J Biomed Mater Res*. 59:349-356.

O'Sullivan, J., S. D'Arcy, F.P. Barry, J.M. Murphy, and C.M. Coleman. 2011. Mesenchymal chondroprogenitor cell origin and therapeutic potential. *Stem Cell Res Ther*. 2:8.

Oh, S.H., T.H. Kim, G.I. Im, and J.H. Lee. 2010. Investigation of pore size effect on chondrogenic differentiation of adipose stem cells using a pore size gradient scaffold. *Biomacromolecules*. 11:1948-1955.

Olde Riekerink, M.B., M.B. Claase, G.H. Engbers, D.W. Grijpma, and J. Feijen. 2003. Gas plasma etching of PEO/PBT segmented block copolymer films. *J Biomed Mater Res A*. 65:417-428.

Ovsianikov, A., M. Gruene, M. Pflaum, L. Koch, F. Maiorana, M. Wilhelmi, A. Haverich, and B. Chichkov. 2010. Laser printing of cells into 3D scaffolds. *Biofabrication*. 2:014104.

Ovsianikov, A., M. Malinauskas, S. Schlie, B. Chichkov, S. Gittard, R. Narayan, M. Lobler, K. Sternberg, K.P. Schmitz, and A. Haverich. 2011. Three-dimensional laser micro- and nano-structuring of acrylated poly(ethylene glycol) materials and evaluation of their cytoxicity for tissue engineering applications. *Acta Biomater*. 7:967-974.

Pachence, J.M. 1996. Collagen-based devices for soft tissue repair. *J Biomed Mater Res*. 33:35-40.

Palmer, G.D., A. Steinert, A. Pascher, E. Gouze, J.N. Gouze, O. Betz, B. Johnstone, C.H. Evans, and S.C. Ghivizzani. 2005. Gene-induced chondrogenesis of primary mesenchymal stem cells in vitro. *Molecular therapy : the journal of the American Society of Gene Therapy*. 12:219-228.

Park, J.S., D.G. Woo, H.N. Yang, H.J. Lim, H.M. Chung, and K.H. Park. 2008. Heparin-bound transforming growth factor-beta3 enhances neocartilage formation by rabbit mesenchymal stem cells. *Transplantation*. 85:589-596.

Park, J.S., H.J. Yang, D.G. Woo, H.N. Yang, K. Na, and K.H. Park. 2010. Chondrogenic differentiation of mesenchymal stem cells embedded in a scaffold by long-term release of TGF-beta 3 complexed with chondroitin sulfate. *Journal of biomedical materials research. Part A*. 92:806-816.

Park, K.M., S.Y. Lee, Y.K. Joung, J.S. Na, M.C. Lee, and K.D. Park. 2009. Thermosensitive chitosan-Pluronic hydrogel as an injectable cell delivery carrier for cartilage regeneration. *Acta Biomater*. 5:1956-1965.

Patel, A.A., R.G. Thakar, M. Chown, P. Ayala, T.A. Desai, and S. Kumar. 2010. Biophysical mechanisms of single-cell interactions with microtopographical cues. *Biomedical microdevices*. 12:287-296.

Patel, P., S. Irvine, J.R. McEwan, and S.N. Jayasinghe. 2008. Bio-protocols for directly forming active encapsulations containing living primary cells. *Soft Matter*. 4:1219-1229.

Pearce, A.I., R.G. Richards, S. Milz, E. Schneider, and S.G. Pearce. 2007. Animal models for implant biomaterial research in bone: a review. *Eur Cell Mater*. 13:1-10.

Portner, R., C. Goepfert, K. Wiegandt, R. Janssen, E. Ilinich, H. Paetzold, E. Eisenbarth, and M. Morlock. 2009. Technical strategies to improve tissue engineering of cartilage-carrier-constructs. *Adv Biochem Eng Biotechnol*. 112:145-181.

Reinholz, G.G., L. Lu, D.B. Saris, M.J. Yaszemski, and S.W. O'Driscoll. 2004. Animal models for cartilage reconstruction. *Biomaterials*. 25:1511-1521.

Richardson, T.P., M.C. Peters, A.B. Ennett, and D.J. Mooney. 2001. Polymeric system for dual growth factor delivery. *Nat Biotechnol*. 19:1029-1034.

Riesle, J., A.P. Hollander, R. Langer, L.E. Freed, and G. Vunjak-Novakovic. 1998. Collagen in tissue-engineered cartilage: types, structure, and crosslinks. *J Cell Biochem*. 71:313-327.

Rosso, F., G. Marino, A. Giordano, M. Barbarisi, D. Parmeggiani, and A. Barbarisi. 2005. Smart materials as scaffolds for tissue engineering. *Journal of cellular physiology*. 203:465-470.

Sachlos, E., and J.T. Czernuszka. 2003. Making tissue engineering scaffolds work. Review: the application of solid freeform fabrication technology to the production of tissue engineering scaffolds. *Eur Cell Mater*. 5:29-39; discussion 39-40.

Saha, K., J.F. Pollock, D.V. Schaffer, and K.E. Healy. 2007. Designing synthetic materials to control stem cell phenotype. *Curr Opin Chem Biol*. 11:381-387.

Sailor, L.Z., R.M. Hewick, and E.A. Morris. 1996. Recombinant human bone morphogenetic protein-2 maintains the articular chondrocyte phenotype in long-term culture. *Journal of orthopaedic research : official publication of the Orthopaedic Research Society*. 14:937-945.

Santourlidis, S., P. Wernet, F. Ghanjati, N. Graffmann, J. Springer, C. Kriegs, X. Zhao, J. Brands, M.J. Arauzo-Bravo, R. Neves, G. Koegler, and M. Uhrberg. 2011. Unrestricted somatic stem cells (USSC) from human umbilical cord blood display uncommitted epigenetic signatures of the major stem cell pluripotency genes. *Stem Cell Res*. 6:60-69.

Sarazin, P., and B.D. Favis. 2003. Morphology control in co-continuous poly(L-lactide)/polystyrene blends: a route towards highly structured and interconnected porosity in poly(L-lactide) materials. *Biomacromolecules*. 4:1669-1679.

Sarazin, P., X. Roy, and B.D. Favis. 2004. Controlled preparation and properties of porous poly(L-lactide) obtained from a co-continuous blend of two biodegradable polymers. *Biomaterials*. 25:5965-5978.

Scharstuhl, A., H.L. Glansbeek, H.M. van Beuningen, E.L. Vitters, P.M. van der Kraan, and W.B. van den Berg. 2002. Inhibition of endogenous TGF-beta during experimental osteoarthritis prevents osteophyte formation and impairs cartilage repair. *Journal of immunology*. 169:507-514.

Schoof, H., J. Apel, I. Heschel, and G. Rau. 2001. Control of pore structure and size in freeze-dried collagen sponges. *J Biomed Mater Res*. 58:352-357.

Schulz, R.M., and A. Bader. 2007. Cartilage tissue engineering and bioreactor systems for the cultivation and stimulation of chondrocytes. *Eur Biophys J*. 36:539-568.

Schulze-Tanzil, G. 2009. Activation and dedifferentiation of chondrocytes: implications in cartilage injury and repair. *Ann Anat*. 191:325-338.

Schuurman, W., D. Gawlitta, T.J. Klein, W. ten Hoope, M.H. van Rijen, W.J. Dhert, P.R. van Weeren, and J. Malda. 2009. Zonal chondrocyte subpopulations reacquire zone-specific characteristics during in vitro redifferentiation. *The American journal of sports medicine*. 37 Suppl 1:97S-104S.

Sekiya, I., B.L. Larson, J.T. Vuoristo, R.L. Reger, and D.J. Prockop. 2005. Comparison of effect of BMP-2, -4, and -6 on in vitro cartilage formation of human adult stem cells from bone marrow stroma. *Cell and tissue research*. 320:269-276.

Sengupta, S., D. Eavarone, I. Capila, G. Zhao, N. Watson, T. Kiziltepe, and R. Sasisekharan. 2005. Temporal targeting of tumour cells and neovasculature with a nanoscale delivery system. *Nature*. 436:568-572.

Shah, R.N., N.A. Shah, M.M. Del Rosario Lim, C. Hsieh, G. Nuber, and S.I. Stupp. 2010. Supramolecular design of self-assembling nanofibers for cartilage regeneration. *Proc Natl Acad Sci U S A*. 107:3293-3298.

Shapiro, L., and S. Cohen. 1997. Novel alginate sponges for cell culture and transplantation. *Biomaterials*. 18:583-590.

Sharma, B., C.G. Williams, M. Khan, P. Manson, and J.H. Elisseeff. 2007. In vivo chondrogenesis of mesenchymal stem cells in a photopolymerized hydrogel. *Plast Reconstr Surg*. 119:112-120.

Sherwood, J.K., S.L. Riley, R. Palazzolo, S.C. Brown, D.C. Monkhouse, M. Coates, L.G. Griffith, L.K. Landeen, and A. Ratcliffe. 2002. A three-dimensional osteochondral composite scaffold for articular cartilage repair. *Biomaterials*. 23:4739-4751.

Shiragami, N., and H. Unno. 1994. Effect of shear stress on activity of cellular enzyme in animal cell. In *Bioprocess and Biosystems Engineering*. Vol. 10. Springer-Verlag. 53-45.

Singh, V. 1999. Disposable bioreactor for cell culture using wave-induced agitation. *Cytotechnology*. 30:149-158.

Solchaga, L.A., K.J. Penick, and J.F. Welter. 2011. Chondrogenic differentiation of bone marrow-derived mesenchymal stem cells: tips and tricks. *Methods in molecular biology*. 698:253-278.

Sproule, T.L., J.A. Lee, H.B. Li, J.J. Lannutti, and D.L. Tomasko. 2004. Bioactive polymer surfaces via supercritical fluids. *Journal of Supercritical Fluids*. 28:241-248.

Steadman, J.R., W.G. Rodkey, and J.J. Rodrigo. 2001. Microfracture: surgical technique and rehabilitation to treat chondral defects. *Clin Orthop Relat Res*:S362-369.

Steinert, A.F., S.C. Ghivizzani, A. Rethwilm, R.S. Tuan, C.H. Evans, and U. Noth. 2007. Major biological obstacles for persistent cell-based regeneration of articular cartilage. *Arthritis Res Ther*. 9:213.

Steinert, A.F., U. Noth, and R.S. Tuan. 2008. Concepts in gene therapy for cartilage repair. *Injury*. 39 Suppl 1:S97-113.

Steinert, A.F., G.D. Palmer, C. Pilapil, U. Noth, C.H. Evans, and S.C. Ghivizzani. 2009. Enhanced in vitro chondrogenesis of primary mesenchymal stem cells by combined gene transfer. *Tissue engineering. Part A*. 15:1127-1139.

Stevens, M.M., R.P. Marini, D. Schaefer, J. Aronson, R. Langer, and V. Prasad Shastri. 2005. *In vivo* engineering of organs: The bone bioreactor. *PNAS*. 102.

Stoop, R., D. Albrecht, C. Gaissmaier, J. Fritz, T. Felka, M. Rudert, and W.K. Aicher. 2007. Comparison of marker gene expression in chondrocytes from patients receiving autologous chondrocyte transplantation versus osteoarthritis patients. *Arthritis Res Ther*. 9:R60.

Suciati, T., D. Howard, J. Barry, N.M. Everitt, K.M. Shakesheff, and F.R. Rose. 2006. Zonal release of proteins within tissue engineering scaffolds. *Journal of materials science. Materials in medicine*. 17:1049-1056.

Taboas, J.M., R.D. Maddox, P.H. Krebsbach, and S.J. Hollister. 2003. Indirect solid free form fabrication of local and global porous, biomimetic and composite 3D polymer-ceramic scaffolds. *Biomaterials*. 24:181-194.

Tang, Y., X. Ye, E.O. Klineberg, S. Curtiss, S. Maitra, and M.C. Gupta. 2011. Temporal and Spatial Expression of BMPs and BMP Antagonists During Posterolateral Lumbar Fusion. *Spine*. 36:E237-244.

Terada, S., H. Yoshimoto, J.R. Fuchs, M. Sato, I. Pomerantseva, M.K. Selig, D. Hannouche, and J.P. Vacanti. 2005. Hydrogel optimization for cultured elastic chondrocytes seeded onto a polyglycolic acid scaffold. *J Biomed Mater Res A*. 75:907-916.

Terraciano, V., N. Hwang, L. Moroni, H.B. Park, Z. Zhang, J. Mizrahi, D. Seliktar, and J. Elisseeff. 2007. Differential response of adult and embryonic mesenchymal progenitor cells to mechanical compression in hydrogels. *Stem Cells*. 25:2730-2738.

Thakar, R.G., M.G. Chown, A. Patel, L. Peng, S. Kumar, and T.A. Desai. 2008. Contractility-dependent modulation of cell proliferation and adhesion by microscale topographical cues. *Small*. 4:1416-1424.

Thorpe, S.D., C.T. Buckley, T. Vinardell, F.J. O'Brien, V.A. Campbell, and D.J. Kelly. 2008. Dynamic compression can inhibit chondrogenesis of mesenchymal stem cells. *Biochem Biophys Res Commun*. 377:458-462.

Tran, S.C., A.J. Cooley, and S.H. Elder. 2011. Effect of a mechanical stimulation bioreactor on tissue engineered, scaffold-free cartilage. *Biotechnol Bioeng*.

Uhrich, K.E., S.M. Cannizzaro, R.S. Langer, and K.M. Shakesheff. 1999. Polymeric systems for controlled drug release. *Chem Rev*. 99:3181-3198.

van Beuningen, H.M., H.L. Glansbeek, P.M. van der Kraan, and W.B. van den Berg. 2000. Osteoarthritis-like changes in the murine knee joint resulting from intra-articular transforming growth factor-beta injections. *Osteoarthritis and cartilage / OARS, Osteoarthritis Research Society*. 8:25-33.

Van Blitterswijk, C.A., P. Thomsen, L. A., J.A. Hubbell, D. Williams, R. Cancedda, J.D. De Bruijn, and J. Sohier. 2008. Tissue Engineering. *Academic Press Series in Biomedical Engineering*.

van Blitterswijk, C.A., J. van den Brink, H. Leenders, and D. Bakker. 1993. The effect of PEO ratio on degradation, calcification and bone bonding of PEO/PBT copolymer (PolyActive). *Cell and Materials*. 3:23-26.

van Dijkhuizen-Radersma, R., S. Metairie, J.R. Roosma, K. de Groot, and J.M. Bezemer. 2005. Controlled release of proteins from degradable poly(ether-ester) multiblock copolymers. *J Control Release*. 101:175-186.

van Dijkhuizen-Radersma, R., F.L. Peters, N.A. Stienstra, D.W. Grijpma, J. Feijen, K. de Groot, and J.M. Bezemer. 2002. Control of vitamin B12 release from poly(ethylene glycol)/poly(butylene terephthalate) multiblock copolymers. *Biomaterials*. 23:1527-1536.

van Dijkhuizen-Radersma, R., J.R. Roosma, P. Kaim, S. Metairie, F.L. Peters, J. de Wijn, P.G. Zijlstra, K. de Groot, and J.M. Bezemer. 2003. Biodegradable poly(ether-ester) multiblock copolymers for controlled release applications. *J Biomed Mater Res A*. 67:1294-1304.

van Dijkhuizen-Radersma, R., J.R. Roosma, J. Sohier, F.L. Peters, M. van den Doel, C.A. van Blitterswijk, K. de Groot, and J.M. Bezemer. 2004. Biodegradable poly(ether-ester) multiblock copolymers for controlled release applications: An in vivo evaluation. *J Biomed Mater Res A*. 71:118-127.

Vasiliadis, H.S., B. Danielson, M. Ljungberg, B. McKeon, A. Lindahl, and L. Peterson. 2010. Autologous chondrocyte implantation in cartilage lesions of the knee: long-term evaluation with magnetic resonance imaging and delayed gadolinium-enhanced magnetic resonance imaging technique. *The American journal of sports medicine*. 38:943-949.

Vihola, H., A. Laukkanen, L. Valtola, H. Tenhu, and J. Hirvonen. 2005. Cytotoxicity of thermosensitive polymers poly(N-isopropylacrylamide), poly(N-vinylcaprolactam) and amphiphilically modified poly(N-vinylcaprolactam). *Biomaterials*. 26:3055-3064.

Vunjak-Novakovic, G., N. Searby, J. De Luis, and L.E. Freed. 2002. Microgravity studies of cells and tissues. *Ann N Y Acad Sci*. 974:504-517.

Waese, E.Y., and W.L. Stanford. 2011. One-step generation of murine embryonic stem cell-derived mesoderm progenitors and chondrocytes in a serum-free monolayer differentiation system. *Stem Cell Res*. 6:34-49.

Wang, J., H.Q. Mao, and K.W. Leong. 2001a. A novel biodegradable gene carrier based on polyphosphoester. *J Am Chem Soc*. 123:9480-9481.

Wang, P.Y. 1989. Compressed poly(vinyl alcohol)-polycaprolactone admixture as a model to evaluate erodible implants for sustained drug delivery. *J Biomed Mater Res*. 23:91-104.

Wang, S., A.C. Wan, X. Xu, S. Gao, H.Q. Mao, K.W. Leong, and H. Yu. 2001b. A new nerve guide conduit material composed of a biodegradable poly(phosphoester). *Biomaterials*. 22:1157-1169.

Wang, X., E. Wenk, X. Zhang, L. Meinel, G. Vunjak-Novakovic, and D.L. Kaplan. 2009. Growth factor gradients via microsphere delivery in biopolymer scaffolds for osteochondral tissue engineering. *Journal of controlled release : official journal of the Controlled Release Society*. 134:81-90.

Williams, G.M., E.F. Chan, M.M. Temple-Wong, W.C. Bae, K. Masuda, W.D. Bugbee, and R.L. Sah. 2010. Shape, loading, and motion in the bioengineering design, fabrication, and testing of personalized synovial joints. *J Biomech*. 43:156-165.

Williamson, A.K., A.C. Chen, K. Masuda, E.J. Thonar, and R.L. Sah. 2003a. Tensile mechanical properties of bovine articular cartilage: variations with growth and relationships to collagen network components. *J Orthop Res*. 21:872-880.

Williamson, A.K., A.C. Chen, and R.L. Sah. 2001. Compressive properties and function-composition relationships of developing bovine articular cartilage. *J Orthop Res*. 19:1113-1121.

Williamson, A.K., K. Masuda, E.J. Thonar, and R.L. Sah. 2003b. Growth of immature articular cartilage in vitro: correlated variation in tensile biomechanical and collagen network properties. *Tissue Eng*. 9:625-634.

Wilson, A., L.A. Shehadeh, H. Yu, and K.A. Webster. 2010. Age-related molecular genetic changes of murine bone marrow mesenchymal stem cells. *BMC Genomics*. 11:229.

Woodfield, T.B., J.M. Bezemer, J.S. Pieper, C.A. van Blitterswijk, and J. Riesle. 2002. Scaffolds for tissue engineering of cartilage. *Crit Rev Eukaryot Gene Expr*. 12:209-236.

Woodfield, T.B., J. Malda, J. de Wijn, F. Peters, J. Riesle, and C.A. van Blitterswijk. 2004. Design of porous scaffolds for cartilage tissue engineering using a three-dimensional fiber-deposition technique. *Biomaterials*. 25:4149-4161.

Woodfield, T.B., C.A. Van Blitterswijk, J. De Wijn, T.J. Sims, A.P. Hollander, and J. Riesle. 2005. Polymer scaffolds fabricated with pore-size gradients as a model for studying the zonal organization within tissue-engineered cartilage constructs. *Tissue Eng*. 11:1297-1311.

Xu, D., Z. Gechtman, A. Hughes, A. Collins, R. Dodds, X. Cui, L. Jolliffe, L. Higgins, A. Murphy, and F. Farrell. 2006. Potential involvement of BMP receptor type IB activation in a synergistic effect of chondrogenic promotion between rhTGFbeta3 and rhGDF5 or rhBMP7 in human mesenchymal stem cells. *Growth Factors*. 24:268-278.

Yamamoto, M., Y. Ikada, and Y. Tabata. 2001. Controlled release of growth factors based on biodegradation of gelatin hydrogel. *Journal of biomaterials science. Polymer edition*. 12:77-88.

Yamane, S., N. Iwasaki, T. Majima, T. Funakoshi, T. Masuko, K. Harada, A. Minami, K. Monde, and S. Nishimura. 2005. Feasibility of chitosan-based hyaluronic acid hybrid biomaterial for a novel scaffold in cartilage tissue engineering. *Biomaterials*. 26:611-619.

Yang, S., K.F. Leong, Z. Du, and C.K. Chua. 2002. The design of scaffolds for use in tissue engineering. Part II. Rapid prototyping techniques. *Tissue Eng*. 8:1-11.

Yeatts, A.B., and J.P. Fisher. 2011. Bone tissue engineering bioreactors: dynamic culture and the influence of shear stress. *Bone*. 48:171-181.

Yeong, W.Y., C.K. Chua, K.F. Leong, and M. Chandrasekaran. 2004. Rapid prototyping in tissue engineering: challenges and potential. *Trends in biotechnology*. 22:643-652.

Zhang, D., and J. Chang. 2008. Electrospinning of three-dimensional nanofibrous tubes with controllable architectures. *Nano letters*. 8:3283-3287.

Zhang, D.M., and J. Chang. 2007. Patterning of electrospun fibers using electroconductive templates. *Advanced Materials*. 19:3664-+.

Zhang, X., M. Hirai, S. Cantero, R. Ciubotariu, L. Dobrila, A. Hirsh, K. Igura, H. Satoh, I. Yokomi, T. Nishimura, S. Yamaguchi, K. Yoshimura, P. Rubinstein, and T.A. Takahashi. 2011. Isolation and characterization of mesenchymal stem cells from human umbilical cord blood: reevaluation of critical factors for successful isolation and high ability to proliferate and differentiate to chondrocytes as compared to mesenchymal stem cells from bone marrow and adipose tissue. *Journal of cellular biochemistry*. 112:1206-1218.

# 11

# Cartilage Regeneration from Bone Marrow Cells Using RWV Bioreactor and Its Automation System for Clinical Application

Toshimasa Uemura[1], Masanori Nishi[1],
Kunitomo Aoki[2] and Takashi Tsumura[3]
[1]*National Institute of Advanced Industrial Science and Technology (AIST), Ibaraki,*
[2]*Industrial Technology Institute of Ibaraki Prefecture, Higashi Ibaraki, Ibaraki,*
[3]*JTEC Corporation, Kobe, Hyogo*
*Japan*

## 1. Introduction

Articular cartilage covers the end of bones in joints and determines the load-bearing characteristics and mobility of joints. It has a thin, smooth, low friction surface with a remarkable resiliency to compressive forces. In general, chondrocytes occupy lacunae in the matrix, and produce cartilaginous ECM (extracellular matrix), which consists of type II collagen (13%), proteoglycans (7%), and water (80%).

Cartilage defects result from aging, joint injury, and developmental disorders, causing joint pain and loss of mobility. Articular cartilage is metabolically active, however, the chondrocytes have a slow turnover rate. Thus, articular cartilage might suffer progressive damage and degeneration with a limited spontaneous repair capability. Total joint arthroplasty is the final choice of treatment, however, it is not suitable for young patients because of the limited life span of the artificial joint. Marrow-stimulating techniques such as microfracturing, multiple drilling, mosaicplasty and autologous chondrocyte implantation are clinically available for young patients, but have some limitations (Ikada, 2006). Marrow-stimulating techniques result in a fibrocartilage with less mechanical strength than hyaline cartilage and only limited repair capacity. The major problems with mosaicplasty are a limited availability of autologous tissue and donor site morbidity, the destruction of healthy non-weight-bearing tissue to repair diseased tissue. Autologous chondrocyte transplantation with a periosteal graft has shown encouraging results, however, predictability and reliability are still questionable.

Ochi et al. (2002) showed a clinical advantage of transplanting autologous chondrocytes cultured in collagen gel for the treatment of full-thickness defects of cartilage in 28 knees over a minimum period of 25 months. Arthroscopic assessment indicated that 26 knees (93%) had a good or excellent outcome. Wakitani et al. (2002) applied cell transplantation to repair human articular cartilage defects in osteoarthritis knee joints. The study group comprised 24 patients with knee OA. Adherent cells expanded from bone marrow aspirates were embedded in collagen gel and transplanted into the articular cartilage defects of 12

knees, with the other 12 knees serving as cell-free controls. Arthroscopic and histological grading scores were better in the cell-transplanted group than cell-free group.
In spite of these successful clinical results to expand the clinical treatment of cartilage diseases, we need to establish a three dimensional culture technique for regenerating large cartilage tissue in vitro. One solution is to use an RWV (rotating wall vessel ) bioreactor.

## 2. Regeneration of cartilaginous tissue from rabbit bone marrow cells under three dimensional culture by RWV bioreactor

### 2.1 RWV (rotating wall vessel) bioreactor

Recently, three-dimensional cell culture techniques have attracted much attention among not only cell and developmental biologists but also clinicians who have an interest in tissue engineering (Abbott, 2003). The limitations of two-dimensional culture using conventional flasks or dishes are becoming clear. In the field of tissue engineering, the chondrocyte cell is a typical example of the major difference between a flat layer of cells and a complex, three-dimensional tissue (Holtzer, 1960; Passaretti et al., 2001). Matured chondrocytes in the two-dimensional condition dedifferentiated without maintaining their phenotype and lost their original phenotype after four rounds of subculture. Clinically, a method of regenerating cartilage tissue needs to be established to treat diseases such as osteoarthritis. Thus, the development of a cell culture system for the growth of three-dimensional cartilage is important. However, problems such as necrosis due to high-density cell culture and shear stress have not yet been solved using conventional stirred fermentors. We examined the use of a rotating wall vessel (RWV) bioreactor that simulates a microgravity environment with low shear stress for cartilage tissue regeneration (Fig.1). This bioreactor generates stress by the horizontal rotation of a cylindrical vessel equipped with a gas exchange membrane. The RWV bioreactor compensates for the effect of gravity, resulting in homogenous cell growth and differentiation without sinking, and cells aggregate and form a three-dimensional tissue. The advantage of using an RWV bioreactor for tissue formation was first reviewed by Unsworth and Lelkes (1998), who discussed the benefits of growing tissues in microgravity and simulated microgravity. The formation of tissue by, for example, endothelial cells (Sanfold et al., 2002), colon carcinoma cell lines (Goodwin et al., 1992), ovarian cancer cells (Goodwin et al., 1997), osteoblasts (Qiu et al., 1999), and erythroid cells (Sytkowski and Davis, 2001), has been reported. In particular, the RWV bioreactor has been shown to stimulate chondrogenesis (Baker and Goodwin, 1997; Duke et al., 1993). Moreover, a comparison of chondrocyte cells cultured in rotating bioreactors in space (Mir space station) and on earth was reported by Freed et al. (1997a). They performed rotating cultures of bovine chondrocytes in polyglycolic acid (PGA) scaffolds and concluded that the culture on earth produced cartilage tissues closer to the natural form than that in space. In this case, the culture period for obtaining tissue was 7 months. Very recently, the chondrogenesis of human cartilage by an RWV bioreactor has been reported (Marlovits et al., 2003). Good quality cartilage tissue was formed by rotating culture from aged human articular cartilage after 90 days of cultivation. In spite of numerous studies on the formation of cartilage tissue from chondrocytes, our report (Ohyabu et al., 2006) was the first on the production of cartilaginous tissue from bone marrow-derived cells by rotating culture. A three-dimensional cell culture technique was established for the construction of large and homogenous cartilage tissues without a scaffold using bone marrow-derived cells and an RWV bioreactor.

Fig. 1. RWV Bioreactor

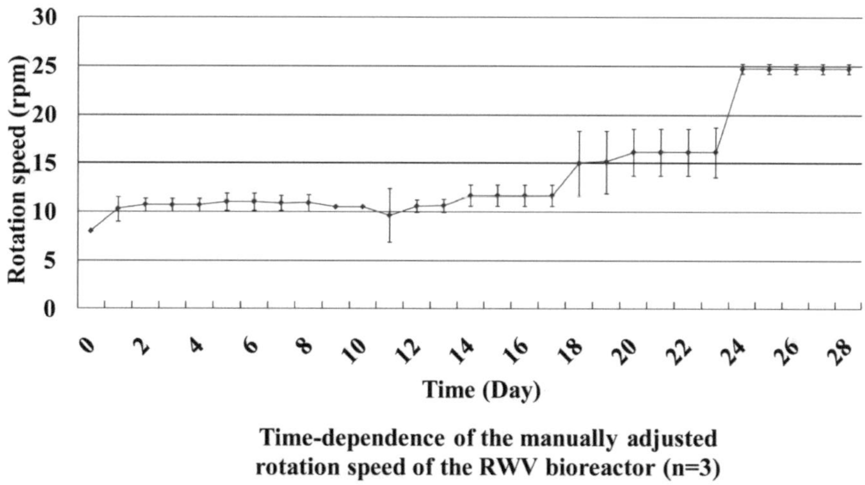


Fig. 2. Time-dependence of the rotation speed

### 2.2 Regeneration of cartilage tissue in vitro using rabbit bone marrow cells and an RWV bioreactor

Bone marrow cells were collected from the femora of six 10-day-old Japanese white rabbits and cultured in a standard medium consisting of Dulbecco's Modified Eagle's Medium(DMEM) containing 10% fetal bovine serum for 3 weeks. The cells were resuspended in a chondrogenic differentiation medium comprising DMEM containing 10% FBS and 10 ng/mL of TGF-β (Johnstone et al., 1998) and seeded in the discoidal vessels of an RWV reactor in a $CO_2$ incubator. A rotatory culture was performed for 4 weeks. The rotation speed was adjusted manually in order to keep cell aggregates freely suspended within the vessel. As a control, a tube culture was performed, a kind of three dimensional culture technique developed by Holtzer and Manning (Holtzer, 1960; Manning and Bonner, 1967) and established by Johnstone et al. (1998). After the rotating culture in the RWV vessel and static culture in the conical tube, the aggregates were harvested and prepared for histochemical and biochemical analysis.

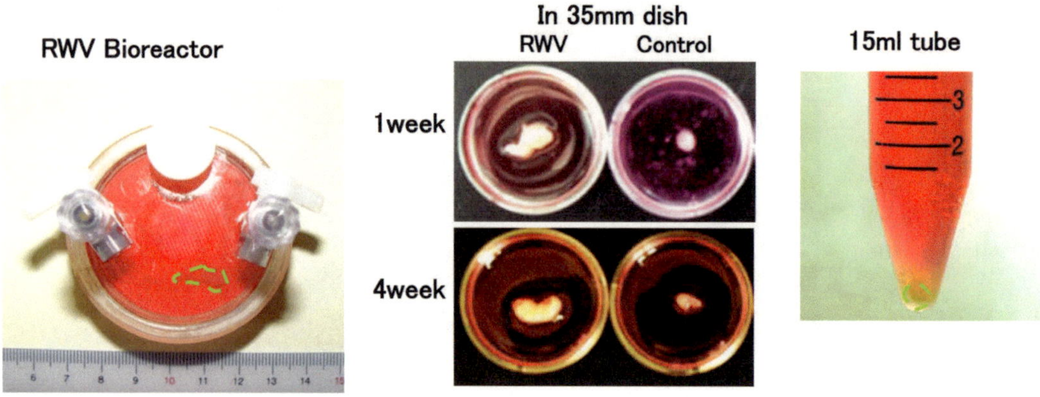


Fig. 3. In vitro cartilage tissue regeneration from rabbit bone marrow cells using RWV bioreactor

The rotation speed of the RWV was adjusted manually to prevent the cell aggregates from sinking in the RWV vessel. The speed was varied between 11 and 25 rpm and increased steadily for 28 days (Fig.2). The change in speed originated from the steady increase in the mass of tissue formed in the vessel. Figure 3 shows images of the tissue formed in the RWV vessel and in the 15-mL conical tube after 1 and 4 weeks of culture. The tissues are of a cylindrical shape. The size (height/diameter) of the tissue in the RWV vessel was 1.00/0.48 cm at 1 week; 1.28/0.53 cm at 2 weeks; and 1.25/0.60cm at 4 weeks. This kind of single tissue formed in the reactor reproducibly in the same experimental conditions. By contrast, the tissue formed in the 15-mL conical tube was smaller: 0.20/0.90 cm at 1 week; 0.20/0.55 cm at 2 weeks; and 0.28/0.74 cm at 4 weeks. The qualities of the tissues as cartilage were evaluated by histochemical methods including immunostaining of collagen type I and collagen type II and safranin-O and toluidine blue staining(Fig.4). The staining of collagen type II was more intense in the RWV than tube culture The results of safranin-O and toluidine blue staining are clearer as shown in Figure 4c and d. The time course of safranin-O and toluidine blue staining in the tissues formed in the tubes shows that the tissue gradually became chondrogenic. However, this change occurred faster in the RWV tissues. Even at 1 week, chondrogenesis occurred in the RWV tissues, but no sign of chondrogenesis was detected in the tissues in the tubes. At 2 weeks, the difference between the two was clearer, and at 4 weeks, chondrogenesis of the RWV tissues was confirmed by the strong and homogeneous matachromasy of safranin-O staining. Comparing the results of safranin-O staining, the RWV tissue at 1 week appeared to be in a similar stage of differentiation as the tube tissue at 4 weeks. In conclusion, we succeeded in the rapid regeneration of three-dimensional large and homogeneous cartilaginous tissue from rabbit bone marrow cells without a scaffold using a RWV bioreactor. Bone marrow cells cultured for 3 weeks were resuspended and cultured for 4 weeks in the chondrogenic medium within the vessel. Large cylindrical cartilaginous tissue 1.25 cm in height and 0.60cm in diameter formed. Their cartilaginous properties were demonstrated by immunohistochemistry of collagen types I and II, mRNA expression of aggrecan, collagen types I and II, GAG/DNA ratio, toluidine blue, and safranin-O staining, and polarization.

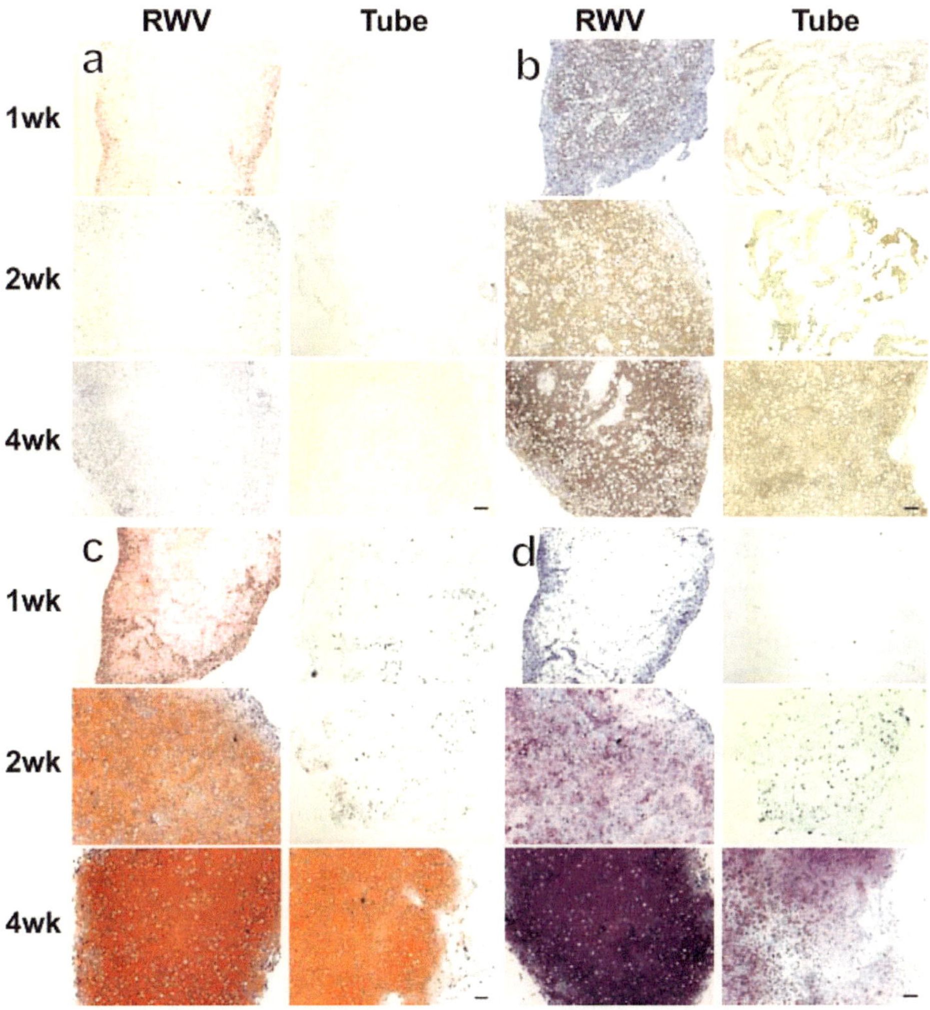


Fig. 4. Immunostaining of collagen type I (a) and collagen type II (b), safranin-O staining (c), and toluidine blue staining (d) of the tissues formed in the RWV vessels and in the conical tubes at 1, 2 and 4 weeks. Bar.1 mm. (Ohyabu et al., 2006)

Despite numerous studies on the formation of cartilage tissue from chondrocytes (Baker and Goodwin, 1997; Duke et al., 1993; Freed and Vunjak-Novakovic, 1997b; Freed et al., 1997a; Marlovits et al., 2003), no report has been published on the production of cartilage tissue from bone marrow-derived cells in rotating cultures. Stem cell-derived tissue formation is a basic concept in tissue engineering. However, three-dimensional cartilage tissue has not yet to be produced in rotating cultures. Our study showed that (1) in vitro chondrogenesis was observed in 3D culture of bone marrow stromal cells and that (2) RWV culture yielded cartilaginous tissues that was superior to static culture. As shown in Figure 4, the cartilaginous tissue is homogeneous and showed no necrotic cells of large size 1.25/0.6 cm. These results are encouraging and promising for cartilage tissue engineering, and an experiment in which the tissue formed by an RWV bioreactor as transplanted to large osteochondral defects described in the next paragraph. It is true that it is the first study to utilize bone marrow stromal cells, but we can gain a lot of benefit by using bone marrow cells. First, the number of autologous cultured chondrocytes is limited and applicable to only a limited area of cartilage damage, while a greater number of chondrocytes can be cultured from bone marrow cells (Wakitani et al., 2004). From this point of view, the clinical applications of RWV using bone marrow cells are wider than those using autologous chondrocytes. Second, bone marrow cells might be suitable for cartilaginous tissue formation by RWV culture. It has been reported that the exposure to RWV at an early stage of chondrogenesis severely limits the ability for cartilage growth, however, at a late stage of chondrogenesis, the RWV environment is beneficial and enhances growth and development using embryonic mouse pre-bone tissues (Klement et al., 2004). These results seem to contradict our results obtained using bone marrow stromal cells. Bone marrow stromal cells might be more suitable for RWV culture than embryonic cells. There are two major differences between the RWV culture and tube culture. One is the way in which the cells aggregate gathered manually in the RWV, whereas they gathered due to centrifugal force in the tube. The other point is that the unique mechanical stress due to the medium flow and gravity continued to apply to the tissue in the RWV (Klaus, 2001; Klement et al., 2004). We speculate that the major reason that superior tissue formed in the RWV was that the appropriate mechanical force was applied to the naturally gathered cells. To study the methodology of cartilage regeneration from bone marrow cells, a comparison of the tissue engineered in the RWV bioreactor with and without a scaffold would be useful. This kind of study was published by our group (Ohyabu et al. 2009).

### 2.3 Regeneration of cartilage tissue in vivo using rabbit bone marrow cells and an RWV bioreactor

As described in the last section, the application of a rotating wall vessel (RWV) bioreactor that simulates a micro-gravity environment with low shear stress for regenerating cartilage tissue and successfully established a three-dimensional (3-D) cell culture technique for the formation of large and homogenous cartilaginous aggregates from bone marrow-derived cells without using a scaffold. This bioreactor generates stress through the horizontal rotation of a cylindrical vessel equipped with a gas exchange membrane. The bioreactor compensates for the effect of gravity, resulting in homogenous growth and differentiation without sinking, and the cells aggregate and form 3-D tissue. This section describes the usefulness of transplanting cartilaginous aggregates formed from rabbit bone marrow-derived cells using an RWV bioreactor (Yoshioka et al., 2007).

Cylindrical defects 5x5 mm in area and 4 mm in depth were created on the patellar groove of the rabbit femur with a hand drill. For the control, the defects were left empty (Group-C: control group, n=18), whereas 3-D cartilage aggregates of 10 mm x 5 mm (height x diameter) were placed into the defects without any flap (Group-T: transplanted group, n=18) as shown in Fig.5. The rabbits were then caged and allowed to move freely without any splinting before being sacrificed 4 (n=6), 8 (n=6) and 12 (n=6) weeks after the operation. Fig.5 shows a macroscopic view of defects of the patella groove of the knee and histological view of a section of osteochondral tissue 4 weeks after transplantation. Large osteochondral defects in the knee joints were successfully repaired with hyaline-like cartilage after the transplantation of allogeneic cartilaginous aggregates formed from bone marrow-derived cells using the RWV bioreactor. As early as 4 weeks after the operation, the defects were filled with reparative tissue that resembled hyaline cartilage. The reparative tissue had a smooth surface and there were no fibrous tissues between the reparative tissue and adjacent normal cartilage. At 8 weeks, enchondral bone had formed in the deeper portion of the reparative tissue. At 12 weeks, in some cases the intensity of staining with safranin-O was slightly reduced in comparison to that in the adjacent normal cartilage, but the reparative tissue retained its thickness. This is the first report of the rapid regeneration of critical osteochondral defects with allogeneic cartilaginous aggregates formed from bone marrow-derived cells without any scaffold using the RWV bioreactor.

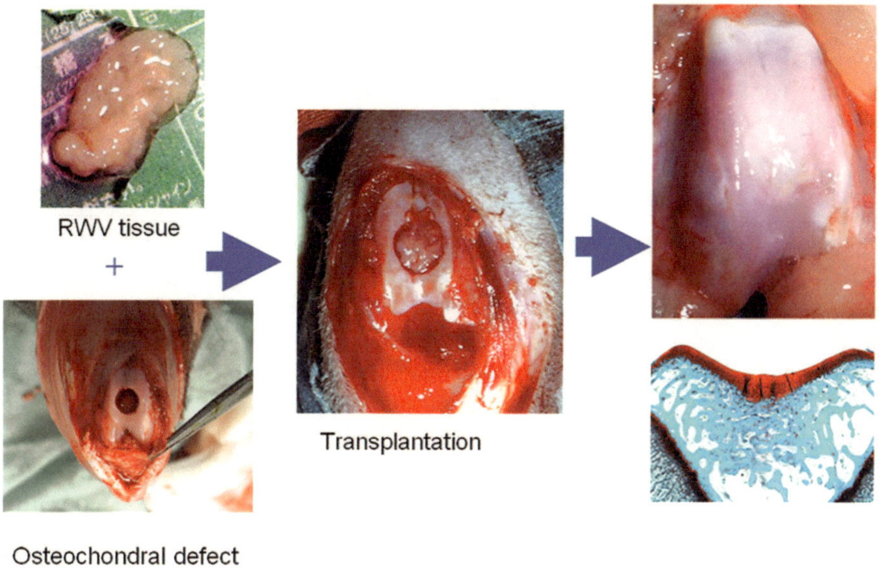


Fig. 5. Transplantation of RWV tissue into osteochondral defects in rabbit knees

There are two important advantages. First, the cells are derived from bone marrow and mesenchymal stem cells (MSCs). Therefore it is possible for the cells to proliferate in a monolayer culture and then to differentiate into cartilage and bone (Pittenger et al.1999, Caplan 1991) . An additional advantage to using bone marrow-derived cells in a clinical context is that they can be collected through aspiration of both sides of the iliac crest under partial anesthesia, a procedure that is very easy to perform and minimally invasive. Second, we used the RWV bioreactor as the 3-D culture system which stimulates chondrogenesis in a simulated micro-gravity environment (Unsworth et al. 1998). In previous reports (Wakitani et al. 1994, Im et al. 2001), engineered cartilage from MSCs required a scaffold for the cells to remain in the defect and to act as a support for inducing the formation of hyaline cartilage. However, using our technique, it is possible to form large and homogenous cartilaginous aggregates without any scaffold which can thus be transplanted into large osteochondral defects, which do not spontaneously regenerate in the rabbit (Shapiro et al. 1993). These aggregates have already produced an abundance of extracellular matrix and type-II collagen which was used to identify the chondrogenic phenotype *in vitro*. This suggests that the aggregates formed in the RWV bioreactor have the characteristics of hyaline cartilage. Rich extracellular matrix embedded chondrocytes to maintain their phenotype protecting them from dedifferentiation. In addition, we had a specific reason to choose aggregates at as early as 1 week after culture in the RWV bioreactor for transplantation. According to our in vitro study(Ohyabu et al.2006), the cells of cartilaginous aggregates after 1 week of culture are not mature, but consist of undifferentiated or chondrogenic precursor cells, which might exhibit plasticity to differentiate into another phenotype such as osteoblasts etc. The aggregates of such cells were influenced by various biological factors from the host bone marrow side (Engstrand 2003, Wozney et al. 1990) and interacted with adjacent cartilage (Tognana et al. 2005, Zhang et al. 2005). A suitable distribution of mechanical stress and synovial factors (Serink et al., 1997, Yanai et al. 2005) also influence the lineage of these cells. These characteristics may thus make it possible to rapidly and suitably regenerate cartilage-bone structure *in vivo*.

## 3. Rotating three-dimensional dynamic culture of adult human bone marrow-derived cells for tissue engineering of hyaline cartilage

The method of constructing cartilage tissue from bone marrow-derived cells in vitro is considered a valuable technique for hyaline cartilage regenerative medicine. Using a rotating wall vessel (RWV) bioreactor originally developed to simulate a microgravity environment, we attempted to efficiently construct hyaline cartilage tissue from human bone marrow-derived cells without using a scaffold (Sakai et al., 2009).

Bone marrow aspirates were taken from the iliac crests of 9 patients using an 11G Bone Marrow Harvest Needle during orthopaedic surgery (mean age of 36 years, range of 23-62 years; Table1) as shown in Fig.6. Briefly, 10 ml of bone marrow sample with anticoagulant was mixed with 25 ml of phosphate-buffered saline, and 35 ml aliquots of bone marrow suspension were overlaid onto a poly-sucrose gradient and centrifuged at 1500 *g* for 15 min at room temperature. The cell layer was carefully removed and suspended and cultured in a growth medium. The growth medium consisted of Dulbecco's Modified Eagle's Medium (DMEM) supplemented with 10% fetal bovine serum (FBS) and antibiotics. Approximately 2-3 weeks later, confluent cells were subcultured and seeded on two representative culture systems (RWV culture and pellet culture) for chondrogenic differentiation using chondrogenic differentiation medium(described in section 2.2)..

| Donor No | Age (years) | sex | predisposing factor | Nucleated cells in Peripheral blood (×1000cells/μl) | Nucleated cells in Bone marrow blood (×1000cells/μl) |
|---|---|---|---|---|---|
| Donor 30 | 28 | Male | Posttraumatic osteonecrosis of the femoral head | 3.8 | 27.5 |
| Donor 31 | 23 | Male | Posttraumatic osteonecrosis of the femoral head | 8.4 | 74.1 |
| Donor 32 | 28 | Female | Posttraumatic osteonecrosis of the femoral head | 10.7 | 22.1 |
| Donor 33 | 26 | Male | Idiopathic osteonecrosis of the femoral head | 7.1 | 36.2 |
| Donor 35 | 42 | Female | Osteoarthritis of hip joint | 7.7 | 32.3 |
| Donor 36 | 37 | Female | Steroid-induced osteonecrosis of the femoral head, SLE | 8.5 | 16.2 |
| Donor 37 | 28 | Female | Steroid-induced osteonecrosis of the femoral head, SLE | 9.9 | 28.3 |
| Donor 38 | 53 | Female | Posttraumatic osteonecrosis of the femoral head | 4.9 | 18.0 |
| Donor 39 | 62 | Female | Idiopathic osteonecrosis of the femoral head | 4.2 | 12.6 |

Table 1. Summary of data from 9 samples (Sakai et al. 2009)

In spite of some studies on the formation of cartilage tissue from chondrocytes cells (Freed et al. 1997a; Marlovits et al. 2003), no report has been published on the production of cartilage tissue from adult human bone marrow-derived cells including mesenchymal stem cells by rotating three-dimensional dynamic culture. After 2 weeks of differentiation and induction of adult human bone marrow-derived cells to form cartilage, the two culture methods were compared using the RWV bioreactor and the pellet culture as shown in Figs 7 and 8. The RWV system produced a larger tissue rich in GAG and collagen II, which are specific components of the cellular matrix of hyaline cartilage. These results suggest that the culture under hydrodynamic conditions using the RWV bioreactor provides a more suitable environment for the induction of cartilage differentiation from adult human bone marrow-derived cells than a conventional static culture system (pellet culture).

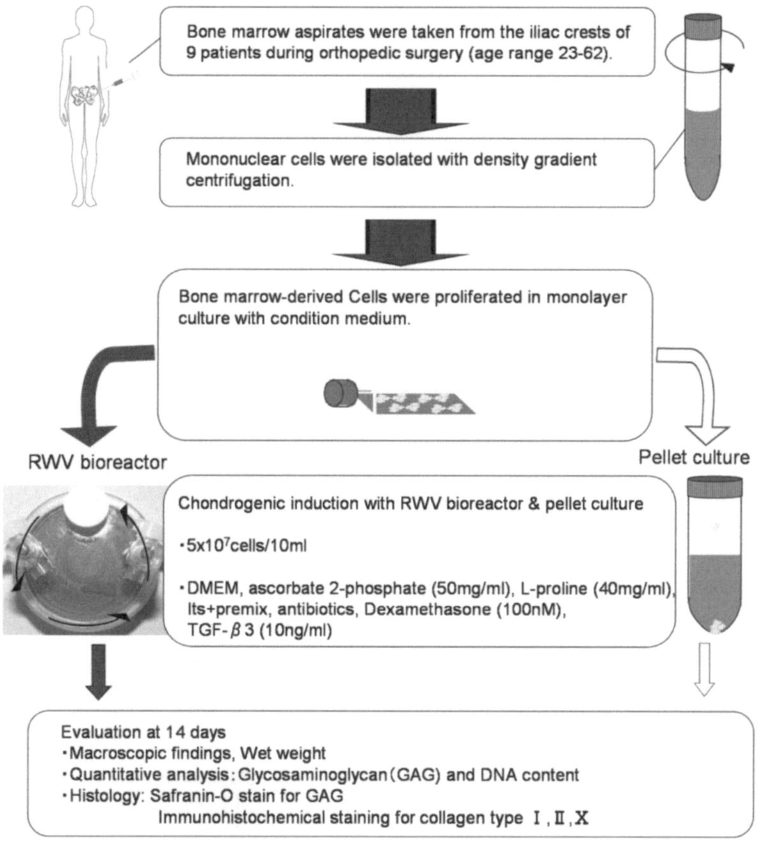


Fig. 6. Brief schematic of experimental protocol (Sakai et al., 2009)

In RWV cultures, rotating the culture medium in the vessel provides an ideal three-dimensional system to achieve high-density cell agglutination and obtain a spatially extended extracellular matrix. Decreased shear stress due to the rotating culture fluid makes high-density cell agglutination possible and affords effective nutrition supply and excretion owing to the circulating fluid. As a result, this system provides high tissue production and high maintenance. Cells tend to retain their differentiated phenotype in vitro only if cultured under conditions that resemble their natural environment in vivo. Therefore, it is reasonable to think that the tissue engineering of cartilage must be conducted in a physiological and mechanical environment similar to that during the *in vivo* ontogeny of the embryo. The formation of cartilage starts with cell agglutination, after early agglutination, an important event for maturation and skeletal patterning (Hall et al., 1992). The formation of bone and cartilage in the fetus occurs in a mechanical environment very similar to an environment with microgravity because the amniotic fluid provides buoyancy to the cells. Many studies have shown the effects of gravity and mechanical load on tissue, and a microgravity environment has considerable effect on tissue formation (Klement et al., 1994, 2004; Freed et al., 1999). Furthermore, some studies have shown an effect of gravity on embryonic bone and cartilage formation. It has been confirmed that the gravity load on mesenchymal stem cells during the early embryonic stages is an important factor determining the orientation of differentiation, and that the effect of a microgravity environment on cells differs between stages; however, the effects of microgravity on the differentiation of mesenchymal stem cells into bone, cartilage and adipose tissue remain to be determined. From the results of this study, the pseudo-microgravity environment provided by the RWV bioreactor may facilitate the differentiation of mesenchymal stem cells into cartilage.

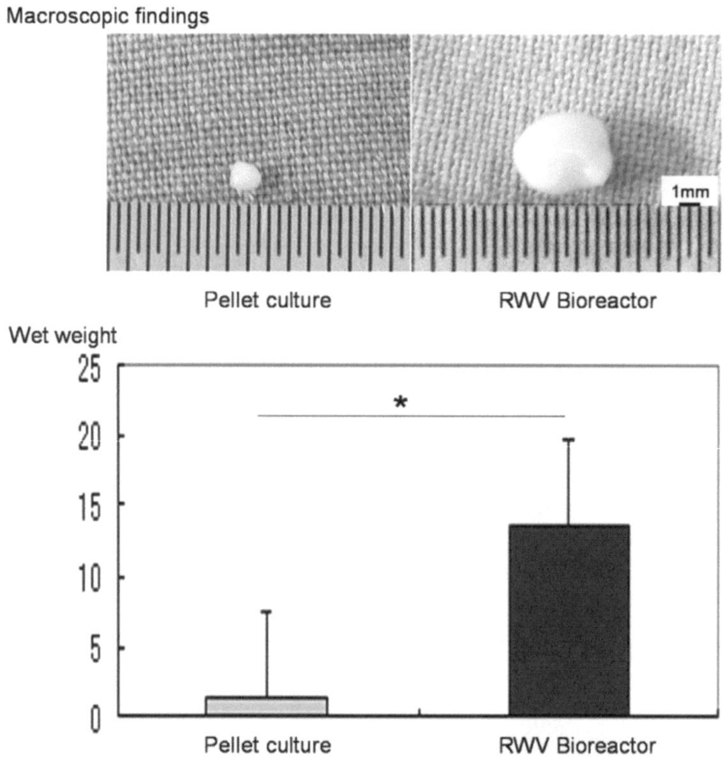


Fig. 7. (A) Comparison of the tissue formed in the pellet culture and RWVculture. Macroscopic photograph after 14 days of pellet culture (left), using an RWV bioreactor (right). (b) Wet weight of the tissue construct after 14 days. Data are expressed as the mean SD (n=9). : p<0.05 (Sakai et al., 2009)

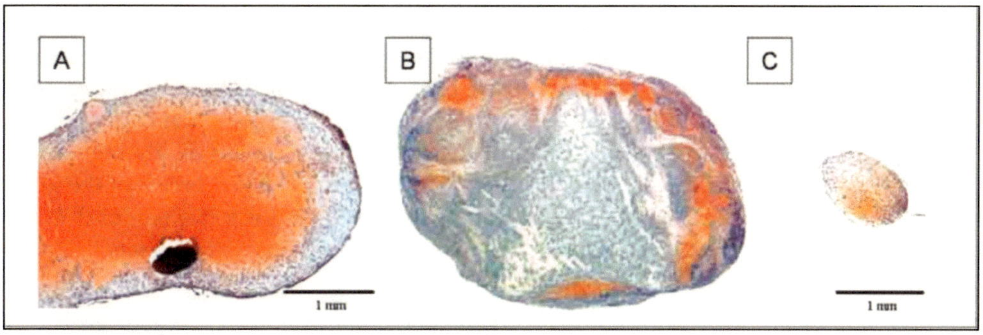


Fig. 8. Safranin-O/fast green staining of cell aggregates cultured using an RWV bioreactor(A, B) and pellet culture system (C). (Original magnification x40). Cell aggregates from 3 donors showed distinct staining with safranin-O (A). Cell aggregates from 6 donors cultured using an RWV bioreactor showed partial staining with safranin-O (B). Construct cultured using the pellet culture system showed faint and heterogeneous staining (C). (Sakai et al. 2009)

## 4. Automatic rotation control system for RWV bioreactors

The RWV bioreactor proved quite useful for the three-dimensional culture of mesenchymal cells as described above. This system is expected to be of laboratory use as well as clinical use, however, it currently requires manual control of the rotation speed to match the increase of mass of the growing tissue as shown in Fig. 2. To overcome this inconvenience, we developed an automatic rotation control system for the RWV bioreactor as shown in Fig.9. A method of catching the position of the cell aggregates in the rotating vessel is key.

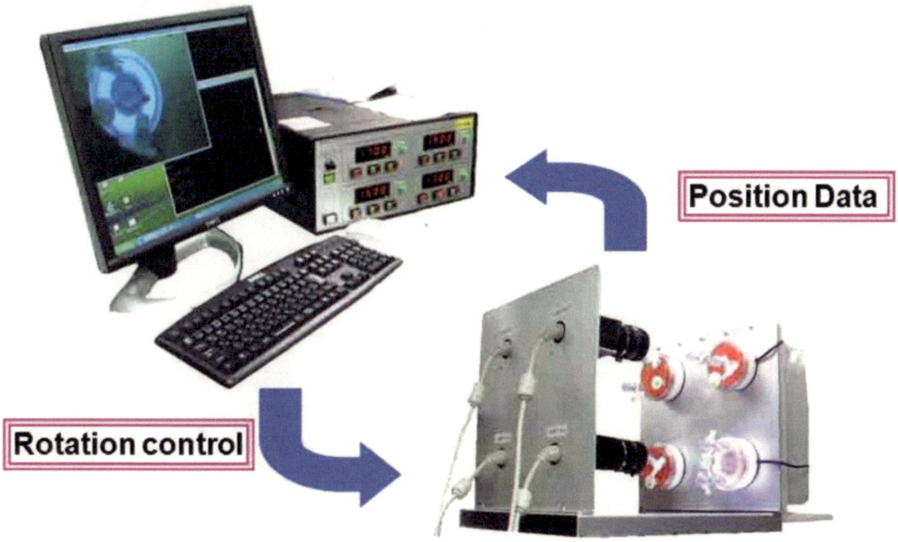


Fig. 9. Automatic Rotation Control System for RWV Bioreactors (for laboratory use)

An image of the vessel containing a medium and a cell aggregate is shown in Fig.8. As commercially available culture mediums are colored pink or red, it is difficult to visualize aggregates in a rotating vessel equipped with valves etc. The complex structure of the vessel prevent simple visualization of the cell aggregate with a CCD camera. Consequently, a tricky method was developed using a dual illumination system as shown in Fig.10. The tissue growing in the RWV vessel was observed using the CCD camera in front of the vessel, and visualized by two illumination systems, cold blue light and white light at the back and front of the vessel respectively, which enabled the visualization of only the tissue. By regulating the RGB signals from the CCD camera, it would be possible to get a masked image of the growing tissue. The control system calculates the position of the tissue and sends a feedback signal to adjust the rotation speed. This automatic system frees up technicians and students.

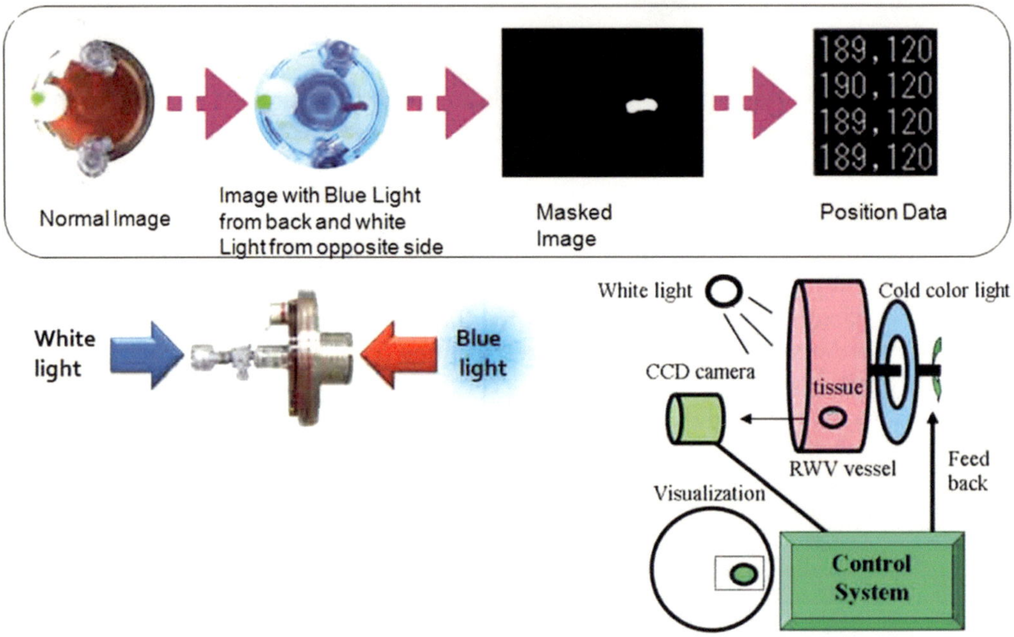


Fig. 10. Principle of visualization of a tissue in RWV vessel: The visualized tissue is judged by the program as to whether it is in the preset rectangular area in the vessel or not, and the control system sends a feedback signal to the rotation system in the RWV bioreactor to increase or decrease the rotation speed if it is outside of this area

## 5. GMP grade automatic RWV cell culture system

For using an RWV bioreactor, for instance for cartilage regeneration therapy, a rough scheme of a series of procedures could be drawn as shown in Fig.11. Bone marrow-derived cells aspirated from the iliac crest are cultured two dimensionally for expansion then three dimensionally by RWV bioreactor, after while they are transplanted into the damaged area. The procedures for cell treatment and cell culture are core processes to produce engineered tissues with structural integrity and functionality. Strict measures to prevent contamination and human error are needed due to the direct use of unsterile products and laborious nature of culture operations. For ensuring a reliable process and good quality products (engineered cartilage), a processing system for RWV culture is necessary and was developed, in which an automatic rotation control (described in the last section) and an automatic medium exchange process are available under GMP level as shown in Fig.12. The automatic RWV cell culture system mainly consists of three blocks, a $CO_2$ incubator and two refrigerators. The $CO_2$ incubator is equipped with an RWV system which rotates two vessels independently, stops their rotation when a change of medium is necessary, and moves the vessels outside of the incubator, where two robot arms grasp two syringes from the refrigerators ( one with fresh medium from the refrigerator beneath the incubator, the other for wasted medium from the refrigerator on the incubator) and moves them to the top and bottom of the vessel, inserts their needles into the caps of the vessel and changes the medium(Fig.13). After the medium exchange procedure, the robot arms deliver each syringe to each refrigerator, the vessel moves to the normal position and the rotation culture starts. In the rotation culture, the rotation speed is controlled by the method described in the last section.

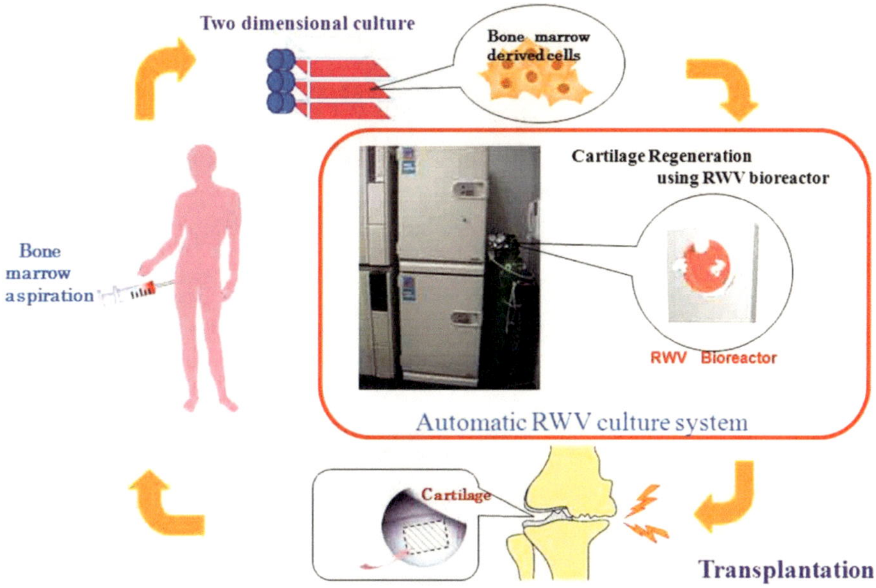


Fig. 11. Procedure for cartilage regeneration therapy using an RWV bioreactor and bone marrow cells

This system limits its function in rotation culture, however, such automatic processing systems are generally inevitable for safety processing for future therapeutic applications in tissue engineering and cell therapy (Kino-oka et al., 2009).

## 6. Acknowledgments

We thank Ms. Naoko Kida, Prof. Pi-chao Wang, Dr. Hajime Mishima, Dr.Shinsuke Sakai, Dr.Tomokazu Yoshioka, and Prof. Naoyuki Ochiai of Tsukuba University, Dr. Yoshimi Ohyabu, Dr. Toshiyuki Ikoma, and Prof. Junzo Tanaka of Tokyo Institute of Technology, Dr. Tetsuya Tateishi of NIMS(National Institute of Materials Sciences), Prof. Duke and Dr. Montufar-Solis of Univ. Texas, Prof. Atsushige Sato of Showa University, Prof. Yoshito Ikada of Nara Medical University, Mr. Hiroyuki Uematsu, and Mr. Nobuyuki Tsuji of Tsuji Electronics CO., Ltd., Dr. Hiromi Okada, and Mr. Tatsuo Masunaga of J-TEC CO. Ltd. , Dr. Hiroko Kojima of AIST(National Institute of Advanced Industrial Science and Technology), and Prof. Masahiro Kino-oka of Osaka University.

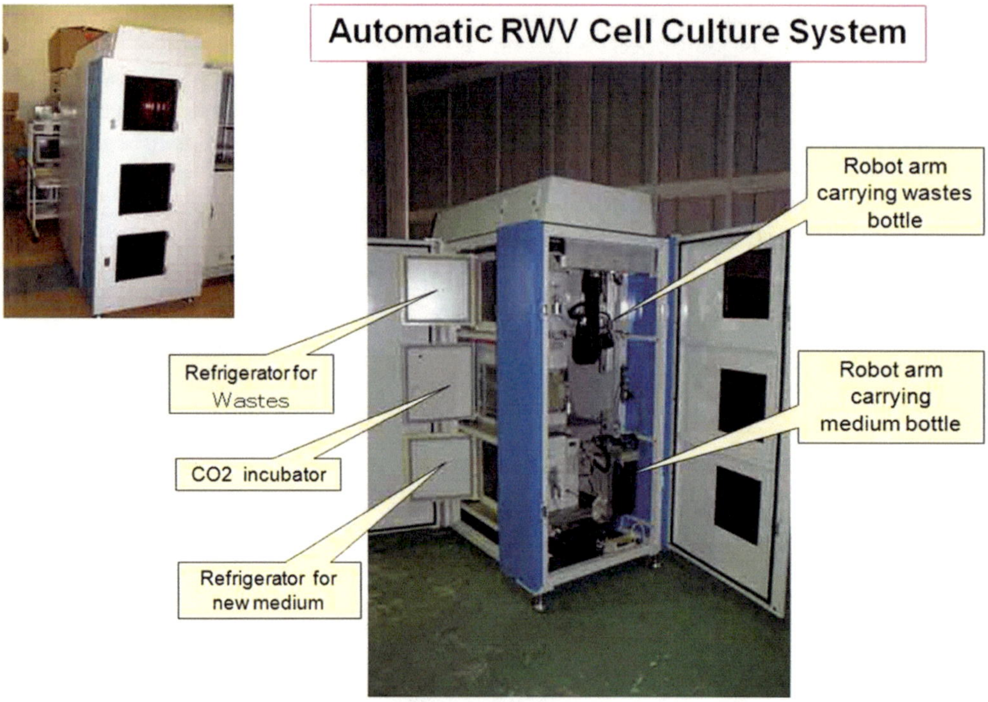


Fig. 12. Automatic RWV Cell Culture System

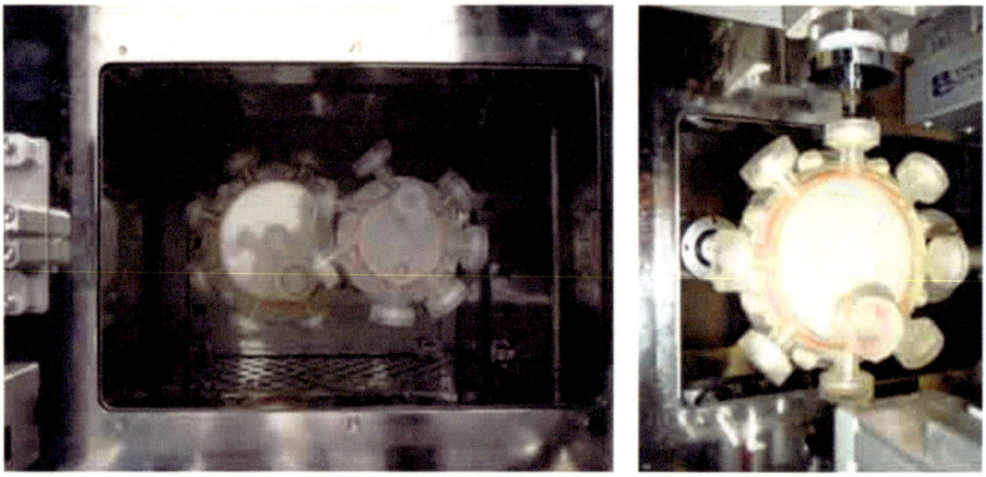

Fig. 13. Two vessels in the $CO_2$ incubator in the automatic RWV culture system (left) and the vessel out of the incubator with two attached syringes (right)

## 7. References

Abbott A. 2003. Biology's new dimension. *Nature* 424:870-872.

Baker TL, Goodwin TJ. 1997. Three dimensional culture of bovine chondrocytes in rotating-wall vessels. *In Vitro Cell Dev Biol Anim* 33:358–365.

Caplan AI. 1991. Mesenchymal stem cells. *J Orthop Res* 9:641-650.

Duke PJ, Daane EL, Montufar-Solis D. 1993. Studies of chondrogenesis in rotating systems. *J Cell Biochem* 51(3):274–282.

Engstrand T. 2003. Molecular biologic aspects of cartilage and bone: potential clinical applications. *Ups J Med Sci* 108:25-35.

Freed LE, Langer R, Martin I, Pellis NR, Vunjak-Novakovic G. 1997a. Tissue engineering of cartilage in space. *Proc Natl Acad Sci USA* 94:13885-13890.

Freed LE, Vunjak-Novakovic G. 1997b. Microgravity tissue engineering. *In Vitro Cell Dev Biol Anim* 33:381–385.

Freed LE, Pellis N, Searby N, de Luis J, Preda C, Bordonaro J, Vanjal-Novakovic G. 1999. Microgravity cultivation of cells and tissues. *Gravit Space Biol Bull* 12:57-66.

Goodwin TJ, Jessup JM, Wolf DA. 1992. Morphologic differentiation of colon carcinoma cell lines HT-29 and HT-29KM in rotating-wall vessels. *In Vitro Cell Dev Biol* 28A(1):47-60.

Goodwin TJ, Prewett TL, Spaulding GF, Becker JK. 1997. Three dimensional culture of a mixed mullerian tumor of the ovary: expression of in vivo characteristics. *In Vitro Cell Dev Biol Anim* 33(5):366-374.

Hall BK, Miyake T. 1992. The membranous skeleton: the role of cell condensations in vertebrate skeletogenesis. *Anat Embryol* (Berl) 186:107-124.

Holtzer H, Abbott J, Lash J, Holtzer S. 1960. The loss of phenotypic traits by differentiated cells in vitro. I. Dedifferentiation of cartilage cells. Proc Natl Acad Sci 46:1533-1542.

Ikada Y. 2006. Tissue Engineering:Fundamentals and Applications, *INTERFACE SCIENCE AND TECHNOLOGY, Volume 8,* Elsevier

Im GI, Kim DY, Shin JH, et al. 2001. Repair of cartilage defect in the rabbit with cultured mesenchymal stem cells from bone marrow *J Bone Joint Surg* Br 83:289-294.

Jessup JM, Goodwin TJ, Spaulding G. 1993. Prospects for use of microgravity-based bioreactors to study three-dimensional host-tumor interactions in human neoplasia. *J Cell Biochem* 51:290-300.

Johnstone B, Hering TM, Caplan AI, Goldberg UM, Yoo JU. 1998. In vitro chondrogenesis of bone marrow-derived mesenchymal progenitor cells. *Exp Cell Res* 238:265-272.

Kino-oka M, Taya M. 2009, Recent development in processing systems for cell and tissue cultures toward therapeutic application. *J Bioscience Bioengineering* 108, 267-276.

Klaus DM. 2001. Clinostats and bioreactors. *Gravit Space Biol Bull* 14: 55–64.

Klement BJ, Young QM, George BJ, Nokkaew M. 2004. Skeletal tissue growth, differentiation and mineralization in the NASA Rotating Wall Vessel. Bone 34:487-498.

Klement BJ, Spooner BS. 1994. Pre-metatarsal skeletal development in tissue culture at unit- and microgravity. *J Exp Zool* 269:230-241.

Klement BJ, Young QM, George, BJ, Nokkaew M. 2004. Skeletal tissue growth, differentiation and mineralization in the NASA rotating wall vessel. *Bone* 34: 487-498.

Marlovits S, Tichy B, Truppe M, Gruber D, Vecsei V. 2003. Chondrogenesis of aged human articular cartilage in a scaffold-free bioreactor. *Tissue Eng* 9(6):1215–1226.

Manning WK, Bonner WM. 1967. Isolation and culture of chondrocytes from human adult articular cartilage. *Arthritis Rheum* 10(3):235–239.

Ochi M, Uchio Y, Kawasaki K et al. 2002. Transplantation of cartilage-like tissue made by tissue engineering in the treatment of cartilage defects of the knee. *J Bone Joint Surg.,* 84:571.

Ohyabu Y, Kida N, Kojima H, Taguchi T, Tanaka J, Uemura T. 2006. Cartilaginous Tissue Formation From Bone Marrow Cells Using Rotating Wall Vessel (RWV) Bioreactor. *Biotechnol. Bioeng.* 95(5):1003-1008

Ohyabu Y, Tanaka J, Ikada Y, Uemura T. 2009. Cartilage Tissue Regeneration from Bone Marrow Cells by RWV Bioreactor Using Collagen Scaffold. *Mater Sci Eng* C29:1150-1155.

Passaretti D, Silverman RP, Huang W, Kirchhoff CH, Ashiku S, Randolph MA, Yaremchuk MJ. 2001. Cultured chondrocytes produce injectable tissue-engineered cartilage in hydrogel polymer. *Tissue Eng* 7(6):805-815.

Pittenger MF, Mackay AM, Beck SC, et al. 1999. Multilineage potential of adult human mesenchymal stem cells. *Science* 284:143-147.

Qiu QQ, Ducheyne P, Ayyaswamy PS. 1999. Fabrication, characterization and evaluation of bioceramic hollow microspheres used as microcarriers for 3-D bone tissue formation in rotating bioreactors. *Biomaterials* 20:989-1001.

Sakai S, Mishima H, Ishii T, Akaogi H, Yoshioka T, Ohyabu Y, Chang F, Ochiai N, Uemura T. 2009. Rotating three-dimensional dynamic culture of adult human bone marrow-derived cells for tissue engineering of hyaline cartilage. *J Orthop Res.*; 27(4):517-21.

Sanfold GL, Ellerson D, Melhado-Gardner C, Sroufe AE, Harris-Hooker S. 2002. Three-dimensional growth of endothelial cells in the microgravity- based rotating wall vessel bioreactor. *In Vitro Cell Dev Biol Anim* 38(9):493–504.

Shapiro F, Koide S, Glimcher MJ. 1993. Cell origin and differentiation in the repair of full-thickness defects of articular cartilage. *J Bone Joint Surg Am* 75:532-553.

Serink MT, Nachemson A, Hansson G. 1977. The effect of impact loading on rabbit knee joints. *Acta Orthop Scand* 48:250-262.

Sytkowski AJ, Davis KL. 2001. Erythroid cell growth and differentiation in vitro in the simulated microgravity environment of the NASA rotating wall vessel bioreactor. *In Vitro Cell Dev Biol Anim* 37:79–83.

Tognana E, Chen F, Padera RF, et al. 2005. Adjacent tissues (cartilage, bone) affect the functional integration of engineered calf cartilage in vitro. *Osteoarthritis Cartilage* 13:129-138.

Unsworth BR, Lelkes PI. 1998. Growing tissues in microgravity. *Nat Med* 4:901–907.

Wakitani S, Goto T, Pineda SJ, et al. 1994. Mesenchymal cell-based repair of large, full-thickness defects of articular cartilage. *J Bone Joint Surg Am* 76:579-592

Wakitani S, Imoto K, Yamamoto T et al. 2002. Human autologous culture expanded bone marrow mesenchymal cell transplantation for repair of cartilage defects in osteoarthritic knees. *Osteoarthritis and Cartilage* 10:199-206.

Wakitani S, Mitsuoka T, Nakamura N, Toritsuka Y, Nakamura Y, Horibe S. 2004. Autologous bone marrowstromal cell transplantation for repair of full-thickness articular cartilage defects in human patellae: Two case reports. *Cell transplant* 13:595–600.

Wozney JM, Rosen V, Byrne M, et al. 1990. Growth factors influencing bone development. *J Cell Sci Suppl* 13:149-156.

Yanai T, Ishii T, Chang F, et al. 2005. Repair of large full-thickness articular cartilage defects in the rabbit: the effects of joint distraction and autologous bone-marrow-derived mesenchymal cell transplantation. *J Bone Joint Surg* Br 87:721-729.

Yoshioka T, Mishima H, Ohyabu Y, Sakai S, Akaogi H, Ishii T, Kojima H, Tanaka J, Ochiai N, Uemura T. 2007. Repair of large osteochondral defects with allogeneic cartilaginous aggregates formed from bone marrow-derived cells using RWV bioreactor. *J Orthop Res.* 25:1291-1298.

Zhang Z, McCaffery JM, Spencer RG, et al. 2005. Growth and integration of neocartilage with native cartilage in vitro. *J Orthop Res* 23:433-439.

# 12

# Cartilage Tissue Engineering: the Application of Nanomaterials and Stem Cell Technology

Adelola O. Oseni, Claire Crowley, Maria Z. Boland,
Peter E. Butler and Alexander M. Seifalian
*Centre for nanotechnology & regenerative Medicine,*
*UCL Division of Surgery & Interventional Science,*
*University College London,*
*Royal Free Hampstead NHS Trust*
*U.K.*

## 1. Introduction

Replacement and reconstruction of pathological or absent cartilage within the human body has been a clinical challenge for many years. The avascular nature of cartilage tissue in all areas of the human body means it has little capacity for regeneration or repair beyond the production of functionally inferior fibrocartilage. Cartilage is injured in a number of ways; in the joint region, repetitive stress can cause irreparable damage, eventually resulting in Osteoarthritis, a debilitating disorder managed only with pain medication or joint replacement. A rise in the incidence of cancer has increased the prevalence of tracheal and nasal cancers, both frequently requiring radical resection as part of aggressive treatment regimes. Congenital disorders, such as Treacher Collins syndrome and Aperts syndrome can cause severe malformation of the ear and nose. It is evident that each of these clinical scenarios involves extensive damage to crucial skeletal cartilage and it is for these reasons that a drive for advancements in cartilage tissue engineering exists.

Tissue engineering uses principles of cell biology, engineering and medicine to generate constructs that can successfully recapitulate the function of native tissues in terms of histology, mechanics and morphology. There is a need for a suitable scaffold that can provide a 3D environment for cells to proliferate and adhere. Debate still continues over the key characteristics needed for the ideal scaffold, but they are likely to differ according to the type and location of cartilage to be engineered. Should it be biodegradable/non biodegradable, natural/synthetic, and what impact do these features have on the flexibility and strength of neocartilage constructs produced? There are many scaffolds that have been extensively investigated in cartilage tissue engineering research from natural collagen and alginate, to the synthetic Polyhydroxyacids and PEG hydrogels. Nonetheless, despite advancements in scaffold design, neocartilage constructs are still mechanically inferior to their natural counterparts, and in vivo problems of poor biointegration, and deterioration in tissue quality over time limit there translation into clinical use.

Nanomaterial science has introduced new methods for improving scaffold quality. Scaffolds can now be engineered on the nanoscale, using techniques such as electrospinning and 3D fibre deposition. Likewise, the incorporation of nanoparticles into polymeric material has allowed the addition of nanoscale features into the matrix structure. Both of these methods produce scaffolds that more closely replicate the extra cellular matrix environment found in native cartilage. It is hoped that this will increase cellular interaction with the scaffold and improve the quality of constructs produced.

With regards to the cell population to be used for engineering these constructs, there is a continued excitement over the possible application of stem cell technology. Stem cells are highly replicative and have multi lineage differentiation capacity. The traditional source of mesenchymal stem cells (precursor of chondrocytes) is bone marrow and various adjuncts to their propagation and differentiation have been explored in detail, such as growth hormones, biomaterials and environmental factors such as shear stress and oxygen tension that are important for culture techniques and bioreactor design. The discovery of new sources of mesenchymal stem cells, such as blood, adipose tissue or the synovium opens up a plethora of possibilities for clinical application, where methods of isolation and differentiation are being optimized.

In light of the numerous advancements that have been made in the last decade, this chapter aims to give a detailed account of cartilage tissue engineering strategy, looking with particular focus at the effect of scaffolds on cell growth, the evolution of stem cell technology and the expansion of bioreactor design and application . We will also explore how an integration of this revolutionary and innovative bench work can be translated into much needed clinical application in the not too distant future.

## 2. Cartilage in the human body

### 2.1 Cartilage tissue biology

Cartilage is a flexible connective tissue found in many areas of the human body, including the joints, ribs, nose, ear, trachea and intervertebral discs (Fig 1). In these regions cartilage can act as structural support, maintain shape or absorb shock during physical exercise. Unlike most other connective tissues, cartilage is largely avascular leading to hypoxic environments that limit the rate of cellular growth and tissue regeneration (115; 116). This in turn limits the capacity of cartilage to repair itself in the event of damage. The main cellular component of cartilage are chondrocytes, highly specialized cells that lie within spaces called lacunae and secrete the extracellular matrix (ECM) of cartilage . As with most connective tissues, the ECM of cartilage consists of a meshwork of macromolecules including collagens, elastin, glycoproteins and proteoglycans, each of which is present in varying amounts, depending on the type and function of cartilage. There are several cell surface receptors that allow chondrocytes to bind these proteins including the integrins, CD44, and the proteoglycan family of receptors e.g. syndecan (144).

The three types of cartilage are hyaline, elastic and fibrocartilage. Hyaline is the most abundant type, white-blue in colour and macroscopically smooth on its surface. It is present on the articular surfaces of joints and in the nasal septum. Hyaline cartilage is covered externally by a fibrous membrane known as the perichondrium, and in the joint especially, it is diffusion from the synovial fluid that provides this tissue with nutrition. It is rich in collagen type II, which forms a meshwork that encases giant aggregates of proteoglycans (Proteins with glycosaminoglycan (GAG) side chains e.g. aggrecan, biglycan, decorin in the extracellular matrix; syndecan, CD44 and fibroglycan as cell surface receptors; serglycan in intracellular tissues) (20; 21). These GAG side chains, keratan and chondroitin sulphate are able to retain water. Cyclical pressures from joint loading are crucial for normal hyaline cartilage function, and encourage the passage of water and nutrients between cartilage and synovial fluid. Elastic cartilage however, is more flexible, due to its rich elastin fibre content that is woven into a collagen mesh (20; 21). Elastin is an insoluble protein polymer that when cross linked with desmosine and isodesmosine make up the elastin fibres themselves. This type of cartilage is also surrounded by perichondrium and is more commonly found in the pinna, Eustachian tube, larynx and epiglottis, providing crucial structural support and flexibility. The third type of cartilage is fibrocartilage, which contains abundant thick collagen type I in addition to type II, that are interlaced into a mesh work of longitudinal and circumferential fibres (20; 21). These collagen bundles impart a great ability for this type of cartilage to withstand high tensile stresses. Fibrocartilage is usually found with the intervertebral discs, sacroiliac joints, pubic symphysis and costochondral joints.

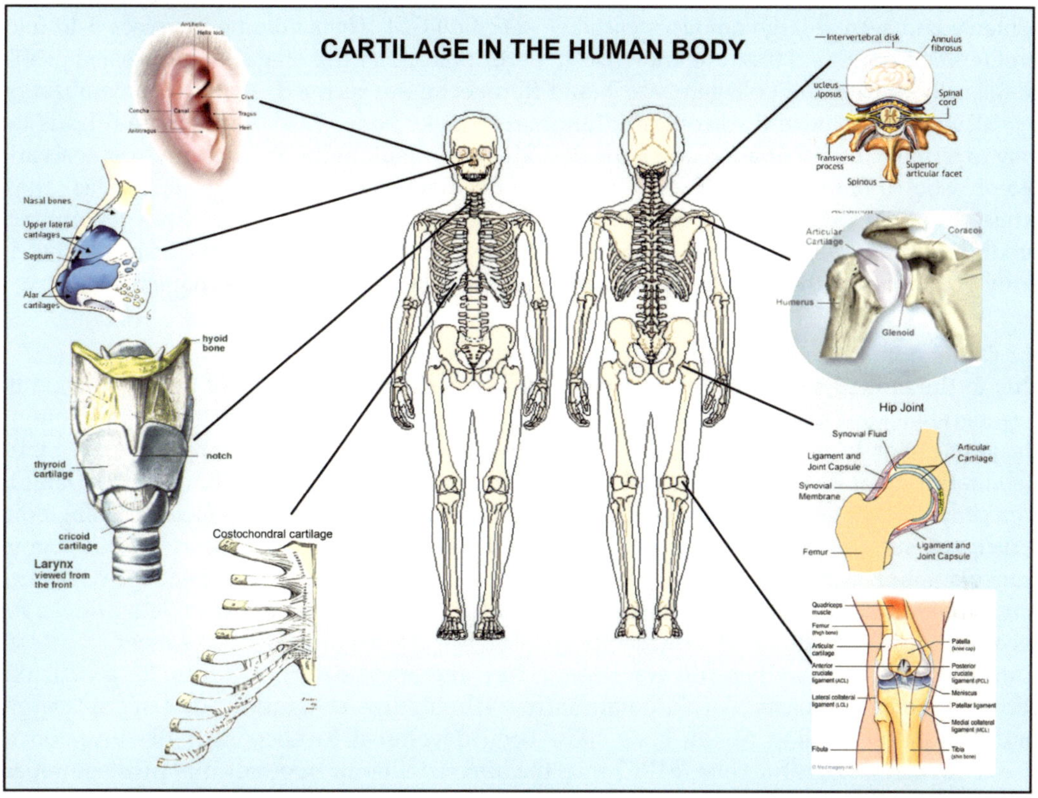


Fig. 1. A diagrammatic representation of cartilaginous regions in the human body

### 2.2 Development of cartilage

Central to effective tissue engineering practice is the understanding of tissue origin and development. This is based on the widely accepted hypothesis that natural tissue regeneration recapitulates developmental processes (118); hence embryological study can give an insight into the regulatory processes and patterns that govern tissue type and function, in addition to forming a foundation for understanding the degeneration and damage seen in tissues of the human body. We can only give a brief outline of the development of cartilaginous tissue specifically, however interested parties are advised to consult specific reviews (17; 70; 121) and books that devote entire chapters to this topic.

One of the epicentres of skeletal cartilage development is the growth plate which produces long bones via the cartilage template in a process known as endochondral ossification. It is important to note that this process is specific to the articular regions of bones and is followed by the eventual replacement by bony tissue. A milieu of hormones and paracrine factors govern a complex interplay of chondrocyte proliferation and differentiation, and the process can be divided into five stages, with the first three mainly being crucial for cartilage formation (144). Mesenchymal stem cells (MSCs) are first committed to becoming chondrocytes by paracrine factors that induce the expression of key transcription factors Pax 1 + scleraxis, which in turn activate cartilage specific genes. During the second stage, the committed MSCs condense into compact nodules and differentiate into chondrocytes. Chondrocytes then proliferate rapidly during the third stage, increase their cytoplasmic contents and secrete large amounts cartilage specific ECM. Their volume increases 5-10 fold, proliferation slows and they are known as hypertrophic. After this stage, the expression profile of the cells change and collagen type X and fibronectin are secreted, enabling mineralization by calcium carbonate and osteoblast infiltration to make bone. Vascular infiltration leads the way to terminal differentiation and bone development, resulting in chondrocyte apoptosis and osteoblastic differentiation. Facial cartilage development is very different, as it is embryologically derived from the cranial neural crest cells that originate from the anterior hindbrain. These cells migrate to specific locations and differentiate under the instruction of an array of Hox genes, the complexities of which are outside the scope of this chapter.

### 2.3 Clinical need for cartilage

Due to the limited self healing capacity of human cartilage, the repair of defects caused by degenerative joint diseases, cancer or trauma gives rise to a challenging clinical problem. In the joint region in particular, lesions of the articular cartilage are frequently associated with debilitating pain and reduced functionality. If not successfully treated long term disability can only be averted by total replacement of the joint. Damage to facial cartilaginous structures such as the nose or ear are only resolved with a prosthesis or autologous transplantation surgery that results in the formation of a donor site and frequently requires a number of revision surgeries. Large scale damage to the trachea has even less options for reconstruction with stents and tracheotomy tubes being the mainstay of treatment.

Cartilage regeneration has always been a key therapeutic target for treating articular cartilage damage in particular. Popular marrow stimulating techniques using micro-fracture or subchondral drilling of the bone have been developed to encourage the invasion of mesenchymal progenitor cells (MPC) into the affected site for spontaneous cartilage repair (94; 109). Unfortunately the outcome of such techniques varies greatly due to the lack of biological instructions for the MPCs to follow. This results in the formation of fibrocartilage which compared with hyaline tissue, has reduced durability and functionality (87; 140). The later invention of cell based therapies such as Autologous Chondrocyte Implantation (ACI) provided an important breakthrough treatment of articular cartilage damage and paved the way perhaps for more complex tissue engineering approaches with matrix assisted ACI introduced later on (16). ACI involves harvesting and propagating a population of autologous mature articular chondrocytes in vitro and re-introducing them into the defect site in cell suspension or in a supported matrix. They are then expected to lay down ECM to repair the site of injury (12; 102). Clinically the results of such procedures are good as they appear to provide symptomatic relief for patients. However, histologically the tissue produced is far inferior to native hyaline, being fibrotic in nature, again with limited

functionality and durability (19; 66; 120; 141). Further evaluations of such techniques has evidenced a strong correlation between the quality of tissue produced and the symptomatic relief of the patient from swelling and pain, once again highlighting the importance of tissue quality in cartilage regeneration. It can be postulated that these clinical breakthroughs buttressed the drive for advancements in cartilage tissue engineering technique.
It is also essential to note that the desired characteristics of engineered cartilage depend heavily on the site to be reconstructed. For instance, in tracheal constructs mechanical integrity, strength, flexibility and durability are all crucial for function, whereas in the facial region the aesthetic properties of the constructs may be equally as important. Taking the specific requirements of the tissue to be regenerated into consideration informs the tissue engineering strategy and the expected out comes of such undertakings.

### 2.4 Tissue engineering cartilage

Cartilage tissue engineering paradigm is based on the isolation of chondrocytes/ chondrocyte precursors from a tissue biopsy, expanding the cell number in culture, seeding them onto 3D scaffold, incubating for a period of time before placing the construct inside a patient. Many studies over the last decades have demonstrated that animal cells, when utilised in this way can produce tissues approaching the biomechanical and histological properties of native cartilage, even after implantation in vivo (3; 8; 24; 48; 58; 64; 77; 99; 110; 113; 152). However challenges do arise regarding the translation of such academic success into the clinical scenario. Challenges include isolating and propagating primary human cells, gaining relevant and reproducible construct morphology and size and ensuring good durability of the construct in vivo. Cell phenotype regulation, in vitro expansion of cell numbers, scaffold design and suitability, bioreactor design are all crucial components of the tissue engineering process that need to be optimized to advance cartilage tissue engineering from a mere academic prologue, into a clinical reality and success. These challenges will be discussed at length in the rest of this chapter.

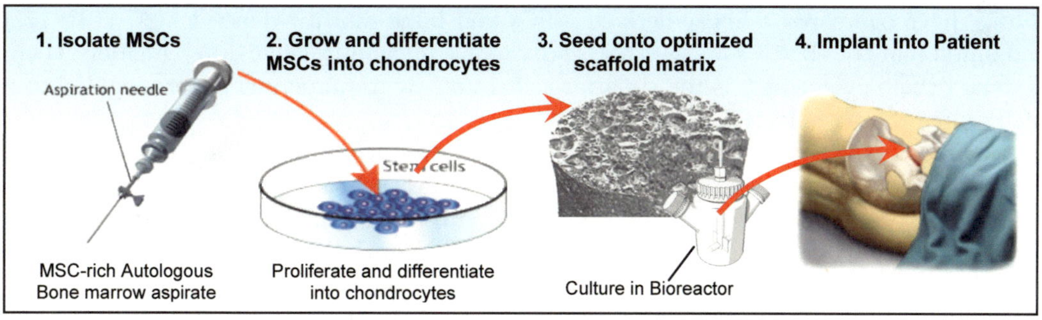


Fig. 2. A diagrammatic outline of tissue engineering strategy in cartilage tissue engineering for articular joint repair (an example of clinical usage)

## 3. Stem cell technology

Stem cells are unspecialised cells with a very high replication capacity. For cartilage regeneration, mesenchymal stem cells are the cell type of choice as they are multipotent stem cells, capable of differentiating into a number of lineages of the musculoskeletal system, including osteoblasts (bone cells), chondrocytes (cartilage cells) and adipocytes (fat cells). Although not immortal, these cells are capable of expanding through many passages in culture while retaining their growth and multi-lineage potential.

MSCs can originate from various tissues including the bone marrow (11), skeletal muscle, adipose (106; 159) synovium (134), the embryo and periosteum. The optimal cell source for cartilage tissue engineering is still being identified. When selecting an ideal cell source, it is important to achieve a number of criteria, including: (i) easy access to/harvesting of the source of MSCs, (ii) extensive self-renewal or expansion capability of the cells (to generate sufficient quantities of cells for large scale tissue engineering, (iii) the ability to readily differentiate into the chondrocytic lineage when induced, and (iv) a lack of or minimal immunogenicity or 'tumourigenic' tendencies. The two most commonly used MSC sources are adipose tissue and bone marrow. Unlike other sources such as embryonic tissue, there are few ethical issues associated with harvesting and using these tissues in research and development. Additionally, bone marrow MSCs (BMSCs) and adipose derived (ADMSCs) are relatively easy to source compared with synovium-derived- or periostium-derived MSCs.

Interestingly, bone marrow is the only organ in which at least two types of stem cells exist; hematopoietic and mesenchymal stem cells (137). The MSCs are found arrayed around the central sinus in the bone marrow. The cells can be isolated from the marrow using standardised techniques and expanded in culture through many generations, while retaining their capacity to differentiate along these pathways when exposed to appropriate culture conditions. Adipose tissue is an abundant, readily available source of MSCs. The cells can be isolated from fat that has been excised or 'liposuctioned' (lipoaspirate). There are advantages and disadvantages to both techniques. Anecdotally, it is thought that excised fat provides a higher yield of MSCs compared with lipoaspirate. This is due minimal mechancal impact upon cell membranes, which would ordinarily cause cell rupture, during the isolation process. Conversely, lipoaspirate is accessible without creating a large donor site defect, a major reason for pursuing tissue engineering methods at the outset. Some studies have compared adipose-derived MSCs and bone marrow-derived MSC (107; 122) and found that both BMSCs and ADASCs are capable of chondrogenic differentiation. There is some debate over which is the superior cell source, with numerous papers highlighting each source at optimum (107).

Mesenchymal stem cells can be identified using a number of methods. These include i) examination of cell morphology, ii) FACS (fluorescence activated cell sorting) analysis to detect the expression of MSC specific markers and iii) proving their differentiation capacity by differentiating the cells into a number of lineages, namely; osteoblastic, adipocytic, and chondrocytic. For FACS analysis, the presence of MSC-specific cell surface proteins such as the following are sought; CD 105 (SH2), CD 90 (THY1) and CD 73 (SH3/4). Similarly, negative markers are used to mark and remove cells expressing cell surface proteins not typically seen on MSCs, such as CD 45, CD 34, and CD 14 (9).

### 3.1 Stem cell differentiation to chondrocytes

Chondrogenesis is the term used to describe the process by which a stem cell is differentiated into a mature chondrocyte and is one of the earliest morphogenetic events of embryonic development (112). The stages where introduced earlier in the section on cartilage tissue biology. They include: MSC condensation, the rise of chondroprogenitors, chondrogenesis, terminal differentiation of progenitor cells and in skeletal development ossification (29).

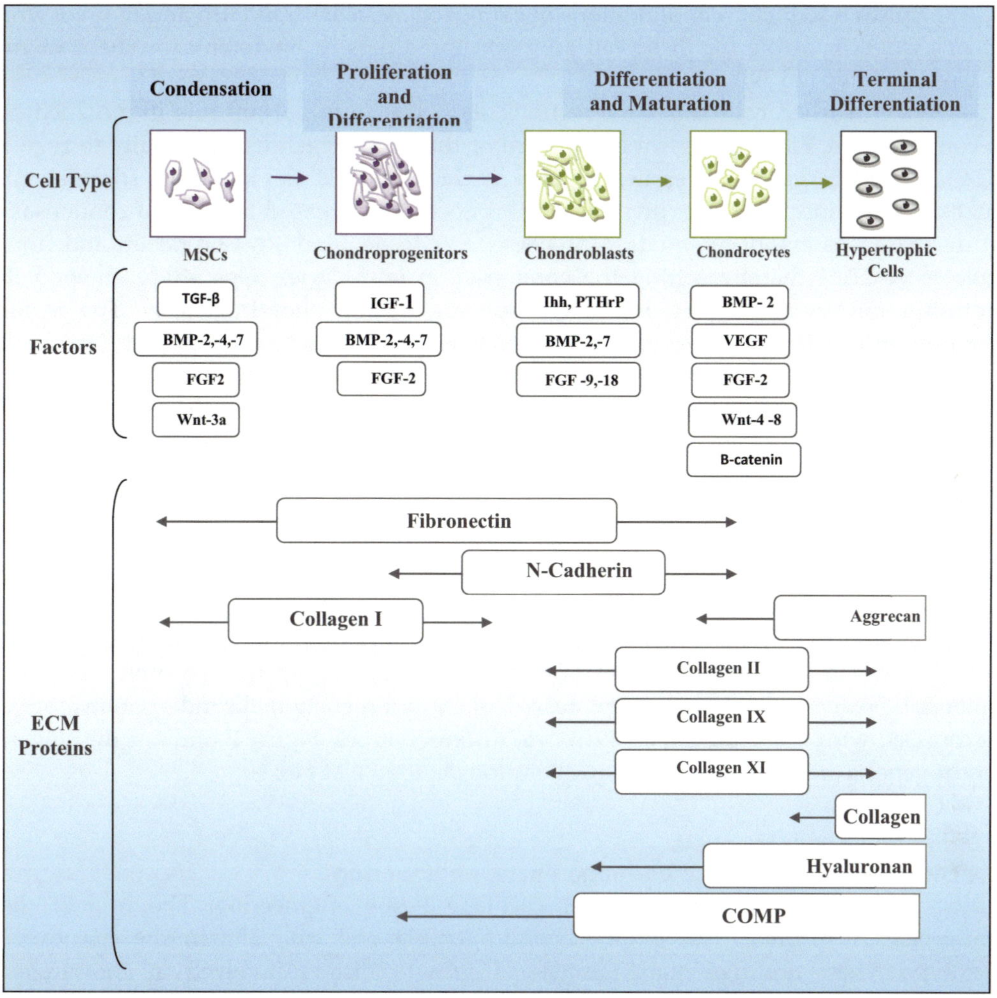


Fig. 3. Schematic diagram of the stages of chondrogenesis, the main growth factors involved in each stage and the accompanying alterations in ECM. (Adapted from Vinatier C. 2009 (144))

Condensation is directed by cell-cell and cell-matrix interactions as well as secreted factors interacting with their related receptors. Prior to condensation, the prechondrocytic MSCs secrete an extracellular matrix (ECM) which is rich in hyaluronan and collagen type I that prevents intimate cell-cell interaction (60). When condensation is initiated, there is an increase in hyaluronidase activity, thus causing a decrease in hyaluronan in the ECM. It is thought that hyaluronan facilitates cell movement, therefore, the increase in hylauronidase and subsequent decrease in hyaluronan allows for close cell-cell interactions. The establishment of the cell-cell interactions is thought to be involved in triggering one or more of the signal transduction pathways that initiate chondrogenic differentiation. Cell adhesion molecules: N-cadherin (neural cadherin) and N-CAM (neural cell adhesion molecule) are also expressed in the condensing mesenchyme, but disappear in the differentiated cartilage, and later can be found only in the perichondrium. Perturbing the function of N-cadherin or N-CAM causes a reduction or alterations in chondrogenesis both in vitro and in vivo, which further evidences a role for these cell adhesion molecules in mediating the mesenchymal condensation step (46; 60).

As mentioned earlier, cell-matrix interactions also play an important role in mesenchymal cell condensation. Fibronectin is a component of the ECM which has the ability to regulate N-CAM. TGF-β1, one of the earliest signals in chondrogenic condensation, stimulates the synthesis of fibronectin. The expression of fibronectin is increased in areas of condensation and decreased as cytodifferentiation proceeds. Syndecan binds to fibronectin and down-regulates N-CAM thereby setting the condensation boundaries. One study showed that fibronectin mRNA undergoes alternative splicing during chondrogenesis. The isoform containing exon EIIIA is present during condensation but disappears once differentiation begins. This suggests that this isoform switching is important for cytodifferentiation to occur. A later study by the same group determined that the function of the fibronectin EIIIA exon seems to regulate mesenchymal cell spreading, therefore permitting and/or promoting adequate cell-cell interaction to take place during the condensation phase of chondrogenesis (46).

The differentiation of chondroprogenitors is characterized by the deposition of cartilage matrix containing collagens II, IX, XI and aggrecan. SOX9 is one of the earliest markers expressed in cells undergoing condensations. It is required for the expression of the type II collagen gene (Col2a1) and other cartilage-specific matrix proteins, including Col11a2 and CD-RAP prior to other matrix deposition in the cartilage anlagen. Two other Sox family members L-Sox5 and Sox6, are not expressed in early mesenchymal condensations, but are co-expressed with Sox 9 during chondrocyte differentiation. Figure 2 outlines the stages of chondrogenesis and the accompanying alterations to the ECM (46; 60).

### 3.2 Inducing chondrogenesis

#### 3.2.1 Biochemical stimuli for cartilage tissue engineering

Specific biomolecules are essential for cartilage tissue engineering. The role of these biomolecules is primarily to induce chondrogenesis and to maintain the chondrocyte phenotype. There are five main families of growth factors involved in chondrogenic differentiation. These are: the transforming growth factor-β super-family (TGFβ), the fibroblast growth factor family (FGF), the insulin-like growth factor family (IGF), the wingless family (Wnt) and the hedgehog family(HH) (144). Below is a brief introduction to each growth family. However, for a more detailed description, the following references are recommended (46; 144). Figure 3 outlines the sequence in which the transcription factors are involved in each stage of chondrogenesis.

The transforming growth factor beta super-family is a family of proteins which have been shown to play a huge role in cartilage formation (11). Members include TGF-β, bone morphogenetic proteins (BMPs), inhibins and activins. All members have been shown to regulate cell growth, differentiation and apoptosis of a large number of different cell types including osteoblasts, chondrocytes, neural and epithelial cells. TGF-β is a secreted protein and exists in five isoforms TGF- β1-5. TFG- β 1,2 and 3 are thought to stimulate proteoglycans and type II collagen synthesis in chondrocytes as well as to induce chondrogenic differentiation of MSCs (144) . Studies have also shown that TGF- β isoforms differ in their effects on various cell types. For example, TGF- β1 has been shown to be responsible for the initial cell-cell interactions between condensing progenitor cells and TGF-β2 mediates hypertrophic differentiation (32).

BMPs are also members of the TGF-β super-family and comprise of a group of 20 proteins each one playing an important role in chondrogenesis and osteogenesis during skeletal development. BMP -2,-4,-6,-7 and -13 have all proven their ability to stimulate chondrogenesis in MSCs (144) . BMP2 in particular, has been found to be expressed in the condensing mesenchyme of the developing limb (101). It regulates chondrogenic development of mesenchymal progenitors (25) as well as stimulates the synthesis of chondrocyte matrix components by articular cartilage in vitro. Even combinations of many growth factors have been to enhance chondrogenesis, for example, BMP-2 with TGF-β3 and BMP-6 with TGF-β3 have been proven to stimulate chondrogenic differentiation and result in chondrogenic lineage development (73; 127).

The FGF family is a group of growth factors consisting of 22 members. Most FGFs are secreted, except for FGF1, 2 and FGF 11 and 14 (144). Signalling by FGF18 and FGF receptor 3 have demonstrated regulation, proliferation, differentiation and matrix production of articular cartilage and growth plate chondrocytes in vivo and in vitro (45).

The IGF family is a group of proteins which have a high similarity to insulin. IGF-1 has the ability to mediate chondrogenesis by increasing proteoglycan and collagen type II production (144). Combining TGF-β3 and IGF-1, has been shown to enhance chondrogenic induction (72). One study examined the effect that IGF-1 has on the chondrogenesis of bone marrow MSCs in the presence and absence of TGF-β signalling. It showed that IGF-1 could modulate MSC chondrogenesis by stimulating proliferation, regulating cell apoptosis and inducing expression of chondrocyte markers. In addition, it proved that the chondroinductive actions of IGF-1 were equally potent to TGF-β1 and independent from the TGF-beta signalling (98). Another similar study investigated the effects of IGF-1 on TGF-β1 induced chondrogenesis. It was found that the combination of IGF-1 and TGF-β1 produced higher amounts of glycosaminoglycan than TGF-beta1 alone at 8 weeks (124).

### 3.2.2 Mechanotransduction in cartilage tissue engineering

In vivo, articular cartilage experiences a variety of stresses and strains on a daily bases. Thus, many groups have extensively researched into the various mechanical stimulation methods for enhancing chondrogenic differentiation and cartilage tissue engineering (67; 68; 135).Examples of the types of mechanical stimuli examined include: hydrostatic pressure (4), cyclic mechanical compression (4), shear stress (139; 143), pulsed ultrasound (130) and dynamic compressive strain (23).

It has been found that exposing differentiating MSCs to various mechanical stimuli results in a shift in the types of protein expressed during chondrogenesis. For example, the application of cyclic, mechanical compression has been shown to result in an increase in proteoglycan and collagen contents as well as a higher amount of proteoglycan-rich, extracellular matrix production. Similarly, the application of shear stress by perfusion to differentiating BM-MSCs results in an enhanced ECM deposition and an increased collagen type II production (143). Low intensity pulsed ultrasound treated cell scaffold constructs show a significant increase of chondrogenic marker gene expression and extracellular matrix deposition in differentiating human BM-MSCs (130).

Collectively, the research shows that MSCs are mechanically sensitive and the chondrogenic differentiation can be modulated and enhanced by mechanical stimulation.

## 4. Biomaterials

It is well established that cells reside, proliferate and differentiate inside a complex 3-dimensional (3D) ECM environment. In cartilage, chondrocytes are surrounded by a highly hydrated matrix of proteins which informs many of their phenotypic states. For example, research has shown that isolated chondrocytes will loose their differentiated phenotypes if cultured in 2-dimension (42). These chondrocytes display a shift towards a fibroblastic phenotype, evident on protein assays and histological evaluations. Type I collagen expression is increased and the typical rounded morphology of the chondrocyte becomes spindle in shape (128). This process has been shown to be reversible upon relocation to 3D matrix environments such as pellet and micro-mass culture systems, which mimic the high cell density phenomenon seen during MSC condensation, a crucial stage of cartilage development (47; 74; 92).

It is in light of this that biomaterials have been proposed as engineered 3D environments in which chondrocytes can reside. For years, material scientists along with cell biologists have worked to optimize the tissue engineering characteristics of various biomaterials. There are a number of characteristics that are thought to be necessary for general tissue engineering attempts. These include a need for biomaterials to allow adequate cell adhesion and migration, with subsequent proliferation and differentiation. The overall architecture of the scaffold should guide and frame tissue formation, whilst providing mechanical support akin to that of native tissue. The scaffold should be porous, as porosity is thought to be crucial in maintaining the phenotype of the differentiated chondrocytes, considering their preference for 3D environments. It would also allow for mass transfer of nutrients and waste products. The scaffold should also be biocompatible with the ability to integrate into surrounding native tissue.

Biomaterial scaffolds can be broadly divided into natural and synthetic scaffolds. In this section we will give a brief overview of existing natural and synthetic scaffolds used for cartilage tissue engineering research, focusing on the regulatory influence these scaffolds have on cell behaviour and the potential application of nanomaterial science to this research.

### 4.1 Natural

Natural materials used as bioactive scaffolds include agarose, alginate, collagen, Hyaluronic acid and acellular cartilage matrix (Table.1). The potential for clinical use of these scaffold matrices is hampered however by poor mechanical strength and flexibility, in addition to a potential for disease transfer and immune system reactivity if allogenically sourced. Their biochemical make-up leaves them prone also to host-related degradations.

*Agarose*: a linear polysaccharide consisting of repeating units of agarobiose, derived from Asian seaweeds and capable of supporting the chondrogenic phenotype. Its ability to form hydrogels allows it to encapsulate chondrocytes providing a 3D matrix for their growth and development. A group in Germany performed allograft transplants of chondrocytes in agarose gel into osteochondral defects in the knee of rabbits. There we no graft versus host rejections, and after 18 months, 47% of grafts had morphologically stable hyaline -like cartilage (117).

| Ref. | Fabrication | Method | Cells Source | Outcome |
|---|---|---|---|---|
| **Natural** | | | | |
| ***ALGINATE*** | | | | |
| (49) | Chondrocytes in suspension with 2% sodium-alginate | *In vivo;* 500µl of suspension injected subcutaneously into dorsa of nude mice. Calcium chloride then injected into this area to stimulate cross linking of the scaffold. Cartilage harvested from 14 to 38 weeks | Human nasal septal chondrocytes | Gross analysis showed that 14/15 constructs resembled native human cartilage. 6 of the explants had histologically homogenous resemblance to native cartilage. The neo-constructs stained positively for Col II. |
| (97) | 3D alginate scaffold prepared by freeze drying | *In vitro;* Cells were cultured in the alginate for 1-4 weeks in a bioreactor | Porcine articular chondrocytes | RT-PCR analysis showed the cells maintained their differentiated phenotype for up to 4 weeks. The cell also proliferated increasing from $5 \times 10^5$ cells to $3 \times 10^7$. |
| (151) | 3D alginate gels | *In vitro;* cell/gel constructs were cultured for 0, 6, 12, 18 and 24 days | Human MSCs | Results of qRT-PCR analysis provided a temporal analysis for marker expression during chondrogenesis. Stage I (days 0–6): Col I and VI, Sox 4, and BMP-2. Stage II (days 6–12): Cartilage oligomeric matrix protein, HAPLN1, Col XI, and Sox 9. Stage III (days 12 -18): Matrilin 3, Ihh, Hbx 7, chondroadherin, and WNT 11. Stage IV (days 18–24): aggrecan, collagen IX, II, and X, osteocalcin, fibromodulin, PTHrP and alkaline phosphatise. |
| (91) | Alginate gel layer | *In vitro;* To evaluate the effect of low-intensity ultrasound (LIUS) on cell viability during chondrogenic differentiation | Human MSCs | When the cell/alginate construct was cultured with TGF-β1, cell viability decreased. However, addition of LIUS enhanced viability and inhibited apoptosis under the same conditions. Demonstrated by the expression profiles of apoptosis genes, p53, bax and bcl-2. |
| (30) | Hydrogel | *In vitro;* Chondrocytes were seeded onto alginate after 1, 2 and 3 passages in a monolayer. | Human nasal septal chondrocytes | Alginate stimulated GAG and Col I deposition supporting the chondrocytic phenotype. Results did not also support other research showing that culture with alginate beads can redifferentiate cells. |

| | | | | |
|---|---|---|---|---|
| ***CHITOSAN*** | | | | |
| (114) | Fibrous scaffold vs. sponge | *In vitro;* constructs analysed 3 days, 10 days and 21 days after cell seeding | Mouse BMSC line | At 10 and 21 days the cells were embedded but did not aggregate, with fibrous scaffolds containing more ECM. The cells had a round morphology. Histology revealed cell and ECM distribution was not homogenous. mRNA expression for Col II was 3 times greater for the fibrous scaffold compared with the sponge at 21 days |
| (22) | Chitosan scaffold and Chitosan microspher es | *In vitro;* Scaffold and microspheres used asTGF-β1 carrier to see the effect of this growth factor on chondrogenic potential | Rabbit articular chondrocytes | Encapsulation efficiency of TGF-β1 was 90.1%. TGF-β1 was released from chitosan in a multiphase fashion. TGF-β1 loaded microspheres significantly improved cell proliferation rate and Col II production, compared with controls with no microspheres or controlled TGF-β1 release. |
| (61) | Chitosan scaffold synthesized via freeze drying | *In vitro;* cells seeded on to chitosan of varying porosity; <10μm, 10-50μm and 70-120μm. Cultured for 28days in a rotating bioreactor | Porcine articular chondrocytes | Chitosan scaffolds remained intact compar4ed with the positive control PGA. However cartilage specific DNA levels and GAG were lower in the Chitosan groups compared with PGA. Chitosan also had the largest pores, with more Chondrocytes, but on histological analysis, the composition of cartilage produced on PGA resembled the structure of native cartilage more than chitosan constructs. |
| ***COLLAGEN*** | | | | |
| (38) | PLGA mesh and Collagen sponge | *In vitro;* hybrid disks of PLGA/Collagen scaffold with different structures. *In vivo;* week old cultured constructs implanted into dorsa of athymic nude mice and harvested after 2, 4 and 8 weeks | Bovine articular chondrocytes | Homogenous cell distribution with natural chondrocyte morphology. Abundant ECM production. Levels of GAG and Collagen II DNA, and aggrecan mRNA increased on the scaffolds with more collagen (semi, collagen on one side of PLGA and sandwich, collagen on both sides). Semi and sandwich compared with natural articular cartilage, had a Young's modulus of 54.8% and 49.3% respectively. 76.8% and 62.7% in stiffness. |

| | | | | |
|---|---|---|---|---|
| (156) | Hydrogel | *In vivo*: comparison of collagen hydrogel and collagen-alginate hydrogel. Gel injected subcutaneously into rabbit backs. | BM-MSC | Homogenous distribution of cells with chondrocyte characteristics demonstrated the chondrogenic differentiation of BM-MSCs. Both collagen hydrogel and collagen alginate hydrogel may induce chondrogenesis. Expression profile of cartilage specific genes differed between collagen hydrogel and collagen alginate, indicating that induction of chondrogenesis is materials dependent. |
| (153) | 3D collagen sponged | *In vitro;* Cells seeded onto collagen sponges and cultured in either standard or serum free culture conditions for 1, 2 and 4 weeks | Bovine articular chondrocytes | Overall chondrogenesis in serum free culture (Nutridoma replacement) was equivalent or better than control cultures in serum. Insulin-transferrinselenium (ITS+3) serum replacement cultures were poor due to decreased cell viability. The porous 3D collagen sponges were able to maintain chondrocyte viability, shape, and synthetic activity with evidence from quantitative assays for cartilage-specific gene expression and biochemical measures of chondrogenesis. |
| ***FIBRIN*** | | | | |
| (83) | Fibrin gel | In vivo: ACI on 30 patients using minimally invasive injection techniques. Mix of fibrin gel and chondrocytes. | Autologous adult chondrocytes | Patients evaluated 24 months post operatively using the Cincinnati knee ligament rating scores, for which 10 patients had excellent result, 17 with good results, two fair and one poor result. Further arthroscopy in 10 patients demonstrated good fill and integration in grafted areas. |
| (131) | PLGA/Fibrin hybrid scaffold | *In vitro;* PLGA scaffold soaked in chondrocyte-fibrin suspension (polymerized by thrombin $CaCl_2$ solution), Constructs were cultured for a maximum of 21 days. | Rabbit articular chondrocytes | Cell proliferation increased steadily until day 14, but declined by day 21. Cartilage formation evident at day 14, confirmed by the presence of cartilaginous cells embedded in basophilic ECM filled lacunae. Proteoglycan and GAG presence was confirmed. Suppression of the cart |

| | | | | |
|---|---|---|---|---|
| | | | | dedifferentiation marker Col 1 observed after 2 and 3 weeks in culture. sGAG production greater in fibrin/PLGA compared with PLGA control. |
| (28) | PLGA/Fibrin hybrid scaffold | *In vivo;* PLGA scaffold soaked in chondrocyte-fibrin suspension (polymerized by thrombin $CaCl_2$ solution) and constructs implanted subcutaneously into dorsum of nude mice for 4 weeks after culture for 3 weeks. Analysis performed at 1, 2 and 4 weeks. | Rabbit articular chondrocytes | Constructs maintained their shape and there was no significant difference between fibrin/PLGA and control PLGA. All exhibited smooth cartilage like properties 1, 2 and 4 weeks after implantation. Presence of proteoglycans and GAG was confirmed. The constructs were also strongly positive for Col II. Notably, sGAG production was greater on fibrin/PLGA scaffold than the control. Overall, both fibrin/PLGA and PLA showed comparable potential in sustaining the chondrogenic phenotype. |
| ***HYALURONIC ACID(HA)*** | | | | |
| (33) | ***Hyaff®-11, biodegradable polymer, nonwoven mesh | *In vitro;* Chondrocytes were harvested from OA patients and seeded onto Hyaff®. Constructs remained in culture for 28 days, analysed on day 0, 7, 14, 21 and 28. | Human Autologous chondrocytes | Viability and proliferation of OA chondrocytes similar to cells from normal subjects. Immunohistochemistry showed no signs of ageing or degeneration in cartilage produced by OA cells. The experimental groups and controls both had significantly raised Col II, Sox 9 and aggrecan. Suggests OA cells benefit from the HA rich environment. |
| (154) | Hydrogel (*in vitro*), beads(*in vivo*) | *In vitro* and *In vivo;* implanted into nude mice. Constructs were cultured in vitro for 2 weeks prior to implantation. Constructs remained implanted for 2 weeks. | Human MSC | Both *in vitro* and *in vivo* cultures of MSC-laden HA hydrogels enabled chondrogenesis. This was measured by the early gene expression and production of cartilage specific matrix proteins (aggrecan, Col II). HA hydrogels were compared to relatively inert poly(ethylene glycol) (PEG) hydrogels, and showed enhanced expression of cartilage specific markers |

| | | | | |
|---|---|---|---|---|
| (107) | HA immobilize d on surface of PLGA scaffold | *In vitro;* biodegradable macroporous PLGA scaffolds chemically conjugated to the surface exposed amine groups of the PLGA. Incubation times varied for each assay. | Bovine articular chondrocytes | Enhanced cellular attachment was observed compared with PLGA controls. GAG and total Col synthesis was significantly increased for HA/PLGA compared to the control. The HA/PLGA constructs exhibited morphological characteristics of cartilage and had cartilage specific Col II expression. |
| **Synthetic** | | | | |
| ***PLGA- Poly(lactic-co-glycolic) acid*** | | | | |
| (107) | PLGA scaffolds | *In vivo:* PLGA scaffolds were seeded with AD-MSC, cultured in TGFβ1 containing medium for 3 weeks, prior to implantation in the subcutaneous pockets of nude mice for 8 weeks. | Human AD-MSC | RT-PCR demonstrated the increased expression profiles of chondrospecific marker mRNA, compared with control samples after 3 weeks in vitro and 8 weeks in vivo. |
| (150) | HA modified porous PLGA scaffold | *In vitro;* cells seeded onto HA/PLGA scaffolds and cultured for a total of 5 days. | Human AD-MSC | The AD-MSC cultured in HA coated wells showed enhanced expression of cartilage specific mRNA. HA-modified PLGA did not affect cell adherence and viability, but did enhance gene expression after 1, 3 and 5 days in culture. GAG and Col I production enhanced after 4 weeks in culture compared with PLGA control. |
| (10) | PLGA scaffolds | *In vivo;* cells were pre-cultured on poly-HEMA coated dish, then seeded onto PLGA. The construct was implanted into the subcutaneous pockets of nude mice for 16 weeks. | Chondrocytes | Macroscopic signs of neo cartilage formation appeared at 8 weeks, and completed by 16 weeks. All constructs showed viable chondrocytes with normal lacunae and ECM. They stained positively for Col II. Control was a cell-free scaffold implanted into the other side of the dorsum on the same mouse. |

| | | | | |
|---|---|---|---|---|
| (75) | PLGA microspheres | *In vivo;* PLGA microsphere seeded with rabbit Chondrocytes injected subcutaneously into dorsa of athymic female mice | Autologous rabbit Chondrocytes | The PLGA microsphere permitted cell adhesion. 4 and 9 weeks post-implantation there was macroscopic and histological evidence of cartilage formation on the seeded PLGA microsphere compared with nothing on the PLGA and chondrocyte controls. |
| (142) | PLGA porous scaffold disks | *In vivo;* MSC seeded PLGA scaffold disks implanted into 36 week old Japanese white rabbits. Constructs were harvested after 4 and 12 weeks. | Rabbit BM-MSC | Engineered cartilage from autologous BM-MSC and PLGA scaffold filled the defects in the rabbit knees. The constructs were macroscopically and histologically similar to hyaline cartilage at 12 weeks post transplantation. |
| ***PCL- Poly(carprolactone)*** | | | | |
| (82) | 3 porous PCL scaffold types investigated (1)PCL/Pluronic F127, (2)PCL collagen and (3)PCL/Pluronic F127/collagen, in addition to (4) PCL only | *In vitro;* 3 porous PCL scaffold modifications investigated (1) PCL/Pluronic F127, (2) PCL collagen and (3) PCL/Pluronic F127/collagen, in addition to (4) PCL only. Cultured for 3 weeks. | Human BM-MSC | The 3 surface treated scaffolds had higher chondrospecific DNA content than the PCL only. GAG concentrations were also higher than in the PCL only, and RT-PCR showed that Sox 9 and Col IIA1 were remarkably elevated in the modified PCLs. Notably, Col IA1 and ColI0A1 mRNA levels were lower in the three modified scaffolds than in the PCL, suggestion prevention of the dedifferentiated phenotype. |
| (95) | Electrospun 3D nanofibrous scaffold | *In vitro;* MSC seeded onto pre-fabricated nanofibrous scaffold for 21 days | Human BM-MSC | Histological analysis was congruent with cartilage formation when cells were grown in medium containing TGFβ1. The cartilage specific gene profile (Aggrecan, Col II and Col X) was low, but improved significantly in chondrogenic medium with TGFβ1. Col X levels were paradoxically down regulated. There was positive immunohistochemistry for cartilage specific ECM molecules. |

| | | | | |
|---|---|---|---|---|
| ***PGA-Polyglycolic acid*** | | | | |
| (158) | Porous PGA and high density polyethylene composite scaffold | *In vivo;* High-density polyethylene carved into cylindrical rods (internal support), with non-woven PGA sheets wrapped around the rods to form the scaffold. Implanted subcutaneously into nude mice. | Porcine BMSC | 8 weeks post-implantation the constructs had formed mature cartilage with an abundant deposition of ECM on SEM. The experimental groups showed a positive histological likeness to cartilage with large number of lacunae and good expression of Col II. |
| (157) | PGA-HA composite scaffold | *In vivo;* MSC were seeded onto the PGA-HA and co-cultured for 72hours. There were then implanted into full thickness cartilage defects in the intercondylar fossa of rabbit femurs. Constructs were then harvested after 16 or 32 weeks of surgery. | Rabbit MSC | Grossly, the constructs demonstrated hyaline cartilage formation and at 16 weeks, there appeared to be integration with surrounding normal cartilage and subchondral bone. At 32 weeks there was no sign of degradation of the neoconstruct. |
| (160) | PGA vs. PLA bio-resorbable nonwoven scaffolds | *In vivo;* Cells seeded onto scaffolds and cultured for 7 days in serum free media, before implantation into subcutaneous nude mice for 6 and 12 weeks | Human articular chondrocytes | Aggrecan synthesis always higher in the PGA groups. mRNA gene expression for Col II significantly higher in the PGA groups after 6 and 12 weeks. Expression of Col X and cartilage oligomeric matrix protein increased on both scaffolds. |
| ***PEG- Poly (ethylene glycol)*** | | | | |
| (125) | PEG-peptide copolymer gels | *In vitro;* RGD and KLER sequences chosen as motifs to modify PEG gels. (KLER is a binding site from decorin protein, known to bind strongly to Col II, RGD promotes survival of encapsulated cells). Cells were encapsulated in the PEG peptide gel and cultured for 6weeks | Human MSCs | After 14 days, cells in RGD and KLER functionalized gels produced 2.5 times as much GAG as those only containing RGD. hMSCs also produced 27x as much hydroxyproline (a major component of collagen) than scrambled sequence gel controls. Col II was more prominent in KLER gels on immunostaining and RT-PCR analysis demonstrated higher levels of Col II and aggrecan synthesis. |

| | | | | |
|---|---|---|---|---|
| (111) | Hydrogel | *In vitro;* cells were encapsulated in the PEG hydrogel and allowed to free swell for 24hrs. | Bovine temporoman d-ibular chondrocytes | Condylar chondrocyte viability was maintained within the constructs during cell culture. RTPCR analysis showed the expression of cartilage specific markers, namely Col II, aggrecan and Col I was maintained |

Table 1. Summary of *in vitro* and *in vivo* studies that have used various scaffolds to engineer cartilage (2005-2010). Abbreviations: AD-MSC, adipose derived mesenchymal stem cells. BMSC, bone marrow stromal cells. BM-MSC, bone marrow derived mesenchymal stem cells. Col, collagen. ECM, extra cellular matrix. GAG, glycosaminoglycan. Hbx, homeobox. Ihh, Indian hedgehog. OA, osteoarthritis. PTHrp, parathyroid hormone replacement hormone. RT-PCR, real-time polymerase chain reaction. SEM, scanning electron microscopy. sGAG, sulphated glycosaminoglycan

*Alginate*: derived from brown marine algae and is consists of 1, 4-linked β-D-mannuronnic and α-L-guluronic residues, which are soluble in aqueous solutions. Cross-linking with bivalent cations such as $Ba^{2+}$ or $Ca^{2+}$ allows it to form stable gels.
*Chitin*: a polysaccharide based analogue of GAG found in the exoskeleton of arthropods. Relatively unexplored bioactive scaffold for tissue engineering, perhaps because it is degraded in vivo by lysozyme; an enzyme found in many human bodily fluids.
*Collagen 1 and II*: As the principle ECM components of cartilage, seeded chondrocytes can bind using inherent cell-surface receptors and use standard signalling pathways to regulate proliferation and growth. Can be fabricated as a sponge, foam, or gel, but like chitin, is subject to enzymatic breakdown.
*Fibrin*: Can be derived from autologous blood samples, and has a comprehensive history of biocompatibility in its clinical use as a wound adhesive. Chondrocytes have integrins that can bind directly to fibrin, much like with collagen.
*Gelatin*: A porous substance derived from hydrolysis of collagen. Its application as a scaffold for cartilage tissue engineering is relatively unchartered.
*Hyaluronic Acid*: a non-sulphated GAG, found in abundantly within the cartilaginous ECM. It is crucial for maintaining the biophysical properties of the cartilage ECM for optimum chondrocyte growth and proliferation.

### 4.2 Synthetic

The main aim of biomimetic materials (synthetic biomaterials) is to generate 3D scaffolds that support essential cell functions in addition to mimicking the biomechanical properties of host tissues, whilst avoiding host immune responses (Table.1). These are two characteristics more difficult to find in natural scaffold alternatives. When considering clinical applications, susceptibility to vascular invasion is a key consideration and there is continued debate between groups about the need for biodegradation. Persistence and stability have been the focal aims for tissue engineering cartilage with the mechanical and biochemical properties of synthetic materials being more amenable to modification than natural scaffolds.
*Polyhydroxyacids*: polyhydroxyacids such as PLLA [poly (L-lactic acid)], PCL [poly (L-lactide-ε-caprolactone)] and PGA [poly (glycolic acid)] have been well studied as potential

cartilage scaffold matrices, where they are easily extruded into fibrous or open lattice sponges. PGA is reportedly highly biodegradable (5 weeks); PLLA can stay *in vivo* up to 3 years. PCL and PGA used to fabricate ear templates for tissue engineering auricular cartilage (133).

*Elastomeric polyurethanes*: Well documented history of use in a variety of biomedical instruments, ranging from urinary and vascular catheters to intra-aortic balloons and mammary implants. Can be fabricated in a biodegradable form, and have been shown to support chondrocyte attachment and growth.

*PEG [poly (ethylene glycol)]*: FDA (Food and Drugs Administration) approved, and extensive research into its ability to promote chondrogenesis

### 4.3 Regulatory influence of scaffolds on cell behaviour

It is widely appreciated that soluble biochemicals such as cytokines, growth factors and chemokines affect the growth and development of all tissues including cartilage. Transforming growth factor beta (TGFβ) and bone morphogenic proteins (BMPs) have been evidenced as highly potent stimulators of cartilage tissue generation (78; 96; 118). In addition to such signalling mechanisms, ECM proteins such as collagens, glycosaminoglycans and proteoglycans exert an array of instructions on cells via transmembrane receptors that affect expression and therefore, cell behaviour. Much of this instruction will crosstalk with growth factor signalling (37; 44). Additional studies have also shown chondrocytes to be particularly receptive to mechanical loading, with this parameter having been evidenced as a crucial factor in the chondrogenic differentiation of MSCs during critical cartilage development. Repetition of these loads and varying the duration and force of the load has positive effects on the structural organization of cartilage ECM (7; 62). The effects of mechanobiology on chondrogenesis have been discussed in detail in the section on stem cells.

In recent times, tissue engineering research had broadened its horizons to understand the effect of scaffold physical properties on cell behaviour. Properties considered include; roughness (88; 89; 147), micro and nanotopography (reviewed in (132)), porosity (155) and surface energy (80; 147). The stiffness of the substrate (scaffold matrix) has been demonstrated as a crucial regulator of stem cell behaviour (15; 48; 52; 119). It is thought that the stiffness or elasticity of a matrix can act as a 'passive' cue for cell processes via a phenomenon known as mechano-transduction. This is a method by which cells convert mechanical stimuli into a chemical response, thus affecting their own behaviour. For detailed reviews see (2; 54). Cells bind to the matrix using integrins. The intra cellular domain connects to the actin and myosin (contractile) cytoskeleton of the cell, and the extracellular domain to the biomaterial. When cells are bound, they apply mechanical forces onto the matrix using their contractile cytoskeleton. Integrins cluster which in turn recruits structural and signalling proteins at the site of contact between cell and matrix, known as a focal adhesion. If a matrix is relatively hard, there is more resistance to the forces applied by the cellular cytoskeleton. This results in a more organized cytoskeleton, more integrin clustering and thus focal adhesions that are greater in maturity. Comparatively, if cells are seeded onto a soft matrix, there is little resistance to counterbalance the cell forces, therefore reduced development of the actin-myosin cytoskeleton. This phenomenon is fundamental considering that changes in cytoskeletal organization affect signalling, thereby translating mechanical processes into chemical responses.

So how can this trend be used in cartilage tissue engineering technology? Let us consider the application of stem cells in tissue engineering cartilage. Stem cells extracted from human or animal sources are frequently expanded in culture. Culturing stem cells on traditional tissue culture plastic could result in preconditioning of the cells in accordance with the stiffness of the plate (51; 52). Depending on the experimental aims, it may be wiser to culture and expand on softer substrates with stiffness comparable to that of native tissue. However conflicting data has shown that stiffer substrates increase the rate of proliferation, whereas soft substrates promote the dedifferentiation of cells (15).This suggests the stiffness of the material used for cartilage tissue engineering is an important parameter not just in terms of mechanical support but also in terms of propagating chondrocyte growth and matrix deposition.

### 4.4 Nanomaterials

Cell coverage over a matrix layer is directly correlated to the spread of microscale ECM proteins over its surface, irrespective of the geometric patterning of such proteins (93). This theory applies at the microscale level of tissue engineering, but at the nanoscale, there is increasing evidence to indicate that cells are able to alter their behaviour differentially in response to changes in nanotopographical surfaces. These changes can be cytoskeletal or a change in morphology, focal adhesions, motility, gene expression and differentiation. Much like the mechano-transduction discussed earlier, there is support for some sort of topography-dependent transduction that communicates independent of chemical signalling from ECM molecules (35). Studies have since demonstrated that this cellular response is heavily related to the pattern and spacing of adhesive ligands (36;41;79;148).

In light of the revelation that nanotopography plays a major role in the governance of cell-matrix interactions, many physical and chemical methods have been developed to engineer geometrically defined nanopatterns on biocompatible scaffolds. Crude methods of acid treatment (85), bonding with calcium cations (63), and coating with nanoparticles (reviewed in (126)) allowed scientists to introduce nanofeatures into the surface topography of scaffolds. Surface modifications with Lanthanum phosphate ($LaPO_4$) nanoparticles increased osteoblast adhesion to traditional bioceramics; Hydroxyapatite and Tricalcium phosphate (53). Likewise with chondrocytes, the levels of adhesion increased on 70%/30% (wt) PLGA/titanium composite scaffolds manufactured to have a nanosurface (76). However the advancement of nanoscience allows more precise pattering of various nanofeatures to further affect cell behaviour. Nanofeatures now come in many forms ranging from nanopits and grooves, to nanopillars, nanodots and traditional nanoparticles. The pattern in which they are arranged is also on the nanoscale. The latest techniques used for nanosurface pattering (reviewed in (132)) include photolithography, electron beam lithography (40), Dip-pen lithography (71) and imprint lithography.

Nanopatterning to mimic the surface density and arrangement of integrin-binding epitopes as seen in the ECM has been a challenge not yet beaten. Studies have shown that integrin mediated signalling operates with a minimum surface density however; the exact spatial organization of these ligands in vivo has not been elucidated. The nearest estimate has come from a group that developed a block-copolymer micelle nanolithography technique to label surfaces with hexagonal arrays of gold nanodots coated with one RGD peptide (found in adhesive glycoproteins such as fibronectin and vitronectin) (26;27). Upon cell seeding they found that only 28nm and 58nm spacing between the nanodots would allow adequate clustering of integrins, which are approximately 8-12nm in size (138). Additional studies on RGD-coated gold nanoparticles have shown that the velocity of migrating cells decreases with an increased particle density, with a peak velocity at circa 120nm, suggesting the boost in particle density increased levels of adhesion (6;65). Interestingly enough, research has also shown the MSC osteoblastic differentiation can be hampered by regularly arranged hexagonal nanopits arrays compared with arrays with a slight irregularity (39). Similar results were found by Biggs et al 2007, where highly ordered nanopits resulted in decreased formation and length of focal adhesions, compared with controlled disorder increasing focal adhesion formation and size (13).

With more research being conducted into the in vitro effects of surface nanopatterning on cell behaviour, there are implications for cartilage tissue engineering research. Data shows that nanostructured PLGA can accelerate chondrocyte attachment, growth and proliferation in addition to improving ECM production (76). In our lab, chondrocytes seeded onto nanocomposite polymer POSS-PCU (UCL nanoBio™) have a faster rate of proliferation compared with controls lacking the nano modification (unpublished data). And though the current research into nanomaterials and cartilage tissue engineering is just evolving, there are many lessons to be learnt from bone (80;81), skin (31) and vascular (100;108) tissue engineering research.

## 5. Bioreactors

Bioreactors are devices in which biological and/or biochemical processes develop under controlled and monitored environmental and operating conditions (104). It is the exceptional control over environmental conditions that makes bioreactor use particularly pertinent in tissue engineering research where specific factors need to be controlled in order to optimise tissue growth. Bioreactors can maintain physiological boundaries at desired levels, enhance nutrient and waste transport rates, and provide specific stimuli to promote optimum growth.

The use of bioreactors has provided a promising method for tackling some causes for poor research outcomes in tissue engineering practice. Restricted, unspecific, or impermanent cell differentiation and poor tissue formation/ remodelling in cartilage tissue engineering largely results from a lack of correct physical stimulation in vitro (86). For example, mechanotransduction, the transduction of mechanical stresses into biochemical signals, affects chondrocyte function. Modifying the mechanical stressors applied to cells in vitro may therefore improve the quality of tissue constructs produced. In early parts of this chapter, the effect of cyclical loading, especially within the articular region has been shown to improve the ECM content of constructs, and therefore the overall construct viability. Mimicking some of these forces in bioreactor systems could also dramatically improve tissue growth. Studies which evidenced the effect of adaptive physical stimulation on mechanotransduction, led to the development of bioreactor devices that transmit forces including shear stress, hydrostatic pressure and compression to articular cartilage in vitro (129).

### 5.1 Mechanical forces

Key to tissue engineering in the joint region specifically is the use of exogenous mechanical forces to simulate loading forces (exerted during daily movement and exercise), which in turn increases the metabolic activity of and ECM production by chondrocytes. Shear stress, compressive forces, tensile forces and hydrostatic pressures are all parameters that can be modulated to influence the quality of cartilaginous constructs engineered. The effect of these mechanical forces on chondrogenesis, have been described earlier in the chapter. We will examine briefly the bioreactors that have been used to study the effects of shear stress however, the different types of bioreactors available for exerting other forces are expertly reviewed in Schulz RM 2007.

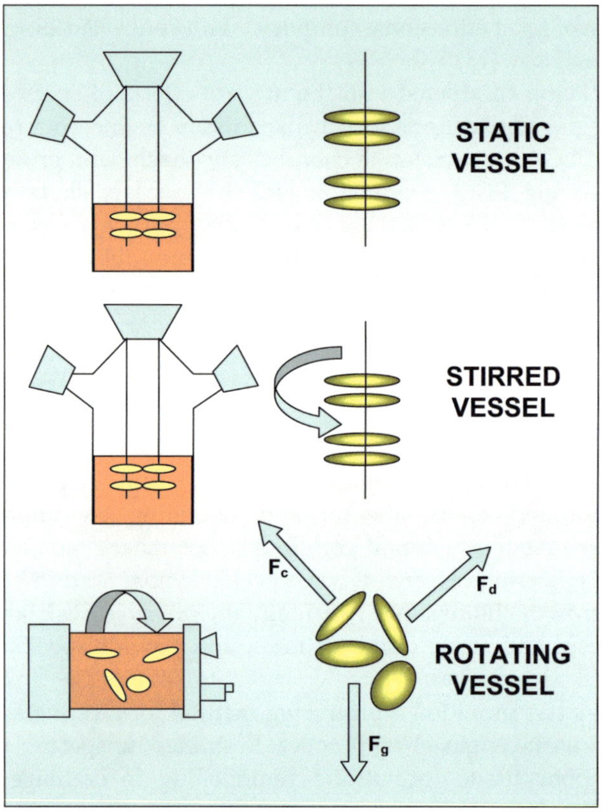


Fig. 4. Experimental appraoches to bioreactor tissue engineering. Representation of static, stirred and rotating vessels. Fc, Fd, Fg refer to centrifugal, drag and net forces respectively. (Adapted from Vunjak-Novakovic G 1999 (145))

The easiest method for examining the effects of shear stress in bioreactor systems is by placing constructs in culture either on a petri dish or in a dynamic or orbital shaker (56;57). Other methods developed have included using spinner flasks or vessels with magnetic stirrers (14;18;136). Extensive study in the nineties looked at the application of shear stress by comparing static and orbital shakers, stirred vessels with rotating vessels (Fig.4) (145). Rotating vessels are more advanced systems where constructs float freely within culture medium, whilst the whole vessel rotates around a central axis at a constants speed. Chondrocytes were seeded onto 97% porous scaffold discs and cultured in the aforementioned vessels for 8 weeks. Results showed that freely cultured constructs were larger than those cultured in static or stirred vessels. They formed cartilaginous ECM with the greatest concentration of GAGs and collagen. Their mechanical properties where also shown to be superior.

### 5.2 Oxygen tension

Optimizing $O_2$ tension within culture systems is an area of great importance in bioreactor design. $O_2$ is the partial pressure of oxygen dissolved in a liquid such as blood. Cells in culture require nutrients and oxygen to proliferate and this is usually achieved through mass transport (net movement of mass from one location to another). When oxygen and nutrients are limiting factors, larger grafts tend to contain a hypoxic, necrotic centre, surrounded by a rim of viable cells (Martin I 2004). In a tissue graft, the density of cells may be higher than the distance oxygen can freely diffuse across by mass transport to provide sufficient oxygen for the inner cells; therefore they are starved of oxygen. Limited $O_2$ diffusion can also affect the spatial distribution of cells and as the $O_2$ concentration gradient decreases from the surface of the tissue compartment to its centre (34). In humans, this problem is solved by the circulatory system and thus nutrients are provided to all cells via a complex network of vessels, slowly decreasing in size the deeper into tissues they enter. It is the proximity of capillaries to somatic (body) cells that allows their mass transfer requirements to be met (105). So how is this problem solved in tissue engineering practice? The introduction of simple stirred flask bioreactors enables the mixing of oxygen and nutrients throughout the medium. So not only does it provide a shear stress which is known to be beneficial for chondrocyte growth and proliferation, but it also reduces the concentration boundary layer of oxygen at the construct surface (14;18;104). In a static culture environment, oxygen would diffuse into cells and carbon dioxide out. The medium in closest proximity to the cells would have a steadily decreasing $O_2$ tension with a conversely increasing $CO_2$ tension. This in turn limits the overall rate diffusion as $O_2$ moves from areas of high tension to areas of low tension. Thus if the culture medium is not circulated or replenished the rate of diffusion will decrease and eventually cease at the point where there is no longer a concentration gradient, leading to cell death.

Studies have also used bioreactors to investigate the effect of different partial pressures of $O_2$ and pH levels on gene and protein expression, as well as the metabolic activity of chondrocytes. Results showed chondrocyte sensitivity to acidic conditions where reduced expression of Coll Type 1, SOX9 and VEGF (vascular endothelial growth factor) were observed. Conversely in hypoxic conditions, VEGF levels were found to be higher, with a pH dependent reduction in Coll Type 1 (43). Culture in bioreactors at low oxygen tension increases the production and retention of glycosaminoglycan (GAG) within the cartilage matrix without affecting chondrocyte proliferation or collagen deposition which typically would requires higher partial pressures of $O_2$ (123). These studies highlight the twofold applications of bioreactors, in maximizing cell growth and tissue generation for clinical use and in research and development to investigate the effect of different biological factors on cell growth.

### 5.3 Growth factors

It has also bee suggested that bioreactors provide suitable environments to add growth stimulating factors to constructs to improve chondrogenesis. For example, transducing human MSCs with an adenoviral vector containing SOX9 and subjecting the construct to mechanical stimulation could increase GAG synthesis (90). Growth factor application of

BMP-2, IGF-1 and TGF-β1 in a bioreactor system can increase the compressive and tensile biomechanical properties of engineered tissue (50). The efficiency of chondrocyte proliferation from low initial seeding densities can also be enhanced by adding various growth factor combinations in to automated bioreactors systems (55). Chitosan scaffolds were used to engineer articular cartilage with the aid of a chondrogenic differentiation factor, BMP-6. Results showed that proliferated cells contained a higher value of GAG, Coll type II and DNA indicating improved chondrogenesis (1). Alternatively, inhibiting the expression of some factors, namely interleukin 6, has also been investigated with the aim of improving tissue growth in bioreactors. In 2010, Wang P et al demonstrated how high levels of interleukin - 6 have been found in osteoarthritic cartilage and suggested that inhibiting this expression may improve cartilage construct culturing in bioreactors (146).

## 6. Challenges for the clinical application of regenerated cartilage

Over the past two decades the amount of data on cartilage tissue engineering strategies has risen exponentially. There is now a plethora of exciting in vitro data evaluating chondrocyte/MSC seeded biomaterial constructs. Perhaps one of the most iconic studies in cartilage tissue engineering research was produced by Cao and Vacanti's group in 1996, when they implanted an auricular shaped cartilaginous construct onto the back of a mouse (24). Even with all the advancements in stem cell and biomaterial technology, the invention of various bioreactor systems, little has progressed beyond this scientifically historic event. Most constructs fail to develop beyond immature, inflexible neocartilage that lacks the durability essential to most clinical applications.

There are a number of reasons for the stagnation in translation to clinical practice. Many of which have been discussed throughout the course of this chapter. On a cellular level, reasons for poor research outcomes could also include; (i)Regenerative cells being lost through leakage of the cell suspension (149), (ii) inflammatory cytokine, matrix metalloproteinase, nitric oxide mediated apoptosis and necrosis at the site of injury. These biochemical factors are released as part of the normal inflammatory and wound healing process, especially at the interface between host and repair tissue, which can also adversely affect biointegration of the neo tissue. The use of anti-apoptotic factors would be crucial in maintaining cell numbers but also in creating a favourable environment for biointegration (5). The poor migration capacity of chondrocytes could also be responsible for hampered infiltration of repair tissue into the host environment. The naturally slow rates of chondrocyte ECM production could slow down integration as well disparities in the organization of neocartilage matrix compared with the zonal arrangement of native cartilage tissue (69;84). Dedifferentiation of chondrogenic cells is another problem, and is likely responsible to the highly fibrotic nature of neocartilage produced, suggesting that over time, cells may have dedifferentiated into fibroblasts or incompletely differentiated into chondrocytes. Solutions would include seeding with cells that have been fully differentiated in vitro, but again there would be difficulties with motility, proliferation and shelf life.

In addition to the cell based scientific problems associated with cartilage engineering tissue research, ambiguous regulatory guidelines currently hamper the flow of development from laboratories to clinics and operating theatres. The EU regulation on Advanced Therapy Medicinal Products (ATMP), which includes tissue engineered constructs, is still in its infancy having only been formally established in December 2008. ATMP regulation aims to proved a coherent and tailored framework for tissue engineered products, however the nascent and fast growing nature of the tissue engineering field means that there is a constant threat of irrelevance over the guidelines developed under this regulation. Tissue engineering technology needs to reach a level of quality controlled and quality assured reproducibility to allow for not just clinical efficiency, but also commercial viability. Methods of stem cell differentiation, cell seeding, scaffold fabrication and bioreactor development/implementation all need to be governed by Good Manufacturing Practice (GMP). Additionally, methods of commercialization ought to be better established, to avoid uncertainty in the markets, improve regulatory approval and clinical uptake/use (103).

## 7. References


[1] Akman, A. C., Seda, T. R., Gumusderelioglu, M., Nohutcu, R. M. Bone morphogenetic protein-6-loaded chitosan scaffolds enhance the osteoblastic characteristics of MC3T3-E1 cells. *Artif. Organs* 34: 65-74, 2010.

[2] Alenghat, F. J., Ingber, D. E. Mechanotransduction: all signals point to cytoskeleton, matrix, and integrins. *Sci. STKE.* 2002: e6, 2002.

[3] Alhadlaq, A., Mao, J. J. Tissue-engineered neogenesis of human-shaped mandibular condyle from rat mesenchymal stem cells. *J Dent. Res* 82: 951-956, 2003.

4. Angele, P., Schumann, D., Angele, M. et al. Cyclic, mechanical compression enhances chondrogenesis of mesenchymal progenitor cells in tissue engineering scaffolds. *Biorheology* 41: 335-346, 2004.

[5] Archer, C. W., Redman, S., Khan, I., Bishop, J., Richardson, K. Enhancing tissue integration in cartilage repair procedures. *J Anat.* 209: 481-493, 2006.

[6] Arnold, M., Hirschfeld-Warneken, V. C., Lohmuller, T. et al. Induction of cell polarization and migration by a gradient of nanoscale variations in adhesive ligand spacing. *Nano. Lett.* 8: 2063-2069, 2008.

[7] Arokoski, J. P., Jurvelin, J. S., Vaatainen, U., Helminen, H. J. Normal and pathological adaptations of articular cartilage to joint loading. *Scand. J Med. Sci. Sports* 10: 186-198, 2000.

[8] Atala, A. Engineering tissues, organs and cells. *J Tissue Eng Regen. Med* 1: 83-96, 2007.

[9] Augello, A., Kurth, T. B., De Bari, C. Mesenchymal stem cells: a perspective from in vitro cultures to in vivo migration and niches. *Eur. Cell Mater.* 20: 121-133, 2010.

[10] Baek, C. H., Ko, Y. J. Characteristics of tissue-engineered cartilage on macroporous biodegradable PLGA scaffold. *Laryngoscope* 116: 1829-1834, 2006.

[11] Barry, F., Boynton, R. E., Liu, B., Murphy, J. M. Chondrogenic differentiation of mesenchymal stem cells from bone marrow: differentiation-dependent gene expression of matrix components. *Exp. Cell Res* 268: 189-200, 2001.

[12] Bentley, G., Biant, L. C., Carrington, R. W. et al. A prospective, randomised comparison of autologous chondrocyte implantation versus mosaicplasty for osteochondral defects in the knee. *J Bone Joint Surg Br* 85: 223-230, 2003.

[13] Biggs, M. J., Richards, R. G., Gadegaard, N., Wilkinson, C. D., Dalby, M. J. Regulation of implant surface cell adhesion: characterization and quantification of S-phase primary osteoblast adhesions on biomimetic nanoscale substrates. *J Orthop Res* 25: 273-282, 2007.

[14] Bouchet, B. Y., Colon, M., Polotsky, A., Shikani, A. H., Hungerford, D. S., Frondoza, C. G. Beta-1 integrin expression by human nasal chondrocytes in microcarrier spinner culture. *J Biomed Mater Res* 52: 716-724, 2000.

[15] Breuls, R. G., Jiya, T. U., Smit, T. H. Scaffold stiffness influences cell behavior: opportunities for skeletal tissue engineering. *Open. Orthop J* 2: 103-109, 2008.
[16] Brittberg, M., Lindahl, A., Nilsson, A., Ohlsson, C., Isaksson, O., Peterson, L. Treatment of deep cartilage defects in the knee with autologous chondrocyte transplantation. *N. Engl. J Med.* 331: 889-895, 1994.
[17] Brochhausen, C., Lehmann, M., Zehbe, R. et al. [Tissue engineering of cartilage and bone : growth factors and signaling molecules]. *Orthopade* 38: 1053-1062, 2009.
[18] Brown, A. N., Kim, B. S., Alsberg, E., Mooney, D. J. Combining chondrocytes and smooth muscle cells to engineer hybrid soft tissue constructs. *Tissue Eng* 6: 297-305, 2000.
[19] Brun, P., Dickinson, S. C., Zavan, B., Cortivo, R., Hollander, A. P., Abatangelo, G. Characteristics of repair tissue in second-look and third-look biopsies from patients treated with engineered cartilage: relationship to symptomatology and time after implantation. *Arthritis Res Ther.* 10: R132, 2008.
[20] Buckwalter, J. A., Mankin, H. J. Articular cartilage: tissue design and chondrocyte-matrix interactions. *Instr. Course Lect.* 47: 477-486, 1998.
[21] Buckwalter, J. A., Mankin, H. J. Articular cartilage: degeneration and osteoarthritis, repair, regeneration, and transplantation. *Instr. Course Lect.* 47: 487-504, 1998.
[22] Cai, D. Z., Zeng, C., Quan, D. P. et al. Biodegradable chitosan scaffolds containing microspheres as carriers for controlled transforming growth factor-beta1 delivery for cartilage tissue engineering. *Chin Med. J (Engl. )* 120: 197-203, 2007.
[23] Campbell, J. J., Lee, D. A., Bader, D. L. Dynamic compressive strain influences chondrogenic gene expression in human mesenchymal stem cells. *Biorheology* 43: 455-470, 2006.
[24] Cao, Y., Vacanti, J. P., Paige, K. T., Upton, J., Vacanti, C. A. Transplantation of chondrocytes utilizing a polymer-cell construct to produce tissue-engineered cartilage in the shape of a human ear. *Plast. Reconstr. Surg* 100: 297-302, 1997.
[25] Carlberg, A. L., Pucci, B., Rallapalli, R., Tuan, R. S., Hall, D. J. Efficient chondrogenic differentiation of mesenchymal cells in micromass culture by retroviral gene transfer of BMP-2. *Differentiation* 67: 128-138, 2001.
[26] Cavalcanti-Adam, E. A., Micoulet, A., Blummel, J., Auernheimer, J., Kessler, H., Spatz, J. P. Lateral spacing of integrin ligands influences cell spreading and focal adhesion assembly. *Eur. J Cell Biol.* 85: 219-224, 2006.
[27] Cavalcanti-Adam, E. A., Tomakidi, P., Bezler, M., Spatz, J. P. Geometric organization of the extracellular matrix in the control of integrin-mediated adhesion and cell function in osteoblasts. *Prog. Orthod.* 6: 232-237, 2005.
[28] Cavallo, C., Desando, G., Facchini, A., Grigolo, B. Chondrocytes from patients with osteoarthritis express typical extracellular matrix molecules once grown onto a three-dimensional hyaluronan-based scaffold. *J Biomed Mater Res A* 93: 86-95, 2010.
[29] Chen, W. H., Lai, M. T., Wu, A. T. et al. In vitro stage-specific chondrogenesis of mesenchymal stem cells committed to chondrocytes. *Arthritis Rheum.* 60: 450-459, 2009.
[30] Chia, S. H., Homicz, M. R., Schumacher, B. L. et al. Characterization of human nasal septal chondrocytes cultured in alginate. *J Am. Coll. Surg* 200: 691-704, 2005.

[31] Chong, E. J., Phan, T. T., Lim, I. J. et al. Evaluation of electrospun PCL/gelatin nanofibrous scaffold for wound healing and layered dermal reconstitution. *Acta Biomater.* 3: 321-330, 2007.

[32] Chung, C., Burdick, J. A. Engineering cartilage tissue. *Adv. Drug Deliv. Rev.* 60: 243-262, 2008.

[33] Chung, C., Burdick, J. A. Influence of three-dimensional hyaluronic acid microenvironments on mesenchymal stem cell chondrogenesis. *Tissue Eng Part A* 15: 243-254, 2009.

[34] Curcio, E., Macchiarini, P., De, B. L. Oxygen mass transfer in a human tissue-engineered trachea. *Biomaterials* 31: 5131-5136, 2010.

[35] Curtis, A. S., Dalby, M., Gadegaard, N. Cell signaling arising from nanotopography: implications for nanomedical devices. *Nanomedicine. (Lond)* 1: 67-72, 2006.

[36] Curtis, A. S., Gadegaard, N., Dalby, M. J., Riehle, M. O., Wilkinson, C. D., Aitchison, G. Cells react to nanoscale order and symmetry in their surroundings. *IEEE Trans. Nanobioscience.* 3: 61-65, 2004.

[37] Czyz, J., Wobus, A. Embryonic stem cell differentiation: the role of extracellular factors. *Differentiation* 68: 167-174, 2001.

[38] Dai, W., Kawazoe, N., Lin, X., Dong, J., Chen, G. The influence of structural design of PLGA/collagen hybrid scaffolds in cartilage tissue engineering. *Biomaterials* 31: 2141-2152, 2010.

[39] Dalby, M. J., Gadegaard, N., Tare, R. et al. The control of human mesenchymal cell differentiation using nanoscale symmetry and disorder. *Nat. Mater* 6: 997-1003, 2007.

[40] Dalby, M. J., Giannaras, D., Riehle, M. O., Gadegaard, N., Affrossman, S., Curtis, A. S. Rapid fibroblast adhesion to 27nm high polymer demixed nano-topography. *Biomaterials* 25: 77-83, 2004.

[41] Dalby, M. J., Riehle, M. O., Sutherland, D. S., Agheli, H., Curtis, A. S. Use of nanotopography to study mechanotransduction in fibroblasts--methods and perspectives. *Eur. J Cell Biol.* 83: 159-169, 2004.

[42] Darling, E. M., Athanasiou, K. A. Rapid phenotypic changes in passaged articular chondrocyte subpopulations. *J Orthop Res* 23: 425-432, 2005.

[43] Das, R. H., van Osch, G. J., Kreukniet, M., Oostra, J., Weinans, H., Jahr, H. Effects of individual control of pH and hypoxia in chondrocyte culture. *J Orthop Res* 28: 537-545, 2010.

[44] Datta, N., Holtorf, H. L., Sikavitsas, V. I., Jansen, J. A., Mikos, A. G. Effect of bone extracellular matrix synthesized in vitro on the osteoblastic differentiation of marrow stromal cells. *Biomaterials* 26: 971-977, 2005.

[45] Davidson, D., Blanc, A., Filion, D. et al. Fibroblast growth factor (FGF) 18 signals through FGF receptor 3 to promote chondrogenesis. *J Biol. Chem.* 280: 20509-20515, 2005.

[46] DeLise, A. M., Fischer, L., Tuan, R. S. Cellular interactions and signaling in cartilage development. *Osteoarthritis. Cartilage.* 8: 309-334, 2000.

[47] Denker, A. E., Nicoll, S. B., Tuan, R. S. Formation of cartilage-like spheroids by micromass cultures of murine C3H10T1/2 cells upon treatment with transforming growth factor-beta 1. *Differentiation* 59: 25-34, 1995.

[48] Discher, D. E., Janmey, P., Wang, Y. L. Tissue cells feel and respond to the stiffness of their substrate. *Science* 310: 1139-1143, 2005.

[49] Dobratz, E. J., Kim, S. W., Voglewede, A., Park, S. S. Injectable cartilage: using alginate and human chondrocytes. *Arch. Facial. Plast. Surg* 11: 40-47, 2009.

[50] Elder, B. D., Athanasiou, K. A. Systematic assessment of growth factor treatment on biochemical and biomechanical properties of engineered articular cartilage constructs. *Osteoarthritis. Cartilage.* 17: 114-123, 2009.

[51] Engler, A. J., Rehfeldt, F., Sen, S., Discher, D. E. Microtissue elasticity: measurements by atomic force microscopy and its influence on cell differentiation. *Methods Cell Biol.* 83: 521-545, 2007.

[52] Engler, A. J., Sen, S., Sweeney, H. L., Discher, D. E. Matrix elasticity directs stem cell lineage specification. *Cell* 126: 677-689, 2006.

[53] Ergun, C., Liu, H., Webster, T. J. Osteoblast adhesion on novel machinable calcium phosphate/lanthanum phosphate composites for orthopedic applications. *J Biomed Mater Res A* 89: 727-733, 2009.

[54] Forgacs, G., Yook, S. H., Janmey, P. A., Jeong, H., Burd, C. G. Role of the cytoskeleton in signaling networks. *J Cell Sci.* 117: 2769-2775, 2004.

[55] Francioli, S. E., Martin, I., Sie, C. P. et al. Growth factors for clinical-scale expansion of human articular chondrocytes: relevance for automated bioreactor systems. *Tissue Eng* 13: 1227-1234, 2007.

[56] Freed, L. E., Martin, I., Vunjak-Novakovic, G. Frontiers in tissue engineering. In vitro modulation of chondrogenesis. *Clin. Orthop Relat Res* S46-S58, 1999.

[57] Freed, L. E., Vunjak-Novakovic, G., Langer, R. Cultivation of cell-polymer cartilage implants in bioreactors. *J Cell Biochem.* 51: 257-264, 1993.

[58] Fuchs, J. R., Hannouche, D., Terada, S., Vacanti, J. P., Fauza, D. O. Fetal tracheal augmentation with cartilage engineered from bone marrow-derived mesenchymal progenitor cells. *J Pediatr. Surg* 38: 984-987, 2003.

[59] Gabay, O., Sanchez, C., Taboas, J. M. Update in cartilage bio-engineering. *Joint Bone Spine* 77: 283-286, 2010.

[60] Goldring, M. B., Tsuchimochi, K., Ijiri, K. The control of chondrogenesis. *J Cell Biochem.* 97: 33-44, 2006.

[61] Griffon, D. J., Sedighi, M. R., Schaeffer, D. V., Eurell, J. A., Johnson, A. L. Chitosan scaffolds: interconnective pore size and cartilage engineering. *Acta Biomater.* 2: 313-320, 2006.

[62] Grodzinsky, A. J., Levenston, M. E., Jin, M., Frank, E. H. Cartilage tissue remodeling in response to mechanical forces. *Annu. Rev. Biomed Eng* 2: 691-713, 2000.

[63] Hanawa, T., Kon, M., Ukai, H., Murakami, K., Miyamoto, Y., Asaoka, K. Surface modifications of titanium in calcium-ion-containing solutions. *J Biomed Mater Res* 34: 273-278, 1997.

[64] Hildner, F., Albrecht, C., Gabriel, C., Redl, H., van, G. M. State of the art and future perspectives of articular cartilage regeneration: a focus on adipose-derived stem cells and platelet-derived products. *J Tissue Eng Regen. Med.* 2011.

[65] Hirschfeld-Warneken, V. C., Arnold, M., Cavalcanti-Adam, A., Lopez-Garcia, M., Kessler, H., Spatz, J. P. Cell adhesion and polarisation on molecularly defined spacing gradient surfaces of cyclic RGDfK peptide patches. *Eur. J Cell Biol.* 87: 743-750, 2008.

[66] Horas, U., Pelinkovic, D., Herr, G., Aigner, T., Schnettler, R. Autologous chondrocyte implantation and osteochondral cylinder transplantation in cartilage repair of the knee joint. A prospective, comparative trial. *J Bone Joint Surg Am.* 85-A: 185-192, 2003.

[67] Huang, A. H., Farrell, M. J., Mauck, R. L. Mechanics and mechanobiology of mesenchymal stem cell-based engineered cartilage. *J Biomech.* 43: 128-136, 2010.

[68] Huang, A. H., Stein, A., Tuan, R. S., Mauck, R. L. Transient exposure to transforming growth factor beta 3 improves the mechanical properties of mesenchymal stem cell-laden cartilage constructs in a density-dependent manner. *Tissue Eng Part A* 15: 3461-3472, 2009.

[69] Hunter, C. J., Levenston, M. E. Maturation and integration of tissue-engineered cartilages within an in vitro defect repair model. *Tissue Eng* 10: 736-746, 2004.

[70] Hunziker, E. B. Mechanism of longitudinal bone growth and its regulation by growth plate chondrocytes. *Microsc. Res Tech.* 28: 505-519, 1994.

[71] Huo, F., Zheng, Z., Zheng, G., Giam, L. R., Zhang, H., Mirkin, C. A. Polymer pen lithography. *Science* 321: 1658-1660, 2008.

[72] Indrawattana, N., Chen, G., Tadokoro, M. et al. Growth factor combination for chondrogenic induction from human mesenchymal stem cell. *Biochem. Biophys. Res. Commun.* 320: 914-919, 2004.

[73] Indrawattana, N., Chen, G., Tadokoro, M. et al. Growth factor combination for chondrogenic induction from human mesenchymal stem cell. *Biochem. Biophys. Res. Commun.* 320: 914-919, 2004.

[74] Johnstone, B., Hering, T. M., Caplan, A. I., Goldberg, V. M., Yoo, J. U. In vitro chondrogenesis of bone marrow-derived mesenchymal progenitor cells. *Exp. Cell Res* 238: 265-272, 1998.

[75] Kang, S. W., Jeon, O., Kim, B. S. Poly(lactic-co-glycolic acid) microspheres as an injectable scaffold for cartilage tissue engineering. *Tissue Eng* 11: 438-447, 2005.

[76] Kay, S., Thapa, A., Haberstroh, K. M., Webster, T. J. Nanostructured polymer/nanophase ceramic composites enhance osteoblast and chondrocyte adhesion. *Tissue Eng* 8: 753-761, 2002.

[77] Kayakabe, M., Tsutsumi, S., Watanabe, H., Kato, Y., Takagishi, K. Transplantation of autologous rabbit BM-derived mesenchymal stromal cells embedded in hyaluronic acid gel sponge into osteochondral defects of the knee. *Cytotherapy.* 8: 343-353, 2006.

[78] Kessler, M. W., Grande, D. A. Tissue engineering and cartilage. *Organogenesis.* 4: 28-32, 2008.

[79] Khang, D., Kim, S. Y., Liu-Snyder, P., Palmore, G. T., Durbin, S. M., Webster, T. J. Enhanced fibronectin adsorption on carbon nanotube/poly(carbonate) urethane: independent role of surface nano-roughness and associated surface energy. *Biomaterials* 28: 4756-4768, 2007.

[80] Khang, D., Lu, J., Yao, C., Haberstroh, K. M., Webster, T. J. The role of nanometer and sub-micron surface features on vascular and bone cell adhesion on titanium. *Biomaterials* 29: 970-983, 2008.

[81] Khang, D., Sato, M., Price, R. L., Ribbe, A. E., Webster, T. J. Selective adhesion and mineral deposition by osteoblasts on carbon nanofiber patterns. *Int. J Nanomedicine.* 1: 65-72, 2006.

[82] Kim, H. J., Lee, J. H., Im, G. I. Chondrogenesis using mesenchymal stem cells and PCL scaffolds. *J Biomed Mater Res A* 92: 659-666, 2010.
[83] Kim, M. K., Choi, S. W., Kim, S. R., Oh, I. S., Won, M. H. Autologous chondrocyte implantation in the knee using fibrin. *Knee Surg Sports Traumatol Arthrosc.* 18: 528-534, 2010.
[84] Klein, T. J., Schumacher, B. L., Schmidt, T. A. et al. Tissue engineering of stratified articular cartilage from chondrocyte subpopulations. *Osteoarthritis. Cartilage.* 11: 595-602, 2003.
[85] Knabe, C., Berger, G., Gildenhaar, R., Klar, F., Zreiqat, H. The modulation of osteogenesis in vitro by calcium titanium phosphate coatings. *Biomaterials* 25: 4911-4919, 2004.
[86] Korossis, S. A., Wilcox, H. E., Watterson, K. G., Kearney, J. N., Ingham, E., Fisher, J. In-vitro assessment of the functional performance of the decellularized intact porcine aortic root. *J Heart Valve Dis.* 14: 408-421, 2005.
[87] Kreuz, P. C., Steinwachs, M. R., Erggelet, C. et al. Results after microfracture of full-thickness chondral defects in different compartments in the knee. *Osteoarthritis. Cartilage.* 14: 1119-1125, 2006.
[88] Kunzler, T. P., Drobek, T., Schuler, M., Spencer, N. D. Systematic study of osteoblast and fibroblast response to roughness by means of surface-morphology gradients. *Biomaterials* 28: 2175-2182, 2007.
[89] Kunzler, T. P., Huwiler, C., Drobek, T., Voros, J., Spencer, N. D. Systematic study of osteoblast response to nanotopography by means of nanoparticle-density gradients. *Biomaterials* 28: 5000-5006, 2007.
[90] Kupcsik, L., Stoddart, M. J., Li, Z., Benneker, L. M., Alini, M. Improving chondrogenesis: potential and limitations of SOX9 gene transfer and mechanical stimulation for cartilage tissue engineering. *Tissue Eng Part A* 16: 1845-1855, 2010.
[91] Lee, H. J., Choi, B. H., Min, B. H., Park, S. R. Low-intensity ultrasound inhibits apoptosis and enhances viability of human mesenchymal stem cells in three-dimensional alginate culture during chondrogenic differentiation. *Tissue Eng* 13: 1049-1057, 2007.
[92] Lefebvre, V., Smits, P. Transcriptional control of chondrocyte fate and differentiation. *Birth Defects Res C. Embryo. Today* 75: 200-212, 2005.
[93] Lehnert, D., Wehrle-Haller, B., David, C. et al. Cell behaviour on micropatterned substrata: limits of extracellular matrix geometry for spreading and adhesion. *J Cell Sci.* 117: 41-52, 2004.
[94] Levy, A. S., Lohnes, J., Sculley, S., LeCroy, M., Garrett, W. Chondral delamination of the knee in soccer players. *Am. J Sports Med.* 24: 634-639, 1996.
[95] Li, W. J., Tuli, R., Huang, X., Laquerriere, P., Tuan, R. S. Multilineage differentiation of human mesenchymal stem cells in a three-dimensional nanofibrous scaffold. *Biomaterials* 26: 5158-5166, 2005.
[96] Lieberman, J. R., Daluiski, A., Einhorn, T. A. The role of growth factors in the repair of bone. Biology and clinical applications. *J Bone Joint Surg Am.* 84-A: 1032-1044, 2002.
[97] Lin, Y. J., Yen, C. N., Hu, Y. C., Wu, Y. C., Liao, C. J., Chu, I. M. Chondrocytes culture in three-dimensional porous alginate scaffolds enhanced cell proliferation, matrix synthesis and gene expression. *J Biomed Mater Res A* 88: 23-33, 2009.

[98] Longobardi, L., O'Rear, L., Aakula, S. et al. Effect of IGF-I in the chondrogenesis of bone marrow mesenchymal stem cells in the presence or absence of TGF-beta signaling. *J. Bone Miner. Res.* 21: 626-636, 2006.
[99] Lu, H. H., Subramony, S. D., Boushell, M. K., Zhang, X. Tissue engineering strategies for the regeneration of orthopedic interfaces. *Ann. Biomed Eng* 38: 2142-2154, 2010.
[100] Lu, J., Rao, M. P., MacDonald, N. C., Khang, D., Webster, T. J. Improved endothelial cell adhesion and proliferation on patterned titanium surfaces with rationally designed, micrometer to nanometer features. *Acta Biomater.* 4: 192-201, 2008.
[101] Lyons, K. M., Pelton, R. W., Hogan, B. L. Organogenesis and pattern formation in the mouse: RNA distribution patterns suggest a role for bone morphogenetic protein-2A (BMP-2A). *Development* 109: 833-844, 1990.
[102] Marlovits, S., Zeller, P., Singer, P., Resinger, C., Vecsei, V. Cartilage repair: generations of autologous chondrocyte transplantation. *Eur. J Radiol.* 57: 24-31, 2006.
[103] Martin, I., Smith, T., Wendt, D. Bioreactor-based roadmap for the translation of tissue engineering strategies into clinical products. *Trends Biotechnol.* 27: 495-502, 2009.
[104] Martin, I., Wendt, D., Heberer, M. The role of bioreactors in tissue engineering. *Trends Biotechnol.* 22: 80-86, 2004.
[105] Martin, Y., Vermette, P. Bioreactors for tissue mass culture: design, characterization, and recent advances. *Biomaterials* 26: 7481-7503, 2005.
[106] Mehlhorn, A. T., Niemeyer, P., Kaiser, S. et al. Differential expression pattern of extracellular matrix molecules during chondrogenesis of mesenchymal stem cells from bone marrow and adipose tissue. *Tissue Eng* 12: 2853-2862, 2006.
[107] Mehlhorn, A. T., Schmal, H., Kaiser, S. et al. Mesenchymal stem cells maintain TGF-beta-mediated chondrogenic phenotype in alginate bead culture. *Tissue Eng* 12: 1393-1403, 2006.
[108] Miller, D. C., Thapa, A., Haberstroh, K. M., Webster, T. J. Endothelial and vascular smooth muscle cell function on poly(lactic-co-glycolic acid) with nano-structured surface features. *Biomaterials* 25: 53-61, 2004.
[109] Mitchell, N., Shepard, N. The resurfacing of adult rabbit articular cartilage by multiple perforations through the subchondral bone. *J Bone Joint Surg Am.* 58: 230-233, 1976.
[110] Mizuta, H., Kudo, S., Nakamura, E., Otsuka, Y., Takagi, K., Hiraki, Y. Active proliferation of mesenchymal cells prior to the chondrogenic repair response in rabbit full-thickness defects of articular cartilage. *Osteoarthritis. Cartilage.* 12: 586-596, 2004.
[111] Nicodemus, G. D., Villanueva, I., Bryant, S. J. Mechanical stimulation of TMJ condylar chondrocytes encapsulated in PEG hydrogels. *J Biomed Mater Res A* 83: 323-331, 2007.
[112] Pelttari, K., Steck, E., Richter, W. The use of mesenchymal stem cells for chondrogenesis. *Injury* 39 Suppl 1: S58-S65, 2008.
[113] Peretti, G. M., Pozzi, A., Ballis, R., Deponti, D., Pellacci, F. Current Surgical Options for Articular Cartilage Repair. *Acta Neurochir. Suppl* 108: 213-219, 2011.
[114] Ragetly, G. R., Slavik, G. J., Cunningham, B. T., Schaeffer, D. J., Griffon, D. J. Cartilage tissue engineering on fibrous chitosan scaffolds produced by a replica molding technique. *J Biomed Mater Res A* 93: 46-55, 2010.

[115] Raghunath, J., Rollo, J., Sales, K. M., Butler, P. E., Seifalian, A. M. Biomaterials and scaffold design: key to tissue-engineering cartilage. *Biotechnol. Appl. Biochem.* 46: 73-84, 2007.
[116] Raghunath, J., Salacinski, H. J., Sales, K. M., Butler, P. E., Seifalian, A. M. Advancing cartilage tissue engineering: the application of stem cell technology. *Curr. Opin. Biotechnol.* 16: 503-509, 2005.
[117] Rahfoth, B., Weisser, J., Sternkopf, F., Aigner, T., von der, M. K., Brauer, R. Transplantation of allograft chondrocytes embedded in agarose gel into cartilage defects of rabbits. *Osteoarthritis. Cartilage.* 6: 50-65, 1998.
[118] Reddi, A. H. Cartilage morphogenetic proteins: role in joint development, homoeostasis, and regeneration. *Ann. Rheum. Dis.* 62 Suppl 2: ii73-ii78, 2003.
[119] Rehfeldt, F., Engler, A. J., Eckhardt, A., Ahmed, F., Discher, D. E. Cell responses to the mechanochemical microenvironment--implications for regenerative medicine and drug delivery. *Adv. Drug Deliv. Rev.* 59: 1329-1339, 2007.
[120] Roberts, S., McCall, I. W., Darby, A. J. et al. Autologous chondrocyte implantation for cartilage repair: monitoring its success by magnetic resonance imaging and histology. *Arthritis Res Ther.* 5: R60-R73, 2003.
[121] Robertson, W. W., Jr. Newest knowledge of the growth plate. *Clin. Orthop Relat Res* 270-278, 1990.
[122] Ronziere, M. C., Perrier, E., Mallein-Gerin, F., Freyria, A. M. Chondrogenic potential of bone marrow- and adipose tissue-derived adult human mesenchymal stem cells. *Biomed. Mater. Eng* 20: 145-158, 2010.
[123] Saini, S., Wick, T. M. Effect of low oxygen tension on tissue-engineered cartilage construct development in the concentric cylinder bioreactor. *Tissue Eng* 10: 825-832, 2004.
[124] Sakimura, K., Matsumoto, T., Miyamoto, C., Osaki, M., Shindo, H. Effects of insulin-like growth factor I on transforming growth factor beta1 induced chondrogenesis of synovium-derived mesenchymal stem cells cultured in a polyglycolic acid scaffold. *Cells Tissues. Organs* 183: 55-61, 2006.
[125] Salinas, C. N., Anseth, K. S. The enhancement of chondrogenic differentiation of human mesenchymal stem cells by enzymatically regulated RGD functionalities. *Biomaterials* 29: 2370-2377, 2008.
[126] Schmidt, R. C., Healy, K. E. Controlling biological interfaces on the nanometer length scale. *J Biomed Mater Res A* 90: 1252-1261, 2009.
[127] Schmitt, B., Ringe, J., Haupl, T. et al. BMP2 initiates chondrogenic lineage development of adult human mesenchymal stem cells in high-density culture. *Differentiation* 71: 567-577, 2003.
[128] Schnabel, M., Marlovits, S., Eckhoff, G. et al. Dedifferentiation-associated changes in morphology and gene expression in primary human articular chondrocytes in cell culture. *Osteoarthritis. Cartilage.* 10: 62-70, 2002.
[129] Schulz, R. M., Bader, A. Cartilage tissue engineering and bioreactor systems for the cultivation and stimulation of chondrocytes. *Eur. Biophys. J* 36: 539-568, 2007.
[130] Schumann, D., Kujat, R., Nerlich, M., Angele, P. Mechanobiological conditioning of stem cells for cartilage tissue engineering. *Biomed Mater Eng* 16: S37-S52, 2006.

[131] Sha'ban, M., Kim, S. H., Idrus, R. B., Khang, G. Fibrin and poly(lactic-co-glycolic acid) hybrid scaffold promotes early chondrogenesis of articular chondrocytes: an in vitro study. *J Orthop Surg Res* 3: 17, 2008.

[132] Shekaran, A., Garcia, A. J. Extracellular matrix-mimetic adhesive biomaterials for bone repair. *J Biomed Mater Res A* 2010.

[133] Shieh, S. J., Terada, S., Vacanti, J. P. Tissue engineering auricular reconstruction: in vitro and in vivo studies. *Biomaterials* 25: 1545-1557, 2004.

[134] Shirasawa, S., Sekiya, I., Sakaguchi, Y., Yagishita, K., Ichinose, S., Muneta, T. In vitro chondrogenesis of human synovium-derived mesenchymal stem cells: optimal condition and comparison with bone marrow-derived cells. *J. Cell Biochem.* 97: 84-97, 2006.

[135] Smith, R. L., Carter, D. R., Schurman, D. J. Pressure and shear differentially alter human articular chondrocyte metabolism: a review. *Clin. Orthop Relat Res* S89-S95, 2004.

[136] Stading, M., Langer, R. Mechanical shear properties of cell-polymer cartilage constructs. *Tissue Eng* 5: 241-250, 1999.

[137] Sterodimas, A., De Faria, J., Correa, W. E., Pitanguy, I. Tissue engineering in plastic surgery: an up-to-date review of the current literature. *Ann. Plast. Surg.* 62: 97-103, 2009.

[138] Takagi, J., Springer, T. A. Integrin activation and structural rearrangement. *Immunol. Rev.* 186: 141-163, 2002.

[139] Tarng, Y. W., Casper, M. E., Fitzsimmons, J. S. et al. Directional fluid flow enhances in vitro periosteal tissue growth and chondrogenesis on poly-epsilon-caprolactone scaffolds. *J Biomed Mater Res A* 95: 156-163, 2010.

[140] Temenoff, J. S., Mikos, A. G. Review: tissue engineering for regeneration of articular cartilage. *Biomaterials* 21: 431-440, 2000.

[141] Tins, B. J., McCall, I. W., Takahashi, T. et al. Autologous chondrocyte implantation in knee joint: MR imaging and histologic features at 1-year follow-up. *Radiology* 234: 501-508, 2005.

[142] Uematsu, K., Hattori, K., Ishimoto, Y. et al. Cartilage regeneration using mesenchymal stem cells and a three-dimensional poly-lactic-glycolic acid (PLGA) scaffold. *Biomaterials* 26: 4273-4279, 2005.

[143] ves da Silva, M. L., Martins, A., Costa-Pinto, A. R. et al. Chondrogenic differentiation of human bone marrow mesenchymal stem cells in chitosan-based scaffolds using a flow-perfusion bioreactor. *J Tissue Eng Regen. Med.* 2010.

[144] Vinatier, C., Mrugala, D., Jorgensen, C., Guicheux, J., Noel, D. Cartilage engineering: a crucial combination of cells, biomaterials and biofactors. *Trends Biotechnol.* 27: 307-314, 2009.

[145] Vunjak-Novakovic, G., Martin, I., Obradovic, B. et al. Bioreactor cultivation conditions modulate the composition and mechanical properties of tissue-engineered cartilage. *J Orthop Res* 17: 130-138, 1999.

[146] Wang, P., Zhu, F., Lee, N. H., Konstantopoulos, K. Shear-induced interleukin-6 synthesis in chondrocytes: roles of E prostanoid (EP) 2 and EP3 in cAMP/protein kinase A- and PI3-K/Akt-dependent NF-kappaB activation. *J Biol. Chem.* 285: 24793-24804, 2010.

[147] Washburn, N. R., Yamada, K. M., Simon, C. G., Jr., Kennedy, S. B., Amis, E. J. High-throughput investigation of osteoblast response to polymer crystallinity: influence of nanometer-scale roughness on proliferation. *Biomaterials* 25: 1215-1224, 2004.

[148] Webster, T. J., Ahn, E. S. Nanostructured biomaterials for tissue engineering bone. *Adv. Biochem. Eng Biotechnol.* 103: 275-308, 2007.

[149] Wood, J. J., Malek, M. A., Frassica, F. J. et al. Autologous cultured chondrocytes: adverse events reported to the United States Food and Drug Administration. *J Bone Joint Surg Am.* 88: 503-507, 2006.

[150] Wu, S. C., Chang, J. K., Wang, C. K., Wang, G. J., Ho, M. L. Enhancement of chondrogenesis of human adipose derived stem cells in a hyaluronan-enriched microenvironment. *Biomaterials* 31: 631-640, 2010.

[151] Xu, J., Wang, W., Ludeman, M. et al. Chondrogenic differentiation of human mesenchymal stem cells in three-dimensional alginate gels. *Tissue Eng Part A* 14: 667-680, 2008.

[152] Yanaga, H., Imai, K., Yanaga, K. Generative Surgery of Cultured Autologous Auricular Chondrocytes for Nasal Augmentation. *Aesthetic Plast. Surg* 2009.

[153] Yates, K. E., Allemann, F., Glowacki, J. Phenotypic analysis of bovine chondrocytes cultured in 3D collagen sponges: effect of serum substitutes. *Cell Tissue Bank.* 6: 45-54, 2005.

[154] Yoo, H. S., Lee, E. A., Yoon, J. J., Park, T. G. Hyaluronic acid modified biodegradable scaffolds for cartilage tissue engineering. *Biomaterials* 26: 1925-1933, 2005.

[155] Zeltinger, J., Sherwood, J. K., Graham, D. A., Mueller, R., Griffith, L. G. Effect of pore size and void fraction on cellular adhesion, proliferation, and matrix deposition. *Tissue Eng* 7: 557-572, 2001.

[156] Zheng, L., Fan, H. S., Sun, J. et al. Chondrogenic differentiation of mesenchymal stem cells induced by collagen-based hydrogel: an in vivo study. *J Biomed Mater Res A* 93: 783-792, 2010.

[157] Zhou, X. Z., Leung, V. Y., Dong, Q. R., Cheung, K. M., Chan, D., Lu, W. W. Mesenchymal stem cell-based repair of articular cartilage with polyglycolic acid-hydroxyapatite biphasic scaffold. *Int. J Artif. Organs* 31: 480-489, 2008.

[158] Zhu, L., Wu, Y., Jiang, H., Liu, W., Cao, Y., Zhou, G. Engineered cartilage with internal porous high-density polyethylene support from bone marrow stromal cells: A preliminary study in nude mice. *Br J Oral Maxillofac Surg* 2009.

[159] Zuk, P. A., Zhu, M., Mizuno, H. et al. Multilineage cells from human adipose tissue: implications for cell-based therapies. *Tissue Eng* 7: 211-228, 2001.

[160] Zwingmann, J., Mehlhorn, A. T., Sudkamp, N., Stark, B., Dauner, M., Schmal, H. Chondrogenic differentiation of human articular chondrocytes differs in biodegradable PGA/PLA scaffolds. *Tissue Eng* 13: 2335-2343, 2007.

# Part 5

## Hollow Organs

# 13

# Bioengineering of Colo-Rectal Tissue

Roman Inglin, Lukas Brügger, Daniel Candinas and Daniel Eberli
*Department of Visceral Surgery and Medicine, Bern University Hospital and Labor for Tissue Engineering and Stem Cell Therapy, Clinic of Urology, University Hospital Zurich*
*Switzerland*

## 1. Introduction

About 30 years ago, first reports have appeared in the literature on successful substitution of gastrointestinal wall defects by synthetic materials. An either absorbable or non-absorbable polymer patch sutured into a full-thickness wall defect of stomach, small intestine, and even colon of different animal species generally showed to be initially integrated and overgrown by neomucosa, and eventually absorbed or intra- or extraluminally extruded (Thompson et al., 1986; Harmon et al., 1979; Contieri et al., 1980; Smyrnis, 1982). Though this approach has been revived in several publications reporting good results (Oh et al., 2002; Uzun et al., 2010), treatment of damaged bowel by implantation of synthetic material has never gained general acceptance in the surgical community, and is, therefore, out of common clinical practice. This may be due to the general concern regarding the use of synthetic material in tissue which is naturally colonized with bacteria, as it is particularly the case in the colon. More recently, also natural materials like collagen sponge or acellular matrix have been tested for their ability either to support healing of an intestinal wall damage or to entirely substitute a full thickness defect. Implantation of patches of these natural biomaterials yielded good results with respect to histological and, to some extent, even functional reconstitution (Wang et al., 2005; Wang et al., 2003; Demirbilek et al., 2003; Badylak, S. et al., 2000; Isch et al., 2001; Kajitani et al., 2001; Mutter et al., 1996). As far as concerns the interposition of a synthetic or natural scaffold of tubular shape into esophagus or small intestine, results are less promising. Poor mucosal regeneration, stricture formation or high mortality rate of the experimental animals due to anastomotic leaks were reported, no matter whether the implanted biomaterial was synthetic absorbable (Thompson et al., 1986), non-absorbable (Fukushima et al., 1983; Watson et al., 1980), or of natural origin (Badylak, S. et al., 2000; Badylak, S.F., 2005; Chen & Badylak, 2001). There are no results reported in the current literature dealing with implantation of a tubular scaffold into the large intestine. In recent years there is a growing effort being done to apply tissue engineering methods for the complete, thus tubular, reconstruction of gastrointestinal organs.
A driving force of the endeavours of gastrointestinal tissue engineering is the intention to circumvent transplantation and the associated lifetime need for immunosuppression as the sole alternative to the complete absence of the respective organ. Since loss of the entire colon is absolutely compatible with life, this is not a condition that would require transplantation, indeed. However, total colectomy leads to important changes in enterohepatic circulation,

microbiology, and in water and sodium absorption (Papa et al., 1997). Furthermore, a considerable number of patients suffer from a deficient or missing anorectal continence organ due to intractable neuro-muscular dysfunction, oncological surgery or congenital malformation. The impact on patient's quality of life and the associated health care costs are considerable. Substitutes of the anorectal continence organ such as artificial bowel sphincter or musculoplasty still go along with a high morbidity and questionable functional results. Tissue engineering might have the potential to avoid some of the shortcomings related to these devices and reconstruction techniques. However, current gastrointestinal tissue engineering is clearly focused on esophagus, stomach and small intestine, whereas research on bioengineering of colon is scarce (Penkala & Kim, 2007). In this chapter we will address the different approaches that have been taken hithterto, and discuss new ground that might be broken to substitute colo-rectal tissue.

## 2. Different approaches for colo-rectal tissue engineering

As a matter of fact, colo-rectal tissue engineering has been struggling for its existence up to date. Given that most hollow organs are organized in a similar fashion, consisting of epithelium or endothelium on the luminal side surrounded by a collagen rich connective tissue and muscle layer, some of the knowledge gained in other fields of hollow organ tissue engineering might be successfully applied to bioengineering of large intestine.

### 2.1 Biomaterials

Studies on materials used with intent to bioengineer colonic tissue do not exist. However, in an attempt to prevent the feared complication of disruption of an anastomosis (a condition where two ends of the colon are sutured together in order to re-establish gastrointestinal continuity after resection of a part of the colon) many synthetic or natural biomaterials have been tested. Despite encouraging results, these biomaterials have not found their way into common clinicial practice yet.

#### 2.1.1 Synthetic polymers

Technical aids such as the SBS-tube (Buch et al., 2002), a synthetic compound consisting of polyethylene glycol and a mixture of palmitic and stearic acid (nonattached intraluminal degradable tube) or Coloshield (Cuilleret et al., 1991), a nondegradable intraluminal Latex or silicone tube sewn to the bowel, were developped in order to support or even enhance regeneration of colonic tissue during the wound healing process. These synthetic materials act through diversion of the intraluminal fecal stream from the zone of the anastomosis rather than by exerting a direct promoting effect on the wound healing. Though, they showed to decrease anastomotic leakage and dehiscence, complications such as erosion of the tube through the bowel wall and obstruction were described(Egozi et al., 1993). Henne-Bruns and colleagues enwrapped (Henne-Bruns et al., 1990) the anastomosis with polyglycolic acid (PGA) meshes. This approach turned out to be associated with motility disorders, delayed healing and leakage followed by peritonitis. Admittedly, these materials have not been tested for colon tissue engineering purposes. However, given the considerable complication rate, they do not qualify for being further assessed in this context. Implanted expanded polytetrafluoroethylene (ePTFE) graft (Uzun et al., 2010), or Dacron (Contieri et al., 1980) used to cover full thickness wall defects in the colon showed to function as a guide for ingrowing tissue that originated from the wound edges, thus, enabling closure of the respective tissue gap. They have nevertheless not been integrated into the regenerating tissue, but were eventually rejected. Given the inert nature of their surfaces, it remains doubtful whether they were suitable for acting as „real" tissue engineering scaffolds that would facilitate engraftment and promote growth of host tissue cells.

PGA scaffolds are successfully being used for small intestinal tissue engineering (Mooney et al., 1994; Sala et al., 2009; Sala et al., 2011). These principles were applied to bioengineering of the colon. Grikscheit and colleagues constructed 2-mm thick nonwoven PGA tubes. They were then seeded with previously harvested mesenchymal - epithelial cell compounds, derived from dissection of autologous full thickness (sigmoid) colon wall and referred to as "oragnoid units", and eventually implanted into the omentum of recipient animals. Within 4 weeks, these constructs grew to neo-colon tissue that exhibited a mucosal architechture and electrophysiological parameters comparable to that of native colon (Grikscheit et al., 2002; Grikscheit et al., 2003). A big achievement beyond doubt. However, the so bioengineered colon derives from syngeneic tissue. This would currently exclude its use for the treatment of patients with diseased autologous donor cells. Furthermore, this technique is substantially limited of what concerns customization of size and shape of the bioengieneered tissue. Indeed, tissue engineering of organoid units is the creation of a premature tissue which has to be further processed in order to become a definite, functional organ.

### 2.1.2 Natural scaffolds

Since interests in bowel tissue engineering are focused on small intestine this is the field where natural scaffolds have mostly been evaluated. Natural scaffolds are synonymous with extracellular matrix (ECM), consisting of interstitial matrix and the basement membrane. ECM of different origin have been evaluated for their potential to serve as scaffold for regeneration of intestinal tissue. Extensive resection of small bowel followed by implantation of lyophilized dura mater was shown to induce neomucosa on the patch, with histological characteristics similar to normal. However, the authors stated that although the creation of neomucosa on dural patches is feasible and conditions a slight improvement in the animal's nutritive status the high perioperative mortality of the experimental animals casts its applicability in intestinal tissue engineering into doubt (Hernandez Bermejo et al., 1993).

Different natural biomaterials have been studied with respect to their influence on healing of colon wall injuries. Small intestinal submucosa (SIS) was shown to be largely successful in promoting healing of a 1 $cm^2$ full thickness wall defect in unprepared large bowel and serving as a bioscaffold for regeneration of the native colonic tissue (Ueno et al., 2007). Another group used as what they referred to as reconstituted connective tissue patches consisting of elastin, fibronectin and collagen. Using these patches applied with fibrin sealant to the edges of a 1 cm diameter colonic defect, complete reconstitution of the mucosal as well as the circular and longitudinal muscle layer after 40 days was obtained. No retraction and inflammatory reaction were found, and the patch was slowly resorbed (Marescaux et al., 1991). Accordingly, placental-derived tissue matrix applied to a 1.5 x 2.0 cm parietal whole in the large intestine healed significantly better than fibrin glue repair alone (Alam et al., 1998).

There is a growing body of evidence that surface characteristics of the biomaterial has a determining influence on cell adhesion patterns and the ability to support and maintain differentiated phenotypes and their functions. Brown and colleagus demonstrated that three different ECMs, including SIS, urinary bladder matrix (UBM) and liver extracellular matrix (LECM) displayd distinct surface properties. These differences were reflected in their ultrastructure, as demonstrated by scanning electron microscopy (SEM) as well the distinct surface pattern of proteins and their fragments. While luminal and abluminal surface topography differs in ECM origninating from hollow organs like urinary bladder (UBM) or intestine (SIS), such a sidedness could not be observed in ECM of solid organs, like LECM (Brown et al., 2010). It remains, however, unclear whether the molecular composition or the ultrastructure of these scaffolds determine to a greater extent the outcome of the cell-scaffold interaction. The authors suggested that it was likely the diverse combination of both the structural and the functional components that accounts for the success of ECM based materials in the respective organ system (Barnes et al., 2011). Given the aforementioned aspects SIS may appear most suitable for intestinal tissue engineering. However, the ideal natural scaffold for colo-rectal constructs has not yet been defined.

### 2.2 Cells

Generally, the combination of biomaterial with cells was shown to improve graft integration and mucosal regeneration compared to when biomaterial is used alone. Cell seeding has been found to induce smooth muscle regeneration with less inflammatory reaction and fibroblast ingrowth as well as reduced contraction and shrinkage than is seen in unseeded grafts (Oberpenning et al., 1999; Yoo et al., 1998). Cell-cell- and cell-matrix interaction play an important role in growth and differentiation of various epithelial cell types, including urothelium (Zhang et al., 2004), oral mucosa (Izumi et al., 2003) and intestine (Kalabis et al., 2003; Rubin, 2007). Furthermore, epithelial and mesenchymal cells (i.e. smooth muscle cells [SMCs]) co-cultured on a biomaterial were shown to mutually influence their proliferation (Zhang et al., 2000; Baskin et al., 2001; Master et al., 2003). The principle of mesenchymal-epithelial interaction has also been substantiated in gut tissue (Pinchuk et al., 2010; Kosinski et al., 2010; van der Flier & Clevers, 2009; Barker et al., 2008). Apart from that, in vitro seeded cells, whether of mesenchymal or epithelial origin, might also function as a place-holder to facilitate ingrowth of host's natural epithelium on the biomaterial in vivo. This might prevent luminal bacteria from getting into contact with the underlying biomaterial, thus avoiding scar tissue formation and contraction of the graft due to infection and inflammation.

Taking into account the layered composition of hollow organs such as large intestine, smooth muscle cells and epithelial cells appear to be of pivotal interest for colo-rectal tissue engineering. Since particularly culture of intestinal epithelial cells still signifies a considerable challenge, alternative epithelial cell sources may keep on being necessary for successful colon bioengineering. It is of utmost importance that the cells used for tissue engineering do not grow in an uncontrolled fashion. Therefore, neither cancer nor immortalized cells qualify for being adopted in this context. On the other hand, primary cell cultures and stem cells seem currently to be the most reasonable of available cell sources for tissue engineering purposes.

### 2.2.1 Primary cell lines

#### 2.2.1.1 Smooth muscle cells (SMCs)

Isolation and culture of SMCs were established in different species and hollow organ systems (Nakase et al., 2006; Raghavan et al., 2011; Raghavan et al., 2010a; Zhang et al., 2004; Kao et al., 1988). Raghavan and colleagues were able to successfully bioengineer a ring-shaped SMC construct, that, once subcutaneously implanted remained viable until harvest at 28 days. The muscle-ring further maintained its alignment, phenotype as well as some physiological properties of SMCs (Raghavan et al., 2010b).

Very recently, the same research group has co-cocultured SMCs originating from human internal anal sphincter with immortomouse fetal enteric neurons, which they eventually implanted in mice (Raghavan et al., 2011). This construct was neovascularized after implantation without signs of inflammation. The bioengineered muscle showed physiological function similar to normal. The cells exhibited a spontaneous myogenic basal tone, and adequately respondeded to direct electrical and chemical stimuli, such as relaxation after application of vasoactive intestinal peptide, or contraction response to cholinergic stimulation with acetylcholine. Such functional bioengineered muscle might be incorporated into the concept of a possible tissue engineered ano-rectal continence organ.

Furthermore, a physiological model of longitudinal smooth muscle tissue was developed (Raghavan et al., 2010a). In this model, the SMCs isolated from rabbit sigmoid colon and arranged in the longitudinal axis formed a highly aligned cell sheet after being seeded at high densities onto laminin-coated Sylgard surfaces with defined wavy microtopographies. The so established longitudinal muscle construct not only maintained smooth muscle phenotype and characteristics of calcium-dependence, but also electrophysiological properties similiar to its archetype.

#### 2.2.1.2 Colon epithelial cells

Though long-term cultures of primary colon epithelial cells (CEC) have been described, they remain a challenge (Chopra et al., 2010; Bartsch et al., 2004; Booth et al., 1995). Traditionally, the method referred to as organ culture was employed to study physiology and intestinal lineage outside of an organism. Small intestinal explants of fetal tissue (few $mm^2$ in size) can be maintained with normal tissue organization for up to 3 weeks, allowing physiological studies on growth, differentiation, and ion transport across the epithelium (Quaroni, 1985; Pyke & Gogerly, 1985). On the other hand, adult intestinal mucosa in organ culture displayed an intense regenerative activity following an initial phase of epithelial cell loss (Ferland & Hugon, 1979; Moorghen et al., 1996). Using adjuncts like embedding on collagen gel or addition of promoting factors to the medium enabled maintaining viable tissue in organ culture for up to 28 days (Ootani et al., 2009). Yet, since organ culture techniques do not allow propagation of a specific cell (i.e. epithelial) lineage, it has limited appeal in tissue engineering applications. Attempts to generate epithelial cell cultures originating from passaged outgrowths of small tissue explants did not work well for intestinal mucosa (Chopra et al., 2010).

Tissue dissociation approaches using chelating and/or proteolytic enzymes appear to qualify better for bioengineering purposes, since specific cell lines can be obtained, held in culture, and successfully passaged. However, certain prerequisites have to be followed if culture of intestinal epithelial cells should be fruitful. After preparation with the generally used chelating agent ethylenediaminetetraacetic acid (EDTA), colonic epithelial cells normally do not attach to the unprepared plastic dish and rapidly degenerate (Deveney et al., 1996). This is presumably due to the disruption of extracellular matrix molecules that normally would facilitate cell-cell and cell-matrix interaction. These epithelial cells much more require plating on a suitable substrate like collagen matrix or a fibroblast feeder layer, whether chemically pretreated or not, to prevent apoptosis and allow cells to replicate (Kalabis et al., 2003; Wildrick et al., 1997). If isolated crypts are prepared from colonic tissue, they also contain stem cells positive for Lgr5, an approved intestinal stem cell marker (Barker et al., 2007). This enables these cell cultures to actively replicate and to form new crypts whose cells differentiate along the crypt-villous axis. Ootani et al. could also demonstrate that the crypt-villous structure is self-organizing, and that it can be generated from a single Lgr5-positive stem cell while any non-epithelial cell type is absent (Ootani et al., 2009). Proteolytic enzymes, including trypsin, dispase, DNAse and collagenase are being applied in order to circumvent the limitations of EDTA. Among a diversity of protocols of enzymatic isolation of epithelial cells, trypsin and collagen are predominantly used (Chopra et al., 2010).

Apart from employing proteolysis, a variety of compositions of culture medium, application of growth factors and additional substances, like hydrocortisone or cholera toxin, culture conditioning with non-epithelial cells as well as the use of antimicrobial agents have been proposed in order to obtain several primary epithelial colon cell cultures of different species (Birkner et al., 2004; Deveney et al., 1996; Chopra & Yeh, 1981; Bartsch et al., 2004; Booth et al., 1995; Yeh & Chopra, 1980). Yet, since suggested concentrations of the respective substrates vary considerably between different authors, Chopra et al. stated in their recent comprehensive review that the precise combination of each substrate and additive must be defined for each system (Chopra et al., 2010).

While even long-term culture of colon epithelial cells were successfully established, their use for tissue engineering purposes have not been published up to date. Given that in vitro seeding of different types of epithelial cells on a scaffold was reported to improve integration of the implanted biomaterial in vivo (Nakase et al., 2008; Oberpenning et al., 1999; Yoo et al., 1998), we speculate that this might also apply to colonic epithelial cells. Furthermore, cell-populated biomaterials may be less prone to inflammation with subsequent fibrosis and scaffold shrinkage due to the fact that the commensal flora is prevented to get into contact with the biomaterial itself. These presumptions, however, have yet to be clarified.

#### 2.2.1.3 Alternative epithelial cells - oral keratinocytes

Assuming that culture of colonic epithelium remains a challenge, alternative cells sources for bioengineering of large intestine are of substantial interest. Oral keratinocytes have already been successfully used in tissue engineering of epithelial cell-lined constructs. Nakase et al. (Nakase et al., 2008) evaluated healing of interposed tissue engineered esophagus in dogs. They prepared tubular composite scaffolds consisting of a PGA felt containing SMCs and, towards its luminal side, human amniotic membrane which was either populated with oral keratinocytes and a feeder layer of autologous fibroblasts or not. These scaffolds were rolled around a polypropylene tube that was 3 cm in length and 2 cm in diameter and wrapped with the omentum of the dog. Three weeks after, the now soft tubular tissue was harvested from the abdomen, and elevated into the right thoracic space as a pedicle graft to replace a 3 cm resection of the hosts's esophagus. Absence of oral keratinocytes on the amniotic membrane side was associated with stricture formation followed by almost complete obstruction after 2 to 3 weeks. Where keratinocyte containing constructs were implanted, however, the in situ tissue-engineered esophagus showed good distensibility and the dogs remained without feeding problems through 420 days. Esophageal peristalsis transferred food to the stomach, despite the absence of peristaltic activity in the tissue-engineered esophagus graft itself. The implanted tissue-engineered esophagus displayed a histological architecture similar to that of the adjacent native esophagus. The capacity of buccal mucosa to prevent stricture formation and, thus, provide patency of bioengineered tubular constructs has been confirmed for the urethra when it was applied as either a patch graft (Li et al., 2008; Bhargava et al., 2008) or an entire tubular implant (Raya-Rivera et al., 2011). Buccal mucosa may, therefore, be a valuable alternative cell source for colon tissue engineering.

### 2.2.2 Organoid units

One of the few research groups who dealt with bioengineering of large intestine loaded mesenchymal - epithelial cell compounds, derived from dissection of autologous full thickness (sigmoid) colon wall and referred to as "oragnoid units", into biodegradable polymer tubes that were eventually implanted into the omentum of recipient animals. After 4 weeks, the constructs have grewn to cysts consisting of many representative elements of an intact colonic architecture including normal epithelium, vascularization, present ganglion cells, and muscularis propria. Furthermore, this neo-colonic tissue exhibited both absorptive and secretory functions. Those animals who have undergone anastomosis of the bioengineered cyst to the native colon suffered no electrolyte imbalances, showed fewer symptoms of dehydration, and had evidence of bile acid recycling as well as raised content of short fatty acids in the stool. These findings were reflected in less weight loss, less relative hyponatremia, decreased stool moisture, elevated transit times and more formed stool in animals with implanted neo-colon (Grikscheit et al., 2002). Further studies revealed that bioengineered large intestine out of previously tissue engineered colonic tissue completely retained the architectural and physiological characteristics of its archetype. Evaluation of physiological function using an Üssing chamber suggested adequate vectorial ion transport, barrier function, and viability (Grikscheit et al., 2003).

These results are promising, of course. However, the fact that the so bioengineered colon derives from syngeneic tissue would currently preclude its use for the treatment of patients with abnormal autologous donor cells, i.e. autoimmune diseases like inflammatory bowel syndrome, or types of hereditary cancer including familial polyposis or HNPCC. Furthermore, this technique is substantially limited of what concerns customization of size and shape of the bioengieneered tissue. Finally, apart from the muscularis propria, smooth muscle layers that might render peristalsis possible are completely absent. Indeed, this approach facilitates the creation of a premature tissue which has to be further processed in order to become a definite, functional organ. Despite these limitations, organoids may be successfully used in combination with other techniques/biomaterials for future tissue engineering of large intestine.

### 2.2.3 Progenitor and stem cells

#### 2.2.3.1 Muscle progenitor and stem cells

The transplantation of muscle progenitor cells (MPCs) has been evaluated as for a treatment of genetic and acquired muscle disorders (Gussoni et al., 1999; Leobon et al., 2003; Yiou et

al., 2003b). MPCs are considered quiescent adult stem cells and are located under the membrane surrounding the muscle fibers. After trauma or damage, MPCs drive muscle tissue regeneration by proliferating and differentiating into myoblasts, further fusing and eventually forming new myofibers. While entire muscle naturally contains cells of different origins, including from the vascular and hematopietic compartment, MPCs represent a subset of muscle cells that are generally committed to the myogenic lineage (Benchaouir et al., 2004). Therefore, MPCs are of particular interest for muscle engineering, and may be a valuable alternative to more original smooth muscle cells in order to restore function of defective or absent anal sphincter. The potential use of injectable cultured MPCs for the treatment of stress urinary incontinence has been investigated in experimental models (Chancellor et al., 2000; Yiou et al., 2003a; Yokoyama et al., 2001). Our own results in dogs showed that autologous muscle progenitor cells are able to reconstitute irreversibly damaged urinary sphincter function. The injected cells were able to survive and formed mature tissue within the damaged sphincter region. This approach has recently been adopted to anal fecal incontinence. Three weeks after external anal sphincterotomy, MPCs from quadriceps myofiber explants were auto-grafted by injection into the anal sphincter of rabbits. The so transplanted MPCs displayed the potential for recapitulation of a myogenic program, yielding improved objective anal measures of resting and stimulated pressures as well as of electromyographic profile (Kajbafzadeh et al., 2010).
An Italian research group reported new muscle fiber formation with increased contractility of sphincter muscle strips in rats after having injected bone marrow-derived mesenchymal stem cells into the previously injured and surgically repaired anal sphincter (Lorenzi et al., 2008). Another study suggested increased contractility of the sphincter muscle in rats after cryoinjury followed by injection of autologous muscle-derived stem cells (Kang et al., 2008). While these results did not reach statistical significance, Frudinger and colleagues (Frudinger et al., 2009) were able to demonstrate a significant clinical improvement of fecal incontinence in patients who recieved an injection of muscle-derived stem cells into their anal sphincters. None of these patients showed signs of rejection or other side effects, proving that the injection of autologous cells is feasible and safe. However, the authors described a discrepancy between clinical improvement and unchanged sphincter pressure and, therefore, recommended that further clinical studies should be preceded by bench experiments.

#### 2.2.3.2 Intestinal stem cells

Recent advances in tracking, proper isolation and further processing of intestinal stem cells have promoted the interest in their use as a potential cell source for biotechnological applications (Umar, 2010). The idea to potentially reproduce intestinal tissue out of a single intestinal stem cell (ISC) is very appealing. As part of the aforementioned progress, specific cell culture conditions have been developed that allow the generation of long-lived organoids from either crypts or even from single ISCs of the small intestine. Sato and colleagues reported that the resulting intestinal „mini-guts" display all the characteristics of native gut epithelium (Sato et al., 2009). Although not yet elucidated in detail, ISC differentiation and self-renewal seems to be regulated *in vivo* by adjacent mesenchaymal cells, including endothelial cells, lymphocytes, muscle cells, and particularly subepithelial myofibroblasts (Powell et al., 2005). This microenvironment which controls the stem cell fate is referred to as stem cell niche (Scadden, 2006).

In an *in vitro* culture, however, where the mesenchymal niche elements are lacking, Paneth cells - specialized intestinal daughter stem cells - have been identified to act as a multifunctional guardian of the mother stem cell supplying both bactericidal lysosymes and essential niche signals (Sato et al., 2011). It could be demonstrated that intestinal organoids, consisting of a central lumen lined by villus-like epithelium and several surrounding crypt-like domains, could be generated from one single Lgr5-positive stem cell (Sato et al., 2009; Barker et al., 2010). Alternatively, Ootani and colleagues used neonatal tissue to initiate three-dimensional intestinal organ cultures. In these long-term cultures architecture of the mesenchymal niche as well as the multilineage epithelial lining which is typical for intestinal mucosa was preserved (Ootani et al., 2009). In contrast to the sphere formation of other cultured epithelial cell types, intestinal stem cell-borne tissue arranges in an asymmetry in which proliferative and differentiated cell types are positioned in accordance to their natural locations in the epithelial lining. It has been speculated that this might be due to the presence of an appropriate niche microenvironment - provided by either available mesenchymal structures and/or Paneth cells - that facilitate local morphogen gradients, thereby generating progenitor zones that are seperate from areas with differentiated cells (Snippert & Clevers, 2011).

Current research is giving rise to a more detailed understanding of the stem cell's surrounding microenvironment and of the different signals that regulate self-renewal and maintain the balance between self-renewal and differentiation of ISCs (Umar, 2010). Accordingly, epidermal growth factor (EGF), Wnt3 and Notch have been identified as essential signals for stem cell support (Sato et al., 2011). Previous work of the same research group was able to demonstrate that the addition of EGF, the Wnt agonist R-spondin 1 and the BMP inhibitor noggin induced single Lgr5-positive stem cells plated in a Matrigel-based culture system to grow into crypt-like structures with de novo generated stem cells and Paneth cells at their bottom (Sato et al., 2009).

This fastly growing knowlegde on regulation of intestinal stem cells' growth and differentiation may stimulate new approaches for tissue engineering of both the small and the large intestine. It is a tempting idea to grow complete intestinal tissue out of a single stem cell whose growth and differentiation would be induced and regulated by the respective factors/signals tethered to the biomaterial itself. Therefore, initialization of an optimized microenvironment that immitates the respective stem cell niche as close to its archetype as possible will be one of the challenges of intestinal tissue engineering of the coming years.

#### 2.2.3.3 Adipose-derived stem cells

Another type of adult stem cells being used in tissue engineering are adipose-derived stem cells (ADSC). ADSC have successfully been derived from fat biopsies and differentiated towards distinct mesenchymal and neurogenic cell lineages (Bunnell et al., 2008; Taha & Hedayati, 2010). Few results on generation of epithelial tissue from ADSC, including corneal (Du et al., 2010; Ho et al., 2011) and urothelium-like cells (Liu et al., 2009) as well as tracheal epithelium (Kobayashi et al., 2010) have been reported up to date. Theoretically, induction of ADSC to differentiate towards intestinal epithelial cells ought to be possible by adding the respective cocktails of chemical inducers or cytokines. This approach might be worth being pursued since the use of ADSC - like adult stem cells in general - does not imply the many socioethic issues that accompany the work with embryonic stem cells. Furthermore, ADSC are abundantly available and easily accessible by liposuction that is itself associated with a low morbidity. Using ADSC for tissue engineering purposes also means that potentially both mesenchymal and epithelial cells can be generated out of one single cell source. However, there is still a way to go to achieve bioengineered tissue on the basis of fat-derived stem cells.

### 2.3 Bioreactors

Once tissue engineering has grown out of its fledgling stages initialization of an adquate vascularization of the growing tissue has arguably become its biggest challenge. Currently, construct development is often restricted to smaller than clinically relevant size due to issues of (in-)sufficient oxygen and nutrient supply. Consequently, feasibility of in vivo integration of the bioengineered tissue may be constrained (Lovett et al., 2009). Interposition of bioengineered substitutes organized in a tubular fashion, such as the gut, appear to be particularly prone to complications (Badylak, S. et al., 2000; Chen & Badylak, 2001; Thompson et al., 1986). Given that inadequate oxygen supply and decreased local microperfusion are believed to be at the origin of many adverse events in gastrointestinal surgery (Anegg et al., 2008; Attard et al., 2005; Sheridan et al., 1987), hypoxia and ischemia presumably account for anastomotic leakage and stricture formation after interposition of bioengineered tubular constructs.

In recent years, considerable effort has been done to develop bioreactors that allow nourishment of growing tissue masses in vitro whose size exceeds the largest possible unit accessible to oxygen and nutrients by sole diffusion. If successful, such bioreactors are believed to clear the way for the generation of clinically applicable, tissue engineered constructs tailored with respect to the required size and shape in vitro.

In various fields of hollow organ tissue engineering, such as of the cardiovascular system (Song et al., 2011; Iyer et al., 2011), as well as the respiratory (Tan et al., 2007) and genitourinary tract (Wei et al., 2011) the rapidly growing know-how on bioreactor design and manufacturing has ultimately enhanced their more and more frequent use. The increasing application of these devices has not yet been parallelled in gastrointestinal tissue engineering. The recent report on a perfusion bioreactor with intent to optimize conditions for long-term culture of primary intestinal epithelium can, however, be considered as an initial step towards the development of such devices in gut bioengineering. Kim and colleagues (Kim et al., 2007) designed and fabricated that bioreactor consisting of a multichannel peristaltic pump, culture medium reservoir, oxygenation and gas exchange unit, air trap, and cell-polymer construct housing unit. The appliance allows tuning of tissue oxygenation and $CO_2$ exchange as well as dynamic cell seeding onto tubular scaffolds. The authors could demonstrate that the dynamically seeded epithelial organoid survived for two days under flow conditions. Yet, they state that, despite the promising results, further optimizations are needed to be done regarding the parameters for dynamic seeding and the employed biomaterial itself in order to enhance cellular attachment and survival. In our opinion, if tubular bioengineered tissue of clinically applicable size ought to be successfully generated in vitro, gastrointestinal tissue engineering will not get by without efforts being performed towards development and use of appropriate bioreactor devices.

## 3. Conclusion

Colon tissue engineering is at its very beginning. There has been some important progress in the last decade. Yet, techniques and knowledge gained from other fields of successful hollow organ tissue engineering should be evaluated, and if promising, applied. A lot of patients worldwide suffer from a deficient or absent colo-rectal continence organ and the subsequent impact on the patient's psycho-social condition and the related health care costs are considerable. Therefore, it seems more than justified to give weight to that field of hollow organ tissue engineering. Given the ongoing progress in biomaterial development and refinement, the rapidly growing know-how on (stem) cell culture and programming towards desired daughter cells as well as the future prospects on potential applications, such as bioreactors, we are convinced that tissue engineering of the colon is awaiting a promising future. The achievement of tailored artificial colon tissue has the potential to facilitate a great step forward in the treatment of patients after abdominoperineal resection (total excision of the ano-rectal continence organ) and, if successful, may be applied to other diseases like intractable fecal incontinence or congenital anorectal malformations.

## 4. Acknowledgement

We acknowledge to all of our collaborators in the Labor of Tissue Engineering and Stem Cell Therapy, Department of Urology, University of Zurich.

## 5. References

Alam, H. et al. (1998), A placental-derived tissue matrix as a bowel wall substitute in rats: preliminary study. *Surgery*, Vol. 124, No. 1, pp. 87-91.
Anegg, U. et al. (2008), Influence of route of gastric transposition on oxygen supply at cervical oesophagogastric anastomoses. *Br J Surg*, Vol. 95, No. 3, pp. 344-9.
Attard, J.A. et al. (2005), The effects of systemic hypoxia on colon anastomotic healing: an animal model. *Dis Colon Rectum*, Vol. 48, No. 7, pp. 1460-70.
Badylak, S. et al. (2000), Resorbable bioscaffold for esophageal repair in a dog model. *J.Pediatr.Surg.*, Vol. 35, No. 7, pp. 1097-1103.
Badylak, S.F. (2005), Regenerative medicine and developmental biology: the role of the extracellular matrix. *Anat.Rec.B New Anat.*, Vol. 287, No. 1, pp. 36-41.
Barker, N. et al. (2007), Identification of stem cells in small intestine and colon by marker gene Lgr5. *Nature*, Vol. 449, No. 7165, pp. 1003-1007.
Barker, N. et al. (2008), The intestinal stem cell. *Genes Dev*, Vol. 22, No. 14, pp. 1856-64.
Barker, N. et al. (2010), Lgr5(+ve) stem cells drive self-renewal in the stomach and build long-lived gastric units in vitro. *Cell Stem Cell*, Vol. 6, No. 1, pp. 25-36.
Barnes, C.A. et al. (2011), The surface molecular functionality of decellularized extracellular matrices. *Biomaterials*, Vol. 32, No. 1, pp. 137-43.
Bartsch, I. et al. (2004), Establishment of a long-term culture system for rat colon epithelial cells. *In Vitro Cell Dev.Biol.Anim*, Vol. 40, No. 8-9, pp. 278-284.
Baskin, L. et al. (2001), Mesenchymal-epithelial interactions in bladder smooth muscle development: effects of the local tissue environment. *J.Urol.*, Vol. 165, No. 4, pp. 1283-1288.
Benchaouir, R. et al. (2004), Evidence for a resident subset of cells with SP phenotype in the C2C12 myogenic line: a tool to explore muscle stem cell biology. *Exp.Cell Res.*, Vol. 294, No. 1, pp. 254-268.
Bhargava, S. et al. (2008), Tissue-engineered buccal mucosa urethroplasty-clinical outcomes. *Eur.Urol.*, Vol. 53, No. 6, pp. 1263-1269.

Birkner, S. et al. (2004), Growth and characterisation of primary bovine colon epithelial cells in vitro. *Altern Lab Anim*, Vol. 32, No. 6, pp. 555-71.

Booth, C. et al. (1995), The isolation and culture of adult mouse colonic epithelium. *Epithelial Cell Biol.*, Vol. 4, No. 2, pp. 76-86.

Brown, B.N. et al. (2010), Surface characterization of extracellular matrix scaffolds. *Biomaterials*, Vol. 31, No. 3, pp. 428-37.

Buch, N. et al. (2002), An intraluminal prosthesis may improve healing of a one-layer colonic anastomosis: an experimental study in pigs. *Eur J Surg*, Vol. 168, No. 3, pp. 165-71.

Bunnell, B.A. et al. (2008), Adipose-derived stem cells: isolation, expansion and differentiation. *Methods*, Vol. 45, No. 2, pp. 115-20.

Chancellor, M.B. et al. (2000), Preliminary results of myoblast injection into the urethra and bladder wall: a possible method for the treatment of stress urinary incontinence and impaired detrusor contractility. *Neurourol Urodyn*, Vol. 19, No. 3, pp. 279-87.

Chen, M.K.Badylak, S.F. (2001), Small bowel tissue engineering using small intestinal submucosa as a scaffold. *J.Surg.Res.*, Vol. 99, No. 2, pp. 352-358.

Chopra, D.P.Yeh, K.Y. (1981), Long-term culture of epithelial cells from the normal rat colon. *In Vitro*, Vol. 17, No. 5, pp. 441-449.

Chopra, D.P. et al. (2010), Intestinal epithelial cells in vitro. *Stem Cells Dev.*, Vol. 19, No. 1, pp. 131-142.

Contieri, E. et al. (1980), [The use of prosthetic material in the repair of gastric and colonic gaps in the rat]. *Chir Patol.Sper.*, Vol. 28, No. 2, pp. 80-86.

Cuilleret, J. et al. (1991), [Endoluminal protection of colorectal anastomosis by Coloshield. Apropos of 14 cases]. *J Chir (Paris)*, Vol. 128, No. 8-9, pp. 351-5.

Demirbilek, S. et al. (2003), Using porcine small intestinal submucosa in intestinal regeneration. *Pediatr.Surg.Int.*, Vol. 19, No. 8, pp. 588-592.

Deveney, C.W. et al. (1996), Establishment of human colonic epithelial cells in long-term culture. *J.Surg Res.*, Vol. 64, No. 2, pp. 161-169.

Du, Y. et al. (2010), Adipose-derived stem cells differentiate to keratocytes in vitro. *Mol Vis*, Vol. 16, No. pp. 2680-9.

Egozi, L. et al. (1993), Complication of the intracolonic bypass. Report of a case. *Dis.Colon Rectum*, Vol. 36, No. 2, pp. 191-193.

Ferland, S.Hugon, J.S. (1979), Organ culture of adult mouse intestine. I. Morphological results after 24 and 48 hours of culture. *In Vitro*, Vol. 15, No. 4, pp. 278-87.

Frudinger, A. et al. (2009), Muscle Derived Cell Injection to Treat Anal Incontinence due to Obstetric Trauma: Pilot Study with One-Year Follow-Up. *Gut*, No.

Fukushima, M. et al. (1983), Seven-year follow-up study after the replacement of the esophagus with an artificial esophagus in the dog. *Surgery*, Vol. 93, No. 1 Pt 1, pp. 70-77.

Grikscheit, T.C. et al. (2002), Tissue-engineered colon exhibits function in vivo. *Surgery*, Vol. 132, No. 2, pp. 200-204.

Grikscheit, T.C. et al. (2003), Tissue-engineered large intestine resembles native colon with appropriate in vitro physiology and architecture. *Ann Surg*, Vol. 238, No. 1, pp. 35-41.

Gussoni, E. et al. (1999), Dystrophin expression in the mdx mouse restored by stem cell transplantation. *Nature*, Vol. 401, No. 6751, pp. 390-394.

Harmon, J.W. et al. (1979), Fate of Dacron prostheses in the small bowel of rabbits. *Surg.Forum*, Vol. 30, No. pp. 365-366.

Henne-Bruns, D. et al. (1990), Reinforcement of colon anastomoses with polyglycolic acid mesh: an experimental study. *Eur.Surg Res.*, Vol. 22, No. 4, pp. 224-230.

Hernandez Bermejo, J.P. et al. (1993), [Neoformation of intestinal mucosa on dura mater patches. Application in the surgical treatment of short bowel syndrome. Experimental study in rats]. *Cir Pediatr*, Vol. 6, No. 3, pp. 133-6.

Ho, J.H. et al. (2011), Isolation and characterization of multi-potent stem cells from human orbital fat tissues. *Tissue Eng Part A*, Vol. 17, No. 1-2, pp. 255-66.

Isch, J.A. et al. (2001), Patch esophagoplasty using AlloDerm as a tissue scaffold. *J.Pediatr.Surg.*, Vol. 36, No. 2, pp. 266-268.

Iyer, R.K. et al. (2011), Engineered cardiac tissues. *Curr Opin Biotechnol*, No.

Izumi, K. et al. (2003), Evaluation of transplanted tissue-engineered oral mucosa equivalents in severe combined immunodeficient mice. *Tissue Eng*, Vol. 9, No. 1, pp. 163-174.

Kajbafzadeh, A.M. et al. (2010), Functional external anal sphincter reconstruction for treatment of anal incontinence using muscle progenitor cell auto grafting. *Dis Colon Rectum*, Vol. 53, No. 10, pp. 1415-21.

Kajitani, M. et al. (2001), Successful repair of esophageal injury using an elastin based biomaterial patch. *ASAIO J.*, Vol. 47, No. 4, pp. 342-345.

Kalabis, J. et al. (2003), Stimulation of human colonic epithelial cells by leukemia inhibitory factor is dependent on collagen-embedded fibroblasts in organotypic culture. *FASEB J.*, Vol. 17, No. 9, pp. 1115-1117.

Kang, S.B. et al. (2008), Sphincter contractility after muscle-derived stem cells autograft into the cryoinjured anal sphincters of rats. *Dis Colon Rectum*, Vol. 51, No. 9, pp. 1367-73.

Kao, H.W. et al. (1988), Cultured circular smooth muscle from the rabbit colon. *In Vitro Cell Dev Biol*, Vol. 24, No. 8, pp. 787-94.

Kim, S.S. et al. (2007), A perfusion bioreactor for intestinal tissue engineering. *J Surg Res*, Vol. 142, No. 2, pp. 327-31.

Kobayashi, K. et al. (2010), A tissue-engineered trachea derived from a framed collagen scaffold, gingival fibroblasts and adipose-derived stem cells. *Biomaterials*, Vol. 31, No. 18, pp. 4855-63.

Kosinski, C. et al. (2010), Indian hedgehog regulates intestinal stem cell fate through epithelial-mesenchymal interactions during development. *Gastroenterology*, Vol. 139, No. 3, pp. 893-903.

Leobon, B. et al. (2003), Myoblasts transplanted into rat infarcted myocardium are functionally isolated from their host. *Proc.Natl.Acad.Sci.U.S.A*, Vol. 100, No. 13, pp. 7808-7811.

Li, C. et al. (2008), Urethral reconstruction using oral keratinocyte seeded bladder acellular matrix grafts. *J.Urol.*, Vol. 180, No. 4, pp. 1538-1542.

Liu, J. et al. (2009), Cell-to-cell contact induces human adipose tissue-derived stromal cells to differentiate into urothelium-like cells in vitro. *Biochem Biophys Res Commun*, Vol. 390, No. 3, pp. 931-6.

Lorenzi, B. et al. (2008), Treatment of experimental injury of anal sphincters with primary surgical repair and injection of bone marrow-derived mesenchymal stem cells. *Dis.Colon Rectum*, Vol. 51, No. 4, pp. 411-420.

Lovett, M. et al. (2009), Vascularization strategies for tissue engineering. *Tissue Eng Part B Rev*, Vol. 15, No. 3, pp. 353-70.

Marescaux, J.F. et al. (1991), Prevention of anastomosis leakage: an artificial connective tissue. *Br J Surg*, Vol. 78, No. 4, pp. 440-4.

Master, V.A. et al. (2003), Urothlelium facilitates the recruitment and trans-differentiation of fibroblasts into smooth muscle in acellular matrix. *J.Urol.*, Vol. 170, No. 4 Pt 2, pp. 1628-1632.

Mooney, D.J. et al. (1994), Design and fabrication of biodegradable polymer devices to engineer tubular tissues. *Cell Transplant*, Vol. 3, No. 2, pp. 203-10.

Moorghen, M. et al. (1996), An organ-culture method for human colorectal mucosa using serum-free medium. *J Pathol*, Vol. 180, No. 1, pp. 102-5.

Mutter, D. et al. (1996), Biomaterial supports for colonic wall defect healing. An experimental study in the rat. *Biomaterials*, Vol. 17, No. 14, pp. 1411-1415.

Nakase, Y. et al. (2006), Tissue engineering of small intestinal tissue using collagen sponge scaffolds seeded with smooth muscle cells. *Tissue Eng*, Vol. 12, No. 2, pp. 403-412.

Nakase, Y. et al. (2008), Intrathoracic esophageal replacement by in situ tissue-engineered esophagus. *J.Thorac.Cardiovasc.Surg.*, Vol. 136, No. 4, pp. 850-859.

Oberpenning, F. et al. (1999), De novo reconstitution of a functional mammalian urinary bladder by tissue engineering. *Nat.Biotechnol.*, Vol. 17, No. 2, pp. 149-155.

Oh, D.S. et al. (2002), Repair of full-thickness defects in alimentary tract wall with patches of expanded polytetrafluoroethylene. *Ann.Surg.*, Vol. 235, No. 5, pp. 708-711.

Ootani, A. et al. (2009), Sustained in vitro intestinal epithelial culture within a Wnt-dependent stem cell niche. *Nat Med*, Vol. 15, No. 6, pp. 701-6.

Papa, M.Z. et al. (1997), Avoiding diarrhea after subtotal colectomy with primary anastomosis in the treatment of colon cancer. *J.Am.Coll.Surg*, Vol. 184, No. 3, pp. 269-272.

Penkala, R.A.Kim, S.S. (2007), Gastrointestinal tissue engineering. *Expert.Rev.Med.Devices*, Vol. 4, No. 1, pp. 65-72.

Pinchuk, I.V. et al. (2010), Intestinal mesenchymal cells. *Curr Gastroenterol Rep*, Vol. 12, No. 5, pp. 310-8.

Powell, D.W. et al. (2005), Epithelial cells and their neighbors I. Role of intestinal myofibroblasts in development, repair, and cancer. *Am.J.Physiol Gastrointest.Liver Physiol*, Vol. 289, No. 1, pp. G2-G7.

Pyke, K.W.Gogerly, R.L. (1985), Murine fetal colon in vitro: assays for growth factors. *Differentiation*, Vol. 29, No. 1, pp. 56-62.

Quaroni, A. (1985), Development of fetal rat intestine in organ and monolayer culture. *J Cell Biol*, Vol. 100, No. 5, pp. 1611-22.

Raghavan, S. et al. (2010a), Bioengineered Three-Dimensional Physiological Model of Colonic Longitudinal Smooth Muscle In Vitro. *Tissue Eng Part C.Methods*, No.

Raghavan, S. et al. (2010b), Successful implantation of physiologically functional bioengineered mouse Internal Anal Sphincter. *Am.J.Physiol Gastrointest.Liver Physiol*, No.

Raghavan, S. et al. (2011), Successful Implantation of Bioengineered, Intrinsically Innervated, Human Internal Anal Sphincter. *Gastroenterology*, No.

Raya-Rivera, A. et al. (2011), Tissue-engineered autologous urethras for patients who need reconstruction: an observational study. *Lancet*, Vol. 377, No. 9772, pp. 1175-82.

Rubin, D.C. (2007), Intestinal morphogenesis. *Curr.Opin.Gastroenterol.*, Vol. 23, No. 2, pp. 111-114.

Sala, F.G. et al. (2009), Tissue-engineered small intestine and stomach form from autologous tissue in a preclinical large animal model. *J.Surg Res.*, Vol. 156, No. 2, pp. 205-212.

Sala, F.G. et al. (2011), A multicellular approach forms a significant amount of tissue-engineered small intestine in the mouse. *Tissue Eng Part A*, No.

Sato, T. et al. (2009), Single Lgr5 stem cells build crypt-villus structures in vitro without a mesenchymal niche. *Nature*, Vol. 459, No. 7244, pp. 262-5.

Sato, T. et al. (2011), Paneth cells constitute the niche for Lgr5 stem cells in intestinal crypts. *Nature*, Vol. 469, No. 7330, pp. 415-8.

Scadden, D.T. (2006), The stem-cell niche as an entity of action. *Nature*, Vol. 441, No. 7097, pp. 1075-9.

Sheridan, W.G. et al. (1987), Tissue oxygen tension as a predictor of colonic anastomotic healing. *Dis Colon Rectum*, Vol. 30, No. 11, pp. 867-71.

Smyrnis, S.A. (1982), Dacron patch for closure of experimental stomach defects. *Br.J.Surg.*, Vol. 69, No. 2, pp. 82-83.

Snippert, H.J.Clevers, H. (2011), Tracking adult stem cells. *EMBO Rep*, Vol. 12, No. 2, pp. 113-22.

Song, Y. et al. (2011), Dynamic culturing of smooth muscle cells in tubular poly(trimethylene carbonate) scaffolds for vascular tissue engineering. *Tissue Eng Part A*, Vol. 17, No. 3-4, pp. 381-7.

Taha, M.F.Hedayati, V. (2010), Isolation, identification and multipotential differentiation of mouse adipose tissue-derived stem cells. *Tissue Cell*, Vol. 42, No. 4, pp. 211-6.

Tan, Q. et al. (2007), Accelerated angiogenesis by continuous medium flow with vascular endothelial growth factor inside tissue-engineered trachea. *Eur J Cardiothorac Surg*, Vol. 31, No. 5, pp. 806-11.

Thompson, J.S. et al. (1986), Growth of intestinal neomucosa on prosthetic materials. *J.Surg.Res.*, Vol. 41, No. 5, pp. 484-492.

Ueno, T. et al. (2007), Small intestinal submucosa (SIS) in the repair of a cecal wound in unprepared bowel in rats. *J Gastrointest Surg*, Vol. 11, No. 7, pp. 918-22.

Umar, S. (2010), Intestinal stem cells. *Curr Gastroenterol Rep*, Vol. 12, No. 5, pp. 340-8.

Uzun, M.A. et al. (2010), Salvage repair of anastomotic dehiscence following colon surgery using an expanded polytetrafluoroethylene graft. *Tech.Coloproctol.*, No.

van der Flier, L.G.Clevers, H. (2009), Stem cells, self-renewal, and differentiation in the intestinal epithelium. *Annu Rev Physiol*, Vol. 71, No. pp. 241-60.

Wang, Z.Q. et al. (2003), Experimental assessment of small intestinal submucosa as a small bowel graft in a rat model. *J.Pediatr.Surg.*, Vol. 38, No. 11, pp. 1596-1601.

Wang, Z.Q. et al. (2005), Morphologic evaluation of regenerated small bowel by small intestinal submucosa. *J.Pediatr.Surg.*, Vol. 40, No. 12, pp. 1898-1902.

Watson, L.C. et al. (1980), Small bowel neomucosa. *J.Surg.Res.*, Vol. 28, No. 3, pp. 280-291.

Wei, X. et al. (2011), A novel bioreactor to simulate urinary bladder mechanical properties and compliance for bladder functional tissue engineering. *Chin Med J (Engl)*, Vol. 124, No. 4, pp. 568-73.

Wildrick, D.M. et al. (1997), Isolation of normal human colonic mucosa: comparison of methods. *In Vitro Cell Dev Biol Anim*, Vol. 33, No. 1, pp. 18-27.

Yeh, K.Y.Chopra, D.P. (1980), Epithelial cell cultures from the colon of the suckling rat. *In Vitro*, Vol. 16, No. 11, pp. 976-986.

Yiou, R. et al. (2003a), Restoration of functional motor units in a rat model of sphincter injury by muscle precursor cell autografts. *Transplantation*, Vol. 76, No. 7, pp. 1053-1060.

Yiou, R. et al. (2003b), The regeneration process of the striated urethral sphincter involves activation of intrinsic satellite cells. *Anat Embryol (Berl)*, Vol. 206, No. 6, pp. 429-35.

Yokoyama, T. et al. (2001), Autologous primary muscle-derived cells transfer into the lower urinary tract. *Tissue Eng*, Vol. 7, No. 4, pp. 395-404.

Yoo, J.J. et al. (1998), Bladder augmentation using allogenic bladder submucosa seeded with cells. *Urology*, Vol. 51, No. 2, pp. 221-225.

Zhang, Y. et al. (2000), Coculture of bladder urothelial and smooth muscle cells on small intestinal submucosa: potential applications for tissue engineering technology. *J.Urol.*, Vol. 164, No. 3 Pt 2, pp. 928-934.

Zhang, Y. et al. (2004), Bladder regeneration with cell-seeded small intestinal submucosa. *Tissue Eng*, Vol. 10, No. 1-2, pp. 181-187.

# 14

# Aspects of Urological Tissue Engineering

Arun K. Sharma[1,2,3] and Dorota I. Rozkiewicz[3,4]
[1]*Children's Memorial Hospital, Chicago Division of Pediatric Urology*
[2]*Northwestern University Feinberg School of Medicine, Department of Urology*
[3]*Northwestern University, Institute for BioNanotechnology in Medicine (IBNAM)*
[4]*Northwestern University, Department of Chemistry*
*USA*

## 1. Introduction

Factors affecting the homeostasis of the urological system stem from a multitude of different sources that include developmental miscues to aberrant signaling pathways leading to malignant transformations. In order to create functional tissues that can act as surrogate tissues while simultaneously mimicking endogenous anatomical and physiological functions, strategies utilizing regenerative medicine based methodologies must be employed. The urological system is comprised of the kidneys and their corresponding ureters; the bladder and sphincter as well as aspects of the genitourinary system. The complex architectural makeup and intricate physiological processes of the urological system at the gross and microscopic levels make tissue engineering attempts to recapitulate these tissues a very difficult and arduous process. Although great strides have been made in the field of urological tissue engineering over the last two decades, the clinical needs of patients suffering from urological defects have largely gone unfulfilled. However, the advent of novel approaches that span multiple disciplines including the material and clinical sciences as well as the constantly evolving field of stem cell biology have curtailed these issues to some extent. This Chapter will delve into strategies that are currently utilized for urological based tissue engineering while also providing alternative tactics to prevailing approaches.

## 2. Conditions affecting the urological system

### 2.1 The kidney

A simplistic overview of nephrogenesis proceeds as three sequential yet overlapping phases during human embryogenesis. During the third to fourth week of development, primitive cell populations derived from intermediate mesoderm converge upon the cervical region of the embryo to form one of two transitory structures termed the pronephros. (McCrory, 1974) These converging cell groups form rudimentary excretory units or nephrotomes that join with the pronephric duct. Although the pronephros is non-functional in mammals, it serves as a precursor structure for the slightly more defined mesonephros. (McCrory, 1974) The progressive development of the definitive kidney continues as the pronephros regresses and the development of mesenophric ducts again derived from intermediate mesoderm becomes apparent. (Pelliniemi, 1983) The mesonephros is composed of two distinct structures, the mesenophric ducts and the mesonephric tubules. The mesonephric tubules acquire a bed of capillaries and are analogous in physiological function as well as anatomic structure to adult nephrons in the definitive kidney. These elementary nephrons include the glomerulus, renal corpuscles, and Bowman's capsule. The mesonephric ducts (Wolffian ducts) eventually give rise to the epididymis, the vas deferens, and the seminal vesicles in males and as vestigial organs in females. (McCrory, 1974; Hannema, 2006; Welsh, 2006) An outpocketing from the mesonephric duct arises during the fifth to six week of development to form the ureteric bud which becomes the decisive ureter as well as the collecting system found within the kidney including major and minor calyces, and the renal pelvis within the final stage of kidney formation, the metanephros. (McCrory, 1974; Fischer, 2001) The physiological aspects of kidney function are numerous in nature and non-exclusively include hormone production and regulation, homeostatic balance, pH and blood pressure control, waste removal and nutrient re-absorption. It is therefore imperative that the various physiological aspects of the kidney remain under proper control and maintenance. The introduction of insult in the forms of either trauma or aberrant developmental events can signify a poor outcome for the patient.

Congenital abnormalities of the kidney and urinary tract (CAKUT) occur in 1 out of 500 newborns which constitutes approximately 30% of all prenatal anomalies. (Toka, 2010) These include a plethora of conditions including horse-shoe kidney, ureter malformation/abnormalities, renal hypoplasia, renal agenesis, polycystic kidney disease, and Wilms' tumor formation and can arise at varying stages of the developmental process. (Toka, 2010; Stahl, 2006; Gimpel, 2010; Acien, 2010) Depending upon the severity of the malformation and organs affected, various strategies can been utilized to treat patients afflicted with CAKUT that include surgical and conservative management. However, even with treatment, kidney and urinary tract functions are typically far from ideal. Certain signaling pathways have been elucidated but only scratch the surface of this complex disorder. (Stahl, 2006; Nakanishi, 2003)

Aside from congenital defects affecting the kidney, acute and chronic conditions also pre-dispose patients to poor clinical outcomes. Acute renal failure occurs with rapid onset and can result in loss of kidney function with eventual renal failure. (Bellomo, 2004; de Mendonca, 2000) As there are myriad circumstances that may cause acute kidney failure, it is a life threatening situation requiring great urgency in order to alleviate the underlying cause. Chronic conditions affecting the kidney are also in great number, most notably those involved with chronic hypertension and diabetes mellitus. The destruction of the kidney due to consistently high blood pressure or continual exposure to copious amounts of glucose results in several scenarios in which to manage the disease. Diabetes is the most common cause of kidney failure within the United States. Data from the 2011 National Diabetes Fact Sheet provided by the American Diabetes Association estimates that there are approximately 26 million people living with diabetes within the United States representing approximately 8% of the population. (http://www.diabetes.org/diabetes-basic/diabetes-statistics) Kidney failure as a result of diabetes accounted for 44% of new cases in 2008 and approximately 50,000 people began treatment for end stage kidney disease. Lastly, approximately 200,000 people were on either dialysis or underwent kidney transplantation as a result of chronic diabetes within the United States. A total of $348 billion was spent in 2007 to address diabetes related issues. This stagnating level of evidence suggests a great medical need to overcome current shortcomings involved with kidney disease as tissue engineering and nanofabrication strategies for the kidney become more prevalent.

Kidney transplants and dialysis are the only current means to treat kidney failure. Attempts to recapitulate the kidney involve the transplantation of fetal kidneys, the integration of new nephrons into the kidney, and the use of embryonic stem cells. Fetal kidney transplantation was thought to be a possible means in which to identify tissue for replacement therapies since it was believed that fetal tissue did not present the high levels of major histocompatibility complexes (MHC) I and II as well as other immunostimulatory antigens. Studies performed by Dekel el al demonstrate the use of fetal human kidneys that were transplanted beneath the renal capsule of immune deficient rats. (Dekel, 1997a) Data suggests that fetal tissue displayed growth and development for as long as 4 months post-transplantation without tissue rejection that may have been caused by populations of T-cells previously residing in the fetal tissue. Analogous adult tissue underwent graft rejection upon stimulus with allogeneic injections of peripheral blood mononuclear cells (PMBC). (Dekel, 1997b, 1997c, 2000) Nephron transplantation became an alternative means in which to rejuvenate a faltering kidney. Woolf et al transplanted sections of metanephroi from E13-E16 mice into the kidney cortex of newborn outbred mice. (Woolf, 1991) Relevant kidney micro-architecture formed over time but complete incorporation of transplanted metanephroi was not found in the collecting ducts. A similar study using an analogous system (but not identical) failed to produce glomeruli. (Rogers, 1998) The lack of positive data with regard to kidney replacement methodologies leaves alternatives such as xenotransplantation with embryonic porcine kidneys. This tissue is not pluripotent but rather precursor tissue that has been naturally differentiated along kidney specific pathways. (Hammerman, 2011)

Lastly, highly compelling work by Roy and colleagues appears to be very promising in that they have created a bio-artificial implantable device that may take the place of damaged or deteriorating kidneys. Preliminary data suggests that this microfluidic bioreactor can mimic the physiological conditions found in the kidney with regard to fluid flow rate in the presence of kidney epithelial cells. (Ferrell, 2010) Furthermore, Roy et al has determined that nanofabricated silicon membranes used in this device demonstrate low levels of complement activation, low platelet adhesion, and low blood coagulability. (Muthusubramaniam, 2011) Studies are currently under way to test the feasibility of this device in animal models.

### 2.2 The ureter

Conditions which may warrant the replacement of a new ureter include a myriad of congenital afflictions including congenital atresia, stricture, bifid ureters as well as ureteral cancer and trauma. Arising from the renal pelvis of each kidney, the ureter descends and crosses into the urinary bladder. Damage to the valve found at the distal end of each ureter can result in vesicoureteral refux with eventually kidney damage if not treated. Within this scenario, damage can also manifest itself within the ureters as evidenced by dilated ureters caused by increased internal pressures. Tissue transplant studies employed to replace defective ureters have been promising. Armatys et al report the use of ileum as a ureteral replacement for reconstruction purposes. (Armatys, 2009) A retrospective study of 91 patients who underwent ureteral reconstruction using grafted ileum revealed that only 3.3% and 6.6% of patients endured long-term anastomotic stricture and fistula, respectively, and had attained normal kidney function. Other groups have utilized vestigial tissue such as the appendix as well as buccal tissue with favorable results. (Dagash, 2008) Ureters constructed from synthetic materials have been used in the past to bypass the use of autolgous tissue sources. Desgrandchamps et al published two reports in which ureters constructed from a silicone polytetrafluoroethylene bonded tube was introduced into the pyelocaliceal renal graft cavities and placed subcutaneously into the suprapubic area where it was eventually guided into the bladder within human patients. (Desgrandchamps, 1995, 1998) A mean 2.5 year follow-up demonstrated that all tubes were patent and displayed no evidence of encrustation or obstruction. More recently, xenogenic accellular collagen membranes have also been used with limited success. (Koziak, 2004) The aforementioned studies provide evidence that various tissues derived and artificial matrices can be used for ureteral reconstruction. In order to further improve on the design and physiological function of the ureter, smart scaffolding material possessing antimicrobial attributes or capable of growth factor release, for example, may be incorporated into this setting such as those found in self assembling nanomolecules.

### 2.3 The urinary bladder

The urinary bladder is a hollow, musculocutaneous organ whose primary function is to expel waste products in the form of urine from the body. Damage or insult to this organ leads to bladder dysfunction which can have dire clinical consequences on upper and lower urological tracts including renal insufficiency with eventual renal failure. Patients suffering from severe bladder dysfunction caused by developmental defect, often resort to surgical means to stabilize poor bladder function. The various forms of spina bifida present the clinician and basic scientist alike with a battery of issues stemming from gross anatomical restructuring to the fine aspects of physiological function. The typical procedure to alleviate problems associated with the neurogenic bladder includes bladder augmentation cystoplasty. What is considered the "gold standard" procedure for patients suffering from end stage neurogenic bladder, a portion of the pathological, native bladder undergoes cystectomy. (Zini, 2004; Husmann, 2004) A detubularized portion of the bowel is then affixed to the cystectomized area of neurogenic bladder and hence serves as a neo-bladder in order to increase capacity. As the bladder is considered functional, it still lacks key attributes of a normally functioning bladder including sensory and contractile/expansion responses requiring lifetime self-catheterization. The reconstructed bladder also possess inherent complications including electrolyte imbalances as an absorptive tissue has been used to replace an excretory tissue; the high potential of infection; excess mucus formation; pronounced stone formation; spontaneous perforation, and the increased likelihood of malignant transformation. (Bankhead, 2000) Although the augmentation cystoplasty procedure is considered standard care for this situation, it is still a stopgap measure.

The biomedical applications of regenerated tissue took its roots from the pioneering work of Langer and colleagues. (Vacanti, 1988; Langer, 1993) Within these studies, it was determined that donor derived cells could be seeded upon synthetic, biodegradable scaffold materials and subsequently manipulated in vitro for eventual in vivo applications. Applying these basic principles of cell biology and engineering, Atala and colleagues embarked upon a mission to recapitulate functional urological tissue with the goal of clinical applicability. Initial studies utilizing rabbit derived urological tissue subsequently seeded upon polyglycolic acid (PGA) meshes and implanted into the tissue within and adjacent to the peritoneal cavity of athymic mice yielded poignant results. (Atala, 1992) Rabbit derived urothelium was able to proliferate and populate the PGA scaffold to an approximate 3 cell thickness while retaining key urothelial markers after 30 days in vivo. This initial study set the stage for several other studies that continued to evaluate the roles of specific cell types combined with synthetic scaffolds for bladder regeneration. (Atala, 1993, 1998) Concurrently, studies by Knapp et al and others were evaluating the use of small intestinal submucosa (SIS) as an alternative biological scaffold for bladder tissue engineering. (Knapp, 1994; Kropp, 1995) The results from these studies seemed to suggest that SIS was a suitable biological material for the bladder augmentation process. However, in two separate studies by Ashley et al, unseeded SIS would be found to be highly pro-inflammatory. (Ashley, 2009, 2010) The seminal work that laid the foundation for future bladder regenerative studies came from Oberpenning et al (Oberpenning, 1999) Within this study, autologous sources of canine urothelial and smooth muscle cells were harvested then seeded onto polymers and subsequently expanded ex vivo. Completely cystectomized canines were then transplanted with this new bladder construct and observed for approximately a year. Very compelling data suggested that the engineered bladders demonstrated a normal micro-architecture, mechanical properties and the ability to store and express urine. This study set the foundation for a human clinical trial utilizing neurogenic urothelial and smooth muscle cells from spina bifida patients with myelomeningocele as a model system with the hopes that the bladder milieu could be recreated in human. (Atala, 2006) Following bladder biopsies, these two cell populations were isolated and expanded ex vivo and subsequently seeded upon either a collagen/PGA hybrid scaffolds or collagen alone scaffolds. These composite scaffolds were anastomosed to the native tissue following an initial cruciate incision to the native bladder. An approximate mean 4 year follow-up revealed no statistically improvement in collagen only based scaffolds for a number of different urodynamic functions including capacity, leak point pressure, and compliance. PGA/collagen scaffolds encased in omentum appeared to display a very modest increase in bladder capacity. The approach taken by Atala et al was bold and extremely novel and provided highly relevant data for this specific patient population. However, the data presented within the work failed to acknowledge that utilizing a pathological source of cells at the onset of this work could possibly result in abnormal cellular behavior or possible reformation of a diseased bladder state. Characterization of bladder smooth muscle cells derived from patients with myelomeningocele indicates that cultured bladder smooth muscle cells possess different characteristics than their normal counterparts. (Lin, 2004; Beqaj, 2005) Finally, the aforementioned augmented bladders continued to be non-contractile thus requiring the patients to presumably self catheterize for the remainder of their lifespan. Although groundbreaking in nature, much work needs to be performed in order to fully realize the goal of creating a functional bladder.

### 2.4 The urinary sphincters

The urinary sphincters are composed of a collection of muscles whose primary function is to prevent the premature leakage of urine from the bladder. (Feki, 2007; Delancey, 2004) The internal urethral sphincter is located at the proximal portion of the urethra and is a continuation of the detrusor muscle consisting of smooth muscle cells. The external urethral sphincter is comprised of skeletal muscle and is situated inferior to the prostate in males and at the bladder's distal inferior end in females. (Koyanagi, 1980) Disruption or dysregulation of the sphincter system can lead to incontinence that encompass various forms including stress urinary incontinence, structural incontinence, urge and overflow incontinence as well as mixed incontinence. (Botlero, 2011; Siegler, 2004; Brubaker, 2011) The economic costs of urinary incontinence are quite staggering and account for greater than $20 billion dollars per year within the United States. (Wilson, 2001) Attempts to alleviate the dysfunction associated with incontinence have been manifested in various tissue engineering attempts to rejuvenate or replace poorly functioning sphincters. Yiou et al tracked the fate of muscle precursor cells that were implanted into a rat model of striated urethral sphincter injury. (Yiou, 2003) Muscle precursor cells were able to adapt to their environment and form myotubules in which acetylcholine receptors were present. Urodynamic studies demonstrated a 41% restoration of sphincter function at approximately one month post muscle precursor cell implantation. A second study similarly utilized muscle derived progenitor cells in a denervated female rat urethra model. (Cannon, 2003) Two weeks following cell injection, urethral muscle strips were analyzed under electrical field stimulation. Injected muscle derived progenitor cells greatly improved the amplitude associated with fast-twitch muscle contractions at a rate of approximately 87% as compared to control animals. The aforementioned studies and those that have incorporated the use of synthetic bulking agents such as poly(lactic-co-glycolic acid) (PLGA) microspheres have demonstrated alternative means in which to overcome obstacles associated with urinary incontinence. (Berjukow, 2004; Oh, 2006) The foundation that these works have created has resulted in over 50 clinical trials related to urinary incontinence. (http://clinicaltrials.gov) The field continues to evolve at a rapid pace partially driven by a need to create a solution to this problem as well as the potential for high financial benefits.

### 2.5 The urethra

Urethral narrowing is the most common affliction affecting the urethra. Physical trauma or insult by bacteria or virus can result in an abundance of scar tissue caused the local invasion of inflammatory cells as the body attempts to resolve the situation at hand resulting in stricture. Clinical presentation includes painful urination, decreased urinary output, dark colored urine, pelvic pain, and incontinence among some of the symptoms. (Ghoniem, 1994; Barbagli, 2007; Mangera, 2010) Depending upon the anatomical location and the severity of the stricture, different techniques have been utilized to remedy stricture exacerbation. Urethroplasty has been successfully utilized as a treatment option to repair the stricture. One method is to excise the stricture from the urethra and then rejoin the opposing ends. A second approach requires tissue grafting from typically autologous sources of buccal mucosa, amniotic membrane, or saphenous vein tissue. (Mangera, 2010; Goel, 2011, Levine, 2007, Shaeer, 2006) Complications associated with this type of surgery include fistula formation, infection, and stricture reformation. (Olajide, 2010) Tissue engineering approaches using synthetic and biologic matrices have also been investigated as alternative approaches to urethral repair. Derivatives of PLGA have been created to mimic buccal mucosa with the added benefit of possessing superior mechanical and elastic properties. (Selim, 2011) Other biomaterials such as SIS, bladder acellular matrix (BAM), and acellular corpus spongiosum matrix (ACSM) have also been utilized in studies with favorable results. (Feng, 2010; Powers, 2010) Since host inflammatory responses appear to be a major culprit behind stricture formation and its persistence, means to address this issue should be explored. One practical solution may be the localized delivery of anti-inflammatory agents into the affected area to prevent further stricture formation. This may be accomplished via a number of delivery vehicles including hydrogel releasing anti-inflammatory agents as well as self-assembling nanomolecules capable of down-regulating the cellular inflammatory response through surface expression of anti-inflammatory epitopes. Current research is focusing upon the creation of peptide amphiphiles that may be capable of such feats. Combined with the anti-inflammatory characteristics of mesenchymal stem cells, this one-two punch to either attenuate inflammation or act in a prophylactic manner may help with stricture development and other biologically relevant other pro-inflammatory situations.

## 3. Stem cells and potential applications in urological regenerative medicine

Stem cells can be described in several different embodiments such as those that are derived from embryonic sources or somatic (adult) in nature. The differential capabilities of these cells can further be sub-categorized as being either totipotent, pluripotent, or multipotent. Totipotent stem cells are a result of a fusion of sperm and egg that have undergone limited divisions during early embryological development in mammalian systems. (Mitalipov, 2009; Seydoux, 2006) At this early stage of development, totipotent stem cells have the potential to become any type of cell in the body which also includes extraembryonic tissue. (Mitalipov, 2009; Seydoux, 2006) As a zygote undergoes further differentiation, totipotent stem cells lose their ability to create whole organisms and become more lineage restricted, hence becoming pluripotent. Pluripotent stem cells can give rise to tissues of endoderm, ectoderm, and mesoderm origin, but not extraembryonic tissue and can be derived from both embryonic and somatic cells. (Pittenger, 1999, Thomson, 1998; Takahashi, 2007; Ko, 2009) Multipotent cells are greatly restricted in their ability to generate specific differentiated cell types and subsequently tissue. These cells retain the ability to self-renew and divide into daughter progeny, but these characteristics are greatly limited. A highly characterized population of cells that fits this definition is mesenchymal stem cells that have been described to differentiate into adipocytes, osteocytes, chondrocytes, and others. (Pittenger, 1999, Phadnis, 2011)

### 3.1 Mesenchymal stem cells

The hematopoietic system and its constituents have been extensively characterized at multiple levels via decades of research and clinical experiences. The hematopoietic system found within the bone marrow of long bones in mammals consists of a hierarchy of cells originating from a primitive hematopoietic stem cell. The hematopoietic stem cell is capable of giving rise to all blood derived cells including those of lymphoid and myeloid origin. (Akashi, 2000) Also residing in the bone marrow are non-hematopoietic multipotent mesenchymal stem cells capable of variable cellular differentiation as previously described. (Takahashi, 2007; Ko, 2009; Phadnis, 2011) The frequency of which bone marrow derived mesenchymal stem cells are found within long bones is approximately 0.001% to 0.1% of nucleated cells which include adipocytes, macrophages, osteoblasts, and reticular connective tissue. (Pittenger, 1999) Therefore, fibroblast cells that reside in the bone marrow are not synonymous with mesenchymal stem cells. The plasticity demonstrated by mesenchymal cells can be utilized in a variety of regenerative medicine settings, especially those targeting urologic defects.

In a series of studies utilizing bone marrow derived mesenchymal stem cells from normal human, adult donors, Sharma et al demonstrate the phenotypic and physiological similarities between bladder smooth muscle cells and adult mesenchymal stem cells. (Sharma, 2009) The mesenchymal stem cells utilized throughout this study as well as subsequent studies were an epitope defined population of highly purified mesenchymal stem cells. This study further demonstrated that both populations of cells can respond to agonist stimulation with statistically indistinguishable contractile responses in vitro while possessing similar contractile machinery. A second study by Sharma and colleagues utilizes adult mesenchymal stem cells in a nude rat bladder augmentation model. (Sharma, 2010a) Within the context of this study, bladder regeneration occurred more potently with mesenchymal stem cells than normal human bladder smooth muscle cells and controls. Scaffolds seeded with mesenchymal stem cells expressed high levels of bladder smooth muscle markers along with typical muscle architecture in the form of muscle fascicles. These scaffolds also demonstrated an approximate 1:1 muscle to collagen ratio at the termination of the experiment which was vastly different from bladder smooth muscle cell seeded and unseeded controls. A third and more recent study by Sharma et al again demonstrate the utility of mesenchymal stem cells in a bladder regenerative setting except in this instance, autologous sources of epitope defined mesenchymal stem cells were utilized to establish a novel bladder augmentation model in a non-human primate. (Sharma, 2010b) Data from this study provides evidence that mesenchymal stem cell seeded scaffolds performed exceptionally as defined by physiological measurements and tissue phenotyping studies with a battery of proliferative and smooth muscle markers compared to controls. More importantly, this newly described augmentation model demonstrates potential clinical feasibility as this non-human primate is highly analogous to humans both anatomically and physiologically. Hence the use of autologous sources of non-pathological cells can be successfully utilized in a bladder regenerative setting.

Although the aforementioned studies provided strong evidence for bladder regeneration in hemicystectomized animals, future endeavors will require a great deal of scientific prowess to regenerate a fully functional bladder. Hence, the natural course of action would be the continuance of these studies. One aspect in conjunction with regenerating tissue is to create an environment that can provide a robust blood supply to developing tissue. Genetically modified mesenchymal stem cells carrying pro-angiogenic growth factors under the control of inducible promoters may aid in this function. A second strategy involves the use of nanomolecules capable of delivering much needed growth factors to tissues undergoing regeneration. Rajangam et al demonstrate robust in vivo angiogenesis via the delivery of heparin binding nanostructures. (Rajangam, 2006) The cross-pollination of stem cell biology with nanotechnology derived materials can potentially provide reliable and reproducible results when applied to urologic tissue regeneration which is the basis for future studies.

There are several tissue sources in which mesenchymal stem cells can be identified and isolated. Two of those sources are found within adipose tissue as well as amniotic fluid which bathes the fetus in utero. These cell populations have been shown to function within urological settings. Dyslipidemia has been associated with a variety of poorly compliant physiological conditions including erectile dysfunction. (Huang, 2010) Damage occurs to the cavernous endothelium and associated nerves through mechanisms that are not quite fully understood. In order to facilitate recovery, adipose derived mesenchymal stem cells have been utilized to augment repair in the penis with the understanding that various factors secreted from mesenchymal stem cells promote re-growth of endothelium and improved nerve conduction. Within the context of this study, adipose derived mesenchymal stem cells were injected into the corpus cavernosum of dyslipidemic rats. Following a period of recovery, erectile function was assessed and mean intracavernous pressure/arterial pressures ratios were superior to control samples. Increased levels of nitric oxide synthase and alpha smooth muscle actin were also observed. (Huang, 2010) This study provides preliminary evidence that adipose derived mesenchymal stem cells can be used to ameliorate the urologic complications that arise in urologic settings. In order to obtain continual exposure of these factors in the afflicted areas, seeding these cells upon implantable nanofabricated materials may allow for immediate regeneration accompanied by long term maintenance. Nanofabricated materials can recreate specific cell niches in order to potentially enhance the regenerative effect and this topic will be discussed in greater detail within subsequent sections. (Murtuza, 2009)

Adipose derived mesenchymal stem cells have also been utilized in bladder regenerative settings and provide a means to obtain a multipotent cell source through adipose tissue aspirates. (Jack, 2009) Adipose derived mesenchymal stem cells maintain an approximate 3% of all nucleated cells within a sample of aspirate. Jack et al seeded adipose derived mesenchymal stem cells upon PLGA scaffolds and implanted them in 50% cystectomized athymic nude rats. 12 weeks post-transplantation, physiological measurements were assessed as well histological and immunofluorescent analyses, and organ bath studies. All data derived from this study indicate that the adipose derived mesenchymal stem cells perform similarly or better than control samples in almost every biological aspect. The novelty of this study obviously involves the use of this unique population cells. Future studies involving potentially sub-total cystectomized large animal models will hopefully provide a great deal of physiological information that may be applied to the human condition.

The recently described characterization of amniotic fluid mesenchymal stem cells offers another potential alternative cell source that may be applied to urologic tissue engineering. Initially described by In 't Anker et al, 2ml of second trimester amniotic fluid was obtained through transcervical isolation. (In't Anker, 2003) Subsequent cell populations were cultured in vitro and coerced to undergo terminal differentiation where they demonstrated differentiation into adipocytes and osteocytes. Transplantation studies also provided evidence that these cells could be used in an allogeneic setting, which suggests that these cells may possess a greater immune-privileged status than their adult counterparts. The application of amniotic fluid derived mesenchymal stem cells in a bladder injury model convincingly demonstrates the wound healing potential of these cells. (De Coppi, 2007) The direct injection of amniotic fluid derived mesenchymal stem cells into the cryo-injured bladder resulted in markedly reduced levels of smooth muscle hypertrophy but also demonstrated a limited level of smooth muscle cell regeneration which may have been attributed to inadequate numbers of injected cells since a dose response curve was not performed. Although both experimental procedures describe this source of cells as potent in function, the means in which they are acquired are highly invasive and pose great risk to fetus and mother. It is estimated that 1 in 300 women experience fetal loss, infection, cramping/vaginal bleeding or other complications during or following amniocentesis. (Simpson, 2007) The isolation and subsequent banking of these cells may allow a convenient means for future therapies in both autologous and allogeneic settings.

### 3.2 Endothelial progenitor cells

The lack of a properly vascularized graft is the bane of many tissue engineering studies regardless of the tissue attempting to be regenerated. Inadequate or improper vascularization leads to poor tissue development or graft failure of implanted cell/scaffold composites. This is evident towards the center of grafted composites as these areas are typically inaccessible to nutrient and gas exchange potentially resulting in a necrotic core. A manner in which to avoid the aforementioned scenarios would be to utilize autologous sources of endothelial progenitor cells. These cells could promote vascular growth even at the most cell dense levels of the graft since the endothelial progenitor cells can simultaneously be seeded along with other bladder cell types thereby encouraging vascular growth. Endothelial progenitors can typically be isolated from either the bone marrow or rather innocuously from the peripheral blood as circulating endothelial progenitor cells.

#### 3.2.1 Bone marrow derived endothelial progenitor cells

It was initially speculated that cells that were destined to become constituents of blood vessels as a result of vascular growth were present throughout embryonic and early stages of fetal development. This was later revealed to be somewhat inaccurate with the advent of specific antibodies utilized to phenotype specific cell populations. The $CD34^+$ hematopoietic stem cell has been greatly characterized over the last several decades and it has been demonstrated that subset populations derived from bone marrow sources have the capacity to either form functional vasculature in or contribute to its remodeling vivo. (Abu El-Asrar, 2009; Ruger, 2008; Baum, 1992; Takahashi, 1999) $CD34^+/CD133^+$ cells derived from the bone marrow have provided evidence that this population can give rise to highly purified endothelial cells. (Quirici, 2001) Sharma et al describe the in vivo potential of this cell population as it was seeded within a chicken chorioallantoic membrane model. (Sharma, 2009) Post experimentation analyses of this work clearly demonstrate the angiogenic potential of bone marrow derived $CD34^+/CD133^+$ cells as human specific vasculature was formed utilizing this model. Other derivatives of this cell population including $CD34^+/CD133^+/VEGFR2^+$, $CD34^+/CD133^+/VEGFR3^+$, $CD34^+/VEGFR2^+$, and $CD133^+/VEGFR2^+$ cells maintain similar properties. (Salven, 2003; Luppi, 2010) However some may argue that the CD34 epitope is not required to describe an endothelial progenitor/precursor cell, although those groups appear to be in the minority. (Friedrich, 2006) In all cases, the frequency of bone marrow derived endothelial progenitor cells is quite low, hence other alternatives cell sources should be examined for potential regenerative medicine purposes.

#### 3.2.2 Circulating endothelial progenitor cells

The seminal discovery that endothelial progenitor cells were persistent throughout the vascular system gave a tremendous boost to a number of tissue engineering approaches. Approximately 11 years ago, Asahara et al describes a simplistic and non-invasive means to isolate circulating endothelial progenitor cells. (Asahara, 1997) By utilizing $CD34^+$ cells isolated from peripheral blood, the study demonstrates that this population of cells was able to differentiate into endothelial cells in vitro while concurrently incorporating into sites of active angiogenesis in an animal model of ischemia. Data goes onto describe the concept that circulating endothelial progenitor cells can home to areas of pre-existing angiogenesis and contribute to the angiogenic process. Since this revelation, many studies have been performed that confirm the concept of circulating endothelial progenitor cells. (Dubois, 2011;

Baker, 2011) One recent example demonstrates the construction of a tri-leaf heart valve seeded with circulating endothelial progenitor cells. (Sales, 2010) Data revealed that endothelial progenitor cells were able to colonize the heart valve and provide both interstitial and endothelial functions to the evolving valve. (Sales, 2010) This again demonstrates the potential utility of endothelial progenitor cells. The use of endothelial progenitor cells is pivotal for tissue regeneration and cannot be understated. The goal of utilizing endothelial progenitor cells for tissue regeneration is definitely not simplistic in the sense by just applying a population of endothelial progenitor cells to a graft and assuming angiogenic events. The shear number of molecules that are involved in the angiogenic pathway need to be carefully balanced in order to prevent aberrant results. This can be demonstrated as an over-abundance of localized vascularization leading to excessive bleeding and subsequent pathological issues. The strengths of using circulating endothelial cells are that they can be isolated in large numbers and more importantly, they can be isolated in an autologous fashion. The latter avoids a battery of immunological events and immunosuppressive therapies that may have to be endured for the lifetime of the patient. Unfortunately, works describing the use endothelial progenitor cells in urological tissue engineering is very limited and should be further investigated as this resource is being squandered.

### 3.3 Neural stem/progenitor cells

The greatest issue involved with urological based tissue engineering strategies (in this author's opinion) is the recapitulation of functional nervous tissue. Although attempts have been made to provide neural functionality to urologic organs such as the bladder utilizing nerve re-routing techniques, these results have been met with skepticism and concern. (Xiao, 2005; Kurzrock, 2010; Park, 2010) Thus, alternate avenues should be evaluated to create functional neural circuitry for urologic organs such as the use of stem cells. The initial isolation and identification of neural stem cells occurred in 1992 by Reynolds and Weiss. (Reynolds & Weiss, 1992) Neural stem cells were isolated from the striatum of the mouse brain and under specific culture conditions coerced to differentiate into neurons and astrocytes. Further testing demonstrated morphological features that were indicative of functional neural cells. This seminal work provided compelling evidence that argued against the long established belief that brain derived tissue could not undergo neurogenesis after birth or damaging insult. Subsequent in vivo studies exploited the use of neural stem and more differentiated neural progenitor cells in a variety of in vivo settings as cells were also being isolated from peripheral nervous systems. (Gage, 2000) One study in particular describes the transplantation of neural progenitor cells into the cortical region of the rat brain that underwent lesion induction. (Prajerova, 2010) Electrophysiological assessments of brain slices revealed the generation of action potentials created by the neural progenitor cells that were statistically greater than control samples. Evidence was also provided in which transplanted neural progenitor cells exhibited expression of differentiated neural markers by immuno-staining. (Prajerova, 2010)

As of this writing, attempts to utilize bona fide neural stem or progenitor cells have not come to fruition in a urological regenerative medicine setting. One study does report the use of neuronal-glial precursor cells which facilitated the mild improvement in bladder physiological function. These included various urodynamic parameters such as mean voiding pressure, capacity, bladder pressure, and mean contraction amplitude. These values were marginally better than control animals. (Temeltas, 2009) It is quite apparent that a considerable amount of research is required to address the major concerns regarding urologic tissue neuro-regeneration. Cultivating cross-discipline approaches may provide a suitable means to an end.

### 3.4 Embryonic stem cells

The seemingly limitless therapeutic potential of human embryonic stem cells has commanded the attention of the scientific community since its initial discovery. (Thomson, 1998) Derived from the inner cell mass of the developing blastocyst, pluripotent embryonic stem cells can be isolated and possess several unique features. These are typified by high nuclei to cytoplasm ratios, very pronounced nucleoli, the capability of indefinite cellular propagation, and most significantly, the ability to differentiate into cell types composed of all three germ layers. (Thomson, 1998) The isolation and characterization of embryonic stem cells laid the foundation for literally thousands of studies that now spans multiple disciplines and aspects of regenerative medicine. (Birket, 2011; Nistor, 2005; Farzaneh, 2010) The use of embryonic stem cells is slowly making its way into the realm of urological tissue engineering. One study describes an in vitro culture environment in which conditioned media from primitive ureteric bud cells were used to support the differentiation of murine mesoderm cells initially derived embryonic stem cells. (Ren, 2010) The mesoderm cells were primed with activin and retinoic acid and subsequently forcibly differentiated into cells of renal lineage with the aid of the conditioned media. The strong inductive effect of the secreted molecules from the ureteric bud culture provides a means to elucidate potential mechanisms involved with mesoderm-renal cell lineage differentiation. A second, more recent study sheds light upon some of the molecules involved in the differentiation of murine embryonic cells into urothelium. (Mauney, 2010) As previously utilized in the aforementioned study by Ren et al, all trans-retinoic acid had the profound ability to induce pluripotent embryonic stem cells in to urothelial cells through a GATA4 and GATA6 signaling pathway. Stimulation of the embryonic stem cells with all trans-retinoic acid caused a prominent increase in several uroplakin genes including UP1A, UP1B, UP2, and UP3B with a simultaneous down-regulation in the pluripotency factor OCT-4 compared to control cell populations. Concurrently, transcription factors GATA4 and GATA6 were also greatly up-regulated. GATA4/6 null embryonic stem cell lines demonstrated a marked reduction in uroplakin expression when undergoing similar treatments as compared to wild type controls. This data suggests a partial mechanism with regard to the inductive effect of all trans-retinoic acid and its ability to drive embryonic stem cells into urothelial cells. Several other studies have also utilized embryonic stem cells or embryoid body derived cells to generate bladder tissue to varying capacities. (Thomas, 2008; Kinebuchi, 2008; Lakshmanan, 2005) Unfortunately, a lack of follow-up studies has seemed to put a damper on this potentially ground breaking field of research.

The utility of embryonic stem cells is obviously not without debate. Ethical and moral dilemmas have significant impact on political policy and are of grave concern for future studies utilizing embryonic stem cells that use federal funds to support research. 2001 saw restrictions placed upon the use and creation of embryonic stem cell lines, a ban that was later rescinded in 2009. The recent political atmosphere continues to be in a state of flux witnessed most recently by a recent court injunction that stopped the use of federal dollars to be used for embryonic stem cell research, thus effecting thousands of researchers. Although the injunction was temporarily stayed upon an appeal, it could set a damaging precedent for the future. Privatized monies as found within the California Institute of Regenerative Medicine currently finance studies with embryonic stem cells with other states in pursuit. (http://cirm.ca.gov/for-researchers/researchfunding) Nevertheless, several clinical trials set forth by privately held companies such as Geron and Advanced Cell Technology are currently underway. (Strauss, 2010; http://www.advancedcell.com/news-and-media/press-release) The controversies surrounded embryonic stem cell research in part fueled the scientific community to pursue other avenues of research that would lead to the identification of an alternative source of pluripotent stem cells.

### 3.5 Induced pluripotent stem cells

Preliminary observations involving studies with somatic cell nuclear transfer into recipient oocytes resulted in cell populations that displayed features that were highly analogous to embryonic stem cells in morphology and function. (McGrath, 1983; Campbell, 1996) Via successive rounds of a highly selective molecular screening process, four genes were determined to be essential in the de-differentiation of terminally differentiated cells back to a pluripotent, embryonic-like state. Takahashi et al produced the first generation of induced pluripotent stem cells from human adult fibroblasts with the over-expression of four transcription factors including Sox2, Oct3/4, c-Myc, and Klf4. (Takahashi, 2007) These cells possessed embryonic stem cell-like qualities as they could differentiate into cells of endoderm, mesoderm, and ectoderm origin; were able to form teratomas in vivo; and proliferate indefinitely without losing their pluripotency. A subsequent study from Yu et al utilized a similar procedure derived from the Takahashi study but determined that a slightly different cocktail of genes (or proteins) could induce the same effect. (Yu, 2007) OCT4, SOX2, NANOG, and LIN28 were identified as pluripotency factors that had the same effect as the aforementioned gene set. Since a slightly different set of genes was utilized in the second study, it would be easy to speculate that different combinations of genes or yet to be identified factors that are tissue specific may be utilized to create pluripotent cell lines. This groundbreaking work led to a number of other studies focusing on specific cell type and disease states. (Moretti, 2010; Ebert, 2009) However, with each great discovery come certain drawbacks. It has been recently demonstrated that induced pluripotent stem cells outwardly function similarly to embryonic stem cells, but the gene expression signatures possessed by each cell type are quite different. (Chin, 2009) This is most likely a result of differential promoter binding by the reprogramming factors. Other studies also demonstrate the detrimental effects that coerced cellular reprogramming may induce upon target cell populations. (Hussein, 2011; Lister, 2011) The use of induced pluripotent stem cells in urological settings (excluding spinal and nerve injury scenarios) is still in its infancy with a paltry number of studies. (Morizane, 2009) As more data emerges with regards to the molecular machinery involved with the reprogramming process (along with potential consequences), perhaps these data will be applied to future urological regenerative medicine based studies.

## 4. Polymeric materials and their potential applications in urologic regenerative medicine

The choice of scaffold material is of paramount significance in determining the outcome of implanted cell/scaffold composites. There are a wide variety of scaffold materials that vary greatly in composition, texture, mechanical and surface characteristics that could potentially be applied in a urological regenerative medicine setting. A number of these materials have been previously discussed in great detail such as bladder acellular matrix, small intestinal submucosa, varying embodiments of poly(lactic-co-glycolic acid) and polyglycolic acid, and collagen. Similarly, other chemically synthesized nano- and macro- based structures are also pivotal in creating hospitable environments for tissue growth and development. The focus of the forthcoming section will concentrate on relatively newly described elastic based materials that may be utilized for urological tissue engineering along with bioactive peptides and nanomolecules.

### 4.1 Synthetic elastomeric substrates for urological tissue engineering

#### 4.1.1 Poly(diol citrates)

The mechanical properties that contribute to the overall composition of various urologic organs in part define its ultimate structure and function, specifically with regard to the urinary bladder. The urinary bladder is a very dynamic organ as demonstrated by its wide elastic properties further exemplified by its ability to undergo repeated contractile/expansion cycles without permanent deformation. Characterization studies of human urinary bladder tissue describe the elastic modulus to be within the low-hundreds of kPa (kilopascal) (Dahms, 1998). Commonly used scaffolds for urinary bladder tissue engineering such as non-woven polyglycolic acid and poly(lactic-co-glycolic acid), and collagen typically exhibit elastic moduli in the giga- and megaPa range. These features predispose this materials to possess poor elongation characteristics of <10% hence making them unsuitable for functional urinary bladder replacement strategies. (Middleton, 2000) It has also been demonstrated that degradation byproducts of these polymers contribute to a localized decrease in pH which can have a negative impact on cell function, including decreased collagen synthesis and impaired cellular differentiation capacity. (Kohn, 2002) Cyclical mechanical forces have also been demonstrated necessary for bladder smooth muscle growth and development while static and non-contractile environments are less likely to produce robust contractile responses. (Heise, 2009) POC [poly(1,8-octanediol-co-citrate)] is a relatively newly described member of a family of diols which is a synthetic polymer that has a wide range of applications including those for very elastic tissues such blood vessels, orthopedic applications, and perivascular wraps. (Yang, 2005; Qiu, 2006; Serrano, 2011) The elastic potential of this polymer is dependent upon several polymerization properties including the length and temperature of scaffold crosslinking. This allows for scaffold customization with regard to elastic potential. Sharma et al demonstrate the use of POC in a bladder augmentation model using bone marrow derived mesenchymal stem cells. (Sharma, 2010) Data from this study suggests the synthesis of a POC film that has elastic properties approaching that of a native human bladder and the ability of the POC scaffolds to undergo elongation up to 137% of their initial length. In vivo data suggests a high regenerative potential when utilizing POC/cell seeded composites as demonstrated by superior muscle to collagen ratios 10 weeks post-implantation (as compared to controls). The non-toxic, degradative by-products of POC also appear to limit the level of stone formation with the context of this study, which may aid future bladder regenerative studies. Lastly, data also suggests that POC can be chemically modified to release pro-angiogenic growth factors over-time that contributes to robust levels of angiogenesis in vivo. (Sharma, unpublished data)

### 4.1.2 Elastomeric poly(ester urethane)urea

A second well described elastomeric compound, poly(ester urethane)urea (PEUU), can also be configured into several different forms including those consistent with blood vessels while retaining elastic properties. (Stankus, 2007) By concurrent electrospraying and electrospinning of vascular smooth muscle cells and PEUU, respectively, a tubular conduit containing the cells was formed with uniform cellular distribution in a radial and circumferential manner. Further testing revealed the ability of the material to remain sutured, while maintaining compliancy that was similar to native arteries. Modification of PEUU also demonstrates the ability to control gene expression utilizing a cleverly created gene induction system. (Baraniak, 2011) Further modifications also allow for the release of bioactive molecules of insulin-like growth factor-1 and hepatocyte growth factor. (Nelson, 2011) Although this material has not been utilized in a urological regenerative setting to date, the implications of this material are obvious as this material could be structured for urinary bladder regenerative studies or the creation of synthetic ureters. An overly simplistic study design would include the seeding of porous scaffolds with relevant bladder cell types as previously described by Baraniak et al accompanied by the appropriate functional testing and immunohistochemical analyses. The added benefit of utilizing PEUU is its ability to become modified so that relevant genes such as those found within urologic epithelial stem cells may contribute to regeneration, for example. (Pascal, 2007) It is hopeful that studies involving PEUU and urological tissue engineering will be seen in the not too distant future.

### 4.2 Bioactive peptides and peptide amphiphiles for regeneration and wound healing

Polymeric materials which are enhanced with bioactive peptides show perhaps the greatest degree of promise for application in tissue regeneration scaffolds. Short peptide sequences originating from the cell binding regions of extracellular matrix (ECM) proteins can be incorporated into a graft material to obtain biospecific adhesion of cells. RGD (arginine-glycine-aspartic acid), the most commonly used peptide in cell adhesion studies, originating from fibronectin, collagen, fibrinogen and other ECM proteins, enhances adhesion and spreading of most cell types (fibroblasts, endothelial, and smooth muscle cells) and interacts with several different integrin receptors. (Humphries, 1990; Massia, 1991) It has been shown in the past that migration of smooth muscle cells and fibroblasts is biphasically dependant on the concentration of RGD attached to the migratory surface with intermediate loadings of RGD displaying maximal mobility. (Olbrich, 1996; Mann 2002) Polymers containing RGD sequence have been used in several wound healing in vivo and in vitro applications. For instance, incorporation of RGD peptide into PEG hydrogels partially eliminated inflammatory reaction of a scaffold implanted subcutaneously in immunocompetent in c57bl/6 mice resulting in a reduced presence of macrophages. (Lynn, 2011) Hydrogels containing RGD peptide were able to promote healing of ulcers, partial thickness burn wounds, and encourage ingrowth of glial tissue in the rat. PHSRN (proline-histidine-serine-arginine-glutamate), a cell binding peptide sequence has been found to act synergistically with RGD sequence for cell adhesion. (van der Veen, 2010; Davis, 2001) When applied individually it can stimulate invasion of ECM by keratinocytes and fibroblasts in vitro and it can enhance the re-epithelialization and concentration of dermal wounds in healing-impaired diabetic mice. (Livant, 2000) Other peptides sequences such as laminin-derived YIGSR (tyrosine-isoleucine-glycine-serine-arginine), LRE (leucine-arginine-glutamic acid) or

IKVAV (isoleucine-lysine-valine-alanine-valine) have potential application in nerve regeneration. (Fittkau, 2005; Pittier, 2005) YIGSR, unlike RGD and PHSRN, does not interact with the 67 kDa laminin binding protein. (Graf, 1987) In addition, YIGSR promotes the adhesion and spreading of many cell types including endothelial cells, fibroblasts, and smooth muscle cells. (Massia, 2011) It was found that PEG surfaces functionalized with cell binding peptide RGD and YIGSR, significantly increased microvascular endothelial cell migration rates relative to RGD alone. (Fittkau, 2005)
Another group demonstrated that degradation of endothelial ECMs with bacterial collagenase releases proangiogenic peptide fragments which can stimulate epithelial responses to injury and wound healing in vivo. (Demidova-Rice, 2011) It has been shown that sequences such as Col4-1, Col4-2, Fibr2, Fibr3, Ten1, Ten2, and Comb1 significantly increase the rate of sprout formation, which enhances cell-cell interactions and improves endothelial motility. Comb1 and Ten2 peptides contain the GXXPG sequence, which was previously reported to enhance endothelial migration and tube formation on collagen type I in vitro; both induce morphogenesis in a Matrigel-based assay. Comb1, as a larger combinatorial peptide with more complex tertiary structure, and two GXXPG sequences also stimulates cellular proliferation. The aforementioned sequences can also stimulate endothelial response to injury in complex environments where the cells remain viable for a prolonged time (up to one week). The 3D model of injury repair which was applied in that research consisted of two layers of Matrigel collagen mixture and a layer of endothelial cells sandwiched in between. Full thickness injury was created by aspiration, and the defect was filled with matrices containing serum, proangiogenic growth factors, or peptides. It was demonstrated that in addition to identification of promising proangiogenic or antiangiogenic therapeutics, the 3D model of injury repair can be employed to study the stability of wound healing compounds and to evaluate drug delivery systems. As described in this Chapter bioactive peptides play an important role in a wound healing process and tissue regeneration. Fabrication of scaffolds with a high specificity for the adhesion of a certain type of cells might require the use of a combination of bioactive peptides such as RGD, IKVAV, Ten 2, Comb 1, and biomimetic synthetic polymers, depending on the in vivo application.
Peptide amphiphiles (PA) are small molecules that contain a peptide sequence and a hydrophobic segment such as an alkyl tail. (Silva, 2004; Rajangam, 2006) Hydrophobic collapse of these filament-forming molecules under strong screening conditions leads to the display of biological signals on their surfaces in high density (on the order of $10^{15}$ signals per $cm^2$). (Cui, 2010) In vivo and in vitro studies have shown that certain PA molecules, bearing bioactive epitopes promote regeneration of spinal cord axons, angiogenesis, bone regeneration, cartilage repair, and selective differentiation of neural progenitor cells into neurons. (Sargeant, 2008; Webber, 2010; Silva, 2004) Previously, PAs were used to coat PGA microfiber scaffold for the improved attachment of smooth muscle cells. (Harrington, 2006; Behanna, 2005) This PGA scaffold was submerged in a suspension of smooth muscle cells in media and modified with PA gel mixed with growth factors as a top layer for the entrapment of cells. The opposite face of the scaffold could them be modified using a second component, which in this case was a mixture of PGA and urothelial cells, since these cells could potentially send necessary signals to smooth muscle cells during regeneration. (Beqaj, 2005; Imamura, 2007) When the human bladder smooth muscle cells and urothelial cells where embedded in a PA-scaffold with a growth factor (bFGF) and incubated in vivo in a

subcutaneous nude rat model, it was found that the human bladder smooth muscle cells were retained and composed the majority of the scaffold cellular content. It was also found that the system which uses a combination of PGA, PA, growth factors and cells demonstrated higher levels of phenotypic alpha-smooth muscle actin than control scaffolds made of PGA and cells only. PAs offer a great potential in regenerative medicine due to their nanoscale filamentous assembly resembling the one in ECM, the ability to incorporate any bioactive epitopes into their structure, and propensity to form gels.

### 4.3 Nano- and microstructured surfaces for bladder regeneration

Nano- and microstructured surfaces offer great potential for a range of biomedical applications from biochips, bioarrays and biosensors to functional membranes and cell adhesion substrates. It is much easier to create nano- or micrometer patterns on surfaces such as metal, glass or silicon than on polymeric surfaces since the surface composition and flatness is much better defined. (Qin, 2010; Jackman; 1998) The techniques that are usually employed for nano- and microstructuring of surfaces include nanoimprint lithography (NiL), microcontact printing, shadow-mask lithography, photolithography, chemical etching processes, robotic spotting techniques, and more. Unfortunately only very few of these techniques are suitable for the surface modification and structuring of (bio)polymers commonly used as the bladder regeneration materials such as PGA, PLGA, poly(L-lactic acid) (PLLA), poly(ε-caprolactone) (PCL), and poly(carbonate-urethane) (PCU). (Miller, 2004; Pattison, 2006) Nanoscale roughness is believed to improve the adhesion of proteins due to its higher surface area and surface energy. Surfaces functionalized with protein such as fibronectin, laminin, collagen I or IV, ECM-type of protein and surfaces with nanometer roughness visibly improve bladder cell adhesion and interactions with the substrate. This phenomena is important in bladder regeneration since many implanted polymeric membranes need to be relatively quickly and efficiently replaced by healthy host tissue. Tissue integration into the biomaterial must occur; therefore resident cells need to adhere appropriately to the substrate in order for subsequent cellular functions to occur at the implant site. Nanoscale roughness of biopolymers or synthetic biodegradable polymers used in bladder regeneration research can be achieved by numerous techniques including surface functionalization with nanometer-size fibers such as PAs, electrospinning, chemical etching, nano- and micromolding and more. (Harrington, 2006; Doshi, 1995; Thapa, 2003) The aforementioned techniques may be applied as potential methods for the fabrication of improved scaffolds for bladder regeneration.

Alternative means to achieve nanostructured membranes for bladder regeneration research utilizes micro or nanomolding. PLGA or PU polymers can be easily cast into molds with pre-fabricated nanometer features made of poly(dimenthylsiloxane) (PDMS). (Thapa, 2003) The procedure relies on pouring a solution of a polymer (for example PLGA or PU) onto a PDMS master mold, curing the polymer until is solid and releasing the polymer from the mold simply by peeling the film off from the mold. The minimum size of the features that could be achieved is defined by the master mold. This process is commonly used in the fabrication of microcontact printing stamps for soft lithography purposes. In this process the mold is usually fabricated in silicon through photolithography or e-beam lithography and the polymer which is used for stamps is usually PDMS. (Xia, 1998) PLGA and PU nanostructured polymers showed advantage in ovine bladder smooth muscle cells adhesion over polymers with larger features.

An interesting method for the bladder surface nanostructuring is the application of nanoparticles to introduce nanoscale roughness or improve certain properties such as permeability of the scaffold. One example was recently introduced where hyaluronic acid (HA) modified PLGA nanoparticles (HA PLGA NPs) were applied for the modification of SIS as a scaffold for bladder regeneration to alter its permeability and simultaneously deliver pro-regenerative compounds. (Roth, 2010) When SIS scaffolds were modified with PLGA NPs of the size in the range of 200-500 nm the scaffold showed a decrease in permeability to urea when compared with unmodified scaffolds. In addition, PLGA modified scaffolds demonstrated significantly higher adhesion of endothelial cells in comparison to unmodified scaffolds. Introduction of pro-regenerative compounds into the SIS scaffold via modification of PLGA NPs with HA (which has been recognized for its ability to promote scarless wound healing), improved the SIS scaffold in vivo performance. HA was attached through electrostatic interaction to the positively charged PLGA NPs which were previously modified with polyethylene imine to produce cationic NPs. The attachment of particles to the SIS scaffold was conducted by incubation and absorption of particles from solution. The modified SIS membrane was investigated for angiogenesis and permeability studies. It was found that HA-NP SIS demonstrated a statistically higher number of blood vessels when compared with unmodified. In addition, HA-NP SIS significantly decreased or even eliminated the propensity to form calcifications, decreased urea permeability and improved full-thickness bladder regeneration.

Electrospinning is one of the most commonly used techniques for the (nano)fiber formation. In this process, nanofibers can be formed through an electrically charged jet of polymer solution or polymer melt. The simplest setup for electrospinning is composed of a high voltage power supply, a polymer solution reservoir (e.g. a syringe with a small diameter needle), and two electrodes (one connected to the needle and the other to the surface of the collecting mat or other substrate). The polymer supply is connected with a power supply through an electrode clipped to the metal needle for electrospinning. Flow control pump can be used to precisely control the flow rate of the polymer solution which is spun onto a substrate. The fiber-collecting surface must be conductive in order to attract the electrospun polymer and can be either stationary or rotating plate for collecting nonwoven or aligned fibers, respectively.

Electrospinning permits fabrication of biodegradable elastomers into matrices that can resemble the scale and mechanical behavior of the native extracellular matrix. There is a great selection of polymers available for the electrospinning of nanofibrous scaffolds used in tissue engineering applications. Moreover, this process can be easily scaled up making the transformation from small-scale laboratory research into large-batch processes feasible. (Greiner, 2008; Zhong, 2010) In addition to being used to fabricate nonwoven mats for wound dressings, there is currently much interest in making scaffolds for bladder tissue engineering. (Roth, 2010; Gelain, 2008) Baker et al used a 3D electrospun nanoscaffold of polystyrene to determine whether the phenotype of cells isolated from the stroma of human ureter specimens could be modulated by growing cells within such scaffold. (Baker, 2008) Non-aligned scaffolds made out of fibers with mean diameter of 200 nm were used in that study. Unfortunately the analysis of SM-MHC protein expression in cultures grown in 3D revealed that those cultures failed to undergo differentiation. It was explained that nanostructured polystyrene scaffolds have a high propensity to adsorb serum which maintained stromal cell cultures in a synthetic non-differentiated phenotype. In this case, the approach of growing cell cultures in a serum-free media on the 3D polystyrene scaffold did not bring positive results. It was found that despite of its promising advantages of electrospun scaffold, stromal cells did not differentiate primarily due to the high propensity of the scaffold to the absorption of proteins from the serum. However, serum-free media reduced the absorption of proteins but cells still showed a greatly reduced plating efficiency and cultures failed to survive.

Findings describing the interactions between human bladder urothelial cells (TEU-2) and a scaffold composite consisting of an electrospun fibrous matrix made of PCU and PLLA fused to a thin film for potential use in urothelial tissue regeneration. (Kundu, 2011) Their results suggested that when electrospun scaffolds alone (without the thin film) were used, their fibrous surface structure while allowing good cellular adhesion, inhibited the proliferation of urothelial cells, whereas cells seeded onto thin films with non-fibrous surfaces proliferated quickly. In addition, it was found that cells on electrospun fibers demonstrated numerous cellular extensions along the matrix fibers, rather than the substantial cell-cell integrations. It was explained that this finding may suggest that the fibrous nature of the electrospun materials do not facilitate the formation of an intact urothelium. The addition of thin films onto electrospun scaffolds, enhanced urothelial cell proliferation, surface coverage, cell-cell interactions, and multilayering in vitro. Electrospun scaffolds alone offer highly porous structures and deep cellular penetration, however they are unlikely to provide a sufficient barrier to water and solute transport when not embedded in a biomimetic synthetic polymer. Furthermore, the three-dimensional structure of such scaffolds may affect cell proliferation, differentiation or adhesion.

## 5. Conclusion

The task to create functional organs as a replacement therapy is indeed quite daunting. Tissue engineering studies during the late 1980s and early 1990s laid the groundwork for the current windfall of clinical applications that have been deployed for organ replacement. The continual evolution of materials science influenced materials combined with discoveries in the field of stem cell biology provides hope that goals of regenerative medicine can be achieved. Although there have been many significant advances with regard to urologic tissue engineering, the field is still in its infancy. There is still a great need to pursue basic and clinical science studies in order to benefit those afflicted with urological trauma or disease.

## 6. References

Abu El-Asrar A., Struyf S., Verbeke H., Van Damme J., Geboes K. (2009) Circulating bone-marrow-derived endothelial precursor cells contribute to neovascularization in diabetic epiretinal membranes. *Acta Ophthalmol.* [Epub ahead of print]

Acién P., Acién M. (2010) Unilateral renal agenesis and female genital tract pathologies. *Acta Obstet Gynecol Scand.* 89(11):1424-1431.

Akashi K., Traver D., Miyamoto T., Weissman I. (2000) A clonogenic common myeloid progenitor that gives rise to all myeloid lineages. *Nature.* 404(6774):193-197.

Armatys S., Mellon M., Beck S., Koch M., Foster R., Bihrle R. (2009) Use of ileum as ureteral replacement in urological reconstruction. *J Urol.* 181(1):177-181.

Asahara T., Murohara T., Sullivan A., Silver M., van der Zee R., Li T., Witzenbichler B., Schatteman G., Isner J. (1997) Isolation of putative progenitor endothelial cells for angiogenesis. *Science.* 275(5302):964-967.

Ashley R., Palmer B., Schultz A., Woodson B., Roth C., Routh J., Fung K., Frimberger D., Lin H., Kropp B. (2009) Leukocyte inflammatory response in a rat urinary bladder regeneration model using porcine small intestinal submucosa scaffold. *Tissue Eng Part A.* 15(11):3241-3246.

Ashley R., Roth C., Palmer B., Kibar Y., Routh J., Fung K., Frimberger D., Lin H., Kropp B. (2010) Regional variations in small intestinal submucosa evoke differences in inflammation with subsequent impact on tissue regeneration in the rat bladder augmentation model. *BJU Int.* 105(10):1462-1468.

Atala A. (1998) Autologous cell transplantation for urologic reconstruction. *J Urol.* 159(1):2-3.

Atala A., Bauer S., Soker S., Yoo J., Retik A. (2006) Tissue-engineered autologous bladders for patients needing cystoplasty. *Lancet.* 367(9518):1241-1246.

Atala A., Freeman M., Vacanti J., Shepard J., Retik A. (1993) Implantation in vivo and retrieval of artificial structures consisting of rabbit and human urothelium and human bladder muscle. *J Urol.* 150(2 Pt 2):608-612.

Atala A., Vacanti J., Peters C., Mandell J., Retik A., Freeman M. (1992) Formation of urothelial structures in vivo from dissociated cells attached to biodegradable polymer scaffolds in vitro. *J Urol.* 148(2 Pt 2):658-662.

Baker J., Zhang L., Imadojemu S., Sharpe A., Patil S., Moore J., Mohler E., Von Feldt J. (2011) Circulating endothelial progenitor cells are reduced in SLE in the absence of coronary artery calcification. *Rheumatol Int.* [Epub ahead of print]

Baker S., Southgate J. (2008) Towards control of smooth muscle cell differentiation in synthetic 3D scaffolds. Biomaterials. 29(23):3357-3366.

Bankhead, R., Kropp, B., Cheng, E. (2000) Evaluation and treatment of children with neurogenic bladders. *J Child Neurol.* 15(3):141-149.

Barbagli G., Lazzeri M. (2007) Surgical treatment of anterior urethral stricture diseases: brief overview. *Int Braz J Urol.* 33(4):461-469.

Baraniak P., Nelson D., Leeson C., Katakam A., Friz J., Cress D., Hong Y., Guan J., Wagner WR. (2011) Spatial control of gene expression within a scaffold by localized inducer release. *Biomaterials.* 32(11):3062-3071.

Baum C., Weissman I., Tsukamoto A., Buckle A., Peault B. (1992) Isolation of a candidate human hematopoietic stem-cell population. *Proc Natl Acad Sci U S A.* 89(7):2804-2808.

Behanna H., Donners J., Gordon A., Stupp S. (2005) Coassembly of amphiphiles with opposite peptide polarities into nanofibers. *J Am Chem Soc.* 127(4):1193-1200.

Bellomo R., Ronco C., Kellum J., Mehta R., Palevsky P. (2004) Acute Dialysis Quality Initiative Workgroup. Acute renal failure - definition, outcome measures, animal models, fluid therapy and information technology needs: the Second International Consensus Conference of the Acute Dialysis Quality Initiative (ADQI) Group. *Crit Care.* 8(4):R204-R212.

Beqaj S., Donovan J., Liu D., Harrington D., Alpert S., Cheng E. (2005) Role of basic fibroblast growth factor in the neuropathic bladder phenotype. *J Urol.* 174(4 Pt 2):1699-1703.

Berjukow S., Margreiter E., Marksteiner R., Strasser H., Bartsch G., Hering S. (2004) Membrane properties of single muscle cells of the rhabdosphincter of the male urethra. *Prostate.* 58(3):238-247.

Birket M., Orr A., Gerencser A., Madden D., Vitelli C., Swistowski A., Brand M., Zeng X. (2011) A reduction in ATP demand and mitochondrial activity with neural differentiation of human embryonic stem cells. *J Cell Sci.* 124(Pt 3):348-358.

Botlero R., Davis S., Urquhart D., Bell R. (2011) Incidence and resolution rates of different types of urinary incontinence in women: findings from a cohort study. *J Urol.* [Epub ahead of print]

Brubaker L., Lukacz E., Burgio K., Zimmern P., Norton P., Leng W., Johnson H., Kraus S., Stoddard A. (2011) Mixed incontinence: comparing definitions in non-surgical patients. *Neurourol Urodyn.* 30(1):47-51.

Campbell K., McWhir J., Ritchie W., Wilmut I. (1996) Sheep cloned by nuclear transfer from a cultured cell line. *Nature.* 380(6569):64-66.

Cannon T., Lee J., Somogyi G., Pruchnic R., Smith C., Huard J., Chancellor M. (2003)Improved sphincter contractility after allogenic muscle-derived progenitor cell injection into the denervated rat urethra. *Urology.* 62(5):958-963.

Chin M., Mason M., Xie W., Volinia S., Singer M., Peterson C., Ambartsumyan G., Aimiuwu O., Richter L., Zhang J., Khvorostov I., Ott V., Grunstein M., Lavon N., Benvenisty N., Croce C., Clark A., Baxter T., Pyle A., Teitell M., Pelegrini M., Plath K., Lowry W. (2009) Induced pluripotent stem cells and embryonic stem cells are distinguished by gene expression signatures. *Cell Stem Cell.* 5(1):111-123.

Cui H., Webber M., Stupp S. (2010) Self-assembly of peptide amphiphiles: from molecules to nanostructures to biomaterials. *Biopolymers.* 94(1):1-18.

Dagash H., Sen S., Chacko J., Karl S., Ghosh D., Parag P., Mackinnon A. (2008) The appendix as ureteral substitute: a report of 10 cases. *J Pediatr Urol.* 4(1):14-19.

Dahms S., Piechota H., Dahiya R., Lue T., Tanagho E. (1998) Composition and biomechanical properties of the bladder acellular matrix graft: comparative analysis in rat, pig and human. *Br J Urol.* 82(3):411-419.

Davis S., Eaglstein W., Cazzaniga A., Mertz P. (2001) An octyl-2-cyanoacrylate formulation speeds healing of partial-thickness wounds. *Dermatol Surg.* 27(9):783-788.

De Coppi P., Callegari A., Chiavegato A., Gasparotto L., Piccoli M., Taiani J., Pozzobon M., Boldrin L., Okabe M., Cozzi E., Atala A., Gamba P., Sartore S. (2007) Amniotic fluid and bone marrow derived mesenchymal stem cells can be converted to smooth muscle cells in the cryo-injured rat bladder and prevent compensatory hypertrophy of surviving smooth muscle cells. *J Urol.* 177(1):369-376.

Dekel B., Burakova T., Ben-Hur H., Marcus H., Oren R., Laufer J., Reisner Y. (1997) Engraftment of human kidney tissue in rat radiation chimera: II. Human fetal kidneys display reduced immunogenicity to adoptively transferred human peripheral blood mononuclear cells and exhibit rapid growth and development. *Transplantation.* 64(11):1550-1558.

Dekel B., Burakova T., Marcus H., Shezen E., Polack S., Canaan A., Passwell J., Reisner Y. (1997) Engraftment of human kidney tissue in rat radiation chimera: I. A new model of human kidney allograft rejection. *Transplantation.* 64(11):1541-1550.

Dekel B., Burakova T., Shezen E., Marcus H., Canaan A., Reisner Y. ( 1997) Human renal allograft rejection in the SCID/rat radiation chimera. *Transplant Proc.* 29(4):2255-2256.

Dekel B., Marcus H., Herzel B., Böcher W., Passwell J., Reisner Y. (2000) In vivo modulation of the allogeneic immune response by human fetal kidneys: the role of cytokines, chemokines, and cytolytic effector molecules. *Transplantation.* 69(7):1470-1478.

Delancey J., Ashton-Miller J. (2004) Pathophysiology of adult urinary incontinence. *Gastroenterology.* 126(1 Suppl 1):S23-S32.

de Mendonça A., Vincent J., Suter P., Moreno R., Dearden N., Antonelli M., Takala J., Sprung C., Cantraine F. (2000) Acute renal failure in the ICU: risk factors and outcome evaluated by the SOFA score. *Intensive Care Med.* 26(7):915-921.

Demidova-Rice T., Geevarghese A., Herman I. (2011) Bioactive peptides derived from vascular endothelial cell extracellular matrices promote microvascular morphogenesis and wound healing in vitro. *Wound Repair Regen.*19(1):59-70.

Desgrandchamps F., Cussenot O., Meria P., Cortesse A., Teillac P., Le Duc A. (1995)Subcutaneous urinary diversions for palliative treatment of pelvic malignancies. *J Urol.* 154(2 Pt 1):367-370.

Desgrandchamps F., Paulhac P., Fornairon S., De Kerviller E., Duboust A., Teillac P., Le Duc A. (1998) Artificial ureteral replacement for ureteral necrosis after renal transplantation: report of 3 cases. *J Urol.* 159(6):1830-1832.

Doshi J., Reneker D. (1995) Electrospinning process and applications of electrospun fibers. *Journal of Electrostatics.* 35:151-160.

Dubois S., Stempak D., Wu B., Mokhtari R., Nayar R., Janeway K., Goldsby R., Grier H., Baruchel S. (2011) Circulating endothelial cells and circulating endothelial precursor cells in patients with osteosarcoma. *Pediatr Blood Cancer.* [Epub ahead of print]

Ebert A., Yu J., Rose F., Mattis V., Lorson C., Thomson J., Svendsen CN. (2009) Induced pluripotent stem cells from a spinal muscular atrophy patient. *Nature.* 457(7227):277-280.

Farzaneh Z., Pournasr B., Ebrahimi M., Aghdami N., Baharvand H. (2010) Enhanced functions of human embryonic stem cell-derived hepatocyte-like cells on three-dimensional nanofibrillar surfaces. *Stem Cell Rev.* 6(4):601-610.

Feki A., Faltin D., Lei T., Dubuisson J., Jacob S., Irion O. (2007) Sphincter incontinence: is regenerative medicine the best alternative to restore urinary or anal sphincter function? *Int J Biochem Cell Biol.* 39(4):678-684.

Feng C., Xu Y., Fu Q., Zhu W., Cui L., Chen J. (2010) Evaluation of the biocompatibility and mechanical properties of naturally derived and synthetic scaffolds for urethral reconstruction. *J Biomed Mater Res A.* 94(1):317-325.

Ferrell N., Desai R., Fleischman A., Roy S., Humes H., Fissell W. (2010) A microfluidic bioreactor with integrated transepithelial electrical resistance (TEER) measurement electrodes for evaluation of renal epithelial cells. *Biotechnol Bioeng.* 107(4):707-716.

Fischer E., Verpont M., Garrett-Sinha L., Ronco P., Rossert J. (2001) Klf6 is a zinc finger protein expressed in a cell-specific manner during kidney development. *J Am Soc Nephrol.* 12(4):726-735.

Fittkau M., Zilla P., Bezuidenhout D., Lutolf M., Human P., Hubbell J., Davies N. (2005) The selective modulation of endothelial cell mobility on RGD peptide containing surfaces by YIGSR peptides. *Biomaterials.* 26(2):167-174.

Friedrich E., Walenta K., Scharlau J., Nickenig G., Werner N. (2006) CD34-/CD133+/VEGFR-2+ endothelial progenitor cell subpopulation with potent vasoregenerative capacities. *Circ Res.* 98(3):e20-e25.

Gage F. (2000) Mammalian neural stem cells. *Science* 287:1433-1438.

Gelain F. (2008) Novel opportunities and challenges offered by nanobiomaterials in tissue engineering. *Int J Nanomedicine.* 3(4):415-424.

Ghoniem G., Lapeyrolerie J., Sood O., Thomas R. (1994) Tulane experience with management of urinary incontinence after placement of an artificial urinary sphincter. *World J Urol.* 12(6):333-336.

Gimpel C., Masioniene L., Djakovic N., Schenk J., Haberkorn U., Tönshoff B., Schaefer F. (2010) Complications and long-term outcome of primary obstructive megaureter in childhood. *Pediatr Nephrol.* 25(9):1679-1686.

Goel A., Goel A., Jain A. (2011) Buccal mucosal graft urethroplasty for penile stricture: only dorsal or combined dorsal and ventral graft placement? *Urology.* [Epub ahead of print]

Graf J., Ogle R., Robey F., Sasaki M., Martin G., Yamada Y., Kleinman H. (1987) A pentapeptide from the laminin B1 chain mediates cell adhesion and binds the 67,000 laminin receptor. *Biochemistry.* 26(22):6896-6900.

Greiner A., Wendorff J. Self-Assembled Nanomaterials I: Nanofibers, Vol. 219, 2008:107-171.

Hammerman M. (2011) Xenotransplantation of embryonic pig kidney or pancreas to replace the function of mature organs. *J Transplant.* 2011:501749.

Hannema S., Print C., Charnock-Jones D., Coleman N, Hughes I. (2006). Changes in gene expression during Wolffian duct development. *Horm Res.* 65(4):200-209.

Harrington D., Cheng E., Guler M., Lee L., Donovan J., Claussen R., Stupp S. (2006) Branched peptide-amphiphiles as self-assembling coatings for tissue engineering scaffolds. *J Biomed Mater Res A.* 78(1):157-167.

Heise R., Ivanova J., Parekh A., Sacks M. (2009) Generating elastin-rich small intestinal submucosa-based smooth muscle constructs utilizing exogenous growth factors and cyclic mechanical stimulation. *Tissue Eng Part A.* 15(12):3951-3960.

http://clinicaltrials.gov (utilizing search terms "urinary" and "incontinence")

http://www.advancedcell.com/news-and-media/press-releases

http://www.cirm.ca.gov/for-researchers/researchfunding

http://www.diabetes.org/diabetes-basics/diabetes-statistics

Huang Y., Ning H., Shindel A., Fandel T., Lin G., Harraz A., Lue T., Lin C. (2010) The effect of intracavernous injection of adipose tissue-derived stem cells on hyperlipidemia-associated erectile dysfunction in a rat model. *J Sex Med.* 7(4 Pt 1):1391-1400.

Humphries M. (1990) The molecular basis and specificity of integrin-ligand interactions. *J. Cell Sci.* 97:585-592.

Husmann D., Snodgrass W., Koyle M., Furness P., Kropp B., Cheng E., Kaplan W., Kramer S. (2004) Ureterocystoplasty: indications for a successful augmentation. *J Urol.* 171(1):376-380.

Hussein S., Batada N., Vuoristo S., Ching R., Autio R., Närvä E., Ng S., Sourour M., Hämäläinen R., Olsson C., Lundin K., Mikkola M., Trokovic R., Peitz M., Brüstle O.,

Bazett-Jones D., Alitalo K., Lahesmaa R., Nagy A., Otonkoski T. (2011) Copy number variation and selection during reprogramming to pluripotency. *Nature.* 471(7336):58-62.

Imamura M., Kanematsu A., Yamamoto S., Kimura Y., Kanatani I., Ito N., Tabata Y., Ogawa O. (2007) Basic fibroblast growth factor modulates proliferation and collagen expression in urinary bladder smooth muscle cells. *Am J Physiol Renal Physiol.* 293(4):F1007-F1017.

In 't Anker P., Scherjon S., Kleijburg-van der Keur C., Noort W., Claas F., Willemze R., Fibbe W., Kanhai H. (2003) Amniotic fluid as a novel source of mesenchymal stem cells for therapeutic transplantation. *Blood.* 102(4):1548-1549.

Jack G., Zhang R., Lee M., Xu Y., Wu B., Rodríguez L. (2009) Urinary bladder smooth muscle engineered from adipose stem cells and a three dimensional synthetic composite. *Biomaterials.* 30(19):3259-3270.

Jackman R., Brittain S., Adams A., Prentiss M., Whitesides G. (1998) Design and fabrication of topologically complex, three-dimensional microstructures. *Science.* 280(5372):2089-2091.

Kinebuchi Y., Johkura K., Sasaki K., Imamura T., Mimura Y., Nishizawa O. (2008) Direct induction of layered tissues from mouse embryonic stem cells: potential for differentiation into urinary tract tissue. *Cell Tissue Res.* 331(3):605-615.

Knapp P., Lingeman J., Siegel Y., Badylak S., Demeter R. (1994) Biocompatibility of small-intestinal submucosa in urinary tract as augmentation cystoplasty graft and injectable suspension. *J Endourol.* 8(2):125-130.

Ko K., Tapia N., Wu G., Kim J., Bravo M., Sasse P., Glaser T., Ruau D., Han D., Greber B., Hausdörfer K., Sebastiano V., Stehling M., Fleischmann B., Brüstle O., Zenke M., Schöler H. (2009) Induction of pluripotency in adult unipotent germline stem cells. *Cell Stem Cell.* 5(1):87-96.

Kohn D., Sarmadi., Helman J., Krebsbach P. (2002) Effects of pH on human bone marrow stromal cells in vitro: implications for tissue engineering of bone. *J Biomed Mater Res.* 60(2):292-299.

Koyanagi T. (1980) Studies on the sphincteric system located distally in the urethra: the external urethral sphincter revisited. *J Urol.* 124(3):400-406.

Koziak A., Kania P., Marcheluk A., Dmowski T., Szcześniewski R., Dorobek A. (2004)Reconstruction of long ureteral obstructions using xenogenic acellular collagen membranes. *Ann Transplant.* 9(4):18-120.

Kropp B., Eppley B., Prevel C., Rippy M., Harruff R., Badylak S., Adams M., Rink R., Keating M. (1995) Experimental assessment of small intestinal submucosa as a bladder wall substitute. *Urology.* 46(3):396-400.

Kundu A., Gelman J., Tyson D. (2011) Composite thin film and electrospun biomaterials for urologic tissue reconstruction. *Biotechnol Bioeng.* 108(1):207-215.

Kurzrock E. (2010) Editorial comment. *J Urol.* 184(2):707; discussion 708.

Lakshmanan Y., Frimberger D., Gearhart J., Gearhart J. (2005) Human embryoid body-derived stem cells in co-culture with bladder smooth muscle and urothelium. *Urology.* 65(4):821-826.

Langer R., Vacanti J., Vacanti C., Atala A., Freed L., Vunjak-Novakovic G. (1995) Tissue engineering: biomedical applications. *Tissue Eng.* 1(2):151-161.

Levine L., Strom K., Lux M. (2007) Buccal mucosa graft urethroplasty for anterior urethral stricture repair: evaluation of the impact of stricture location and lichen sclerosus on surgical outcome. *J Urol.* 178(5):2011-2015.

Lin H., Cowan R., Moore P., Zhang Y., Yang Q., Peterson J., Tomasek J., Kropp B., Cheng E. (2004) Characterization of neuropathic bladder smooth muscle cells in culture. *J Urol.* 171(3):1348-1352.

Lister R., Pelizzola M., Kida Y., Hawkins R., Nery J., Hon G., Antosiewicz-Bourget J., O'Malley R., Castanon R., Klugman S., Downes M., Yu R., Stewart R., Ren B., Thomson J., Evans R., Ecker J. (2011) Hotspots of aberrant epigenomic reprogramming in human induced pluripotent stem cells. *Nature.* 471(7336):68-73.

Livant D., Brabec R., Kurachi K, Allen D., Wu Y., Haaseth R., Andrews P., Ethier S., Markwart S. (2000) The PHSRN sequence induces extracellular matrix invasion and accelerates wound healing in obese diabetic mice. *J Clin Invest.* 105(11):1537-1545.

Luppi P., Powers R., Verma V., Edmunds L., Plymire D., Hubel C. (2010) Maternal circulating CD34+VEGFR-2+ and CD133+VEGFR-2 + progenitor cells increase during normal pregnancy but are reduced in women with preeclampsia. *Reprod Sci.* 17(7):643-652.

Lynn A., Blakney A., Kyriakides T., Bryant S. (2011) Temporal progression of the host response to implanted poly(ethylene glycol)-based hydrogels. *J Biomed Mater Res A.* 96(4):621-631.

Mangera A., Chapple C. (2010) Management of anterior urethral stricture: an evidence-based approach. *Curr Opin Urol.* 20(6):453-458.

Mann B., West J. (2002) Cell adhesion peptides alter smooth muscle cell adhesion, proliferation, migration, and matrix protein synthesis on modified surfaces and in polymer scaffolds. *J Biomed Mater Res.* 60(1):86-93.

Massia S., Hubbell J. (1991) An RGD spacing of 440 nm is sufficient for integrin alpha V beta 3-mediated fibroblast spreading and 140 nm for focal contact and stress fiber formation. *J Cell Biol.* 114(5):1089-1100.

Massia S., Rao S., Hubbell J. (1993) Covalently immobilized laminin peptide Tyr-Ile-Gly-Ser-Arg (YIGSR) supports cell spreading and co-localization of the 67-kilodalton laminin receptor with alpha-actinin and vinculin. *J Biol Chem.* 268(11):8053-8059.

Mauney J., Ramachandran A., Yu R., Daley G., Adam R., Estrada C. (2010) All-trans retinoic acid directs urothelial specification of murine embryonic stem cells via GATA4/6 signaling mechanisms. *PLoS One.* 5(7):e11513.

McCrory WW. (1974) The normal embryologic development of the kidney: a basis for understanding structural abnormalities. *Birth Defects Orig Artic Ser.* 10(4):3-11.

McGrath J., Solter D. (1983) Nuclear transplantation in the mouse embryo by microsurgery and cell fusion. *Science.* 220(4603):1300-1302.

Middleton JC., Tipton AJ. (2000) Synthetic biodegradable polymers as orthopedic devices. *Biomaterials.* 21(23):2335-2346.

Miller D., Thapa A., Haberstroh K., Webster T. (2004) Endothelial and vascular smooth muscle cell function on poly(lactic-co-glycolic acid) with nano-structured surface features. *Biomaterials.* 25(1):53-61.

Mitalipov S., Wolf D. (2009) Totipotency, pluripotency and nuclear reprogramming. *Adv Biochem Eng Biotechnol.* 114:185-199.

Moretti A., Bellin M., Welling A., Jung C., Lam J., Bott-Flügel L., Dorn T., Goedel A., Höhnke C., Hofmann F., Seyfarth M., Sinnecker D., Schömig A., Laugwitz K. (2010) Patient-specific induced pluripotent stem-cell models for long-QT syndrome. *N Engl J Med.* [Epub ahead of print]

Morizane R., Monkawa T., Itoh H. (2009) Differentiation of murine embryonic stem and induced pluripotent stem cells to renal lineage in vitro. *Biochem Biophys Res Commun.* 390(4):1334-1339.

Mundy A., Andrich D. (2011) Urethral strictures. *BJU Int.* 107(1):6-26.

Murtuza B., Nichol J., Khademhosseini A. (2009) Micro- and nanoscale control of the cardiac stem cell niche for tissue fabrication. *Tissue Eng Part B Rev.* 15(4):443-454.

Muthusubramaniam L., Lowe R., Fissell W., Li L., Marchant R., Desai T., Roy S. (2011)Hemocompatibility of silicon-based substrates for biomedical implant applications. *Ann Biomed Eng.* [Epub ahead of print]

Nakanishi K., Yoshikawa N. (2003) Genetic disorders of human congenital anomalies of the kidney and urinary tract (CAKUT). *Pediatr Int.* 45(5):610-616.

Nelson D., Baraniak P., Ma Z., Guan J., Mason N., Wagner W. (2011) Controlled release of IGF-1 and HGF from a biodegradable polyurethane scaffold. *Pharm Res.* [Epub ahead of print]

Nistor G., Totoiu M., Haque N., Carpenter M., Keirstead H. (2005) Human embryonic stem cells differentiate into oligodendrocytes in high purity and myelinate after spinal cord transplantation. *Glia.* 49(3):385-396.

Oberpenning F., Meng J., Yoo J., Atala A. (1999) De novo reconstitution of a functional mammalian urinary bladder by tissue engineering. *Nat Biotechnol.* 17(2):149-55.

Oh S., Lee J., Ghil S., Lee S., Yuk S., Lee J. (2006) PCL microparticle-dispersed PLGA solution as a potential injectable urethral bulking agent. *Biomaterials.* 27(9):1936-1944.

Olajide A., Salako A., Aremu A., Eziyi A., Olajide F., Banjo O. (2010) Complications of transverse distal penile island flap: urethroplasty of complex anterior urethral stricture. *Urol J.* 7(3):178-182.

Olbrich K., Andersen T., Blumenstock F., Bizios R. (1996) Surfaces modified with covalently-immobilized adhesive peptides affect fibroblast population motility. *Biomaterials.* 17(8):759-764.

Oottamasathien S., Wang Y., Williams K., Franco O., Wills M., Thomas J., Saba K., Sharif-Afshar A., Makari J., Bhowmick N., DeMarco R., Hipkens S., Magnuson M., Brock J., Hayward S., Pope J., Matusik R. (2007) Directed differentiation of embryonic stem cells into bladder tissue. *Dev Biol.* 304(2):556-566.

Park J. (2010) Editorial comment. *J Urol.* 184(2):708; discussion 708.

Pascal L., Deutsch E., Campbell D., Korb M., True L., Liu A. (2007) The urologic epithelial stem cell database (UESC) - a web tool for cell type-specific gene expression and immunohistochemistry images of the prostate and bladder. *BMC Urol.*7:19.

Pattison M., Webster T., Haberstroh K. (2006) Select bladder smooth muscle cell functions were enhanced on three-dimensional, nano-structured poly(ether urethane) scaffolds. *J Biomater Sci Polym Ed.* 17(11):1317-1332.

Pelliniemi L., Kellokumpu-Lehtinen P., Hoffer A. (1983) Glycogen accumulations in differentiating mesonephric ducts and tubuli in male human embryos. *Anat Embryol* 168(3):445-453.

Phadnis S., Joglekar M., Dalvi M., Muthyala S., Nair P., Ghaskadbi S., Bhonde R., Hardikar A. (2011) Human bone marrow-derived mesenchymal cells differentiate and mature into endocrine pancreatic lineage in vivo. *Cytotherapy.* 13(3):279-293.

Pittenger M., Mackay A., Beck S., Jaiswal R., Douglas R., Mosca J., Moorman M., Simonetti D., Craig S., Marshak D. (1999) Multilineage potential of adult human mesenchymal stem cells. *Science.* 284(5411):143-147.

Pittier R., Sauthier F., Hubbell J., Hall H. (2005) Neurite extension and in vitro myelination within three-dimensional modified fibrin matrices. *J Neurobiol.* 63(1):1-14.

Powers M., Campbell B., Weisse C. (2010) Porcine small intestinal submucosa augmentation urethroplasty and balloon dilatation of a urethral stricture secondary to inadvertent prostatectomy in a dog. *J Am Anim Hosp Assoc.* 46(5):358- 365.

Prajerova I., Honsa P., Chvatal A., Anderova M. (2010) Neural stem/progenitor cells derived from the embryonic dorsal telencephalon of D6/GFP mice differentiate primarily into neurons after transplantation into a cortical lesion. *Cell Mol Neurobiol.* 30(2):199-218.

Qin D., Xia Y., Whitesides G. (2010) Soft lithography for micro- and nanoscale patterning. *Nat Protoc.* 5(3):491-502.

Qiu H., Yang J., Kodali P., Koh J., Ameer G. (2006) A citric acid-based hydroxyapatite composite for orthopedic implants. *Biomaterials.* 27(34):5845-5854.

Quirici N., Soligo D., Caneva L., Servida F., Bossolasco P., Deliliers G. (2001) Differentiation and expansion of endothelial cells from human bone marrow CD133(+) cells. *Br J Haematol.* 115(1):186-194.

Rajangam K., Behanna H., Hui M., Han X., Hulvat J., Lomasney J., Stupp S. (2006) Heparin binding nanostructures to promote growth of blood vessels. *Nano Lett.* 6(9):2086-2090.

Ren X., Zhang J., Gong X., Niu X., Zhang X., Chen P., Zhang X.(2010) Differentiation of murine embryonic stem cells toward renal lineages by conditioned medium from ureteric bud cells in vitro. *Acta Biochim Biophys Sin (Shanghai).* 42(7):464-471.

Reynolds B., Weiss S. (1992) Generation of neurons and astrocytes from isolated cells of the adult mammalian central nervous system. *Science.* 255(5052):1707-1710.

Rogers S., Lowell J., Hammerman N., Hammerman M. (1998) Transplantation of developing metanephroi into adult rats. *Kidney Int.* 54(1):27-37.

Roth C., Donovan B., Tonkin J., Klein J., Frimberger D., Kropp B. (2010) Endoscopic injection of submucosal bulking agents for the management of incontinent catheterizable channels. *J Pediatr Urol.* 5(4):265-268.

Rüger B., Breuss J., Hollemann D., Yanagida G., Fischer M., Mosberger I., Chott A., Lang I., Davis P., Höcker P., Dettke M. (2008) Vascular morphogenesis by adult bone marrow progenitor cells in three-dimensional fibrin matrices. *Differentiation.* 76(7):772-783.

Sales V., Mettler B., Engelmayr G., Aikawa E., Bischoff J., Martin D., Exarhopoulos A., Moses M., Schoen F., Sacks M., Mayer J. (2010) Endothelial progenitor cells as a sole source for ex vivo seeding of tissue-engineered heart valves. *Tissue Eng Part A.* 16(1):257-267.

Sargeant T., Oppenheimer S., Dunand D., Stupp S. (2008) Titanium foam-bioactive nanofiber hybrids for bone regeneration. *J Tissue Eng Regen Med.* 2(8):455-462.

Selim M., Bullock A., Blackwood K., Chapple C., MacNeil S. (2011) Developing biodegradable scaffolds for tissue engineering of the urethra. *BJU Int.* 107(2):296-302.

Serrano M., Vavra A., Jen M., Hogg M., Murar J., Martinez J., Keefer L., Ameer G., Kibbe M. (2011) Poly(diol-co-citrate)s as novel elastomeric perivascular wraps for the reduction of neointimal hyperplasia. *Macromol Biosci.* [Epub ahead of print]

Seydoux G., Braun R. (2006) Pathway to totipotency: lessons from germ cells. *Cell.* 127(5):891-904.

Shaeer O., El-Sadat A. (2006) Urethral substitution using vein graft for hypospadias repair. *J Pediatr Urol.* 2(5):518-521.

Sharma A., Bury M., Marks A., Fuller N., Meisner J., Tapaskar N., Halliday L., Matoka D., Cheng E. (2010) A non-human primate model for urinary bladder regeneration utilizing autologous sources of bone marrow derived mesenchymal stem cells. *Stem Cells.* [Epub ahead of print]

Sharma A., Fuller N., Sullivan R., Fulton N., Hota P., Harrington D., Villano J., Hagerty J., Cheng E. (2009) Defined populations of bone marrow derived mesenchymal stem and endothelial progenitor cells for bladder regeneration. *J Urol.* 182(4 Suppl):1898-1905.

Sharma A., Hota P., Matoka D., Fuller N., Jandali D., Thaker H., Ameer G., Cheng E. (2010) Urinary bladder smooth muscle regeneration utilizing bone marrow derived mesenchymal stem cell seeded elastomeric poly(1,8-octanediol-co-citrate) based thin films. *Biomaterials.* 31(24):6207-6217.

Siegler E., Reidenberg M. (2004) Treatment of urinary incontinence with anticholinergics in patients taking cholinesterase inhibitors for dementia. *Clin Pharmacol Ther.* 75(5):484-488.

Simpson J., Otaño L. (2007) Prenatal genetic diagnosis. In: Gabbe SG, Niebyl JR, Simpson JL, eds. Obstetrics:Normal and Problem Pregnancies. 5th ed. Philadelphia, Pa: Elsevier Churchill Livingstone; chap 7.

Silva G., Czeisler C., Niece K., Beniash E., Harrington D., Kessler J., Stupp S. (2004) Selective differentiation of neural progenitor cells by high-epitope density nanofibers. *Science.* 303(5662):1352-1355.

Stahl D., Koul H., Chacko J., Mingin G. (2006) Congenital anomalies of the kidney and urinary tract (CAKUT): a current review of cell signaling processes in ureteral development. *J Pediatr Urol.* 2(1):2-9.

Stankus J., Soletti L., Fujimoto K., Hong Y., Vorp D., Wagner W. (2007) Fabrication of cell microintegrated blood vessel constructs through electrohydrodynamic atomization. *Biomaterials.* 28(17):2738-2746.

Strauss S. (2010) Geron trial resumes, but standards for stem cell trials remain elusive. *Nat Biotechnol.* 28(10):989-990.

Takahashi K., Tanabe K., Ohnuki M., Narita M., Ichisaka T., Tomoda K., Yamanaka S. (2007)Induction of pluripotent stem cells from adult human fibroblasts by defined factors. *Cell.* 131(5):861-872.

Takahashi T., Kalka C., Masuda H., Chen D., Silver M., Kearney M., Magner M., Isner J., Asahara T. (1999) Ischemia- and cytokine-induced mobilization of bone marrow-derived endothelial progenitor cells for neovascularization. *Nat Med.* 5(4):434-438.

Temeltas G., Dagci T., Kurt F., Evren V., Tuglu I. (2009) Bladder function recovery in rats with traumatic spinal cord injury after transplantation of neuronal-glial restricted precursors or bone marrow stromal cells. *J Urol.* 181(6):2774-2779.

Thapa A., Miller D., Webster T., Haberstroh K. (2003) Nano-structured polymers enhance bladder smooth muscle cell function. *Biomaterials.* 24(17):2915-2926.

Thapa A., Webster T., Haberstroh K. (2003) Polymers with nano-dimensional surface features enhance bladder smooth muscle cell adhesion. *J Biomed Mater Res A.* 67(4):1374-1383.

Thomas J., Oottamasathien S., Makari J., Honea L., Sharif-Afshar A., Wang Y., Adams C., Wills M., Bhowmick N., Adams M., Brock J., Hayward S., Matusik R., Pope J. (2008) Temporal-spatial protein expression in bladder tissue derived from embryonic stem cells. *J Urol.* 180(4 Suppl):1784-1789.

Thomson J., Itskovitz-Eldor J., Shapiro S., Waknitz M., Swiergiel J., Marshall V., Jones J. (1998) Embryonic stem cell lines derived from human blastocysts. *Science.* 282(5391):1145-1147.

Toka H., Toka O., Hariri A., Nguyen H. (2010) Congenital anomalies of kidney and urinary tract. *Semin Nephrol.* 30(4):374-386.

Vacanti J., Morse M., Saltzman W., Domb A., Perez-Atayde A., Langer R. (1988) Selective cell transplantation using bioabsorbable artificial polymers as matrices. *J Pediatr Surg.* 23(1 Pt 2):3-9.

van der Veen V., van der Wal M., van Leeuwen M., Ulrich MM., Middelkoop E. (2010) Biological background of dermal substitutes. *Burns.* 36(3):305-321.

Webber M., Kessler J., Stupp S. (2010) Emerging peptide nanomedicine to regenerate tissues and organs. *J Intern Med.* 267(1):71-88.

Welsh M., Saunders P., Marchetti N., Sharpe R. (2006) Androgen-dependent mechanisms of Wolffian duct development and their perturbation by flutamide. *Endocrinology.* 147(10):4820-4830.

Wilson L., Brown J., Shin G., Luc K., Subak L. (2001) Annual direct cost of urinary incontinence. *Obstet Gynecol* 98:398-406.

Woolf A., Hornbruch A., Fine L. (1991) Integration of new embryonic nephrons into the kidney. *Am J Kidney Dis.* 17(6):611-614.

Xia Y., Whitesides G. (1998) Soft Lithography. In. *Annu. Rev. Mater. Sci.* 28:153–184.

Xiao C., Du M., Li B., Liu Z., Chen M., Chen Z., Cheng P., Xue X., Shapiro E., Lepor H. (2005) An artificial somatic-autonomic reflex pathway procedure for bladder control in children with spina bifida. *J Urol.* 173(6):2112-2116.

Yang J., Motlagh D., Webb A., Ameer G. (2005) Novel biphasic elastomeric scaffold for small-diameter blood vessel tissue engineering. *Tissue Eng.* 11(11-12):1876-1886.

Yiou R., Yoo J., Atala A. (2003) Restoration of functional motor units in a rat model of sphincter injury by muscle precursor cell autografts. *Transplantation.* 76(7):1053-1060.

Yu J., Vodyanik M., Smuga-Otto K., Antosiewicz-Bourget J., Frane J., Tian S., Nie J., Jonsdottir G., Ruotti V., Stewart R., Slukvin I., Thomson J. (2007) Induced pluripotent stem cell lines derived from human somatic cells. *Science.* 318(5858):1917-1920.

Zhang Q., Jiang J., Han P., Yuan Q., Zhang J., Zhang X., Xu Y., Cao H., Meng Q., Chen L., Tian T., Wang X., Li P., Hescheler J., Ji G., Ma Y. (2010) Direct differentiation of atrial and ventricular myocytes from human embryonic stem cells by alternating retinoid signals. *Cell Res.* [Epub ahead of print]

Zhong S., Zhang Y., Lim C. (2010) Wiley Interdisciplinary Reviews-Nanomedicine and Nanobiotechnology 2:510-525.

Zini L., Yiou R., Lecoeur C., Biserte J., Abbou C., Chopin D. (2004) Tissue engineering in urology. *Ann Urol.* 38(6):266-274.

# Part 6

## Craniofacial Tissues

# 15

# Tooth Organ Engineering: Biological Constraints Specifying Experimental Approaches

*Sabine Kuchler-Bopp[1,2], Laetitia Keller[1,2], Anne Poliard[3] and Herve Lesot[1,2]
[1]*INSERM UMR977, Université de Strasbourg,*
[2]*Faculté de Chirurgie Dentaire, Université de Strasbourg,*
[3]*EA 2496, Université Paris Descartes, Faculté de Chirurgie Dentaire, Montrouge, France*

## 1. Introduction

Basically, two types of approaches are currently being developed for tooth engineering. The first one consists in the engineering of dental constituents, such as the periodontium, the pulp/dentin, or the enamel/dentin complexes (Duan et al., 2011; Honda et al., 2009; Huang, 2009; Park et al., 2010). In parallel with these experiments of tissue engineering, attempts are also made to reconstruct a whole tooth (Arany et al., 2009; Honda et al., 2008; Hu et al., 2006a; Komine et al., 2007; Nakao et al., 2007; Ohazama et al., 2004). Most of this chapter will consider the second goal, and only use data from tooth tissue engineering, when they bring information about the cellular or molecular mechanisms that are involved and/or about the specific constraints, which they may illustrate.

Few groups are interested in a biomimetic approach to engineer a whole tooth, including crown, roots, and periodontium by using cultured cell-cell re-associations and trying to recapitulate the successive steps of tooth development (Arany et al., 2009; Honda et al., 2008; Hu et al., 2006a; Nakao et al., 2007; Ohazama et al., 2004). Specific questions arising from this research concern the experimental design, and the search for easily available cell sources. The panel of approaches is progressively restricted by two types of biological constraints: those specifically related to tooth functionality and those related to more general biological aspects such as the maintenance of the cell heterogeneity in the dental mesenchyme and in the periodontium, keeping the gradients of cell differentiation and their 3D-geometry, or the maintenance of the cell kinetic parameters to ensure the differential cusp timing and growth. These complementary points will be discussed in light of parallel approaches being developed by different groups.

## 2. Abbreviations

BMSCs: bone marrow stem cells, CGHT: chondroitin-6-sulfate/gelatin/hyaluronate tri-copolymer, CL: cervical loop, DM: dental mesenchyme, DEJ: dentin-enamel junction, DFPCs: dental follicle precursor cells, DP; dental pulp, DPSCs: dental pulp stem cells, DRG: dorsal root

* Sabine Kuchler-Bopp and Laetitia Keller contributed equally to this work

ganglia, ECM: extracellular matrix, ED: embryonic day, FGF: fibroblast growth factor, GFP: green fluorescent protein, HA/TCP: hydroxyapatite/tricalcium phosphate, IDE: inner dental epithelium, iDPSCs: induced dental pulp stem cells, iPSCs: induced pluripotent stem cells, MSCs: mesenchymal stem cells, NF-PLLA: nanofibrous-poly-Llactic acid, ODE: outer dental epithelium, PDLSCs: periodontal ligament stem cells, PEK: primary enamel knot, PGA: polyglycolic acid, PLGA: poly-L-lactate-co-glycolate, PLLA: poly-L-lactate, SBP: stromal bone producing cells, SCAPs: stem cells from the apical papilla, Shh; Sonic hedgehog, SI: stratum intermedium, SHEDs: stem cells from human exfoliated deciduous teeth, SPCs side population cells, SR: stellate reticulum, SMA: smooth muscle actin, β-TCP: β-tricalcium phosphate.

## 3. Engineering a whole tooth using embryonic dental cells

Embryonic dental cells are commonly used as they allow a full tooth development from cell-cell re-associations (Honda et al., 2008; Hu et al., 2005; Nakao et al., 2007). A two stages methodology has been developed, where cell-cell re-associations are cultured *in vitro* before implantation in adult mice (Fig. 1; Hu et al., 2006a; Nakao et al., 2007). The *in vitro* culture allowed epithelial histogenesis, initiation of crown morphogenesis and odontoblasts to become post-mitotic (Fig. 2A, C, D). Ameloblast functional differentiation as well as root and periodontal tissue development started during the implantation period (Fig. 2B, E). This two steps method is quite well established and has been used with either dental or non-dental cell sources (Arany et al., 2009; Honda et al., 2007; Nakao et al., 2007; Ohazama et al., 2004).

### 3.1 Epithelial histogenesis and crown morphogenesis

During tooth development, the primary enamel knot (PEK), a morphogenetic center controlling crown morphogenesis, transiently forms in the enamel organ (Jernvall & Thesleff, 2000). The histogenesis and functionalization of this structure in cell-cell re-associations is thus an essential step to be achieved. Using embryonic dental cells, the PEK reproducibly formed during early stages of the re-association *in vitro* (Fig. 2C) and was characterized by the condensation of non dividing cells in the epithelium, their expression of Shh and a local concentration of apoptosis (Hu et al., 2005). Four days after this PEK specification, cusps developed, showing that this structure was functional (Fig. 2A). However, up to now, this could not be achieved when using bone marrow cells instead of embryonic dental mesenchymal cells (Nait Lechguer et al., 2009). It was shown that the mesenchyme specifies epithelial histogenesis (Lesot & Brook, 2009; Nait Lechguer et al., 2009) and that the number of mesenchymal cells used for re-association is a critical parameter in determining the number of cusps (Hu et al., 2006a). In contrast with bone marrow cells (Ohazama et al., 2004), the potentialities of the embryonic dental mesenchymal cells to stimulate tooth formation are lost *in vitro* (Keller et al., 2011). Attempts were made to recover it by pre-culturing these cells in the presence of FGF2 (Arany et al., 2009; Keller et al., 2011). Like FGF2, the use of FGF1 or FGF8 or their combination did not allow tooth formation (data not shown). In cell-cell re-associations, the potentialities of cells to engage in tooth development seem to be revealed during the first few hours of co-culture, several days before it became histologically visible. There is no molecular marker to anticipate the fate of cells in re-associations and to predict whether they can engage in tooth formation. On the other hand, there is no published example where cell-cell re-associations could engage in early stage of tooth re-formation and stop thereafter, as observed in several mutants where the bud to cap transition is critical (Jernvall & Thesleff, 2000; Peters & Balling, 1999).

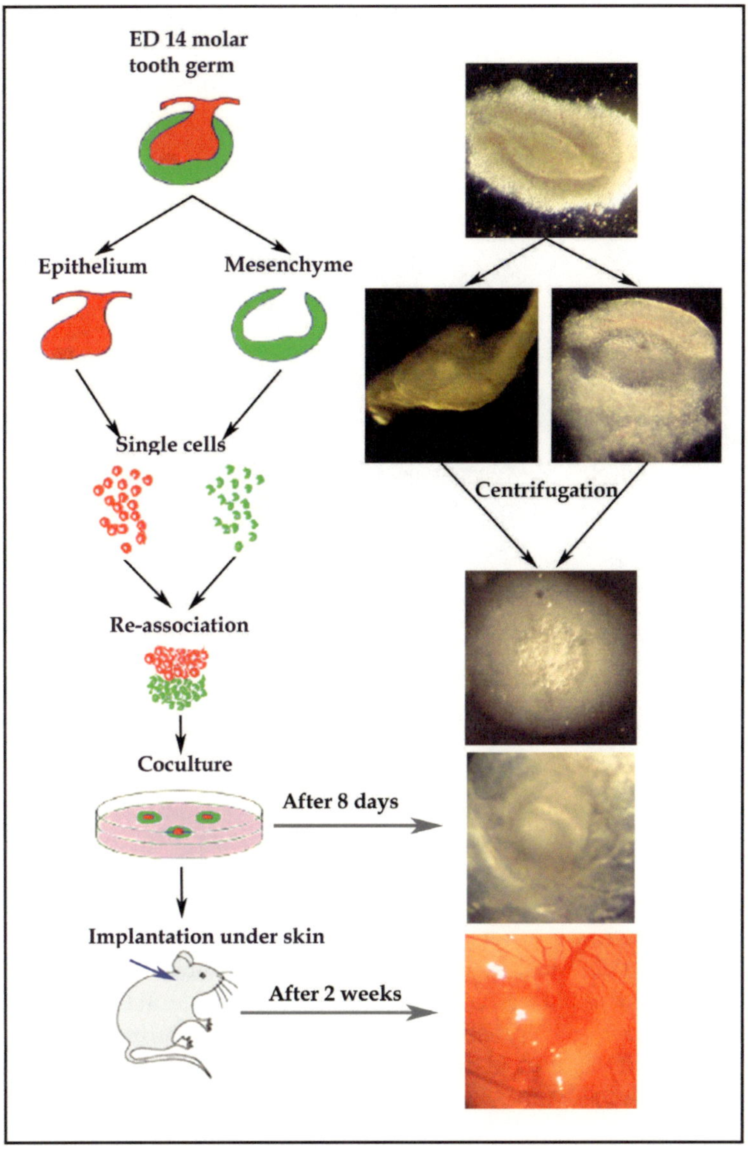


Fig. 1. Schematic representation of the experimental procedures.
The mandibular first molars were dissected from ICR embryos at embryonic day (ED) 14. All procedures with animals were in compliance with the recommendations of the European Economic Community (86/609/CEE) on the use and care of laboratory animals. The dental epithelium and mesenchyme were dissociated by using 1% trypsin in DMEM-F12 at 4°C for 30 min. After separation, the dental epitheliums and mesenchymes were dissociated into cells by centrifugation through 70 µm nylon filters. Epithelial and mesenchymal cells were then pelleted by centrifugation at 9000 g., fragments of each pellet were re-associated and cultured for 8 days. After 8 days *in vitro*, the re-associations were implanted between skin and muscles behind the ears in ICR or GFP adult mice. The implantations were maintained for up to two weeks

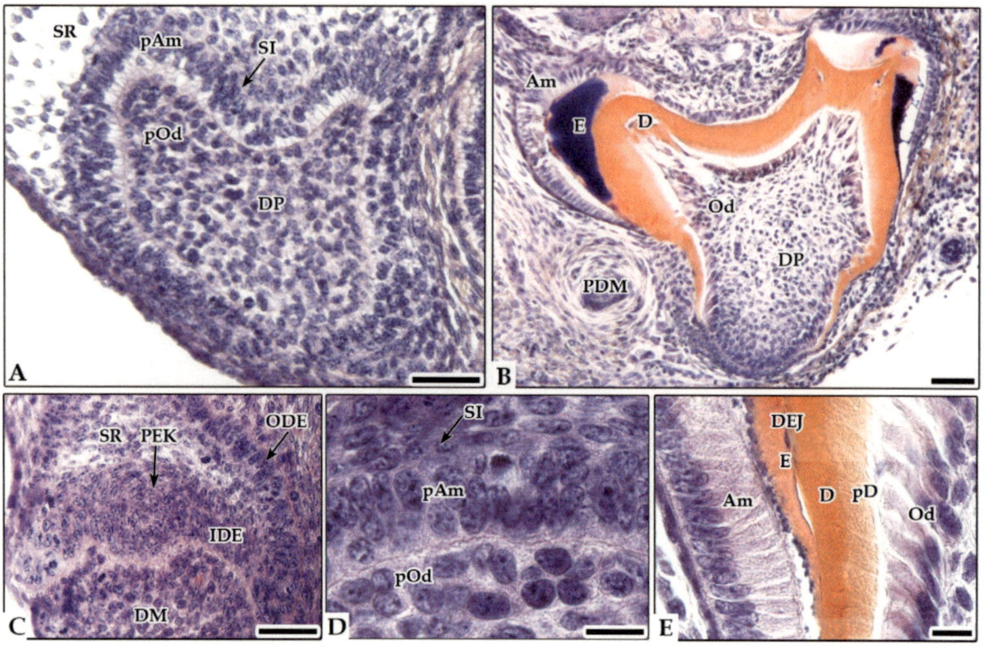


Fig. 2. Histology of dental epithelial and mesenchymal cell-cell re-associations cultured for 4 days (C), 8 days (A, D) and implanted for 2 weeks under skin (B, E).
For histology, samples were fixed in Bouin-Hollande, embedded in paraffin, and 5 µm serial sections were stained with Mallory. Implants were demineralized in 4.3% EDTA before their inclusion in paraffin. At four days in culture, the PEK was visible (C) and cusps developed after 8 days (A). After 8 days in culture, preodontoblasts and postmitotic odontoblasts are facing preameloblasts (D). After implantation (B, E) ameloblasts became functional, gradient of differentiation and mineralization of the dental matrices were achieved. Am: ameloblasts, CL: cervical loop, D: dentin, DEJ: dentin-enamel junction, DM: dental mesenchyme, DP: dental pulp, E: enamel, IDE: inner dental epithelium, ODE: outer dental epithelium, Od: odontoblasts, pAm: preameloblasts, pD: predentin, PDM: peridental mesenchyme, PEK:primary enamel knot, pOd: preodontoblasts, SI: stratum intermedium, SR: stellate reticulum. Bars= 10µm (D, E) and 40 µm (A, B, C)

The *in vitro* culture of embryonic dental cell-cell re-associations allowed the rapid deposition of a basement membrane at the epithelial-mesenchymal junction. A complete epithelial histogenesis could be achieved within 4 to 6 days, including the inner and outer dental epitheliums, the stellate reticulum and, later, the stratum intermedium (Fig. 2A, C; Hu et al., 2005). This stratum intermedium plays an essential role in the functional differentiation of ameloblasts (Lesot & Brook, 2009; Nait Lechguer et al., 2011; Yoshida et al., 2010). Only in one report, non polarized ameloblasts-like cells expressing amelogenin appeared to be generated in the absence of a stratum intermedium (Honda & Hata, 2010). However, it is not known whether these cells, after longer implantation periods might indeed lead to the formation of organized enamel crystals.
However, it remains to be determined whether and how far such potentialities can be maintained at later embryonic stages and also in the adult (Keller et al., 2011). Keratinization, leading to cyst formation, is indeed a major problem arising when culturing

cell-cell re-associations, or even after implantation (Arany et al., 2009; Honda et al., 2008; Komine et al., 2007). Such cysts may then mechanically interfere with tooth morphogenesis. It is not clear yet, whether embryonic dental epithelial cells can be expanded *in vitro*, without loosing their potentialities, as observed with mesenchymal cells (Keller et al., 2011).

### 3.2 Functional cell differentiation

During tooth development, the differentiation of odontoblasts and ameloblasts occurs in a progressive way. Gradients in the cytological and functional differentiation of the two cell types warrant the adhesion of enamel to dentin. It also allows the further growth of the tooth from the late bell stage, while it is already impaired at the tip of the cusps by the matrix accumulated there. In the absence of interposed scaffold, the exchanges between the epithelial and ecto-mesenchymal compartments can occur directly and the planar signaling within the preodontoblast-odontoblast cell layer can have a direct echo on the preameloblast-ameloblast layer (Ruch et al., 1995). It is not known how far this is transmitted to the stratum intermedium. In the absence of scaffold, gradients of differentiation and the progressive polarized deposition and mineralization of the dental matrices could be reproduced in cultured and implanted cell-cell re-associations (Fig. 2B, E). Recently, the organization of these mineralized matrices has been investigated (Honda et al., 2008; Nait Lechguer et al., 2011). Tooth-specific constraints include the necessity to get a correct mineralization of the matrices (dentin, enamel and cementum), which itself requires a good vascularization of the reconstructed organ (Fig. 4A-D; Nait Lechguer et al., 2008, 2011). Ultimately, the quality of the dentin-enamel junction (DEJ) and the orientation of the crystals in each matrix will determine the functionality of the tooth (Katz et al., 2007). The DEJ has been viewed as a connector between two matrices with distinct mechanical properties. The quality of this DEJ, with its own mechanical and bonding specificities, is thus a critical parameter to be preserved in tooth engineering. The risk of interfering with the integrity of the DEJ might then represents a major limitation in the use of scaffolds designed to shape engineered teeth (Xu et al., 2008).
Depending on the region of the dental mesenchyme during tooth development, local specificities were observed, which supported differential timings in cusp formation, their individual kinetic of development, and their final different sizes (Lesot et al., 1999). These were suggested to result from regional variations in cell proliferation kinetics, thus requiring sophisticated mechanisms to coordinate these events in the ecto-mesenchyme and neighbor inner dental epithelium (Lesot et al., 2001; Lisi et al., 2003). These parameters have not been investigated in the case of cultured and implanted cell-cell re-associations, although they would determine the crown shape. To avoid probably very difficult questions, it was suggested to put efforts mainly on root engineering allowing then the implantation of an artificial crown (Volponi et al., 2010).

### 3.3 Root formation

More and more, the root appears as the key point in tooth engineering because its functionality is related to tooth anchoring as well as in the mediation of tooth movement in the jaw. For this reason, it is important to avoid ankylosis. As determining tooth functionality, the development of root in cell-cell re-associations has been investigated as well (Honda et al., 2010; Hu et al., 2006a; Nait Lechguer et al., 2011).
Using embryonic dental cells, root formation could be initiated, but only after implantation of the cultured cell-cell re-associations (Fig. 3A, B; Hu et al., 2006a). In these conditions, the

periodontium formed and elongated cells, similar to periodontal ligament fibers, extended from the root dentin to newly formed bone (Fig. 3C; Hu et al., 2006a). During the two weeks of implantation, the periodontium remained as a non-mineralized zone in between the dentin/cementum (Fig. 3D, E) and newly formed bone (Nait Lechguer et al., 2011). Longer periods of implantation will have to be tested in order to determine whether the absence of mineralization of this area can be preserved. The maintenance of homeostasis after long term implantation will also have to be investigated since the periodontal ligament is a rather complex tissue (for review see Nanci & Bosshardt, 2006).

Attempts to regenerate the periodontium were also performed using postnatal stem cells from the periodontal ligament (Table 3; Seo et al., 2004). However, this potential cell source only represents a few cells. Other approaches have been developed, using bone marrow stem cells (Kawaguchi et al., 2004), or muscle derived-stem cells (Yang et al., 2010). The role of blood vessels and the participation of progenitor/pericytic cells in periodontium development and regeneration have been extensively reviewed (Ripamonti & Petit, 2009). However, with these cells, the goal was very focused and no attempt was made to test how far, in a different context, they could give rise to other dental tissues. Such a hypothesis has been formulated recently (Rothova et al., 2011).

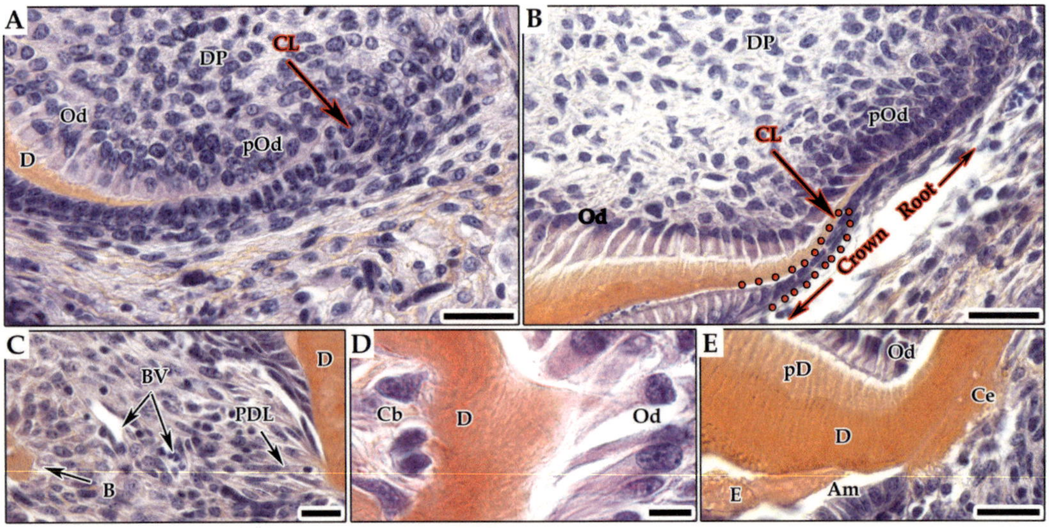


Fig. 3. Histology of root formation in cell-cell re-associations cultured for 8 days and implanted for 1 week (A) and 2 weeks under skin (B, C, D, E).
Between one week (A) and two weeks (B) of implantation, the root was forming. Cementoblasts (D) secreted cementum (E) and the peridontium (C) remained as a non-mineralized zone in between the dentin/cementum and newly formed bone. Am: ameloblasts, B: bone, BV: blood vessel, Cb: cementoblasts, Ce: cementum, CL: cervical loop, D: dentin, DP: dental pulp, E: enamel Od: odontoblasts, pOd: preodontoblasts, pD: predentin, PDL: periodontal ligament. Bars=10μm, (D) 20 μm (C, D) and 30 μm (A, B)

### 3.4 Organ vascularization

The primary role of vascularization is related to oxygenation and metabolic exchanges, thus allowing cell survival. Since it cannot be functional *in vitro*, culturing large sized samples may be rapidly impaired, so that the implantation of the re-associations is absolutely necessary. The interstitial flow, generated by vascularization, may further mediate the extracellular transport of molecules, including growth factors, thus coupling mechanical and chemical signaling (Griffith & Schwartz, 2006). The requirement of vascularization for dental matrices mineralization, explains its specific patterning in time and space during tooth development, including in the enamel organ (Fig. 4C; Manzke et al., 2005; Nait Lechguer et al., 2008). Cultured dental embryonic cell-cell re-associations could be vascularized after they were implanted under the skin or in the jaw of adult mice (Fig. 4A-E; Nait Lechguer et al., 2008, 2011; Nakao et al., 2007). It has been suggested that dental pulp stem cells might give rise to functional blood vessels (Cordeiro et al., 2008). However, cultured embryonic dental cell-cell re-associations from ICR mouse embryos have been implanted in GFP mice, which showed that all newly formed blood vessels entering the dental mesenchyme and the enamel organ, originated from the host (Nait Lechguer et al., 2008).

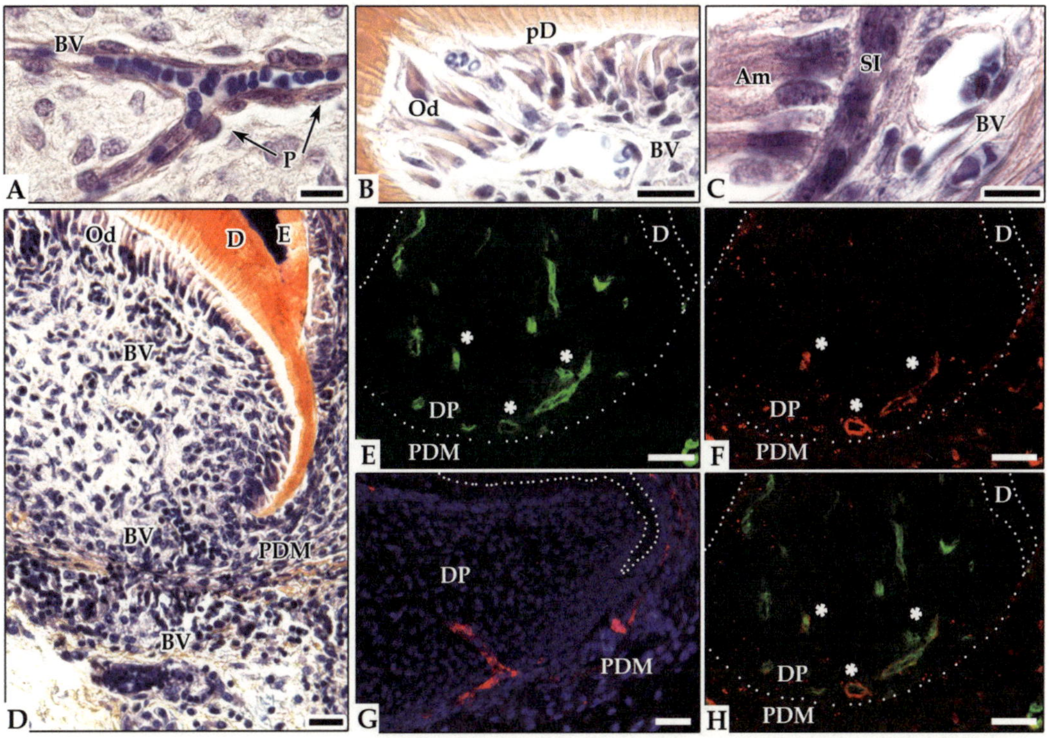


Fig. 4. Vascularization of subcutaneously implanted cultured epithelial/mesenchymal cell-cell re-associations in adult ICR mice (A-H).
Epithelial/mesenchymal cell-cell re-associations were cultured for 8 days *in vitro* prior to implantation for 2 weeks under skin. Histology showed that blood vessels can enter in the dental pulp (D), migrate in the pulp (A), and reach odontoblasts (B). In the enamel organ, blood vessels were in close contact with the stratum intermedium (C). Blood vessels were localized by immunofluorescent staining for CD31 (E, H) and pericytes were detected by immunofluorescence staining for αSMA (G, F, H). Cells detected by αSMA coming from the PDM could reach and enter in the pulp (G). All blood vessels detected by CD31 were not positive for αSMA (H). Implants were embedded in Tissue-Tek (Agar Scientific, Saclay,

France) and frozen at -30°C. Immunostaining was performed on serial frozen sections (7 µm) with rat monoclonal anti-mouse CD31 (1/100) (BD Pharmingen, Evry, France) and polyclonal rabbit anti-human αSMA (1/100) (Abcam, Cambridge, MA, USA). Sections were incubated for 2hrs at room temperature with the primary antibodies and for 1hr with donkey anti-rat secondary antibodies conjugated to Alexa 488 (1/200) and chicken anti-rabbit conjugated to Alexa 594 (1/500) (Molecular Probes, Invitrogen SARL, Cergy Pontoise, France). Am: ameloblasts, BV: blood vessel, D: dentin, DP: dental pulp, E: enamel, Od: odontoblasts, P: pericyte, pD: predentin, PDM: periodontal mesenchyme, SI: stratum intermedium. Bars=10µm, (A, C) 20 µm (B, G) and 60 µm (D, E, F, H)

Vascularization may play a further role, at least in the mesenchymal compartment. Cells with a mesodermal origin participate in the dental pulp formation including the blood vessels, which develop there (Chai et al., 2000; Rothova et al., 2011). Some of the mesenchymal stem cells present in the dental pulp might originate from perycites (Feng et al., 2010; Lovschall et al., 2007). Since the perivascular environment may play an essential role as it remains a potential source of stem cells possibly involved in reparative processes (Shi & Gronthos, 2003), this context could explain the results published by Arany and co-workers (2009). These authors could obtain the formation of a tooth when re-associating a clonal cell line derived from a dental mesenchyme with a competent enamel organ and implanting these re-associations. As far as the implants can be vascularized, external cells might thus participate in the re-formation of the dental mesenchyme and allow restoring the complex cellular heterogeneity of this tissue, which hardly could be expected to originate from a clonal cell line (Keller et al., 2011). Pericytes in implanted cell-cell re-association were visualized after immunostaining for a smooth muscle actin (αSMA) (Nehls & Drenckhahn, 1991) and detected both in the apical part of the dental pulp (Fig. 4G) and peridental mesenchyme (Fig. 4F, H). All blood vessels detected by CD31 were not positive for αSMA (Fig. 4H). The presence of stem cells associated with the blood vessels during the early stages of vascularization of implanted cell-cell re-associations will have to be investigated to better understand the data from Arany et al. (2009).

### 3.5 Innervation

Besides the mediation of pain sensation, tooth innervation is involved in several complementary aspects such as the recruitment of immunocompetent cells (Byers et al., 2003; Fried et al., 2000), the control of odontoblast function (Fristad et al., 1999) and the prevention of tooth ankylosis. Being necessary for tooth functionality, the innervation remains as a major challenge in tooth engineering. The innervation of the tooth is initiated relatively late during development and involves sensory and sympathetic nerves. Complex molecular mechanisms regulate the timing and pattern of tooth innervation (Fried et al., 2007; Kettunen et al., 2005; Luukko et al., 2008). It has been suggested that the epithelium instructs and controls dental axons navigation in the dental mesenchyme (Luukko et al., 2008).

Attempts to implant extracted or engineered teeth in the jaw of adult mice could be achieved (Ferreira et al., 2007; Honda et al., 2006; Nait Lechguer et al., 2008; Nakao et al., 2007). However, it always remained quite far from nerves. Still a correct positioning of the implanted cell-cell re-associations in the jaw cannot be controlled in the mouse or rat, where they are relatively small. It is thus impossible to specifically position the prospective root-forming part of the implant so that it is oriented towards the alveolar nerve (Honda et al., 2006; Nait Lechguer et al., 2008). In parallel, preliminary experiments have been performed to test the possibility to obtain the innervation of cultured cell-cell re-associations after implantation under the skin. For this purpose, embryonic dental cell-cell re-associations were implanted together with DRG (Fig. 5A). After 2 weeks of implantation, immunostaining for peripherin showed the axonal growth in the peridental tissues (Fig. 5C, D), but not in the dental mesenchyme (Fig. 5B, C, D). The impossibility to innervate the dental mesenchyme might be related to the embryonic stage (ED14) that was used to prepare dental cells. Indeed at ED14, innervation remains inhibited in physiological conditions, and only becomes possible postnatally (Luukko et al., 2008). The expression of semaphorins, which regulate innervation, will have to be investigated in implanted cell-cell re-associations.

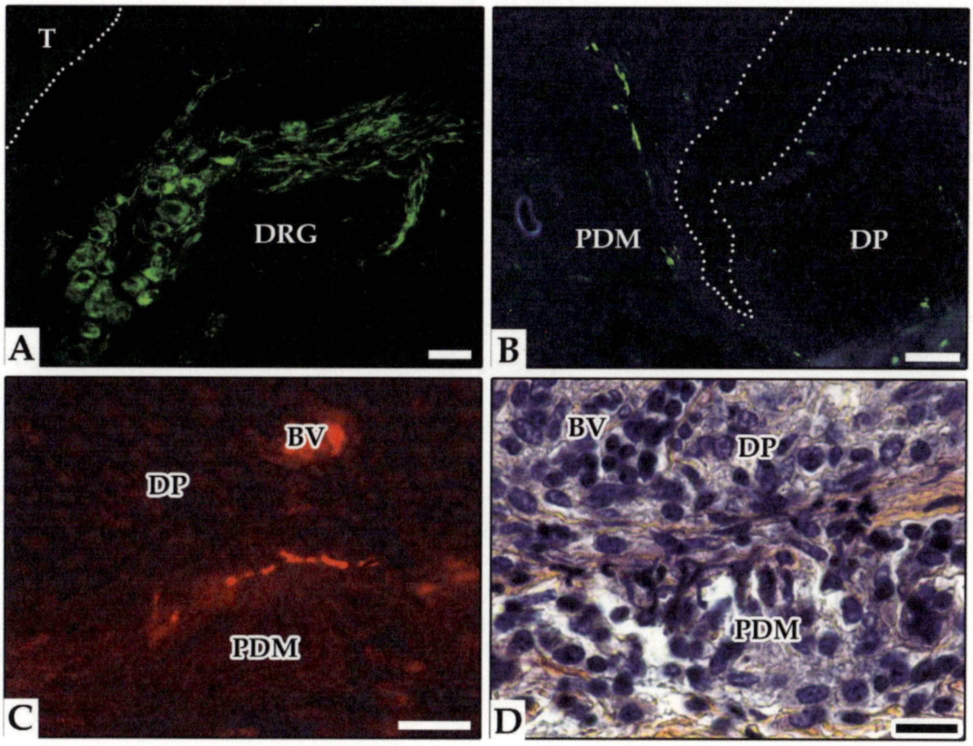


Fig. 5. Innervation of subcutaneously implanted cultured epithelial/mesenchymal cell re-associations with DRG in adult ICR mice (A-D).
Epithelial-mesenchymal cell-cell re-associations were cultured for 8 days *in vitro* prior to co-implantation with DRG from ED12. After 2 weeks of implantation under the skin (A), axons and DRG were detected by immunofluorescent staining for peripherin. Axons growing from DRG in contact with the PDM (A, B) were detected at the limit between PDM and DP (C, D). However, they never entered the DP (B, C). Implants were fixed with 4% paraformaldehyde and embedded in paraffin. Immunostaining was performed on serial paraffin sections (7 μm) with rabbit polyclonal anti-human peripherin (1/100) (Abcam, Cambridge, MA, USA). Sections were incubated for 2hrs at room temperature with the primary antibody and for 1hr with chicken anti-rabbit conjugated to Alexa 594 (1/500) and to Alexa 488 (1/200) (Molecular Probes, Invitrogen SARL, Cergy Pontoise, France). BV: blood vessel, DP: dental pulp, DRG: dorsal root ganglia, PDM: periodontal mesenchyme, T: tooth. Bars= 20 μm (C, D) and 40 μm (A, B)

## 4. Using matrix or artificial polymers instead of cells only

A tricky point in the experimental designs being developed in different laboratories is whether to use scaffolds or not. Several scaffolds with different composition and complementary purposes have been used for tooth engineering (Fig. 6).

### 4.1 Pre-shaped scaffolds

In some instances, scaffolds of different natures have been designed to try to "pre-shape" the tooth to be engineered (Fig. 6; Kim et al., 2010; Young et al., 2005). Despite their potential interest to direct crown morphology, scaffolds also have their own limitations. Indeed, they may impair correct cell-cell interactions, and thus interfere both with the general histogenesis and with epithelial-mesenchymal interactions. This is critical since these interactions control all aspects of tooth development: epithelial histogenesis, crown morphogenesis, and cell differentiation, as well as root development (Jernvall & Thesleff, 2000; Ruch et al., 1982; Slavkin et al., 1984).

The question of tissue interfaces junction has been raised with other models such as the complex ligament-, tendon- or cartilage-bone interfaces. For this specific purpose, attempts are made to develop stratified scaffolds (Lu et al., 2010). Similar challenges exist when trying to restore dentin/enamel, dentin/cementum, or cementum/periodontal fibers interfaces and maintain their specific mechanical properties. During tooth development, tissues interfaces usually show multi-step formation where extensive matrix remodeling takes place (Bosshardt et al., 2005; Ho et al., 2009; Imbeni et al., 2005). In all these interfaces, there is a tight structure-function relationship. In case of multiple bone interfaces, these characteristics were taken into account to design biomimetic stratified scaffolds (Lu et al., 2010). In this specific case however, scaffolds were not designed to try shaping a tissue, but to set up a 3D spatial organization of cells activities. The ultimate goal was to reproduce physiological mechanical properties by patterning matrix heterogeneity. Similarly for periodontium engineering, attempts have been made to design hybrid scaffolds allowing the maintenance of the ligamentous fibers spatial organization (Park et al., 2010). According to these authors, the design will have to be further developed by introducing mechanical forces in order to better adjust the spatial organization of fibers.

Depending on their structure and surface organization, scaffolds may also interfere with horizontal signaling within the odontoblast and/or ameloblast layer(s). The main problem would be related to the odontoblast layer, since the extent of the gradients which they may form will have direct consequences on the patterning of ameloblast differentiation and, ultimately, on enamel organization (Nait Lechguer et al., 2011). The morphology of the epithelial-mesenchymal junction, just before ameloblast differentiate, is thought to represent the physiological scaffold and template for the final crown shape (Skinner & Gunz, 2010). During tooth development, the different cusps show different specific kinetics of cell differentiation (Lisi et al., 2003). Still very little is known about the mechanisms involved in it despite active research to address this question (Salazar-Ciudad & Jernvall, 2004, 2010).

Obviously, the design of a scaffold is strongly related to the tissue to be engineered (Chan & Leong, 2008) and to the purpose of its use: either for organ shaping (Kim et al., 2010; Young et al., 2005), or, more simply to maintain high cell density and cohesion as discussed below (Arany et al., 2009; Nakao et al., 2007).

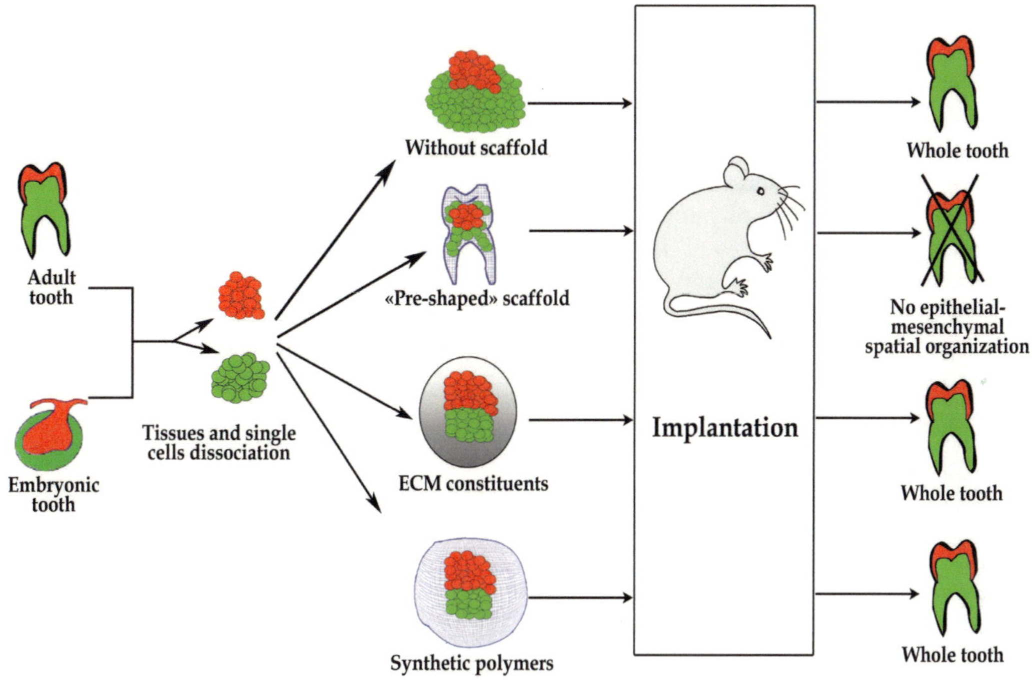


Fig. 6. Schematic representation of experimental procedures developed for tooth organ engineering

### 4.2 Extracellular matrix constituents

For tooth engineering, extracellular matrix (ECM) constituents (Fig. 6) were not used to shape the forming organ but either to maintain cells together and work with high cell density or to retain signaling molecules involved in stimulating morphogenesis and/or cell differentiation (Griffith & Schwartz, 2006; Honda et al., 2005; Huang et al., 2010; Nakao et al., 2007; Sumita et al., 2009). The level of signaling in the mesenchyme was recognized as a potentially limiting factor in tooth engineering (Hu et al., 2006a; Nait Lechguer et al., 2009). The matrix may be involved in signaling either directly through cell surface receptors, or indirectly because of its ability to differentially bind diffusible growth factors, and also because it contains enzymes modulating or tuning the two other ways (Discher et al., 2009; Reilly & Engler, 2010; Rhodes & Simons, 2007). These different molecules may explain why and how the ECM can interfere with distinct cellular activities such as the adhesion, migration, proliferation, survival and differentiation. For this reason also, it is difficult to design the composition of the ECM scaffold. Beside matrix stiffness, which may interfere with cell migration and thus with morphogenesis, Griffith and Schwartz (2006) took into account the formation of molecular gradients, one parameter, which may be very important for tooth formation. The main difficulty in using ECM constituents is related to the physiological complexity of the 3D environment, as it exists in a tissue. However, cells can rapidly modify and re-adjust this environment as far as ECM constituents can initially be populated by the different cell types present in a tissue, so that a physiological homeostasis and histogenesis can then be restored (Keller et al., 2011). ECM scaffold remodeling is thus an important parameter to be investigated when the organ formation progresses (Badylak et al., 2009).

Different matrix constituents have been used for tooth engineering (Table 1). However, collagen, under different forms, was by far the most frequently used (Honda et al., 2007; Nakao et al., 2007; Sumita et al., 2006). Collagen gel has been successfully used for dental cells encapsulation, as obviously neither interfering with the epithelial histogenesis nor with the patterning of cell differentiation (Arany et al., 2009; Honda et al., 2008; Nakao et al., 2007). It would be interesting to analyze the behavior of epithelial cells in such an environment, which they physiologically never meet.
Comparing the potentialities of different scaffolds, possible inhibitory effects of collagen on hard tissue formation have been reported (Ohara et al., 2010). However, other studies ended up with tooth formation (Arany et al., 2009; Nakao et al., 2007). Although their degree of mineralization was not measured, dentin and enamel were secreted in these two independent studies.

| **ECM constituents** | | |
|---|---|---|
| **Collagen gel** | Whole tooth | *Komine et al., 2007*<br>*Nakao et al., 2007*<br>*Arany et al., 2009* |
| | Dentin and enamel-like structures | *Ohara et al., 2010* |
| **Collagen sponge** | Whole tooth | *Sumita et al., 2006* |
| **Fibrin gel** | Dentin and enamel-like structures | *Ohara et al., 2010* |
| **GCHT** | Whole tooth | *Kuo et al., 2011* |

Table 1. Extracellular matrix constituents used for tooth engineering

### 4.3 Synthetic polymers

Several synthetic polymers have been used to embed cells for tooth engineering (Fig. 6), mostly polyglycolic acid (PGA) and poly-lactideco-glycolide (PLGA) (Table 2). They were chosen for the possibility they offer to control their structural and mechanical properties (Hacking & Khademhosseini, 2009). For this reason, they have been widely used for tissue engineering. The possibility to re-create a specific histological organization has been investigated, in the case of lung tissue engineering by growing pluripotent cells seeded onto synthetic polymers, showing that identifiable pulmonary structures could develop (Cortiella et al., 2006). Although PGA is satisfactory to grow cells *in vitro*, this polymer can induce inflammatory reactions after implantation in immunocompetent hosts (Ceonzo et al., 2006; Cortiella et al., 2006).
Different groups have used PGA for tooth engineering (Duailibi et al., 2004, 2008; Iwatsuki et al., 2006; Ohara et al., 2010), but the risk of inflammation was only discussed by Duailibi et al. (2008). Concerning biodegradation, PGA fiber meshes were still present after 10 days

of implantation (Iwatsuki et al., 2006). It will be important to more precisely evaluate what happens at later stages.

As stated above, synthetic polymers might introduce discontinuities in the setting of gradients of cell progression towards differentiation and thus interfere with the geometry of epithelial-mesenchymal interactions. However, Honda et al., (2008) showed convincing images of tooth formation using dental cells seeded in PGA polymers, but coated with collagen.

A comparative study suggested that the potential of PGA scaffold to support tooth formation was not as good as collagen sponges scaffold (Honda et al., 2010). Considering the initial stages of tooth formation, similar conclusions were drawn when comparing collagen or fibrin gels with PGA or ß-tricalcium phosphate (ß-TCP) (Ohara et al., 2010).

| **Synthetic polymers** | | |
|---|---|---|
| **PGA** (Polyglycolate) | Enamel, dentin and pulp tissue<br>Enamel-like structure and dentin | *Duailibi et al., 2004*<br>*Ohara et al., 2010* |
| Collagen coated **PGA** | Ameloblast and dentin tissue | *Honda et al., 2005* |
| | Whole tooth | *Iwatsuki et al., 2006*<br>*Honda et al., 2008* |
| **PLGA** (poly-L-lactate-co-glycolate) | Enamel, dentin and pulp tissue | *Duailibi et al., 2004* |
| | Enamel- and dentin-like tissue | *Duailibi et al., 2008* |
| **PLLA** (poly-L-lactate) | Functional odontoblasts and tubular dentin | *Sakai et al., 2010* |
| **NF-PLLA** (nanofibrous -poly-L-lactic acid) | Odontoblast-like cells | *Wang et al., 2010* |
| **PGA/PLLA** (Polyglycolate/poly-L-lactate) | Enamel- and dentin-like tissue | *Duailibi et al., 2008* |
| **Inorganic scaffolds** | | |
| **β-TCP** (β-tricalcium phosphate) | Odontoblast- like cells | *Ohara et al., 2010* |
| **HA/TCP** (hydroxyapatite/tricalcium phosphate) | Dentin-like matrix and odontoblast -like cells | *Gronthos et al., 2000*<br>*Miura et al., 2003*<br>*Sonoyama et al., 2006* |

Table 2. Synthetic polymers and inorganic scaffolds used for tooth engineering

## 5. Cell sources

Although embryonic cells represent an excellent tool to set up methodologies for tooth engineering, they are not a cell source to be used in a clinical context. One possible alternative would be the use of the stem cells evidenced in numerous adult organs (Bergmann & Steller, 2010). To engineer a complete tooth and be able to reproduce the epithelial-mesenchymal interactions taking place during development, it will be necessary to combine mesenchymal and epithelial adult stem cells with odontogenic potentialities. The identification and selection of these two cell populations/sources is a major challenge.

For the mesenchymal compartment, the use of mesenchymal stem cells (MSCs) certainly appears as first evidence. These cells were initially characterized in the bone marrow (Friedenstein et al., 1970; Pittenger et al., 1999). They appear as a subset of fibroblast-like cells able to form colonies that can be selected by their adherence to plastic and expanded *in vitro*. As classical stem cells, they are capable of self-renewal and of displaying multilineage differentiation ability *in vitro*, not only towards the classical "triumvirate", osteoblasts, adipocytes and chondrocytes, but also towards the endothelial, cardiomyocytes or neural fate depending on *in vitro* induction (for review see Nombela-Arrieta et al., 2011). This multipotentiality together with their easiness of *in vitro* expansion has made these cells promising in regenerative medicine, mainly dealing with bone defects, cardiac repair, and MSCs immunosuppressive properties (for review see Augello et al., 2010; Chanda et al., 2010). Bone marrow stem cells (BMSCs) have also been tested for tooth enginering (Hu et al., 2006b; Nait Lechguer et al., 2009; Ohazama et al., 2004). They could substitute the dental mesenchyme, and engage in tooth formation (Ohazama et al., 2004). Furthermore, they could also be reprogrammed *in situ* to engage in ameloblast functional differentiation (Hu et al., 2006b).

Since their first description, MSC-like cells have been reported in many embryonic and adult connective tissues (da Silva Meirelles et al., 2006), including some of the adult tooth: dental pulp of permanent (Gronthos et al., 2000) or deciduous teeth (Miura et al., 2003), periodontal ligament (Seo et al., 2004), apical papilla (Sonoyama et al., 2008) or dental follicle (Casagrande et al., 2011; Morsczeck et al., 2005). Among dental MSC-like cells, the dental pulp stem cells (DPSCs) are the most easily accessible. They have already demonstrated their plasticity and capacity to participate in tissue repair (for review see Huang, 2009). DPSCs have been described in human (Gronthos et al., 2000; Miura et al., 2003; Sonoyama et al., 2008), rat (Yang et al., 2007) and mouse (Balic et al., 2010). Most dental mesenchymal cells are believed to derive from the cranial neural crest (Chai et al., 2000; Lumsden, 1984), DPSCs appear as candidates of choice to substitute the mesenchymal component of the tooth. Moreover, they might have retained some type of "memory" of their tissue of origin and preferentially differentiate towards an odontogenic program (Batouli et al., 2003; Gronthos et al., 2000 and Fig. 4 therein).

The stem cells of the dental pulp (Table 3, SHEDs, DPSCs or SCAPs) constitute *per se* an heterogenous cell population which, depending on the isolation, selection and culture procedures may display different features and degrees of stemness (Table 3, Bakopoulou et al., 2011; Huang et al., 2006). It is remarkable that although the majority of studies involved cells isolated by enzymatic dissociation of the pulpal tissue, the most immature dental pulp stem cells were obtained from explants culture (Siqueira da Fonseca et al., 2009).

Determining which type of DPSCs is the best suited for tooth engineering is therefore an essential question. The purification of a homogeneous DPSCs population is not possible due to a lack of known specific markers. Pulp stem cells are thus enriched on their ability to form colonies (Gronthos et al., 2000), their expression of cell surface molecules such as STRO-1 (Gronthos et al., 2000), or CD34 (Laino et al., 2006) or on their capacity of exclusion of a DNA binding dye (Iohara et al., 2006). Not surprisingly, the resulting different cell populations display various differentiation potentials, upon appropriate *in vitro* induction (for review see Huang, 2009) and are able to participate in bone (d'Aquino et al., 2009; Seo et al., 2008), neuronal (Arthur et al., 2008) and corneal tissue formation *in vivo* (Table 3; Gomes et al., 2010). In an implanted tooth slice model, heterogeneous populations of DPSCs, SCAPs or SHEDs have already shown their capacity to participate in the formation of a pulp-like tissue with dentin secretion and apposition (Huang et al., 2010; Sakai et al., 2010). However, the potential of these different populations to participate in whole tooth formation has not yet been investigated. For this purpose, it will be necessary to set up models for preclinical evaluations and to standardize the procedures. A major drawback in this field lies in the fact that most of these pulp stem cell sub-populations are described in the human tooth, while experimental strategies to test their competence to participate in tooth formation have been set up with murine models (Arany et al., 2009; Hu et al., 2006a; Nakao et al., 2007; Ohazama et al., 2004).

A hint that post-natal dental pulp cells may maintain such a potential, at least to a certain extent, comes from the work of Masaki Honda and coworkers. These authors have shown that unselected epithelial and mesenchymal cells derived from porcine third molar (6 months after birth), are able to participate in the formation of a tooth-like tissue (Honda et al., 2008). The precise nature of these cells and their possible relationship with dental pulp stem cells will have to be investigated. It will also have to be determined why, in these experimental conditions, tooth formation was not impaired by mechanisms of cell sorting out (Manning et al., 2010; Steinberg, 1962, 1963).

Tooth engineering also requires epithelial stem cells. Dental epithelial stem cells have been described in the continuously growing mouse incisor, which is a very peculiar tooth (Harada et al., 1999). However, all epithelial cells have disappeared in the erupted molar, except for the very restricted population of Malassez rests, when it exists (Huang et al., 2009; Shinmura et al., 2008). Therefore, the search for non-dental epithelial cell sources is ongoing.

The palatal mucosal epithelium isolated from embryos or newborn mice has been shown to be able to substitute for a dental epithelium, and to contribute to tooth formation in a re-association model using embryonic dental mesenchyme. Cultured dissociated palatal epithelial cells can maintain this potential up to one month post-natally, provided they are cultivated as cell sheets to maintain epithelial cell properties (Nakagawa et al., 2009).

Along the same line, young human skin keratinocyte progenitors, cultured in cell sheets, can contribute to tooth-crown like structure formation after chimeric re-association with ED 13.5 mouse molar mesenchyme (Wang et al., 2010). However, in these conditions, ameloblasts do not differentiate. Their differentiation needed to be induced by exogenous FGF8 (Wang et al., 2010).

In the ongoing search for cell sources, an interesting alternative resides in the induced pluripotent stem cells (iPSCs). Since the pioneer work of Takahashi and Yamanaka (2006), it has been demonstrated that mouse adult somatic cells can be reprogrammed to an embryonic-like state, by the introduction of 4 transcription factors, Oct4, Sox2, c-Myc and Klf4 or Lin28. These cells exhibit the morphology and growth properties of murine embryonic stem cells and appear capable of contributing to virtually every cell type in the body (Maherali et al., 2007). Their human counterparts have subsequently been generated (Lowry et al., 2008; Takahashi et al., 2007). Emerging data suggest that iPSCs retain a memory of their tissue of origin (Quatrocelli et al., 2011; Tian et al., 2011) and therefore could differentiate more easily towards their original program (Hu et al., 2010). Human iPSCs have already been generated from dental pulp stem cells (SHEDs, DPSCs and SCAPs) (Duan et al., 2011; Tamaoki et al., 2010) and different types of epithelia (Buchholtz et al., 2009; Novak et al., 2010). The potential of such cells will have to be evaluated when grown in re-association, to determine whether they are indeed valuable candidates for whole tooth engineering.

Whatever the chosen cells, further studies to improve the success rate in tooth formation should focus primarily on the enrichment/purification, of competent cells. This will require the finding of markers permitting this enrichment as well as specific culture protocols allowing their expansion and the maintenance of their differentiation potentials *in vitro*.

## 6. Maintenance of cell potentialities *in vitro*

Adult stem cells are known to reside in defined anatomical compartments specified by cellular and acellular components. This specialized microenvironment is known as a niche and was first described as a concept in the context of hematopoiesis (Schofield, 1978). The existence of niches has then been validated in many other tissues (Weissman, 2000). It integrates systemic and local cues (growth factors, ECM as well as specific cell types) to maintain a balance between self-renewal and differentiation. In the absence of stimuli, adult stem cells remain quiescent but signals generated by inflammation, infection or lesion will lead to their recruitment/activation towards differentiation and repair.

In contrast to the niche of the intestine, hair or blood stem cells whose understanding has progressed over the recent years (for review see Fuchs & Segre, 2000 and Li & Clevers, 2010 and references therein), the dental pulp stem cell niche is still poorly understood. This might be explained by the facts that 1) the pulp is not an actively renewing tissue, 2) there is a lack of specific markers identifying its stem cells and 3) it is technically difficult to work with strongly mineralized tissue. DPSCs are thus essentially defined by their functionalities. DPSCs freshly established in culture, present a phenotype similar to BMSCs (Shi et al., 2001). They express STRO-1 and a series of surface markers, proposed by the International Society for Cell Therapy for the identification of human MSCs (Dominici et al., 2006). However the extent to which these markers reflect the *in vivo* phenotype remains an unresolved question. In normal pulp, the STRO-1 expressing cells are only found in the microvasculature (Shi & Gronthos, 2003), an observation compatible with the hypothesis that DPSCs, as MSCs from other tissues, might correspond to pericytes (Crisan et al., 2008; Feng et al., 2010). Indirect data from BrdU labeling experiments have brought support to this hypothesis (Ishikawa et al., 2010; Téclès et al., 2005). But the fact that DPSCs appear as a heterogeneous cell population raises the question of the existence of other potential stem cell niches in the pulp (Harichane et al., 2011; Mitsiadis et al., 1999, 2003).

| Name | Isolation and selection | *In vitro* differentiation potential | *In vivo* differentiation potential | First description |
|---|---|---|---|---|
| **DPSCs** | Enzymatic dissociation<br>Stro1+ or Colony-forming units/low density inoculation | Odontoblasts<br>Osteoblasts<br>Adipocytes<br>Neurons | Bone-like tissue<br>**Dentin-pulp like complex** | *Gronthos et al., 2000* |
| **SBP/DPSCs** | Enzymatic dissociation<br>CD34+ | Osteoblasts<br>Endotheliocytes<br>Melanocytes | Bone-like tissue<br>Alveolar bone | *Laino et al., 2006* |
| **SPCs** | Enzymatic dissociation<br>Hoecht exclusion | Odontoblasts<br>Chondrocytes<br>Adipocytes<br>Neurons | **Odonto/osteodentinoblasts**<br>**Pulp regeneration** | *Iohara et al., 2006* |
| **iDPSCs** | Outgrowth culture<br>Low density inoculation | Odontoblasts<br>Corneal epithelial cells<br>Muscle cells<br>Osteoblasts<br>Neurons<br>Chondrocytes | Contribution to preterm chimera and to corneal reconstruction | *Kerkis et al., 2006* |
| **SHEDs** | Enzymatic dissociation<br>Colony-forming units/low density inoculation | Odontoblasts<br>Osteoblasts<br>Muscle cells<br>Adipocytes<br>Neurons<br>Endotheliocytes | **Dentin-pulp tissue**<br>**Odontoblast-like cells**<br>Bone | *Miura et al., 2003* |
| **PDLScs** | Enzymatic dissociation<br>Low density inoculation | Cementoblasts<br>Osteoblasts<br>Adipocytes<br>Chondrocytes<br>Neurons | **Cementum-like formation**<br>**PDL-like formation** | *Seo et al., 2004* |
| **DFPCs** | Enzymatic dissociation<br>Low density inoculation | Cementoblasts<br>Osteoblasts<br>Odontoblasts<br>Adipocytes<br>Muscle cells<br>Chondrocytes<br>Neurons | **PDL-like formation**<br>**Cementum matrix** | *Morsczeck et al., 2005* |
| **SCAPs** | Enzymatic dissociation<br>Low density inoculation | Odontoblasts<br>Adipocytes<br>Muscle cells<br>Chondrocytes<br>Neurons | **Dentin-pulp like complex**<br>**Odontoblasts** | *Sonoyama et al., 2008* |

Table 3. Different types of stem cells from the tooth: isolation, culture procedures and differentiation potentials

Working with stem cells requires their growth and expansion *in vitro*. However, it is well known that *in vitro* culture conditions go along with major changes in gene expression. For instance, culture on specific substrates, in different media, serum substitutes, oxygen concentration has been shown to deeply affect the transcriptome/epigenome of ES cells (Allegrucci et al., 2007; Forsyth et al., 2008). Recent studies have further shown that changes in culture conditions may modify the potency of stem cells or reprogram adult stem cells to endow them with a broader differentiation potential (for review see Roobrouck et al., 2011). Similarly, cultured DPSCs have been shown to undergo phenotypic changes, losing markers while acquiring others with subculture (Govindasamy et al., 2010; Patel et al., 2009; Yu et al., 2010). Trying to improve the culture conditions of DPSCs for their use in whole tooth engineering is now a major challenge.

## 7. Conclusion and prospects

Recent studies in tooth engineering led to the analysis of different potential cell sources and to the development of several complementary experimental approaches. The biological constraints, which become better characterized as research progresses, will necessarily restrict the number of protocols and strategies.

Potential human cell sources have been reported (Gronthos et al., 2000; Miura et al., 2003), while banks for autologous cell sources are being developed (Arora et al., 2009; Tamaoki et al., 2010; Woods et al., 2009). The biological potentialities of the different cells types, which recently were identified from human sources, will have to be tested both *in vitro* and *in vivo*. This will allow determining whether they may recapitulate the different developmental stages and lead to the engineering of a functional tooth. The next challenge will be to deal with the engineering of larger teeth, where full vascularization and innervation will become critical.

As far as possible, it may be better to work in the absence of scaffold or matrices to engineer whole teeth (Hu et al., 2005, 2006a; Ohazama et al., 2004). It better fits with the necessity to preserve epithelial-mesenchymal interactions and with the maintenance of gradual cell differentiation. Furthermore, it is important to prevent any risk of delayed immune response, which may occur with collagen (Zippel et al., 2008) or PLGA scaffolds (Tavassol et al., 2010). As a broad question, related to tissue engineering and regeneration in general, the possible negative consequences of using collagen, PGA and other components to embed cells are investigated (Marx et al., 2008; Ohara et al., 2010; Pihlajamäki et al., 2010). Scaffolds made of PGA fibers were used to tooth bud cells from 6 month old miniature pig third molars (Ohara et al., 2010). PGA supported odontoblast-like cell functional differentiation, but these conditions did not seem to be optimal for histogenesis. PGA was also used by the same group to grow embryonic dental cells from molars at ED14 (Iwatsuki et al., 2006). In these conditions, after implantation, teeth formed, which showed all aspects of epithelial histogenesis and gradients of functional odontoblasts and ameloblasts (Iwatsuki et al., 2006). However, the same results could be obtained without scaffold (Nait Lechguer et al., 2008, 2011).

All together, these results indicate that the limitations in tooth engineering are more related to the cell sources and to the difficulty to maintain their potentialities during cell expansion *in vitro*.

## 8. Acknowledgements

The authors thank Hervé Gegout for the histology. This work was funded by the Dental School from Strasbourg University, by the IFRO, by a grant from Fondation des Gueules

Cassées and by a grant supporting Strategic Research of Nihon University, School of Dentistry at Matsudo from MEXT, 2008-2012.

## 9. References

Allegrucci, C.; Wu, Y.Z.; Thurston, A.; Denning, C.N.; Priddle, H.; Mummery, C.L.; Ward-van Oostwaard, D.; Andrews, P.W.; Stojkovic, M.; Smith, N.; Parkin, T.; Jones, M.E.; Warren, G.; Yu, L.; Brena, R.M.; Plass, C. & Young, L.E. (2007). Restriction landmark genome scanning identifies culture-induced DNA methylation instability in the human embryonic stem cell epigenome. *Hum. Mol. Genet.* Vol.16, No.10, (May 2007), pp. 1253-1268, ISSN 1460-2083

Arany, S.; Kawagoe, M. & Sugiyama, T. (2009). Application of spontaneously immortalized odontoblast cells in tooth regeneration. *Biochem. Biophys. Res. Commun.* Vol.381, No.1, (March 2009), pp. 84-89, ISSN 0006-291X

Arora, V.; Arora, P. & Munshi, A.K. (2009). Banking stem cells from human exfoliated deciduous teeth (SHED): saving for the future. *J. Clin. Pediatr. Dent.* Vol.33, No.4, (Summer 2009), pp. 289-294, ISSN 1053-4628

Arthur, A.; Rychkov, G.; Shi, S.; Koblar, S.A. & Gronthos, S. (2008). Adult human dental pulp stem cells differentiate toward functionally active neurons under appropriate environmental cues. *Stem Cells* Vol.26, No.7, (July 2008), pp.1787-1795, ISSN 1549-4918

Augello, A.; Kurth, T.B. & De Bari, C. (2010). Mesenchymal stem cells: a perspective from in vitro cultures to in vivo migration and niches. *Eur. Cell Mater.* Vol.1, No.20, (September 2010), pp.121-133, ISSN 1473-2262

Badylak, S.F.; Freytes, D.O. & Gilbert, T.W. (2009). Extracellular matrix as a biological scaffold material: Structure and function. *Acta Biomater.* Vol.5, No.1, (January 2009), pp. 1-13, ISSN 1742-7061

Bakopoulou, A.; Leyhausen, G.; Volk, J.; Tsiftsoglou, A.; Garefis, P.; Koidis, P. & Geurtsen, W. (2011). Assessment of the impact of two different isolation methods on the osteo/odontogenic differentiation potential of human dental stem cells derived from deciduous teeth. *Calcif. Tissue Int.* Vol.88, No.2, (February 2011), pp.130-141, ISSN 1432-0827

Balic, A.; Aguila, H.L.; Caimano, M.J.; Francone, V.P. & Mina, M. (2010). Characterization of stem and progenitor cells in the dental pulp of erupted and unerupted murine molars. *Bone* Vol.46, No.6, (June 2010), pp.1639-1651, ISSN 8756-3282

Batouli, S.; Miura, M.; Brahim, J.; Tsutsui, T.W.; Fisher, L.W.; Gronthos, S.; Robey, P.G. & Shi, S. (2003). Comparison of stem-cell-mediated osteogenesis and dentinogenesis. *J. Dent. Res.* Vol.82, No.12, (December 2003), pp. 976-981, ISSN 1544-0591

Bergmann, A. & Steller, H. (2010). Apoptosis, stem cells, and tissue regeneration. *Sci. Signal.* Vol.26, No.145, (October 2010), pp.re8, ISSN 1937-9145

Bosshardt, D.D.; Degen, T. & Lang, N.P. (2005). Sequence of protein expression of bone sialoprotein and osteopontin at the developing interface between repair cementum and dentin in human deciduous teeth. *Cell Tissue Res.* Vol.320, No.3, (June 2005), pp. 399-407, ISSN 1432-0878

Buchholz, D.E.; Hikita, S.T.; Rowland, T.J.; Friedrich, A.M.; Hinman, C.R.; Johnson, L.V. & Clegg, D.O. (2009). Derivation of functional retinal pigmented epithelium from induced pluripotent stem cells. *Stem Cells* Vol.27, No.10, (October 2009), pp. 2427-2434, ISSN 1549-4918

Byers, M.R.; Suzuki, H. & Maeda, T. (2003). Dental neuroplasticity, neuro-pulpal interactions, and nerve regeneration. *Microsc. Res. Tech.* Vol.60, No.5, (April 2003). pp. 503-515, ISSN 1097-0029

Casagrande, L.; Cordeiro, M.M.; Nör, S.A. & Nör, J.E. (2011). Dental pulp stem cells in regenerative dentistry. *Odontology* Vol.99, No.1, (January 2011), pp. 1-7, ISSN 1618-1247

Ceonzo, K.; Gaynor, A.; Shaffer, L.; Kojima, K.; Vacanti, C.A. & Stahl, G.L. (2006). Polyglycolic acid-induced inflammation: role of hydrolysis and resulting complement activation. *Tissue Eng.* Vol.12, No.2, (February 2006), pp. 301-308, ISSN 1937-3376

Chai, Y.; Jiang, X.; Ito, Y.; Bringas, P. Jr.; Han, J.; Rowitch, D.H.; Soriano, P.; McMahon, A.P. & Sucov, H.M. (2000). Fate of the mammalian cranial neural crest during tooth and mandibular morphogenesis. *Development* Vol.127, No.8, (April 2000), pp. 1671-1679, ISSN 0146-0404

Chan, B.P. & Leong, K.W. (2008). Scaffolding in tissue engineering: general approaches and tissue-specific considerations. *Eur. Spine J.* Suppl. 4, (December 2008), pp. 467-479, ISSN 1432-0932

Chanda, D.; Kumar, S. & Ponnazhagan, S. (2010). Therapeutic potential of adult bone marrow-derived mesenchymal stemcells in diseases of the skeleton. *J. Cell. Biochem.* Vol.111, No.2, (October 2010), pp. 249-257, ISSN 1097-4644

Cordeiro, M.M.; Dong, Z.; Kaneko, T.; Zhang, Z.; Miyazawa, M.; Shi, S.; Smith, A.J. & Nör, J.E. (2008). Dental pulp tissue engineering with stem cells from exfoliated deciduous teeth. *J. Endod.* Vol.34, No.8, (August 2008), pp. 962-969, ISSN 0099-2399

Cortiella, J.; Nichols, J.E.; Kojima, K.; Bonassar, L.J.; Dargon, P.; Roy, A.K.; Vacant, M.P.; Niles, J.A. & Vacanti, C.A. (2006). Tissue-engineered lung: an in vivo and in vitro comparison of polyglycolic acid and pluronic F-127 hydrogel/somatic lung progenitor cell constructs to support tissue growth. *Tissue Eng.* Vol.12, No.5, (May 2006), pp. 1213-1225, ISSN 1937-3376

Crisan, M.; Yap, S.; Casteilla, L.; Chen, C.W.; Corselli, M.; Park, T.S.; Andriolo, G.; Sun, B.; Zheng, B.; Zhang, L.; Norotte, C.; Teng, P.N.; Traas, J.; Schugar, R.; Deasy, B.M.; Badylak, S.; Buhring, H.J.; Giacobino, J.P.; Lazzari, L.; Huard, J. & Péault, B. (2008). A perivascular origin for mesenchymal stem cells in multiple human organs. *Cell Stem Cell* Vol.3, No.3, (September 2008), pp. 301-313, ISSN 1934-5909

d'Aquino, R.; De Rosa, A.; Laino, G.; Caruso, F.; Guida, L.; Rullo, R.; Checchi, V.; Laino, L.; Tirino, V. & Papaccio, G. (2009). Human dental pulp stem cells: from biology to clinical applications. *J. Exp. Zool. B Mol. Dev. Evol.* Vol.132B, No.5, (July 2009) pp. 408-415, ISSN 1520-541X

da Silva Meirelles, L.; Chagastelles, P.C. & Nardi, N.B. (2006). Mesenchymal stem cells reside in virtually all post-natal organs and tissues. *J. Cell Sci.* Vol.119, No.11, (June 2006), pp. 2204-2213, ISSN 1477-9137

Discher, D.E.; Mooney, D.J. & Zandstra, P.W. (2009). Growth factors, matrices, and forces combine and control stem cells. *Science* Vol.324, No.5935, (June 2009), pp. 1673-1677, ISSN 1095-9203

Dominici, M.; Le Blanc, K.; Mueller, I.; Slaper-Cortenbach, I.; Marini, F.; Krause, D.; Deans, R.; Keating, A.; Prockop, D.J. & Horwitz, E. (2006). Minimal criteria for defining multipotent mesenchymal stromal cells. The International Society for Cellular Therapy position statement. *Cytotherapy* Vol.8, No.4, (2006), pp. 315-317, ISSN 1477-2566

Duailibi, M.T.; Duailibi, S.E.; Young, C.S.; Bartlett, J.D.; Vacanti, J.P. & Yelick, P.C. (2004). Bioengineered teeth from cultured rat tooth bud cells. *J. Dent. Res.* Vol.83, No.7, (July 2004), pp. 523-528, ISSN 1544-0591

Duailibi, S.E.; Duailibi, M.T.; Zhang, W.; Asrican, R.; Vacanti, J.P. & Yelick, P.C. (2008). Bioengineered dental tissues grown in the rat jaw. *J. Dent. Res.* Vol.87, No.8, (August 2008), pp. 745-750, ISSN 1544-0591

Duan, X.; Tu, Q.; Zhang, J.; Ye, J.; Sommer, C.; Mostoslavsky, G.; Kaplan, D.; Yang, P. & Chen, J. (2011). Application of induced pluripotent stem (iPS) cells in periodontal tissue regeneration. *J. Cell Physiol.* Vol.226, No.1, (January 2011), pp. 150-157, ISSN 1097-4652

Feng, J.; Mantesso, A. & Sharpe, P.T. (2010). Perivascular cells as mesenchymal stem cells. *Expert Opin. Biol. Ther.* Vol.10, No.10, (October 2010), pp. 1441-1451, ISSN 1744-7682

Ferreira, C.F.; Magini, R.S. & Sharpe, P.T. (2007). Biological tooth replacement and repair. *J. Oral Rehabil.* Vol.34, No.12, (December 2007), pp. 933-939, ISSN 1365-2842

Forsyth, N.R.; Kay, A.; Hampson, K.; Downing, A.; Talbot, R. & McWhir, J. (2008). Transcriptome alterations due to physiological normoxic (2% O2) culture of human embryonic stem cells. *Regen. Med.* Vol.3, No.6, (November 2008), pp. 817-833, ISSN 1746-0751

Fried, K.; Nosrat, C.; Lillesaar, C. & Hildebrand, C. (2000). Molecular signaling and pulpal nerve development. *Crit. Rev. Oral Biol. Med.* Vol.11, No.3, (2000), pp. 318-332, ISSN 1045-4411

Fried, K.; Lillesaar, C.; Sime, W.; Kaukua, N. & Patarroyo, M. (2007). Target finding of pain nerve fibers: neural growth mechanisms in the tooth pulp. *Physiol. Behav.* Vol.92, No1-2, (September 2007), pp. 40-45, ISSN 0031-9384

Friedenstein, A.J.; Chailakhjan, R.K. & Lalykina, K.S. (1970). The development of fibroblastcolonies in monolayer cultures of guinea-pig bone marrow and spleen cells. *Cell Tissue Kinet.* Vol.3, No.4, (October 1970), pp. 393-403, ISSN 0008-8730

Fristad, I.; Vandevska-Radunovic, V. & Kvinnsland, I.H. (1999). Neurokinin-1 receptor expression in the mature dental pulp of rats. *Arch. Oral Biol.* Vol.44, No.2, (February 1999), pp. 191-195, ISSN 0003-9969

Fuchs, E. & Segre, J.A. (2000). Stem cells: a new lease on life. *Cell* Vol.100, No.1, (January 2000), pp. 143-155, ISSN 0092-8674

Gomes, J.A.; Geraldes Monteiro, B.; Melo, G.B.; Smith, R.L.; Cavenaghi Pereira da Silva, M.; Lizier, N.F.; Kerkis, A.; Cerruti, H. & Kerkis, I. (2010). Corneal reconstruction with tissue-engineered cell sheets composed of human immature dental pulp stem cells. *Invest. Ophthalmol. Vis. Sci.* Vol.51, No.3, (March 2010), pp. 1408-1414, ISSN 1552-5783

Griffith, L.G. & Schwartz, M.A. (2006). Capturing complex 3D tissue physiology in vitro. *Nat. Rev. Mol. Cell. Biol.* Vol.7, No.3, (March 2006), pp. 211-224, ISSN 1471-0072

Gronthos, S.; Mankani, M.; Brahim, J.; Robey, P.G. & Shi, S. (2000). Postnatal human dental pulp stem cells (DPSCs) *in vitro* and in vivo. *Proc. Natl. Acad. Sci. U.S.A.* Vol.97, No.25, (December 2000), pp. 13625-13630, ISSN 0022-0345

Gronthos, S.; Brahim, J.; Li, W.; Fisher, L.W.; Cherman, N.; Boyde, A.; DenBesten, P.; Robey, P.G. & Shi, S. (2002). Stem cell properties of human dental pulp stem cells. *J. Dent. Res.* Vol.81, No.8, (August 2002), pp. 531-535, ISSN 1544-0591

Govindasamy, V.; Abdullah, A.N.; Ronald, V.S.; Musa, S.; Ab Aziz, Z.A.; Zain, R.B.; Totey, S.; Bhonde, R.R. & Abu Kasim, N.H. (2010). Inherent differential propensity of dental pulp stem cells derived from human deciduous and permanent teeth. *J. Endod.* Vol.36, No.9, (September 2010), pp. 1504-1515, ISSN 0099-2399

Hacking, S.A. & Khademhosseini, A. (2009). Applications of microscale technologies for regenerative dentistry. *J. Dent. Res.* Vol.88, No.5, (May 2009), pp. 409-421, ISSN 1544-0591

Harada, H.; Kettunen, P.; Jung, H.S.; Mustonen, T.; Wang, Y.A. & Thesleff, I. (1999). Localization of putative stem cells in dental epithelium and theirassociation with Notch and FGF signaling. *J. Cell Biol.* Vol.147, No.1, (October 1999), pp. 105-120, ISSN 1540-8140

Harichane, Y.; Hirata, A.; Dimitrova-Nakov, S.; Granja, I.; Goldberg, M.; Kellermann, O. & Poliard, A. (2011). Pulpal progenitors and dentin repair *ADR*, (2011), in press, ISSN 1099-5595

Ho, S.P.; Yu, B.; Yun, W.; Marshall, G.W.; Ryder, M.I. & Marshall, S.J. (2009). Structure, chemical composition and mechanical properties of human and rat cementum and its interface with root dentin. *Acta Biomater.* Vol.5, No.2, (February 2009), pp. 707-718, ISSN 1742-7061

Honda, M.J.; Sumita, Y.; Kagami, H. & Ueda. M. (2005). Histological and immunohistochemical studies of tissue engineered odontogenesis. *Arch. Histol. Cytol.* Vol.68, No.2, (June 2005), pp. 89-101, ISSN 1349-1717.

Honda, M.J.; Ohara, T.; Sumita, Y.; Ogaeri, T.; Kagami, H. & Ueda, M. (2006). Preliminary study of tissue-engineered odontogenesis in the canine jaw. *J. Oral Maxillofac. Surg.* Vol.64, No.2, (February 2006), pp. 283-289, ISSN 1698-6946

Honda, M.J.; Tsuchiya, S.; Sumita, Y.; Sagara, H. & Ueda, M. (2007). The sequential seeding of epithelial and mesenchymal cells for tissue-engineered tooth regeneration. *Biomaterials* Vol.28, No.4, (February 2007), pp. 680-689, ISSN 0142-9612

Honda, M.J.; Fong, H.; Iwatsuki, S.; Sumita, Y. & Sarikaya, M. (2008). Tooth-forming potential in embryonic and postnatal tooth bud cells. *Med. Mol. Morphol.* Vol.41, No.4, (December 2008), pp. 183-192, ISSN 1860-1499

Honda, M.J.; Shinmura, Y. & Shinohara, Y. (2009). Enamel tissue engineering using subcultured enamel organ epithelial cells in combination with dental pulp cells. *Cells Tissues Organs* Vol.189, No.1-4, (August 2008), pp. 261-267, ISSN 1422-6421

Honda, M.J. & Hata, K. (2010). Enamel tissue engineering. In : « Tooth engineering » (Ed. D. Eberli). 2010, pp. 281-296, ISBN 978-953-307-079-7

Honda, M.J.; Tsuchiya, S.; Shinohara, Y.; Shinmura, Y. & Sumita, Y. (2010). Recent advances in engineering of tooth and tooth structures using postnatal dental cells. *Japanese Dental Science* Vol.46 (2010), pp. 54-66, ISSN 1882-7616

Hu, B.; Nadiri, A.; Bopp-Kuchler, S.; Perrin-Schmitt, F. & Lesot, H. (2005). Dental epithelial histomorphogenesis in vitro. *J. Dent. Res.* Vol.84, No.6, (June 2005), pp. 521-525. ISSN 0022-0345

Hu, B.; Nadiri, A.; Kuchler-Bopp, S.; Perrin-Schmitt, F.; Peters, H. & Lesot, H. (2006a). Tissue engineering of tooth crown, root, and periodontium. *Tissue Eng.* Vol.12, No.8, (August 2006), pp. 2069-2075, ISSN 1937-3376

Hu, B.; Unda, F.; Bopp-Kuchler, S.; Jimenez, L.; Wang, X.J.; Haïkel, Y.; Wang, S.L. & Lesot, H. (2006b). Bone marrow cells can give rise to ameloblast-like cells. *J. Dent. Res.* Vol.85, No.5, (May 2006), pp. 416-421, ISSN 1544-0591

Hu, Q.; Friedrich, A.M.; Johnson, L.V. & Clegg, D.O. (2010). Memory in induced pluripotent stem cells: reprogrammed human retinal-pigmented epithelial cells show tendency for spontaneous redifferentiation. *Stem Cells* Vol.28, No.11, (November 2010), pp. 1981-1991, ISSN 1549-4918

Huang, G.T.; Sonoyama, W.; Chen, J. & Park, S.H. (2006). In vitro characterization of human dental pulp cells: variousisolation methods and culturing environments. *Cell Tissue Res.* Vol.324, No.2, (May 2006), pp. 225-236, ISSN 1432-0878

Huang, G.T. (2009). Pulp and dentin tissue engineering and regeneration: current progress. *Regen. Med.* Vol.4, No.5, (September 2009), pp. 697-707, ISSN 1746-0751

Huang, G.T.; Gronthos, S. & Shi, S. (2009). Mesenchymal stem cells derived from dental tissues vs. those from other sources: their biology and role in regenerative medicine. *J. Dent. Res.* Vol.88, No.9, (September 2009), pp. 792-806, ISSN 1544-0591

Huang, G.T.; Yamaza, T.; Shea, L.D.; Djouad, F.; Kuhn, N.Z.; Tuan, R.S. & Shi, S. (2010). Stem/progenitor cell-mediated de novo regeneration of dental pulp with newly deposited continuous layer of dentin in an in vivo model. *Tissue Eng. Part A.* Vol.16, No.2, (February 2010), pp. 605-615, ISSN 1937-3376

Imbeni, V.; Kruzic, J.J.; Marshall, G.W.; Marshall, S.J. & Ritchie, R.O. (2005). The dentin-enamel junction and the fracture of human teeth. *Nat. Mater.* Vol.4, No.3, (March 2005), pp. 229-232, ISSN 1476-4660

Iohara, K.; Zheng, L.; Ito, M.; Tomokiyo, A.; Matsushita, K. & Nakashima, M. (2006). Side population cells isolated from porcine dental pulp tissue with self-renewal and multipotency for dentinogenesis, chondrogenesis, adipogenesis, and neurogenesis. *Stem Cells* Vol.24, No.11, (November 2006), pp. 2493-2503, ISSN 1549-4918

Ishikawa, Y.; Ida-Yonemochi, H.; Suzuki, H.; Nakakura-Ohshima, K.; Jung, H.S.; Honda, M.J.; Ishii, Y.; Watanabe, N. & Ohshima, H. (2010). Mapping of BrdU label-retaining dental pulp cells in growing teeth and their regenerative capacity after injuries. *Histochem. Cell Biol.* Vol.134, No.3, (September 2010), pp. 227-241, ISSN 1432-119X

Iwatsuki, S.; Honda, M.J.; Harada, H. & Ueda, M. (2006). Cell proliferation in teeth reconstructed from dispersed cells of embryonic tooth germs in a three-dimensional scaffold. *Eur. J. Oral Sci.* Vol.114, No.4, (August 2006), pp. 310-317, ISSN 1600-0722

Jernvall, J. & Thesleff, I. (2000). Reiterative signaling and patterning during mammalian tooth morphogenesis. *Mech. Dev.* Vol.92, No.1, (March 2000), pp. 19-29, ISSN 0925-4773

Katz, J.L.; Misra, A.; Spencer, P.; Wang, Y.; Bumrerraj, S.; Nomura, T.; Eppell, S.J. & Tabib-Azar, M. (2007). Multiscale mechanics of hierarchical structure/property relationships in calcified tissues and tissue/material interfaces. *Mater. Sci. Eng. A Struct. Mater.* Vol.27, No.3, (April 2007) pp. 450-468, ISSN 0921-5093

Kawaguchi, H.; Hirachi, A.; Hasegawa, N.; Iwata, T.; Hamaguchi, H.; Shiba, H.; Takata, T.; Kato, Y. & Kurihara, H. (2004). Enhancement of periodontal tissue regeneration by transplantation of bone marrow mesenchymal stem cells. *J. Periodontol.* Vol.75, No.9, (September 2004), pp. 1281-1287, ISSN 0022-3492

Keller, L.; Kuchler-Bopp, S.; Acuna-Mendoza, S.; Poliard, A. & Lesot, H. (2011). Tooth engineering: searching for dental mesenchymal cells sources. *Front. Physio.* Vol.2, (March 2011) doi: 10.3389, ISSN 1664-042X

Kerkis, I.; Kerkis, A.; Dozortsev, D.; Stukart-Parsons, G.C.; Gomes Massironi, S.M.; Pereira, L.V.; Caplan, A.I. & Cerruti, H.F. (2006). Isolation and characterization of a population of immature dental pulpstem cells expressing OCT-4 and other embryonic stem cell markers *Cells Tissues Organs.* Vol.184, No.3-4, (2006), pp. 105-116, ISSN 1422-6421

Kettunen, P.; Loes, S.; Furmanek, T.; Fjeld, K.; Kvinnsland, I.H.; Behar, O.; Yagi, T.; Fujisawa, H.; Vainio, S.; Taniguchi, M. & Luukko, K. (2005). Coordination of trigeminal axon navigation and patterning with tooth organ formation: epithelial-mesenchymal interactions, and epithelial Wnt4 and Tgfbeta1 regulate semaphorin 3a expression in the dental mesenchyme. *Development* Vol.132, No.2, (January 2005), pp. 323-334, ISSN 1011-6370

Kim, K.; Lee, C.H.; Kim, B.K. & Mao, J.J. (2010). Anatomically shaped tooth and periodontal regeneration by cell homing. *J. Dent. Res.* Vol.89, No.8, (August 2010), pp. 842-847, ISSN 1544-0591

Komine, A.; Suenaga, M.; Nakao, K.; Tsuji, T. & Tomooka, Y. (2007) Tooth regeneration from newly established cell lines from a molar tooth germ epithelium. *Biochem. Biophys. Res. Commun.* Vol.355, No.3, (April 2007), pp. 758-763, ISSN 0006-291X

Kuo, T.F.; Lin, H.C.; Yang, K.C.; Lin F.H.; Chen, M.H.; Wu, C.C. & Chang, H.H. (2011) Bone Marrow Combined With Dental Bud Cells Promotes Tooth Regeneration in Miniature Pig Model. *Artf. Organs* Vol.35, No.2, (February 2011), pp. 1525-1594, ISSN 1525-1594

Laino, G.; Graziano, A.; d'Aquino, R.; Pirozzi, G.; Lanza, V.; Valiante, S.; De Rosa, A.; Naro, F.; Vivarelli, E. & Papaccio, G. (2006). An approachable human adult stem cell source for hard-tissue engineering. *J. Cell Physiol.* Vol.206, No.3, (March 2006), pp.693-701, ISSN1097-4652

Lesot, H.; Peterková, R.; Schmitt, R.; Meyer, J.M.; Viriot, L.; Vonesch, J.L.; Senger, B.; Peterka, M. & Ruch, J.V. (1999). Initial features of the inner dental epithelium histo-morphogenesis in the first lower molar in mouse. *Int. J. Dev. Biol.* Vol.43, No.3, (May 1999) pp. 245-254, ISSN 0214-6282

Lesot, H. & Brook, A.H. (2009). Epithelial histogenesis during tooth development. *Arch. Oral Biol.* Vol.54, Suppl.1, (December 2009), pp. 25-33, ISSN 0003-9969

Lesot, H.; Lisi, S.; Peterkova, R.; Peterka, M.; Mitolo, V. & Ruch, J.V. (2001). Epigenetic signals during odontoblast differentiation. *Adv. Dent. Res.* Vol.15, (August 2001), pp. 8-13, ISSN 0895-9374

Li, L. & Clevers, H. (2010). Coexistence of quiescent and active adult stem cells in mammals. *Science* Vol.327, No.5965, (January 2010), pp. 542-545, ISSN 1095-9203

Lisi, S.; Peterková, R.; Peterka, M.; Vonesch, J.L.; Ruch, J.V. & Lesot, H. (2003). Tooth morphogenesis and pattern of odontoblast differentiation. *Connect. Tissue Res.* Vol.44, Suppl.1, (2003), pp. 167-170, ISSN 1607-8438

Lovschall, H.; Mitsiadis, T.A.; Poulsen, K.; Jensen, K.H. & Kjeldsen, A.L. (2007). Coexpression of Notch3 and Rgs5 in the pericyte-vascular smooth muscle cell axis in response to pulp injury. *Int. J. Dev. Biol.* Vol.51, No.8, (2007), pp.715-721, ISSN 0214-6282

Lowry, W.E.; Richter, L.; Yachechko, R.; Pyle, A.D.; Tchieu, J.; Sridharan, R.; Clark, A.T. & Plath, K. (2008). Generation of human induced pluripotent stem cells from dermal fibroblasts. *Proc. Natl. Acad. Sci. U.S.A.* Vol.105, No.8, (February 2008), pp. 2883-2888, ISSN 1091-6490

Lu, H.H.; Subramony, S.D.; Boushell, M.K. & Zhang, X. (2010). Tissue engineering strategies for the regeneration of orthopedic interfaces. *Ann. Biomed. Eng.* Vol.38, No.6, (June 2010), pp. 2142-2154, ISSN 1350-4533

Lumsden, A.G. (1984). Tooth morphogenesis : Contribution of the cranial neural crest cells in mammals. In Belcourt A.B. & Ruch J.V. eds, Vol.125, Colloque Inserm Paris (1984), pp. 19-27,

Luukko, K.; Moe, K.; Sijaona, A.; Furmanek, T.; Hals Kvinnsland, I.; Midtbø, M. & Kettunen, P. (2008). Secondary induction and the development of tooth nerve supply. *Ann. Anat.* Vol.190, No.2, (2008), pp. 178-187, ISSN 0940-9602

Maherali, N.; Sridharan, R.; Xie, W.; Utikal, J.; Eminli, S.; Arnold, K.; Stadtfeld, M.; Yachechko, R.; Tchieu, J.; Jaenisch, R.; Plath, K. & Hochedlinger, K. (2007). Directly reprogrammed fibroblasts show global epigenetic remodeling and widespread tissue contribution. *Cell Stem Cell* Vol.1, No.1, (June 2007), pp. 55-70, ISSN 1934-5909

Manning, M.L.; Foty, R.A.; Steinberg, M.S. & Schoetz, E.M. (2010). Coaction of intercellular adhesion and cortical tension specifies tissue surface tension. *Proc. Natl. Acad. Sci. U.S.A.* Vol.107, No.28, (July 2010), pp. 12517-12522, ISSN 1091-6490

Manzke, E.; Katchburian, E.; Faria, F.P. & Freymüller, E. (2005). Structural features of forming and developing blood capillaries of the enamel organ of rat molar tooth germs observed by light and electron microscopy. *J. Morphol.* Vol.265, No.3, (September 2005), pp. 335-342, ISSN 1097-4687

Marx, G.; Hotovely-Salomon, A.; Levdansky, L.; Gaberman, E.; Snir, G.; Sievner, Z.; Klauzner, Y.; Silberklang, M.; Thomas, D.; Hoffman, N.; Luke, S.; Lesnoy, D. & Gorodetsky, R. (2008). Haptide-coated collagen sponge as a bioactive matrix for tissue regeneration. *J. Biomed. Mater. Res. B Appl. Biomater.* Vol.84, No.2, (February 2008), pp. 571-583, ISSN 1552-4981

Mitsiadis, T.A.; Fried, K. & Goridis, C. (1999). Reactivation of Delta-Notch signaling after injury: complementary expression patterns of ligand and receptor in dental pulp. *Exp. Cell Res.* Vol.246, No.2, (February 1999), pp. 312-318, ISSN 0014-4827

Mitsiadis, T.A.; Roméas, A.; Lendahl, U.; Sharpe, P.T. & Farges, J.C. (2003). Notch2 protein distribution in human teeth under normal and pathological conditions. *Exp. Cell Res.* Vol.282, No.2, (January 2003), pp. 101-109, ISSN 0014-4827

Miura, M.; Gronthos, S.; Zhao, M.; Lu, B.; Fisher, L.W.; Robey, P.G. & Shi, S. (2003). SHED: stem cells from human exfoliated deciduous teeth. *Proc. Natl. Acad. Sci. U.S.A.* Vol.100, No.10, (May 2003), pp. 5807-5812, ISSN 1091-6490

Morsczeck, C.; Götz, W.; Schierholz, J.; Zeilhofer, F.; Kühn, U.; Möhl, C.; Sippel, C. & Hoffmann, K.H. (2005). Isolation of precursor cells (PCs) from human dental follicle of wisdom teeth. *Matrix Biol.* Vol.24, No.2, (April 2005), pp. 155-165, ISSN 0945-053X

Nakagawa, E.; Itoh, T.; Yoshie, H. & Satokata, I. (2009). Odontogenic potential of post-natal oral mucosal epithelium. *J. Dent. Res.* Vol.88, No.3, (March 2009), pp. 219-223, ISSN 1544-0591

Nait Lechguer, A.; Kuchler-Bopp, S.; Hu, B.; Haïkel, Y. & and Lesot, H. (2008). Vascularization of engineered teeth. *J. Dent. Res.* Vol.87, No.12, (December 2008), pp. 1138-1143, ISSN 1544-0591

Nait Lechguer, A.; Kuchler-Bopp, S. & Lesot, H. (2009). Crown formation during tooth development and tissue engineering. *J. Exp. Zool. B. Mol. Dev. Evol.* Vol.312B, No.2, (January 2009), pp. 399-407, ISSN 1473-2262

Nait Lechguer, A.; Couble, M.L.; Labert, N.; Kuchler-Bopp, S.; Keller, L.; Magloire, H.; Bleicher, F. & Lesot, H. (2011). Cell Differentiation and Matrix Organization in Engineered Teeth. *J. Dent. Res.* Vol.90, No.5, (May 2011), pp. 583-589, ISSN 1544-0591

Nakao, K.; Morita, R.; Saji, Y.; Ishida, K.; Tomita, Y.; Ogawa, M.; Saitoh, M.; Tomooka, Y. & Tsuji, T. (2007). The development of a bioengineered organ germ method. *Nat. Methods* Vol.4, No.3, (March 2007), pp. 227-230, ISSN 1548-7091

Nanci, A. & Bosshardt, D.D. (2006). Structure of periodontal tissues in health and disease. *Periodontol. 2000* Vol.40, (2006), pp. 11-28, ISSN 0906-6713

Nehls, V. & Drenckhahn, D. (1991). Heterogeneity of microvascular pericytes for smooth muscle type alpha-actin. *J. Cell Biol.* Vol.113, No.1, (April 1991), pp. 147-154, ISSN 1540-8140

Nombela-Arrieta, C.; Ritz, J. & Silberstein, L.E. (2011). The elusive nature and function of mesenchymal stem cells. *Nat. Rev. Mol. Cell Biol.* Vol.12, No.2, (February 2011), pp. 126-131, ISSN 1471-0072

Novak, A.; Shtrichman, R.; Germanguz, I.; Segev, H.; Zeevi-Levin, N.; Fishman, B.; Mandel, Y.E.; Barad, L.; Domev, H.; Kotton, D.; Mostoslavsky, G.; Binah, O. & Itskovitz-Eldor, J. (2010). Enhanced reprogramming and cardiac differentiation of human keratinocytes derived from plucked hair follicles, using a single excisable lentivirus. *Cell. Reprogram.* Vol.12, No.6, (December 2010), pp. 665-678, ISSN 2152-4998

Ohara, T.; Itaya, T.; Usami, K.; Ando, Y.; Sakurai, H.; Honda, M.J.; Ueda, M. & Kagami, H. (2010). Evaluation of scaffold materials for tooth tissue engineering. *J. Biomed. Mater. Res. A.* Vol.94, No.3, (September 2010), pp. 800-805, ISSN 1552-4965.

Ohazama, A.; Modino, S.A.; Miletich, I. & Sharpe, P.T. (2004). Stem-cell-based tissue engineering of murine teeth. *J. Dent. Res.* Vol.83, No.7, (July 2004), pp. 518-522, ISSN 1544-0591

Park, C.H.; Rios, H.F.; Jin, Q.; Bland, M.E.; Flanagan, C.L.; Hollister, S.J. & Giannobile, W.V. (2010). Biomimetic hybrid scaffolds for engineering human tooth-ligament interfaces. *Biomaterials* Vol.31, No.23, (August 2010), pp. 5945-5952, ISSN 0142-9612

Patel, M.; Smith, A.J.; Sloan, A.J.; Smith, G. & Cooper, P.R. (2009). Phenotype and behaviour of dental pulp cells during expansion culture. *Arch. Oral Biol.* Vol.54, No.10, (October 2009), pp. 898-908, ISSN 0003-9969

Peters, H. & Balling, R. (1999). Teeth. Where and how to make them. *Trends Genet.* Vol.15, No.2, (February 1999), pp. 59-65, ISSN: 0168-9479

Pihlajamäki, H.K.; Salminen, S.T.; Tynninen, O.; Böstman, O.M. & Laitinen, O. (2010). Tissue restoration after implantation of polyglycolide, polydioxanone, polylevolactide, and metallic pins in cortical bone: an experimental study in rabbits. *Calcif. Tissue Int.* Vol.87, No.1, (July 2010), pp. 90-98, ISSN 1432-0827

Pittenger, M.F.; Mackay, A.M.; Beck, S.C.; Jaiswal, R.K.; Douglas, R.; Mosca, J.D.; Moorman, M.A.; Simonetti, D.W.; Craig, S. & Marshak, D.R. (1999). Multilineage potential of adult human mesenchymal stem cells. *Science* Vol.284, No.5411 (April 1999), pp. 143-147, ISSN 1095-9203

Quattrocelli, M.; Palazzolo, G.; Floris, G.; Schöffski, P.; Anastasia, L.; Orlacchio, A.; Vandendriessche, T.; Chuah, M.K.; Cossu, G.; Verfaillie, C. & Sampaolesi, M. (2011). Intrinsic cell memory reinforcesmyogenic commitment of pericyte-derived iPSCs. *J. Pathol.* Vol.223, No.5, (April 2011), pp. 593-603, ISSN 0022-3417

Reilly, G.C. & Engler, A.J. (2010). Intrinsic extracellular matrix properties regulate stem cell differentiation. *J. Biomech.* Vol.43, No.1, (January 2010), pp. 55-62, ISSN 0021-9290

Ripamonti, U. & Petit, J.C. (2009). Bone morphogenetic proteins, cementogenesis, myoblastic stem cells and the induction of periodontal tissue regeneration. *Cytokine Growth Factor Rev.* Vol.20, No.5-6, (October-December 2009), pp. 489-499, ISSN 1359-6101

Rhodes, J.M. & Simons, M. (2007). The extracellular matrix and blood vessel formation: not just a scaffold. *J. Cell Mol. Med.* Vol.11, No.2, (March-April 2007), pp. 176-205, ISSN 2136-2133

Roobrouck, V.D.; Vanuytsel, K. & Verfaillie, C.M. (2011). Culture Mediated Changes in Fate and/or Potency of Stem Cells. *Stem Cells* (February 2011), doi: 10.1002, ISSN 1549-4918

Rothova, M.; Feng, J.; Sharpe, P.T.; Peterkova, R. & Tucker, A.S. (2011). Contribution of mesoderm to the developing dental papilla. *Int. J. Dev. Biol.* Vol.55, No.1, (2011), pp. 59–64, ISSN 1696-3547

Ruch, J.V.; Lesot, H.; Karcher-Djuricic, V.; Meyer, J.M. & Olive, M. (1982). Facts and hypotheses concerning the control of odontoblast differentiation. *Differentiation* Vol.21, No.1, (May 1982), pp. 7-12, ISSN 1432-0436

Ruch, J.V.; Lesot, H. & Bègue-Kirn, C. (1995). Odontoblast differentiation. *Int. J. Dev. Biol.* Vol.39, No.1, (February 1995), pp. 51-68, ISSN 0214- 6282

Sakai, V.T.; Zhang, Z.; Dong, Z.; Neiva, K.G.; Machado, M.A.; Shi, S.; Santos, C.F. & Nör, J.E. (2010). SHED differentiate into functional odontoblasts and endothelium. *J. Dent. Res.* Vol.89, No.8, (August 2010), pp. 791-796, ISSN 1544-0591

Salazar-Ciudad, I. & Jernvall, J. (2004). How different types of pattern formation mechanisms affect the evolution of form and development. *Evol. Dev.* Vol.6, No.1, (January-February 2004), pp. 6-16, ISSN 1525-142X

Salazar-Ciudad, I. & Jernvall, J. (2010). A computational model of teeth and the developmental origins of morphological variation. *Nature* Vol. 464, No.7288, (March 2010), pp. 583-586, ISSN 0028-0836

Schofield, R. (1978). The relationship between the spleen colony-forming cell and the haemopoietic stem cell. *Blood Cells* Vol.4, No.1-2, (1978), pp. 7-25, ISSN 1079-9796

Seo, B.M.; Miura, M.; Gronthos, S.; Bartold, P.M.; Batouli, S.; Brahim, J.; Young, M.; Robey, P.G.; Wang, C.Y. & Shi, S. (2004). Investigation of multipotent postnatal stem cells from human periodontal ligament. *Lancet* Vol.364, No.9429, (July 2004), pp. 149-155, ISSN 0140-6736

Seo, B.M.; Sonoyama, W.; Yamaza, T.; Coppe, C.; Kikuiri, T.; Akiyama, K.; Lee, J.S. & Shi, S. (2009). SHED repair critical-size calvarial defects in mice. *Oral Dis.* Vol.14, No.5, (July 2008), pp. 428-434. Erratum in: *Oral Dis.* Vol.14, No.4, (May 2009), pp. 302, ISSN 1601-0825

Shi, S. & Gronthos, S. (2003). Perivascular niche of postnatal mesenchymal stem cells in human bone marrow and dental pulp. *J. Bone Miner. Res.* Vol.18, No.4, (April 2003), pp. 696-704, ISSN 0884-0431

Shi, S.; Robey, P.G. & Gronthos, S. (2001). Comparison of human dental pulp and bone marrow stromal stem cells by cDNA microarray analysis. *Bone* Vol.29, No.6, (December 2001), pp. 532-539, ISSN 1095-9203

Shinmura, Y.; Tsuchiya, S.; Hata, K. & Honda, M.J. (2008). Quiescent epithelial cell rests of Malassez can differentiate into ameloblast-like cells. *J. Cell Physiol.* Vol.217, No.3, (December 2008), pp. 728-738, ISSN 1097-4652

Siqueira da Fonseca, S.A.; Abdelmassih, S.; de Mello Cintra Lavagnolli, T.; Serafim, R.C.; Clemente Santos, E.J.; Mota Mendes, C.; de Souza Pereira, V.; Ambrosio, C.E.; Miglino, M.A.; Visintin, J.A.; Abdelmassih, R.; Kerkis, A. & Kerkis, I. (2009). Human immature dental pulp stem cells contribution to developing mouse embryos: production of human/mouse preterm chimaeras. *Cell Prolif.* Vol.42, No.2, (April 2009), pp. 132-140, ISSN 0960-7722

Skinner, M.M. & Gunz, P. (2010). The presence of accessory cusps in chimpanzee lower molars is consistent with a patterning cascade model of development. *J. Anat.* Vol.217, No.3, (September 2010), pp. 245-253, ISSN 1469-7580

Slavkin, H.C.; Snead, M.L.; Zeichner-David, M.; Jaskoll, T.F. & Smith, B.T. (1984). Concepts of epithelial-mesenchymal interactions during development: tooth and lung organogenesis. *J. Cell Biochem.* Vol.26, No.2, (February 1984), pp. 117-125, ISSN 1097-4644

Sonoyama, W.; Liu, Y.; Fang, D.; Yamaza, T.; Seo, B.M.; Zhang, C.; Liu, H.; Gronthos, S.; Wang, C.Y.; Wang, S. & Shi, S. (2006). Mesenchymal stem cell-mediated functional tooth regeneration in swine. *PLoS One* Vol.1, (December 2006), pp. 79, ISSN 1932-6203

Sonoyama, W.; Liu, Y.; Yamaza, T.; Tuan, R.S.; Wang, S.; Shi, S. & Huang, G.T. (2008). Characterization of the apical papilla and its residing stem cells from human immature permanent teeth: a pilot study. *J. Endod.* Vol.34, No.2, (February 2008), pp. 166-171, ISSN 0099-2399

Steinberg, M.S. (1962). On the mechanism of tissue reconstruction by dissociated cells. I. Population kinetics, differential adhesiveness and the absence of directed migration. *Proc. Natl. Acad. Sci. U.S.A.* Vol.48, (September 1962), pp. 1577-1582, ISSN 1091-6490

Steinberg, M.S. (1963). Reconstruction of tissues by dissociated cells. Some morphogenetic tissue movements and the sorting out of embryonic cells may have a common explanation. *Science* Vol.141 (August 1963), pp. 401-408, ISSN 1095-9203

Sumita, Y.; Honda, M.J.; Ohara, T.; Tsuchiya, S.; Sagara, H.; Kagami, H. & Ueda, M. (2006). Performance of collagen sponge as a 3-D scaffold for tooth-tissue engineering. *Biomaterials* Vol.27, No.17, (June 2006), pp. 3238-3248, ISSN 0142-9612

Sumita, Y.; Tsuchiya, S.; Asahina, I.; Kagami, H. & Honda, M.J. (2009). The location and characteristics of two populations of dental pulp cells affect tooth development. *Eur. J. Oral Sci.* Vol.117, No.2, (April 2009), pp. 113-121, ISSN 1600-0722

Takahashi, K. & Yamanaka, S. (2006). Induction of pluripotent stem cells from mouse embryonic and adult fibroblast cultures by defined factors. *Cell* Vol. 126, No.4, (August 2006), pp. 663-676, ISSN 0092-8674

Takahashi, K.; Tanabe, K.; Ohnuki, M.; Narita, M.; Ichisaka, T.; Tomoda, K. & Yamanaka, S. (2007). Induction of pluripotent stem cells from adult human fibroblasts by defined factors. *Cell* Vol.131, No.5, (November 2007), pp. 861-872, ISSN 0092-8674

Tamaoki, N.; Takahashi, K.; Tanaka. T.; Ichisaka, T.; Aoki, H.; Takeda-Kawaguchi, T.; Iida, K.; Kunisada, T.; Shibata, T.; Yamanaka, S. & Tezuka, K. (2010). Dental pulp cells for induced pluripotent stem cell banking. *J. Dent. Res.* Vol.89, No.8, (August 2010), pp. 773-778, ISSN 1544-0591

Tavassol, F.; Schumann, P.; Lindhorst, D.; Sinikovic, B.; Voss, A.; von See, C.; Kampmann, A.; Bormann, K.H.; Carvalho, C.; Mülhaupt, R.; Harder, Y.; Laschke, M.W.; Menger, M.D.; Gellrich, N.C. & Rücker, M. (2010). Accelerated angiogenic host tissue response to poly(L-lactide-co-glycolide) scaffolds by vitalization with osteoblast-like cells. *Tissue Eng. Part A* Vol.16, No.7, (July 2010), pp. 2265-2279, ISSN 2152-4955

Téclès, O.; Laurent, P.; Zygouritsas, S.; Burger, A.S.; Camps, J.; Dejou, J. & About, I. (2005). Activation of human dental pulp progenitor/stem cells in response to odontoblast injury. *Arch. Oral Biol.* Vol.50, No.2, (February 2005), pp. 103-108, ISSN 0003-9969

Tian, C.; Wang, Y.; Sun, L.; Ma, K. & Zheng, J.C. (2011). Reprogrammed mouse astrocytes retain a "memory" of tissue origin and possess more tendencies for neuronal differentiation than reprogrammed mouse embryonic fibroblasts. *Protein Cell* Vol.2, No.2, (February 2011), pp. 128-140, ISSN 1674-8018

Volponi, A.A.; Pang, Y. & Sharpe, P.T. (2010). Stem cell-based biological tooth repair and regeneration. *Trends Cell Biol.* Vol.20, No.12, (December 2010), pp. 715-722, ISSN 0962-8924

Wang, B.; Li, L.; Du, S.; Liu, C.; Lin, X.; Chen, Y. & Zhang, Y. (2010). Induction of human keratinocytes into enamel-secreting ameloblasts. *Dev. Biol.* Vol.344, No.2, (August 2010), pp. 795-799, ISSN 0012-1606

Weissman, I.L. (2000). Translating stem and progenitor cell biology to the clinic: barriers and opportunities. *Science* Vol.287, No.5457, (February 2000), pp. 1442-1446, ISSN 1095-9203

Woods, E.J.; Perry, B.C.; Hockema, J.J.; Larson, L.; Zhou, D. & Goebel, W.S. (2009). Optimized cryopreservation method for human dental pulp-derived stem cells and their tissues of origin for banking and clinical use. *Cryobiology* Vol.59, No.2, (October 2009), pp. 150-157, ISSN 0011-2240

Xu, W.P.; Zhang, W.; Asrican, R.; Kim, H.J.; Kaplan, D.L. & Yelick, P.C. (2008). Accurately shaped tooth bud cell-derived mineralized tissue formation on silk scaffolds. *Tissue Eng. Part A.* Vol.14, No.4, (April 2008), pp. 549-557, ISSN 2152-4955

Yang, X.; van der Kraan, P.M.; van den Dolder, J.; Walboomers, X.F.; Bian, Z.; Fan, M. & Jansen JA. (2007). STRO-1 selected rat dental pulp stem cells transfected with adenoviral-mediated human bone morphogenetic protein 2 gene show enhanced odontogenic differentiation. *Tissue Eng.* Vol.13, No.11, (November 2007), pp. 2803-2812, ISSN 1937-3376

Yang, Y.; Rossi, F.M. & Putnins, E.E. (2010). Periodontal regeneration using engineered bone marrow mesenchymal stromal cells. *Biomaterials* Vol.31, No.33, (November 2010), pp. 8574-8582, ISSN 0142-9612

Yoshida, T.; Miyoshi, J.; Takai, Y. & Thesleff, I. (2010). Cooperation of nectin-1 and nectin-3 is required for normal ameloblast function and crown shape development in mouse teeth. *Dev. Dyn.* Vol.239, No.10, (October 2010), pp. 2558-2569, ISSN 1097-0177

Young, C.S.; Kim, S.W.; Qin, C.; Baba, O.; Butler, W.T.; Taylor, R.R.; Bartlett, J.D.; Vacanti, J.P. & Yelick, P.C. (2005). Developmental analysis and computer modelling of bioengineered teeth. *Arch. Oral Biol.* Vol.50, No.2, (February 2005), pp. 259-265, ISSN 0003-9969

Yu, J.; He, H.; Tang, C.; Zhang, G.; Li, Y.; Wang, R.; Shi, J. & Jin, Y. (2010). Differentiation potential of STRO-1+ dental pulp stem cells changes during cell passaging. *BMC Cell Biol.* Vol.11, (May 2010), pp. 32, ISSN 1471-2121

Zippel, R.; Wilhelm, L.; Hoene, A.; Walschus, U.; Ueberrueck, T. & Schlosser, M. (2008). Local tissue reaction and differentiation of the prosthesis-specific antibody response following functional implantation of vascular grafts in pigs. *J. Biomed. Mater. Res. B Appl. Biomater.* Vol.85, No.2, (May 2008), pp. 334-342, ISSN 1552-4981

16

# Transplantation of Corneal Stroma Reconstructed with Gelatin and Multipotent Precursor Cells from Corneal Stroma

Tatsuya Mimura[1], Yasuhiko Tabata[2] and Shiro Amano[1]
*[1]Department of Ophthalmology, University of Tokyo Graduate School of Medicine,*
*[2]Institute for Frontier Medical Sciences, Kyoto University*
*Japan*

## 1. Introduction

The cornea is composed of a multilayered epithelium, Bowman's membrane, stroma, Descemet's membrane and endothelium. As compared to the stroma, tissue engineering of the corneal epithelium or endothelium is relatively simple and well established, as the stroma is mainly composed of continuous cells joined by tight junctions with a few subcellular extracellular matrix (ECM) proteins. Corneal stromal engineering, however, has yet to be fully achieved due to difficulties encountered when trying to constitute an artificial corneal stroma, which consists of few fibroblast-like keratocytes, nerve fibers, and a transparent ECM of collagen fibrils and proteoglycans.

Stem cells or progenitor cells are defined by their capacity for self-renewal and the ability to generate different types of cells (multipotentiality), which leads to the formation of mature tissues. In contrast, the precursor cells are unipotent cells with a limited proliferative activity. Regenerative stem cells or precursors can be detected by the sphere-forming assay in various adult tissues, including the central nervous system (Nunes et al., 2003), bone marrow (Krause et al., 2001), skin (Kawase et al., 2004; Toma et al., 2001), retina (Coles et al., 2004), corneal epithelium (Mimura et al., 2010a), corneal stroma (Amano et al., 2006; Mimura et al., 2008a, 2008b; Uchida et al., 2005; Yamagami et al., 2007), and corneal endothelium (Amano et al., 2006; Mimura et al., 2005a, 2005b, 2005c, 2007, 2010b; Yamagami et al., 2006a, 2007; Yokoo et al., 2005).

Despite the many successes achieved in the isolation and characterization of stem cells from various tissues, relatively few studies have investigated the efficacy of stem cell transplantation therapy in animals. While the three-dimensional structure that is responsible for maintaining the cell-to-cell interactions is indispensable for tissue engineering using stem cells, the structural complexity involved in tissue engineering does not easily allow us to extend the investigations of stem cell transplantations.

In this chapter, we introduce our recent work about regenerative medicine and tissue engineering of corneal stroma using multipotent precursor cells. We isolated precursors with the propensity to develop into corneal keratocyte-like cells from the stroma of rabbit corneas and investigated the distribution and proliferative capacity of precursor cells derived from the central and peripheral regions of the cornea by the sphere-forming assay.

Additionally, we constructed corneal stroma using keratocyte precursors and porous gelatin hydrogels *in vitro* by tissue engineering and investigated the feasibility of the engineered corneal stroma with keratocyte precursors in a rabbit model.

## 2. Origin and development of corneal stroma

Corneal stromal cells (keratocytes) are derived from the neural crest (Bahn et al., 1984; Johnston et al., 1979; Mooy et al., 1990). During embryonic development, neural crest cells, from which the keratocytes originate (Bahn et al., 1984; Johnston et al., 1979), show two waves of migration and differentiation during the corneal growth (Liu et al., 1998; Meier, 1982). In the first wave, the corneal epithelium forms and then synthesizes the primary stroma (i.e., periocular mesenchymal cells of neural crest origin). Subsequently, the neural crest cells migrate to the margin of the optic cup, and to the area between the lens and the corneal epithelium. This contributes to development of the corneal stroma and the trabecular meshwork. During the second wave of migration, neural crest cells invade the primary stroma and then undergo differentiation into keratocytes.

In adults, corneal stromal cells are composed of keratocytes and bone marrow-derived cells of the monocyte lineage (Yamagami et al., 2006b). The latter cell population is continuously replaced from the bone marrow (Nakamura et al., 2005). We previously demonstrated that while the bone marrow-derived cells isolated from corneal stroma do not form spheres (Uchida et al., 2005), the keratocytes derived from the corneal stroma do form spheres (Mimura et al., 2008a). This suggests that the precursors isolated from the corneal stroma reside among the keratocytes. If so, these precursors might be able to supply new keratocytes and thus, they could a potentially promote corneal wound healing. Keratocyte precursors may also have a role in the healing of corneal stromal wounds, since proliferation and migration of residual keratocytes from the peripheral part of the stroma appear to promote healing (Wilson & Kaufman, 1990).

## 3. Isolation of sphere colonies

### 3.1 Primary sphere-forming assay

Rabbits were handled in accordance with the ARVO Statement on the Use of Animals in Ophthalmic and Vision Research. New Zealand white rabbits (weighing 2.0-2.4 kg) were used. The animals were anesthetized with intramuscular injections of ketamine hydrochloride (60 mg/kg) and xylazine (10 mg/kg), after which the eyes were enucleated. Subsequently, the eyes were washed three times with sterile saline and then immersed for 5 min in saline containing 10% povidone-iodine and 50 mg/ml gentamicin. After further rinsing with saline, the cornea was excised from each eye along the scleral rim. The epithelium was then carefully removed from the corneal stroma by scraping the outer surface of the cornea, while fine forceps were used to peel away the corneal endothelium and Descemet's membrane in a sheet that covered an area from the periphery to the center of the inner surface of the cornea. Using appropriate trephines and forceps, samples of the corneal stroma were excised from the periphery of the cornea (6.0-10.0 mm in diameter) and from the central region (6.0 mm in diameter) (Fig. 1A). The stromal samples were cut into small pieces that were approximately 1.0 mm in diameter. These were incubated overnight at 37°C in basal medium containing 0.02% collagenase. Subsequently, the tissue pieces were washed with phosphate-buffered saline (PBS), incubated in PBS containing 0.2%

ethylenediaminetetraacetic acid (EDTA) for 5 minutes at 37°C, and then dissociated into single cells by trituration with a fire-polished Pasteur pipette. After centrifugation at 800 g for 5 minutes, the cells were resuspended in the basal medium, which consisted of Dulbecco's modified Eagle's medium (DMEM)/F12 medium supplemented with B27, 20 ng/mL epidermal growth factor (EGF), and 40 ng/mL basic fibroblast growth factor (bFGF).

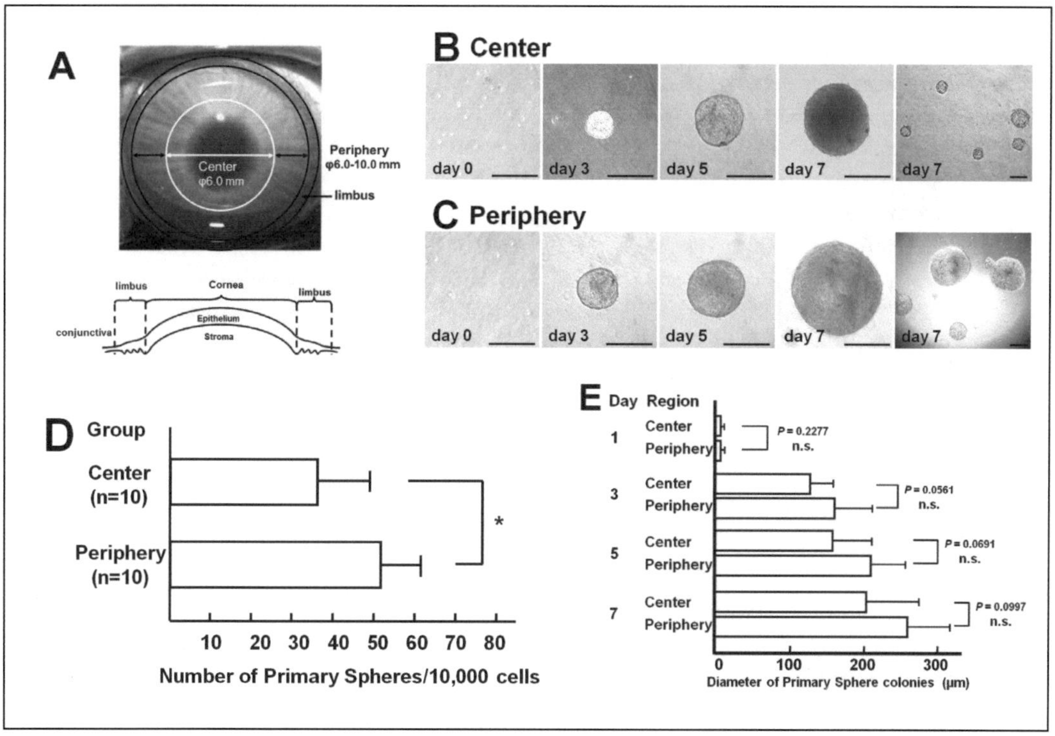


Fig. 1. (A) Anterior view of a rabbit cornea and a diagram of the corneal epithelium and stroma. Stromal keratocytes were isolated from tissue specimens obtained from both the peripheral (6.0-10.0 mm in diameter) and central regions (6.0 mm in diameter). (B, C) Primary sphere formation by the keratocytes from the peripheral and central regions of the rabbit cornea. Scale bar=200 μm. (D) The number of primary spheres derived from stromal tissue was compared between the periphery and the center of the cornea. The number of sphere colonies obtained from samples of the peripheral stroma (n=10) was significantly higher than that for the samples of the central stroma (n=10) after seven days of culture (* $P$=0.00021, unpaired *t*-test ). (E) The sizes of the primary sphere colonies derived from the samples of the peripheral (n=10) and central (n=10) corneal stroma were compared. The mean size of the spheres from both regions gradually increased during the culture period and exceeded 250 μm by day 7 (periphery: 258±63 μm versus center: 203 ± 71 μm after seven days, mean ± SD). n.s.= not significant

Isolated keratocytes were counted using a hemocytometer. Trypan blue staining indicated the viability of the isolated cells was greater than 90%. The sphere-forming assay was employed for the primary culture of the cells (Reynolds & Weiss, 1992). Basal medium containing a methylcellulose gel matrix (0.8%; Wako Pure Chemical Industries) was used to prevent reaggregation of the cells, as per a previously described method (Gritti et al., 1999). Plating was done at a density of 10 viable cells/μL (50,000 cells/well or 2,500 cells/cm$^2$) in the uncoated wells of 60 mm culture dishes. To measure the diameter of sphere colonies, culture dishes were observed under an inverted phase-contrast microscope with a 10× objective lens, followed by analysis of the images using the NIH image program developed at the US National Institutes of Health (n=10). The number of spheres per 10,000 cells was calculated for each well. To distinguish growing spheres from dying cell clusters, only those with a diameter of more than 50 μm were counted.

After keratocytes were disaggregated into single cells and cultured for 7 days, the viable spheres grew larger and the non-proliferating cells were eliminated. To compare the density of the precursors between the peripheral and central regions of the cornea, primary spheres were isolated separately from the peripheral and central stroma. Representative spheres obtained from the peripheral and central regions are shown in the photographs in Figures 1B and 2C. There were significantly more spheres (51.4 ± 10.1 per 10,000 cells, mean ± SD) obtained from the peripheral corneal stroma as compared to the central stroma (35.9 ± 3.0 per 10,000 cells) ($P$=0.00021, unpaired $t$-test; Fig. 1D). Although this result suggests that the stroma from both the peripheral and central regions of the cornea contains precursor cells, there were significantly more precursors in the peripheral stroma versus the central stroma. Additionally, there were no significant differences noted with respect to the size of the primary spheres that were derived from the two regions after culturing for 3, 5, and 7 days. This suggests that there were no differences in the proliferative capacity of the precursors obtained from either region (Fig. 1E).

### 3.2 Secondary sphere formation

To further evaluate the proliferative capacity of the keratocytes, cells from the primary spheres were passaged under the same culture conditions that were used for the initial growth of the spheres. For passaging, primary spheres (day 7) were treated with 0.5% EDTA and after dissociation into single cells, they were then plated into the wells of 60 mm culture dishes at a density of 10 cells/μL. Cultures were continued for 7 days in the basal medium that contained methylcellulose gel matrix to prevent reaggregation. Experiments were performed twice and representative results are shown in Figure 2 (n=10).

Secondary spheres were generated after dissociation of the primary spheres derived from the peripheral or central stroma. Replating to generate secondary sphere colonies was less efficient than the generation of the primary spheres, which indicates that the precursor cells appear to have a limited proliferative capacity. Photographs of representative secondary spheres are shown in Figures 2A and 2B. The number of secondary spheres per 10,000 cells was significantly higher when the primary spheres passaged were derived from the peripheral stroma versus the central stroma (45.6 ± 6.4 vs. 33.4 ± 2.1, respectively; $P$=0.000025; unpaired $t$-test; Fig. 2C). Despite differences in the properties of the cells derived from the keratocyte spheres that were obtained from the peripheral and central regions of the cornea, there were no significant differences noted in the expression of mesenchymal and neural cell markers. These findings imply that keratocyte precursors preferentially reside in the peripheral corneal stroma and have a stronger proliferative capacity as compared to the cells from the central stroma, while the precursors from both regions demonstrate similar multipotentiality.

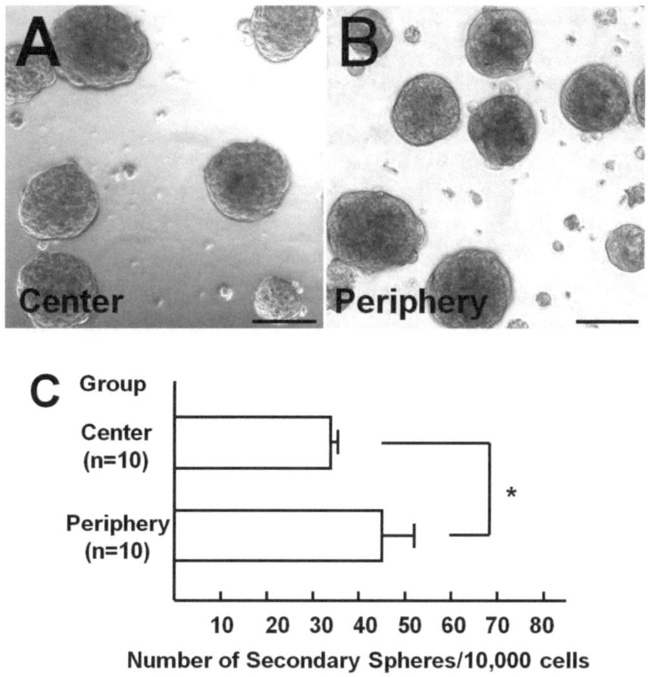


Fig. 2. Formation of secondary sphere. (A, B) Secondary spheres were generated after the dissociation of the primary spheres that were derived from the peripheral or central keratocytes. Scale bar=100 µm. (C) The replating efficiency from the primary to the secondary colonies was higher for the spheres derived from the peripheral stroma as compared to the central stroma (* $P$=0.000025, unpaired $t$-test)

### 3.3 Differentiation of sphere colonies

Individual primary spheres (day 7) were transferred to 13 mm glass coverslips coated with 50 µg/ml poly-L-lysine (PLL) and 10 µg/ml fibronectin in separate wells, as has been previously described (Reynolds & Weiss, 1992). To promote differentiation, 1% FBS was added to the basal medium, with the culture then continued for another 7 days. Immunocytochemical examinations of the 7-day-old spheres and their progeny were performed after 7 days of adherent culture on the glass coverslips.

Nestin has been used as a marker for the detection of immature neural progenitor cells within the multipotential sphere colonies derived from the brain (Gage, 2000), skin (Toma et al., 2001), inner ear (Li et al., 2003), retina (Tropepe et al., 2000), corneal stroma (Uchida et al., 2005), and endothelium (Mimura et al., 2005b, 2005c; Yokoo et al., 2005). In addition, it was recently suggested that the stem cell marker CD34 could be a useful cell surface marker for human keratocytes (Toti et al., 2002). Most cells in the spheres were immunopositive for nestin, CD34, and 5-bromo-2′-deoxyuridine (BrdU) (Fig. 3A). Subsequently, we then examined whether the sphere colonies could give rise to cells expressing neural lineage markers. Some cells in the sphere colonies along with their progeny expressed microtubule-associated protein 2 (MAP2: a neural cell marker) and neuron-specific enolase (NSE: a marker of neural differentiation) (Fig. 3A). Most of the cells in the spheres and their progeny were immunoreactive for vimentin (which is a marker of mesenchymal cells) or for alpha-smooth muscle actin (αSMA: a marker of fibroblasts). However, all of the cells were negative for staining by the control IgG and the differentiated epithelial cell marker cytokeratin 3 (Figs. 3A and 3B). Expression of nestin and vimentin by the spheres and their progeny was confirmed using RT-PCR (Fig. 3C). Both spheres were derived from the peripheral and central regions of the cornea and their progeny displayed the same patterns of immunostaining (data not shown) and mRNA expression (Fig. 3C).

Spheres derived from both the peripheral and central regions of the rabbit cornea expressed a stem cell marker (CD34) and a neural stem cell marker (nestin), while their progeny expressed mesenchymal markers (vimentin and α-SMA) and neural lineage markers (MAP2 and NSE). These findings indicate that spheres isolated from the corneal stroma of rabbits contained bi-potential precursors, with their progeny able to display the morphologic characteristics of keratocytes. Taken together, these results suggest that precursors from the corneal stroma remain close in nature to the tissue of origin and that they undergo differentiation into corneal keratocytes. Thus, since precursors need to be able to efficiently differentiate and produce their tissue of origin, precursors obtained from the corneal stroma may be more appropriate for tissue regeneration or cell transplantation than those that are derived from the multipotential stem cells.

## 4. Tissue engineering of corneal stroma

### 4.1 The organization of the corneal stromal matrix

Adult vertebrate corneal stroma is primarily composed of collagen type 1 fibrils, smaller amounts of other extracellular matrix (ECM) proteins, and a few keratocytes. Therefore, both the cells and the stratified complex of the ECM, which forms a scaffold for the precursors, are necessary in order to be able to reconstruct a three-dimensional corneal stroma. The porous scaffolds that are used in tissue engineering contribute to cell proliferation and differentiation in suitable environments, as well as taking part in the maintenance of the structure and composition of injured tissues. Furthermore, three-dimensional porous scaffolds provide a larger surface for cell attachment, migration and proliferation in contrast to the two-dimensional scaffolds, which facilitate contact inhibition in confluent cells. Because corneal keratocyte proliferation is substrate-dependent, increases in the surface area of the culture substrate are preferable. Several three-dimensional substrates that use collagen have been designed and tested for their capacity for proliferative enhancement (Takahashi et al., 2004, 2005; Yasuda et al., 2004). While their long-term safety, stability, and efficacy in vivo have been adequately established in humans, there are significant drawback of using collagen due to its poor biodegradation and bioabsorption. Biodegradation and biocompatibility are probably the two most important properties that have to be examined when considering the use of insoluble biomaterials. Unfortunately, biological reactions can develop when using these materials in vivo, resulting in tissue opacity. Gelatin, which is a denatured type of collagen, possesses most of the properties of an ideal scaffold and thus, has been clinically applied as an implant material (Waldrop & Semba, 1993). We have also used a biodegradable porous gelatin hydrogel as a carrier of the corneal keratocyte precursors, and found the material to be effective in facilitating cell migration and delivering oxygen and nutrients to the migrated cells.

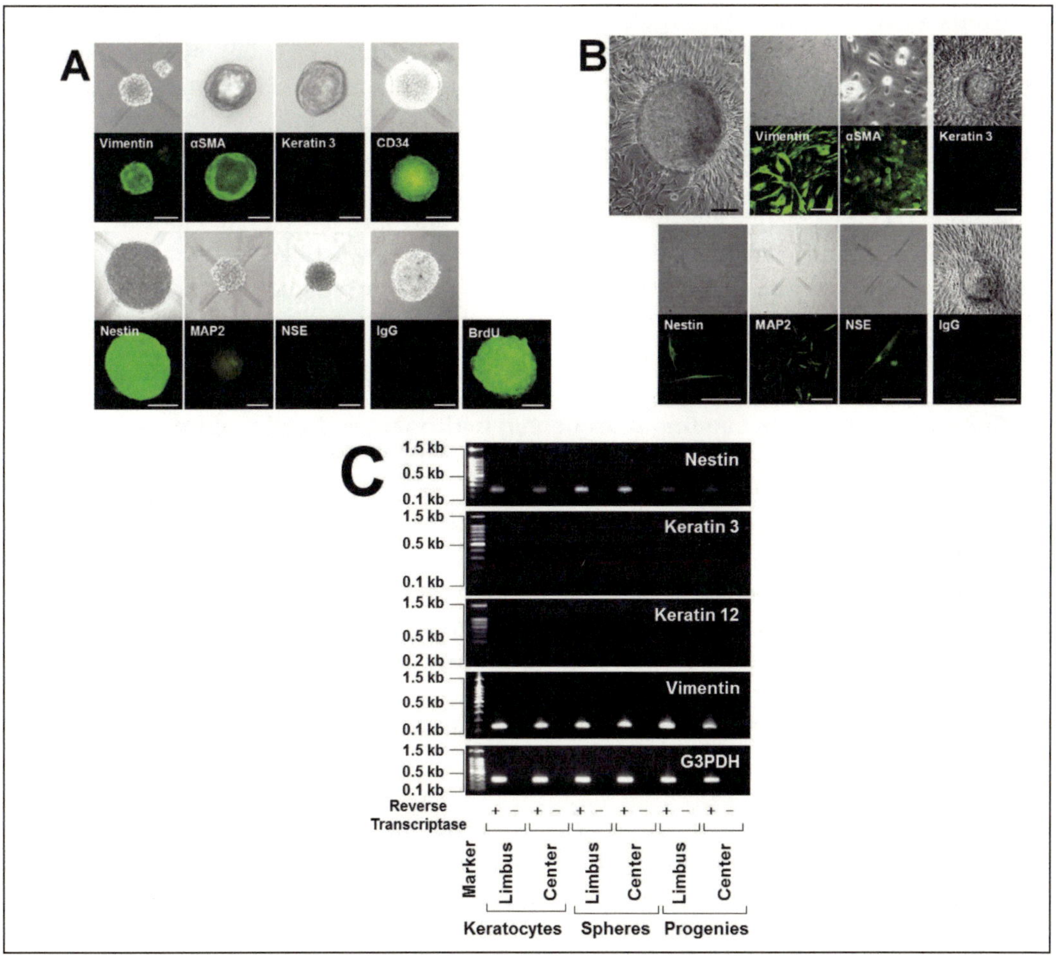


Fig. 3. (A) Immunocytochemical analysis of sphere colonies from the peripheral stroma on day 7. Bright-field images and immunostaining of the spheres are shown. The spheres were stained for vimentin, alpha-smooth muscle actin (α-SMA), cytokeratin 3, nestin, microtubule-associated protein 2 (MAP2), neuron-specific enolase (NSE), and CD34. Each colony was also labeled by BrdU. As a negative control, IgG was used instead of the primary antibody. Scale bar=100 μm. (B) Immunocytochemical analysis of differentiated cells from spheres derived from the peripheral cornea. The cells that migrated out from the spheres expressed α-SMA, MAP2, and NSE, indicating that the colonies contain differentiated mesenchymal and neuronal cells. No staining with the control IgG was noted. Scale bar=100 μm. (C) Reverse-transcription polymerase chain reaction analysis of the corneal stromal tissue, spheres, and sphere progeny. Glyceraldehyde-3-phosphate dehydrogenase (G3PDH) gene expression was detected in all samples except those processed without reverse-transcriptase (RT). Vimentin was expressed by the corneal stromal tissues and the spheres derived from the peripheral or central regions and their progeny. However, it was not detected by PCR of the total RNA without RT. Expression of nestin by the progeny was lower than that seen for the spheres from both the peripheral and central regions of the cornea. No expression of keratin 3 or 12 was detected in any of these samples

### 4.2 Preparation of gelatin hydrogels

Porous gelatin hydrogels were prepared through the chemical crosslinking of an aqueous gelatin solution with glutaraldehyde, using a method described elsewhere (Tabata et al., 1999; Yamamoto et al., 2003). Briefly, an aqueous gelatin solution mixed with glutaraldehyde was cast into a polypropylene dish (138x138 mm$^2$, BIO-BIK), followed by the crosslinking reaction, which was performed at 4°C for 12 h. The hydrogel samples were added to a 100 mM aqueous glycine solution at 37°C and stirred for 1 h to block the residual glutaraldehyde. After washing three times with double-distilled water, the hydrogel samples were freeze-dried and sterilized with ethylene oxide gas and ultraviolet for 2 hours. Before use, the desiccated sheets were trephined with a 5.0 mm diameter trephine and then immersed in sterilized saline for 10 minutes.

### 4.3 Sphere or keratocyte seeding onto gelatin hydrogels and cell culture

The 5.0 mm gelatin hydrogel disks were placed into wells of a 24-well culture dish. Subsequently, 10 primary rabbit corneal keratocyte spheres that had been cultured for 7 days or 100,000 primary cultured rabbit corneal keratocytes were applied to the porous gelatin hydrogels. The reconstructed corneal stroma was then placed in the 24-well plates and centrifuged at 1000 rpm (176 g) for 10 min to promote the attachment of the cells to the porous gelatin hydrogels. To trace the localization in some of the experiments, primary rabbit corneal keratocyte spheres (cultured for 7 days) were labeled with a fluorescent cell tracker, as has been described elsewhere (Mimura et al., 2004). The corneal keratocytes or the corneal keratocyte spheres were cultured in the basal medium containing 10% FBS for 7 days, with the medium changed every 2 to 3 days.

### 4.4 Immunocytochemical analysis of ECM in gelatin hydrogels

Corneal keratocytes or keratocyte precursors were cultured on porous gelatin hydrogels for 7 days, with the degree of ECM production by the cells then evaluated. Vimentin expression was more frequent in the porous gelatin hydrogels that contained keratocyte precursors versus those with only the corneal keratocytes (Fig. 4). Weak expression of laminin was seen in the gelatin hydrogels with keratocyte precursors, while no expression was observed in those with the corneal keratocyte. There was little or no expression of type I and IV collagen found in either the porous gelatin hydrogels with corneal keratocytes or the keratocyte precursors (Fig. 4).

Despite having a total cell number less than 100,000 corneal keratocytes within the 10 corneal keratocyte spheres, when the porous gelatin hydrogels that incorporated the spheres were cultured *ex vivo* for 7 days, there was intense immunostaining for vimentin, a marker of mesenchymal cells. This staining was much more intense than that which was seen for the incorporated 100,000 corneal keratocytes that were also cultured for 7 days. This indicates that the corneal keratocyte precursors have a superior proliferative potential on the gelatin hydrogels as compared to the corneal keratocytes. While weak expression of laminin and collagens was detected in the gelatin hydrogels that incorporated the corneal keratocytes precursors, these ECM components were barely detected in the gelatin that incorporated the corneal keratocytes *ex vivo* before transplantation. These results indicate that the gelatin hydrogel, itself, does not have the ability to induce tissue regeneration either *in vitro* or *ex vivo*.

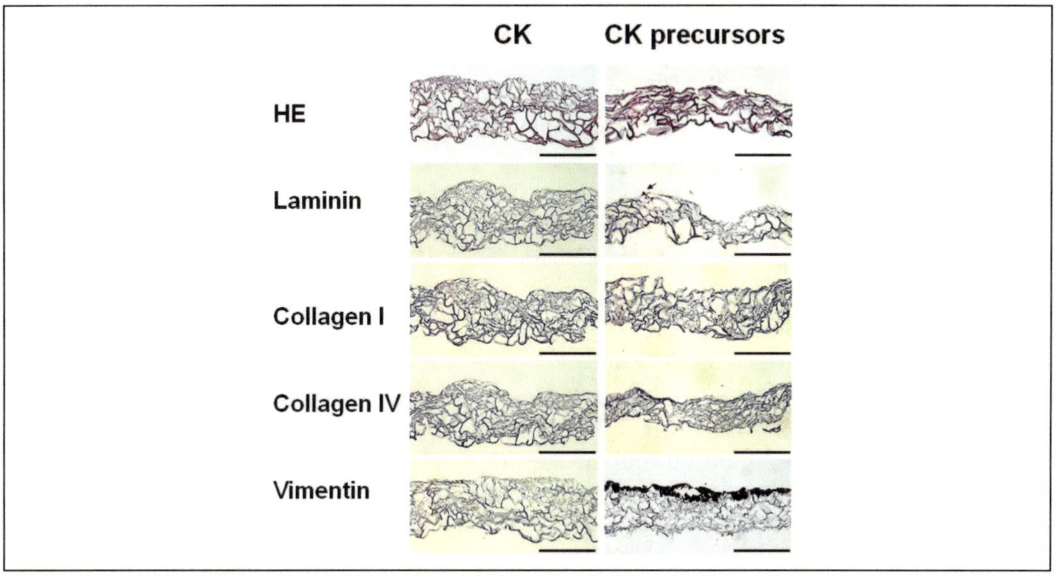


Fig. 4. Immunohistochemical analysis of the extracellular matrix in the porous gelatin hydrogel with corneal keratocytes or keratocyte precursors. Corneal keratocytes or keratocyte precursors were seeded onto porous gelatin hydrogels and cultured for one week. Vimentin staining was more intense in the gelatin hydrogels with corneal fibroblast precursors than that seen with the corneal fibroblasts (arrows). Except for a weak expression of laminin in the gelatin hydrogels with corneal keratocyte precursors (arrow), there was no expression of the other ECM components such as laminin, type I collagen, and type IV collagen in the gelatin hydrogel with the corneal keratocytes or keratocyte precursors before transplantation. Scale bar=200 μm. HE=hematoxylin and eosin staining

## 5. Transplantation of gelatin hydrogels with corneal keratocyte precursors

### 5.1 Surgical procedure

New Zealand white rabbits (weighing 2.0-2.4 kg, n=24) were anesthetized with an intramuscular injection of ketamine hydrochloride (60 mg/kg) and xylazine (10 mg/kg). A diamond knife was used to make a 6 mm incision 1.0 mm inside the superior limbus, at a depth of 150-200 μm (Fig. 5A). A lamellar dissection of the cornea was then performed with a bevel-up crescent knife. The reconstructed corneal stroma with corneal keratocyte precursors and porous gelatin was then implanted into a mid-stroma corneal pocket with a jeweler's forceps (Fig. 5B). No sutures were used for the corneal incision after the implantation (Fig. 5C). An antibiotic ointment (0.3% ofloxacin) was applied to the eye. The 24 rabbits were divided into three groups: the gelatin group (n=8) for which porous gelatin hydrogels alone were transplanted, the keratocyte/gelatin group (n=8) for which corneal keratocytes cultured on porous gelatin hydrogels for 7 days were transplanted, and the precursor/gelatin group (n=8) for which the corneal keratocyte precursors cultured on porous gelatin hydrogels for 7 days were transplanted. Histological examinations and immunohistochemistry were performed in the animals killed at 1 week (n=2 in each group) and at 4 weeks after surgery (n=6 in each group). None of the animals received any immunosuppressive treatment after the transplantation.

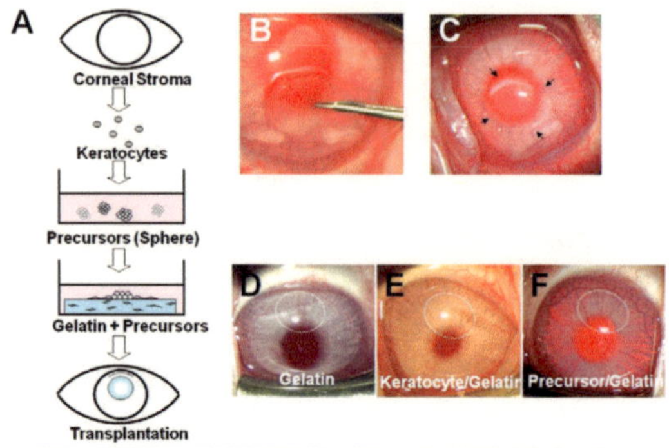


Fig. 5. Schematic illustration and clinical findings. A: Keratocyte precursors were isolate from the rabbit corneal stroma using a sphereforming assay. Corneal stroma was engineered by cultivating precursors in porous gelatin for one week. The engineered corneal stromal sheet with precursors was transplanted in a pocket of rabbit corneal stroma. B,C: Gelatin hydrogels (gelatin group), gelatin hydrogels with corneal keratocyte (keratocyte/gelatin group), or gelatin hydrogels with corneal keratocyte precursors (precursor/gelatin group) were implanted into the corneal stroma (indicated by arrows in C). D-F: Representative photographs of corneas four weeks after transplantation in each group are shown. No corneal opacity and no rejection were observed in any group four weeks after transplantation (indicated by the white circles in D-F)

### 5.2 Clinical observation after surgery

Each eye that underwent the surgical procedure was checked two to three times a week by external examination, in addition to being photographed on postoperative days 7, 14, 21, and 28. As seen in the representative anterior segment photographs from the gelatin group (transplanted with gelatin hydrogels alone, Fig. 5D), the keratocyte/gelatin group (transplanted with gelatin hydrogels with corneal keratocytes, Fig. 5E), and the precursor/gelatin group (transplanted with gelatin hydrogels with corneal keratocyte precursors, Fig. 5F), the corneas were clear, and there was no edema or mononuclear cell infiltration of the stroma in any of the groups. There were also no apparent inflammatory reactions suggestive of immunological rejection observed with a slit lamp microscope nor were there any increases of intraocular pressure, or possible side effects noted in any of the groups during the follow-up period.

### 5.3 Histological examination and analysis of ECM production with immunohistochemistry

The corneas were excised from the eyes 1 and 4 weeks after the transplantations and subjected to histological examinations and immunohistochemical staining. In the gelatin and keratocyte/gelatin groups, the few cells that did migrate into the thick transplanted gelatin hydrogels exhibited no changes in shape. In contrast, large numbers of cells migrated into the gelatin hydrogels of the precursor/gelatin group, which became thinner than the other two groups at 1 and 4 weeks after transplantation. The expressions of vimentin and ECM, such as laminin and type I and type IV collagens were more intense in the precursor/gelatin group versus the other groups after 1 week (data not shown), with their expressions significantly increased at 4 weeks (Fig. 6A). These results suggest that the gelatin hydrogel by itself can induce ECM production to some extent *in vivo*, although the efficacy is not as high as that seen for the gelatin hydrogels with precursors. These findings clearly indicate that transplanted corneal keratocyte precursors possess the ability to generate new corneal stromal tissues *in vivo* by cell proliferation and differentiation properties, and ECM production.

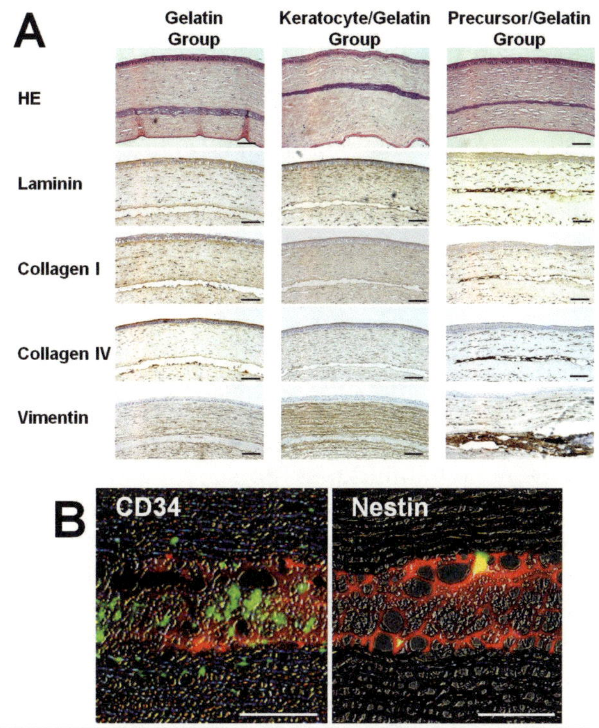


Fig. 6. (A) Histological findings and immunocytochemical analysis of the extracellular matrix at 4 weeks after transplantation. Transplanted gelatin hydrogels were found in the corneal stroma in all groups. HE staining revealed no mononuclear cell infiltration around the gelatin hydrogels in any of the groups. The precursor/gelatin group showed more intense staining for laminin, type I collagen, type IV collagen, and vimentin in the transplanted gelatin as compared to that seen in the gelatin and keratocyte/gelatin groups at 4 weeks after transplantation. Scale bar=100 μm. (B) Immunolocalization of CD34- or nestin-positive cells within the transplanted DiI-positive precursors in the precursor/gelatin group 4 weeks after transplantation of gelatin hydrogels with corneal keratocyte precursors. A computer software program (Adobe Photoshop) was used to superimpose the findings in a single image. In this image, bright light represents the background (black and white), while rhodamine (red color) shows the transplanted DiI-labeled corneal keratocyte precursors in the gelatin hydrogels, and FITC (green color) shows the CD34- or nestin-positive cells. The light red color associated with the many DiI-positive corneal keratocyte precursors indicates the whole transplanted gelatin hydrogel. A few CD34- positive cells or nestin-positive spindle cells are seen scattered within the gelatin hydrogels.
Scale bar=100 μm

### 5.4 Expression of stem or progenitor cell markers in the transplanted gelatin hydrogels

We next examined the expressions of the stem cell marker, CD34, and the neural progenitor cell marker, nestin, in the precursor/gelatin group 4 weeks after transplantation. Many CD34-positive cells were detected in and around the transplanted gelatin hydrogels (Fig. 6B). In addition, a few nestin-positive cells were also seen in the transplanted gelatin hydrogels, whereas nestin-positive cells were barely detected in the corneal stroma around the gelatin hydrogels (Fig. 6B). This indicates that corneal keratocyte precursors with a greater self-renewal potential continue to proliferate even after the transplantation, and thus, are able to supply the keratocytes that are necessary for regeneration of the host stroma. The nestin-positive cells were also present in the transplanted gelatin hydrogels, which indicates that they can contribute to the induction of the nerve regeneration. Large corneal nerves penetrate from the pericorneal nerve plexus into the anterior corneal stroma as thick nerve bundles, which provides sensory innervation of the cornea (Maclver & Tanelian, 1993). Since the transplanted corneal keratocyte precursors would be expected to consistently promote nerve regeneration *in vivo*, they may consequently be able to produce sufficient corneal sensation.

### 5.5 Advantage of transplantation of keratocyte precursors with gelatin hydrogels

The combined approach of transplanting the corneal keratocyte precursors along with the gelatin hydrogels into a corneal stromal pocket has several advantages over the use of penetrating keratoplasty in a full-thickness donor cornea. For example, complications associated with open-sky surgery that involve expulsive hemorrhage and risks of wound dehiscence are essentially eliminated. In addition, several postoperative complications, such as postoperative corneal irregular astigmatism, wound leakage, corneal infection, vascularization, and persistent epithelial defect can be avoided when using the combined approach. In conventional full-thickness human corneal allografting with local and/or systemic immunosuppressants, the leading cause of the failure is allograft rejection (Price et al., 1991; Wilson & Kaufman, 1990). Since histologically there was no apparent inflammatory reaction observed, this suggests that there was no immunological rejection detected in the current corneal keratocyte precursor allotransplantation. Thus, this finding indicates that the transplanted precursors can survive in the corneal stroma without rejection.

## 6. Conclusion

We demonstrated that as compared to the stroma from the central cornea, the stroma from the peripheral region of the rabbit cornea contains a higher density of precursors with a strong proliferative capacity. These keratocyte precursors are able to differentiate into both mesenchymal fibroblasts and neural cells. Furthermore, by using corneal keratocyte precursors and gelatin hydrogels, we were able to establish a new method of three-dimensional reconstruction of the corneal stroma. This new approach that uses corneal keratocyte precursor–based corneal stromal regeneration combined with gelatin hydrogel is a potentially promising new therapy that can be used to attract keratocytes and ECM after the transplantation of the corneal keratocyte precursors. Additionally, the current findings have important implications with regard to the field of regenerative medicine, as this therapeutic tissue engineering approach appears to be applicable to any type of cell. The transplantation of corneal keratocyte precursors into a corneal stromal pocket proved to be a simple and an effective treatment strategy. By being able to use precursors to initiate corneal regeneration, and ECM production to improve wound healing, this new method could potentially be used to replace conventional full-thickness corneal grafting, thereby helping to alleviate the worldwide shortage of donor corneas that exists at the present time.

## 7. Acknowledgment

This work is supported by a Grant-in-Aid for Scientific Research from the Ministry of Education, Culture, Sports, Science and Technology of Japan.

## 8. References

Amano, S., Yamagami, S., Mimura, T., Uchida, S., & Yokoo, S. (2006). Corneal stromal and endothelial cell precursors. *Cornea,* Vol.25, No.10(Suppl 1), (December 2006), pp. S73-77, ISSN 0277-3740

Bahn, C. F., Falls, H. F., Varley, G. A., Meyer, R. F., Edelhauser, H. F., & Bourne, W. M. (1984). Classification of corneal endothelial disorders based on neural crest origin. *Ophthalmology,* Vol.91, No.6, (June 1984), pp. 558-563, ISSN 0161-6420

Coles, B. L., Angenieux, B., Inoue, T., Del Rio-Tsonis, K., Spence, J. R., McInnes, R. R., Arsenijevic, Y., & van der Kooy, D. (2004). Facile isolation and the characterization of human retinal stem cells. *Proc Natl Acad Sci U S A,* Vol.101, No.44, (November 2004), pp. 15772-15777, ISSN 1091-6490

Gage, F. H. (2000). Mammalian neural stem cells. *Science,* Vol.287, No.5457, (February 2000), pp. 1433-1438, ISSN 0036-8075

Gritti, A., Frolichsthal-Schoeller, P., Galli, R., Parati, E. A., Cova, L., Pagano, S. F., Bjornson, C. R., & Vescovi, A. L. (1999). Epidermal and fibroblast growth factors behave as mitogenic regulators for a single multipotent stem cell-like population from the subventricular region of the adult mouse forebrain. *J Neurosci.* Vol.19, No.9, (May 1999), pp. 3287-3297, ISSN 1529-2401

Johnston, M. C., Noden, D. M., Hazelton, R. D., Coulombre, J. L., & Coulombre, A. J. (1979). Origins of avian ocular and periocular tissues. *Exp Eye Res,* Vol.29, No.1, (July 1979), pp. 27-43, ISSN 0014-4835

Kawase, Y., Yanagi, Y., Takato, T., Fujimoto, M., & Okochi, H. (2004). Characterization of multipotent adult stem cells from the skin: transforming growth factor-beta (TGF-beta) facilitates cell growth. *Exp Cell Res,* Vol.295, No.1, (April 2004), pp. 194-203, ISSN 0014-4827

Krause, D. S., Theise, N. D., Collector, M. I., Henegariu, O., Hwang, S., Gardner, R., Neutzel, S., & Sharkis, S. J. (2001). Multi-organ, multi-lineage engraftment by a single bone marrow-derived stem cell. *Cell,* Vol.105, No.3, (May 2001), pp. 369-377, ISSN 0092-8674

Li, H., Liu, H., & Heller, S. (2003). Pluripotent stem cells from the adult mouse inner ear. *Nat Med,* Vol.9, No.10, (October 2003), pp. 1293-1299, ISSN 1078-8956

Liu, C. Y., Shiraishi, A., Kao, C. W., Converse, R. L., Funderburgh, J. L., Corpuz, L. M., Conrad, G. W., & Kao, W. W. (1998). The cloning of mouse keratocan cDNA and genomic DNA and the characterization of its expression during eye development. *J Biol Chem,* Vol.273, No.35, (August 1998), pp. 22584-22588, ISSN 0021-9258

MacIver, M. B., & Tanelian, D. L. (1993). Structural and functional specialization of A delta and C fiber free nerve endings innervating rabbit corneal epithelium. *J Neurosci*, Vol.13, No.10, (October 1993), pp. 4511-4524, ISSN 1529-2401

Meier, S. (1982). The distribution of cranial neural crest cells during ocular morphogenesis. *Prog Clin Biol Res*, Vol.82, (1982), pp. 1-15, ISSN 0361-7742

Mimura, T., Yamagami, S., Yokoo, S., Usui, T., Tanaka, K., Hattori, S., Irie, S., Miyata, K., Araie, M., & Amano, S. (2004). Cultured human corneal endothelial cell transplantation with a collagen sheet in a rabbit model. *Invest Ophthalmol Vis Sci*, Vol.45, No.9, (September 2004), pp. 2992-2997, ISSN 0146-0404

Mimura, T., Yamagami, S., Yokoo, S., Yanagi, Y., Usui, T., Ono, K., Araie, M., & Amano, S. (2005a). Sphere therapy for corneal endothelium deficiency in a rabbit model. *Invest Ophthalmol Vis Sci*, Vol.46, No.9, (September 2005), pp. 3128-3135, ISSN 0146-0404

Mimura, T., Yokoo, S., Araie, M., Amano, S., & Yamagami, S. (2005b). Treatment of rabbit bullous keratopathy with precursors derived from cultured human corneal endothelium. *Invest Ophthalmol Vis Sci*, Vol.46, No.10, (October 2005), pp. 3637-3644, ISSN 0146-0404

Mimura, T., Yamagami, S., Yokoo, S., Araie, M., & Amano, S. (2005c). Comparison of rabbit corneal endothelial cell precursors in the central and peripheral cornea. *Invest Ophthalmol Vis Sci*, Vol.46, No.10, (October 2005), pp. 3645-3648, ISSN 0146-0404

Mimura, T., Yamagami, S., Usui, T., Seiichi, Honda, N., & Amano, S. (2007). Necessary prone position time for human corneal endothelial precursor transplantation in a rabbit endothelial deficiency model. *Curr Eye Res*, Vol.32, No.7-8, (July-August 2007), pp. 617-623, ISSN 1460-2202

Mimura, T., Amano, S., Yokoo, S., Uchida, S., Usui, T., & Yamagami, S. (2008a). Isolation and distribution of rabbit keratocyte precursors. *Mol Vis*, Vol.14, (January 2008), pp. 197-203, ISSN 1090-0535

Mimura, T., Amano, S., Yokoo, S., Uchida, S., Yamagami, S., Usui, T., Kimura, Y., & Tabata, Y. (2008b). Tissue engineering of corneal stroma with rabbit fibroblast precursors and gelatin hydrogels. *Mol Vis*, Vol.14, (January 2008), pp. 1819-1828, ISSN 1090-0535

Mimura, T., Yamagami, S., Uchida, S., Yokoo, S., Ono, K., Usui, T., & Amano, S. (2010a). Isolation of adult progenitor cells with neuronal potential from rabbit corneal epithelial cells in serum- and feeder layer-free culture conditions. *Mol Vis*, Vol.16, (August 2010), pp. 1712-1719, ISSN 1090-0535

Mimura, T., Yamagami, S., Yokoo, S., Usui, T., & Amano, S. (2010b). Selective isolation of young cells from human corneal endothelium by the sphere-forming assay. *Tissue Eng Part C Methods*, Vol.16, No.4, (August 2010), pp. 803-812, ISSN 1937-3384

Mooy, C. M., Clark, B. J., & Lee, W. R. (1990). Posterior axial corneal malformation and uveoretinal angiodysgenesis--a neurocristopathy? *Graefes Arch Clin Exp Ophthalmol*, Vol.228, No.1, (1990), pp. 9-18, ISSN 0721-832X

Nakamura, T., Ishikawa, F., Sonoda, K. H., Hisatomi, T., Qiao, H., Yamada, J., Fukata, M., Ishibashi, T., Harada, M., & Kinoshita, S. (2005). Characterization and distribution of bone marrow-derived cells in mouse cornea. *Invest Ophthalmol Vis Sci*, Vol.46, No.2, (February 2005), pp. 497-503, ISSN 0146-0404

Nunes, M. C., Roy, N. S., Keyoung, H. M., Goodman, R. R., McKhann, G., 2nd, Jiang, L., Kang, J., Nedergaard, M., & Goldman, S. A. (2003). Identification and isolation of multipotential neural progenitor cells from the subcortical white matter of the adult human brain. *Nat Med,* Vol.9, No.4, (April 2003), pp. 439-447, ISSN 1078-8956

Price, F. W., Jr., Whitson, W. E., & Marks, R. G. (1991). Graft survival in four common groups of patients undergoing penetrating keratoplasty. *Ophthalmology,* Vol.98, No.3, (March 1991), pp. 322-328, ISSN 0161-6420

Reynolds, B. A., & Weiss, S. (1992). Generation of neurons and astrocytes from isolated cells of the adult mammalian central nervous system. *Science,* Vol.255, No.5052, (March 1992), pp. 1707-1710, ISSN 0036-8075

Tabata, Y., Hijikata, S., Muniruzzaman, M., & Ikada, Y. (1999). Neovascularization effect of biodegradable gelatin microspheres incorporating basic fibroblast growth factor. *J Biomater Sci Polym Ed,* Vol.10, No.1, (1999), pp. 79-94, ISSN 1568-5624

Takahashi, Y., & Tabata, Y. (2004). Effect of the fiber diameter and porosity of non-woven PET fabrics on the osteogenic differentiation of mesenchymal stem cells. *J Biomater Sci Polym Ed,* Vol.15, No.1, (2004), pp. 41-57, ISSN 1568-5624

Takahashi, Y., Yamamoto, M., & Tabata, Y. (2005). Osteogenic differentiation of mesenchymal stem cells in biodegradable sponges composed of gelatin and beta-tricalcium phosphate. *Biomaterials,* Vol.26, No.17, (June 2005), pp. 3587-3596, ISSN 0142-9612

Toma, J. G., Akhavan, M., Fernandes, K. J., Barnabe-Heider, F., Sadikot, A., Kaplan, D. R., & Miller, F. D. (2001). Isolation of multipotent adult stem cells from the dermis of mammalian skin. *Nat Cell Biol,* Vol.3, No.9, (September 2001), pp. 778-784, ISSN 1465-7392

Toti, P., Tosi, G. M., Traversi, C., Schurfeld, K., Cardone, C., & Caporossi, A. (2002). CD-34 stromal expression pattern in normal and altered human corneas. *Ophthalmology,* Vol.109, No.6, (June 2002), pp. 1167-1171, ISSN 0161-6420

Tropepe, V., Coles, B. L., Chiasson, B. J., Horsford, D. J., Elia, A. J., McInnes, R. R., & van der Kooy, D. (2000). Retinal stem cells in the adult mammalian eye. *Science,* Vol.287, No.5460, (March 2000), pp. 2032-2036, ISSN 0036-8075

Uchida, S., Yokoo, S., Yanagi, Y., Usui, T., Yokota, C., Mimura, T., Araie, M., Yamagami, S., & Amano, S. (2005). Sphere formation and expression of neural proteins by human corneal stromal cells in vitro. *Invest Ophthalmol Vis Sci,* Vol.46, No.5, (May 2005), pp. 1620-25, ISSN 0146-0404

Waldrop, T. C., & Semba, S. E. (1993). Closure of oroantral communication using guided tissue regeneration and an absorbable gelatin membrane. *J Periodontol,* Vol.64, No.11, (November 1993), pp. 1061-1066, ISSN 0022-3492

Wilson, S. E., He, Y. G., Weng, J., Li, Q., McDowall, A. W., Vital, M., & Chwang, E. L. (1996). Epithelial injury induces keratocyte apoptosis: hypothesized role for the interleukin-1 system in the modulation of corneal tissue organization and wound healing. *Exp Eye Res,* Vol.62, No.4, (April 1996), pp. 325-327, ISSN 0014-4835

Wilson, S. E., & Kaufman, H. E. (1990). Graft failure after penetrating keratoplasty. *Surv Ophthalmol,* Vol.34, No.5, (March-April 1990), pp. 325-356, ISSN 0039-6257

Yamagami, S., Mimura, T., Yokoo, S., Takato, T., & Amano, S. (2006a). Isolation of human corneal endothelial cell precursors and construction of cell sheets by precursors. *Cornea,* Vol.25, No.10(Suppl 1), (December 2006), pp. S90-92, ISSN 0277-3740

Yamagami, S., Ebihara, N., Usui, T., Yokoo, S., & Amano, S. (2006b). Bone marrow-derived cells in normal human corneal stroma. *Arch Ophthalmol,* Vol.124, No.1, (January 2006), pp. 62-69, ISSN 0003-9950

Yamagami, S., Yokoo, S., Mimura, T., Takato, T., Araie, M., & Amano, S. (2007). Distribution of precursors in human corneal stromal cells and endothelial cells. *Ophthalmology,* Vol.114, No.3, (March 2007), pp. 433-439, ISSN 0161-6420

Yamamoto, M., Takahashi, Y., & Tabata, Y. (2003). Controlled release by biodegradable hydrogels enhances the ectopic bone formation of bone morphogenetic protein. *Biomaterials,* Vol.24, No.24, (November 2003), pp. 4375-4383, ISSN 0142-9612

Yasuda, K., Inoue, S., & Tabata, Y. (2004). Influence of culture method on the proliferation and osteogenic differentiation of human adipo-stromal cells in nonwoven fabrics. *Tissue Eng,* Vol.10, No.9-10, (September-October 2004), pp. 1587-1596, ISSN 2152-4955

Yokoo, S., Yamagami, S., Yanagi, Y., Uchida, S., Mimura, T., Usui, T., & Amano, S. (2005). Human corneal endothelial cell precursors isolated by sphere-forming assay. *Invest Ophthalmol Vis Sci,* Vol.46, No.5, (May 2005), pp. 1626-31, ISSN 0146-0404

# 17

# Human Ear Cartilage

Lu Zhang, Qiong Li, Yu Liu, Guangdong Zhou, Wei Liu and Yilin Cao
*Department of Plastic and Reconstructive Surgery, Shanghai 9th People's Hospital, Shanghai Jiao Tong University School of Medicine, Shanghai Key Laboratory of Tissue Engineering, Shanghai P.R. China*

## 1. Introduction

The human ear (Fig. 1) is of an ovoid form, with its larger end directed upward. Its lateral surface is irregularly concave, directed slightly forward, and presents numerous eminences and depressions to which names have been assigned (Beahm, Walton, 2002; Walton, Beahm, 2002). The prominent rim of the human ear is called the helix while another curved prominence, parallel with and in front of the helix, is called the antihelix; this divides above into two crura, between which is a triangular depression, the fossa triangularis. The narrow-curved depression between the helix and the antihelix is called the scapha; the antihelix describes a curve around a deep, capacious cavity, the concha, which is partially divided into two parts by the crus or commencement of the helix; the upper part is termed the cymba concha, the lower part the cavum concha. In front of the concha, and projecting backward over the meatus, is a small pointed eminence, the tragus, so called from its being generally covered on its under surface with a tuft of hair, resembling a goat's beard. Opposite the tragus, and separated from it by the intertragic notch, is a small tubercle, the antitragus. Below this is the lobule, composed of tough areolar and adipose tissues, and wanting the firmness and elasticity of the rest of the auricula.

Up to now, total human ear reconstruction for congenital microtia or auricular traumatic amputation still remains one of the greatest challenges for plastic surgeons(Brent, 1999; Nagata, 1993; TANZER, 1959). Although tissue engineering is a promising method for repair and reconstruction of cartilage defects(Chung, Burdick, 2008; Langer, Vacanti, 1993), engineering cartilage with a delicate three dimensional (3D) structure, such as a human ear, remains a great challenge in this field(Ciorba, Martini, 2006; Sterodimas et al., 2009; Zhang, 2010). Since in 1997 Cao *et al.* engineered the cartilage with a shape of human auricle in a nude mouse model(Cao et al., 1997), many researchers have tried to explore further developments of this tissue engineering system, but few of them have succeeded in *in vitro* regeneration of a cartilage construct with a complete and anatomically refined auricle structure(Haisch et al., 2002; Isogai et al., 2004; Kamil et al., 2003; Kamil et al., 2004; Naumann et al., 2003; Neumeister et al., 2006; Shieh et al., 2004; Xu et al., 2005)(Table 1).

One major reason leading to the failure of *in vitro* engineering a cartilage construct with sufficient control over shape is the lack of appropriate scaffolds(Liu et al., 2010). The optimal scaffold used for engineering a cartilage construct with accurate designed shapes should possess at least three characteristics: good biocompatibility for cartilage formation, ease of being processed into a specific shape, and sufficient mechanical strength for retaining the pre-designed shape during chondrogenesis. Polyglycolic acid (PGA) has proven to be one of the most successful scaffolds for cartilage regeneration(Cui et al., 2009; Frenkel, Di, 2004; Heath, Magari, 1996). Cartilage engineered with the PGA scaffold has structure and composition similar to the native tissue, as demonstrated by histological analysis and cartilage specific matrices(Aufderheide, Athanasiou, 2005; Moran et al., 2003; Yan et al., 2009). However, the most widely used form of PGA material in cartilage engineering is unwoven fiber mesh, which is difficult to be initially prepared into a complicated 3D structure and would most likely fail to maintain its original architecture during subsequent *in vitro* chondrogenesis due to insufficient mechanical support(Gunatillake, Adhikari, 2003; Kim, Mooney, 1998; Moran et al., 2003).

| Year | Issue Name | Author | Scaffold | Seeding cells | Shape | Culture |
|---|---|---|---|---|---|---|
| 1997 | Plastic and Reconstructive Surgery | Cao Yilin et al. | PGA+PLA | Bovine chondrocytes | 3-year-old child (partial size) | nude mouse |
| 2002 | European Archives of Oto-Rhino-Laryngology | Andreas Haisch et al. | PGA+PLA+Fibrinogen | human nasal septum chondrocytes | Poor shape (silicon stent) | nude mouse (partial) |
| 2003 | European Archives of Oto-Rhino-Laryngology | A.Naumann et al. | Hyaff 11 | Human nasoseptal chondrocytes | Real size? | Vitro? (partial) |
| 2003 | The Laryngoscope | Syed H. Kamil et al. | PGA+PLA | Newborn calf shoulders chondrocytes | Real size Auricle and Nasal Tip | Vitro+ nude rat |
| 2004 | Tissue engineering | NORITAKA ISOGAI et al. | PLLA+PCL | Newborn calf shoulders chondrocytes | 1-year-old child (partial size) | Nude mouse |
| 2004 | Biomaterials | Shyh-Jou Shieh et al. | PGA ,PCL ,P4HB | Adult sheep chondrocytes; rabbit ear cartilage but failed | Poor shape | Vitro+ nude mouse |
| 2004 | The Laryngoscope | Kamil,S.H. et al. | Calcium alginate, pluronic, PGA | Chondrocytes | Poor shape(gold stent) | Pig, sheep |
| 2005 | Plastic and Reconstructive Surgery | Xu Jianwei et al. | Fibrin | Chondrocytes,perichondrium | Hand sculpted | Nude mouse |
| 2006 | Plastic and Reconstructive Surgery | Michael W. Neumeister et al. | Fibrin glue | Chondrocytes femoral vascular pedicle | Poor shape but vascularized (silicone mold) | Rat |

Table 1.

To overcome these problems, two crucial issues should be addressed. First, the PGA-based scaffold should be prefabricated into the exact shape of a human ear. Second, the mechanical strength of the above-mentioned scaffold should be further enhanced so that it can retain the pre-designed shape during *in vitro* chondrogenesis.

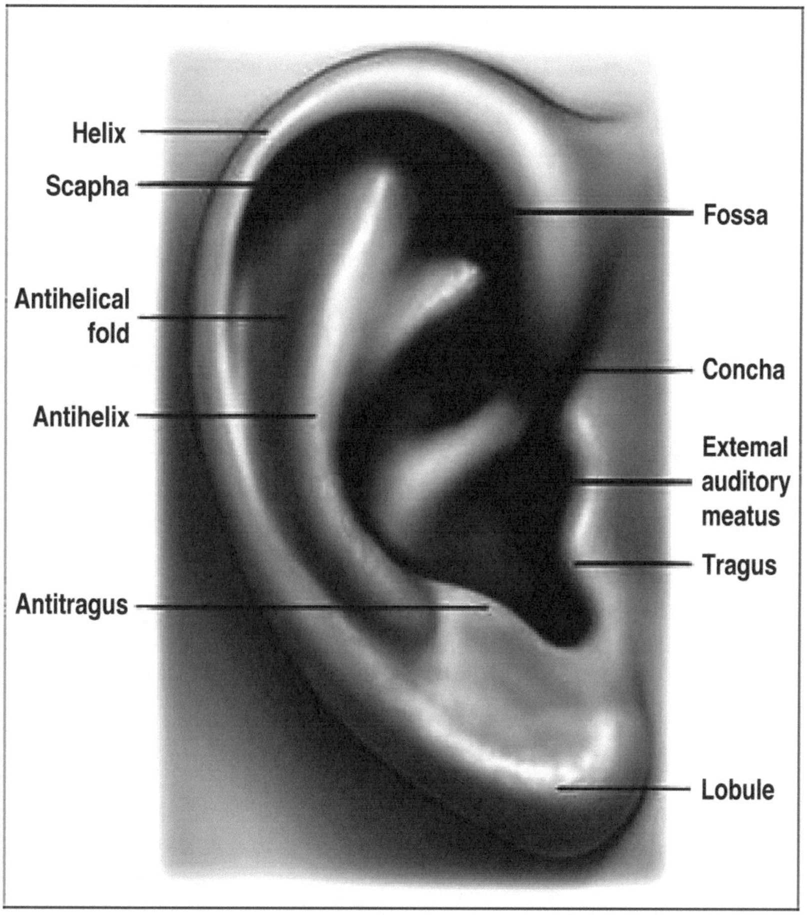


Fig. 1. The outline of a human ear

In order to meet these requirements, in the current study, a computer aided design and manufacturing (CAD/CAM) technique was employed to fabricate a set of negative molds, which was then used to press the PGA fibers into the pre-designed ear structure. Furthermore, the mechanical strength of the scaffold was enhanced by coating the PGA fibers with an optimized amount of PLA. Then, the feasibility of engineering a shape controllable ear cartilage *in vitro* was explored by seeding chondrocytes into the optimized scaffolds. In addition, the exactness of the shape of the ear graft was quantitively evaluated by a 3D laser scanning system.

## 2. Materials and methods

### 2.1 Preparation of scaffolds with different PLA contents

40 mg of unwoven PGA fibers (provided by Dong Hua University, Shanghai, China) were compressed into a cylinder shape of 13mm in diameter and 1.5mm in thickness. A solution

of 0.3 % PLA (Sigma, St. Louis, MO, USA) in dichloromethane was evenly dropped onto the PGA scaffold, dried in a 65 ºC oven, and weighed. The PLA mass ratio was calculated according to the formula: PLA%= (final mass-original mass)/final mass×100%. The above procedures were repeated until the predetermined PLA mass ratios of 0%, 10%, 20% and 30% were achieved.

### 2.2 Mechanical analysis of the scaffolds

The mechanical properties of the scaffolds were analyzed by a biomechanical analyzer (Instron-5542, Canton, MA, USA). The scaffold disks were compressed at a constant compressive strain rate of 0.5 mm/min until a maximum of 10% total strain was reached. The maximum compressive force and Young's modulus were determined from the stress-strain curve.

### 2.3 Biocompatibility evaluation of the scaffolds

**Cell seeding:** Chondrocytes were isolated from the articular cartilage of newborn swine (2-3weeks old) as described(Rodriguez et al., 1999). The harvested chondrocytes were adjusted to a final concentration of $50\times10^6$ cells/mL, and a 200uL cell suspension was pipetted onto each scaffold. The cell-scaffold constructs were then incubated for 5h at 37ºC with 95% humidity and 5% $CO_2$ to allow for complete adhesion of the cells to the scaffolds. Then, the constructs were covered by pre-warmed culture medium and cultured under the same conditions.

**Cell adhesion:** After 24 hours of incubation, the cell-scaffold constructs were gently transferred into a new 6-well plate for subsequent culture to evaluate cartilage formation. The remaining cells were collected and counted. The cell seeding efficiencies of the scaffolds with different PLA contents were calculated based on the formula: (total cell number-remaining cell number)/ total cell number×100%(Moran et al., 2003).

**Scanning electron microscopy (SEM):** The constructs were cultured *in vitro* and the attachment and matrix production of the cells on the scaffolds were examined by SEM (Philips XL-30, Amsterdam, Netherlands) after 2 weeks and 8 weeks.

**Evaluation of cartilage formation:** The constructs were harvested after 8 weeks of culture. The cartilage formation on different scaffolds was evaluated histologically by staining with hematoxylin and eosin (HE) and Safranin O, and immunohistochemically with type II collagen(Zhang, Spector, 2009).

### 2.4 Mold fabrication by CAD/CAM

A patient's normal ear was scanned by CT to obtain the geometric data. These data were further processed by a CAD system to generate the half-sized mirror image data (both positive and negative) of the normal ear, and the resultant data were input into a CAM system (Spectrum 510, Z Corporation) for the fabrication of the resin models by 3D printing. The negative mold was composed of two parts: the outer part and the inner part. In order to make the mold pressure-loadable, the outer part was replaced by a silicon rubber, which was molded according to the inner part of the resin negative mold.

### 2.5 Fabrication of ear shaped scaffold

Two hundred milligrams of unwoven PGA fibers were pressed using the negative mold for over 12 hours. A solution of 0.3 % PLA (Sigma, St. Louis, MO, USA) in dichloromethane was evenly dropped onto the PGA scaffold, dried in a 65 °C oven, weighed, and pressed again with the negative mold. This procedure was repeated until the final PLA mass ratio of 20% was reached. The edge of the scaffold was carefully trimmed according to the shape of the positive mold.

### 2.6 Three-dimensional laser surface scanning

A 3D laser scanning system was used for the shape analysis(Yu et al., 2009). The surface image data were collected from both the positive mold and the ear shaped scaffolds using a Konica Minolta Vivid 910 and Polygen Editing Tools version 2.21 (Konica Minolta, Tokyo, Japan). These data were further processed by RapidForm 2006 (INUS, Seoul, South Korea) and HP xw6200 (Hewlett Packard, Shanghai, China). The resultant data obtained from the ear-shaped scaffolds were compared to those from the positive mold, which served as a standard. Variations in voxels smaller than 1mm were considered similar, and the number of these similar voxels was divided by the number of total voxels to calculate the similarity level.

### 2.7 In vitro construction of ear-shaped cartilage

A 1mL aliquot of chondrocyte suspension with a density of $50\times10^6$ cells/mL was seeded into the ear-shaped scaffold followed by incubating for 5h, according to the cell seeding procedures described above. Then, the construct was gently transferred into a 50mL centrifuge tube for subsequent culture. The culture medium was changed every other day. The constructs were harvested at 4weeks, 8 weeks and 12 weeks for evaluation of shape exactness and cartilage specific histology.

### 2.8 Statistical analysis

The differences of cell seeding efficiencies (n=6), Young's moduli (n=6), and maximum compressive loadings (n=6) among the four PLA content groups were analyzed using the Student's t-test. A $p$-value less than 0.05 was considered statistically significant.

## 3. Results

### 3.1 Mechanical analysis of different scaffolds

The mechanical properties of the scaffolds were analyzed to evaluate the effects of PLA coating with different amount on the mechanical strength. As shown in Figure 2, all the scaffolds had regular cylinder shapes with the same diameter of 13mm (Fig. 2A-2D). No obvious differences in appearance were observed among the PLA/PGA scaffolds with different PLA amounts (Fig. 2B-2D). As expected, the pure PGA group (0% PLA group) showed a flat compressive stress-strain curve close to the X axis, indicating that pure PGA scaffolds had relatively low mechanical strength. With an increase in PLA content, the compressive stress-strain curves became steeper and more linear before the maximum loadings were reached (Fig. 2E), and the compressive moduli (Fig. 2F) as well as maximum loadings (Fig. 2G) also increased. Noticeably, there was a significant increase (over 4 folds) in both compressive moduli and maximum loading in scaffolds fabricated with 20% PLA compared to those with 10% PLA. Furthermore, the scaffold with 20% PLA reached a compressive modulus around 45MPa (45.42±10.52 MPa), which was similar to that of native adult human articular cartilage [19]. As expected, the 30% PLA group achieved the highest maximum loading and Young's modulus in all groups, although no significant difference was observed in Young's modulus between the 20% and 30% groups.

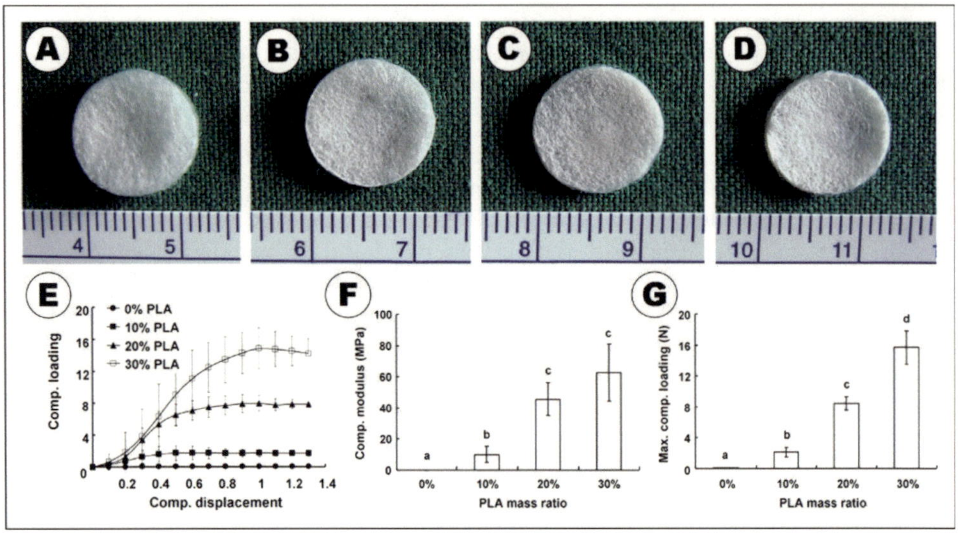


Fig. 2. The influences of PLA contents on mechanical properties. PGA fibers are pressed into a regular cylindrical shape (A). No obvious differences in appearance are observed among the PLA/PGA scaffolds with different PLA ratios of 10% (B), 20% (C), and 30% (D). The scaffolds have different stress-strain curves (E), with significant differences in maximum loading (F) and Young's modulus (G). Different lower-case letters indicate significant differences (p<0.01)

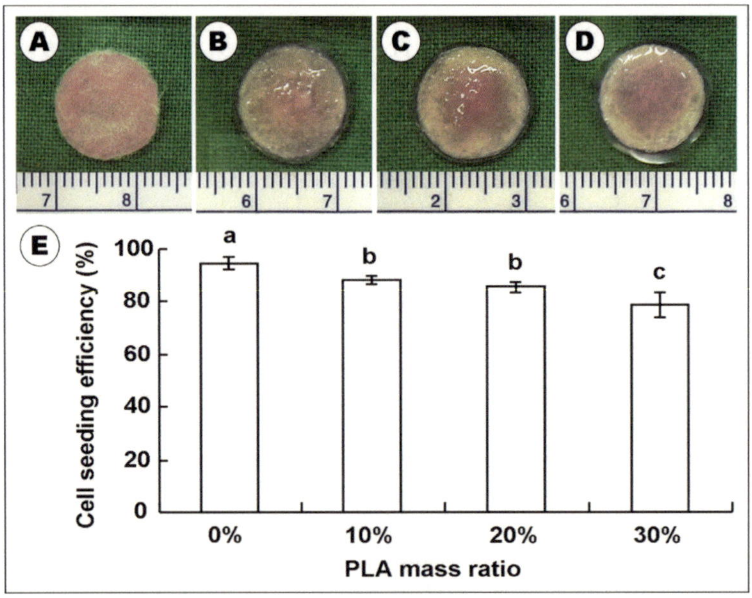


Fig. 3. The influences of PLA contents on cell seeding efficiency. Scaffolds with different PLA contents of 0% (A), 10% (B), 20% (C), and 30% (D) absorb different volumes of the cell suspension. Cell seeding efficiencies decrease with increasing PLA contents in the scaffolds, and a significant decrease is observed in the scaffolds with 30% PLA compared to those with 10% and 20% PLA (E). Different lower-case letters indicate significant differences (p<0.05)

### 3.2 Evaluation of the biocompatibility of the scaffolds with different PLA contents

Cell seeding efficiencies, SEM, and histological examination were performed to evaluate the influence of PLA contents on cell compatibility of the scaffolds and on final cartilage formation. The results showed that the increase in PLA content could lead to the reduction in the ability of the scaffolds to absorb the cell suspensions (Fig. 3A-3D), which may be related to the different pore structures (Fig. 4A-4D) and hydrophobicity of the scaffolds with different PLA contents. Quantitative analysis demonstrated that all the groups with PLA presented significantly lower cell seeding efficiencies compared to the group without PLA ($p<0.05$). Most notably, there was a significant decrease in cell seeding efficiencies in scaffolds with 30% PLA compared to those with 10% and 20% PLA, while no significant differences were observed between the scaffolds with 10% and 20% PLA (Fig. 3E).

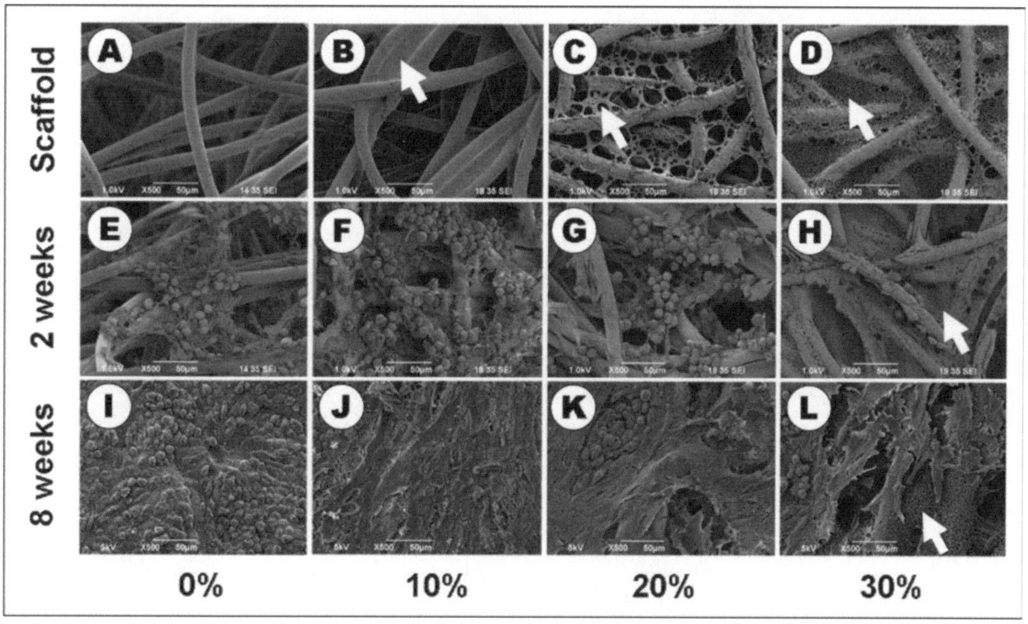


Fig. 4. SEM examination for the influences of PLA contents on cell distribution and ECM production. Scaffolds with different PLA contents show different pore structures (A-D). At 2 weeks, no obvious differences in cell distribution are observed among groups with 0% (E), 10% (F), and 20% (G) PLA, while an obvious decrease in cell number is observed in 30% PLA group (H). At 8 weeks, inferior ECM deposition is observed in 30% PLA group (L) compared to the other groups (I-K). The white arrows indicate the coated PLA

Naturally, the evaluation of final cartilage formation is the most important criterion to determine whether a scaffold can be used for cartilage engineering. As shown in Figure 5, after 8 weeks of *in vitro* culture, homogenous cartilage-like tissue with abundant cartilage-specific extracellular matrices (ECM) was observed in the constructs with 0% (Fig. 5E, 5I, 5M), 10% (Fig. 5F, 5J, 5N), and 20% (Fig. 5G, 5K, 5O) PLA. However, in the group with 30% PLA (Fig. 5H, 5L, 5P), there were high amounts of undegraded scaffold in the constructs, and only sporadic cartilage-like tissues were observed. These findings were consistent with the SEM examinations, which showed an obvious decrease in both cell number and ECM deposition in 30% PLA group (Fig. 4H, 4L) compared to the other groups (Fig. 4E-4G, 4I-4K). Therefore, these results indicate that 20% but not 30% is an acceptable PLA amount for preparing the scaffolds in terms of cell seeding efficiency, ECM production, and cartilage formation.

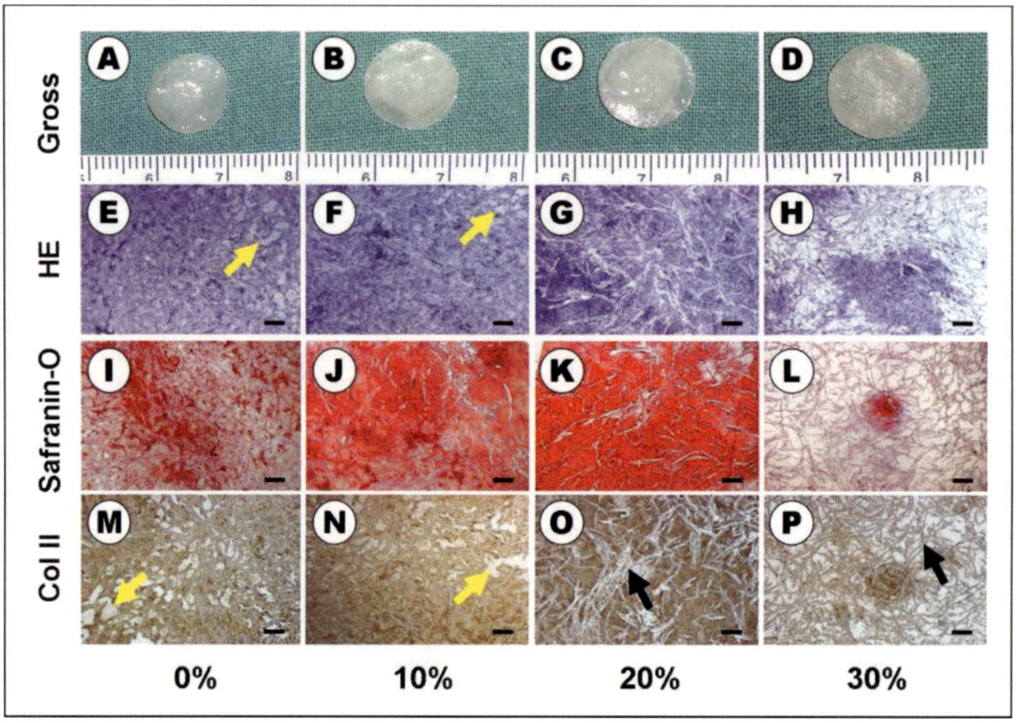


Fig. 5. The influences of PLA contents on cartilage formation. Grossly, the construct without PLA shrinks a little in diameter (A). The constructs that contain PLA basically maintain their original sizes (B-D). Histologically, homogenous cartilage-like tissue is observed in groups with 0% (E, I, M), 10% (F, J, N), and 20% (G, K, O) PLA, except that more compact structures and more undegraded scaffold fibers are observed in 20% PLA group compared with 0% and 10% PLA groups. In the group with 30% PLA (H, L, P), obvious heterogeneous cartilage was observed with an abundance of undegraded scaffolds. The black arrows indicate the undegraded PGA fibers. The yellow arrows indicate void regions caused by fast degradation of the scaffolds. Scale bar = 100μm

### 3.3 Preparation and shape analysis of ear-shaped scaffold

Because sufficient mechanical strength and good biocompatibility could be achieved in the scaffold with 20% PLA, this formulation was further used for the preparation of the human ear-shaped scaffold. In order to prepare the scaffold into a shape that is mirror-symmetrical to the normal ear, a set of negative molds in half size of an ear (Fig. 6F-6G) was fabricated according to the mirror image (Fig. 6B) of the normal ear (Fig. 6A). The resulting ear-shaped scaffold (Fig. 6H-6J; Fig. 7A, 7E) achieved a similarity level of above 97% compared to the positive mold, the standard for comparison, (Fig. 6C-6E) according to the shape analysis. These results indicate that the mold fabricated by CAD/CAM technology is allowed to accurately fabricate a scaffold into an ear-shape mirror-symmetrical to the normal ear.

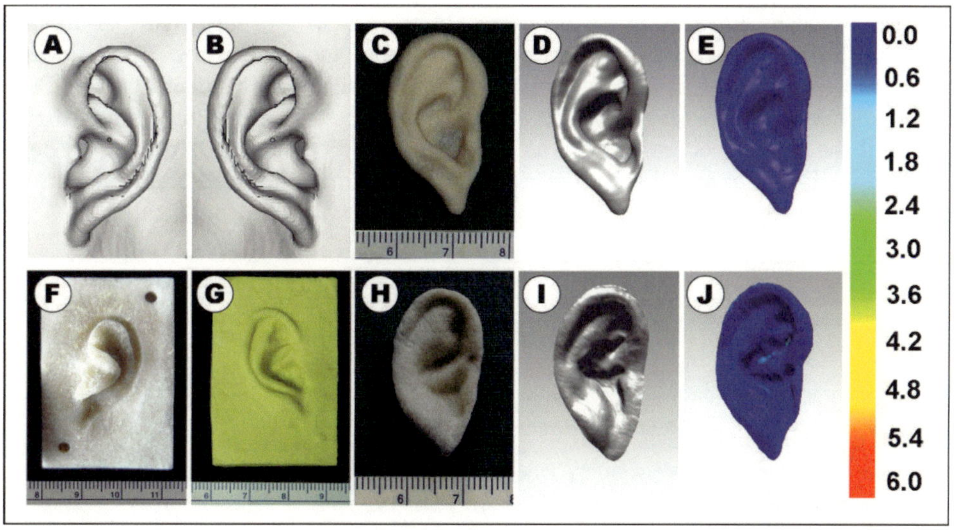


Fig. 6. Preparation and shape analysis of the ear-shaped scaffolds. (A): 3D image of the normal ear; (B): the mirror image of A; (C): The half-sized resin positive mold; (D): laser scan image of C; (E): color map of D; (F): inner part of the resin negative mold fabricated by 3D printing; (G): outer part of the negative mold cast from F with silicon rubber; (H): the ear-shaped PLA/PGA scaffold; (I): laser scan image of H; (J): color map of I compared to D

### 3.4 Construction of ear-shaped cartilage in vitro

The scaffolds were then used to explore the feasibility of engineering an ear-shaped cartilage *in vitro*. Similarly to the cylindrical scaffold containing 20% PLA, the ear-shaped scaffold also had good compatibility with seeded chondrocytes (data not shown). Most importantly, all the cell-scaffold constructs largely maintained their original ear-like shape during *in vitro* culture, and the shape similarity of the engineered ear grafts was retained at a level of 85.2% at 4 weeks (Fig.7 B, F), 84.0% at 8 weeks (Fig.7 C, G), and 86.2% at 12 weeks (Fig.7 D, H) compared to positive mold, indicating that the mechanical strength of the scaffolds was strong enough to maintain the ear-shape throughout the *in vitro* culture period.
Histologically, the structure of the ear-shaped constructs gradually became compact with prolonged culture time. At 4 weeks, cartilage-like tissue was preliminarily formed despite the presence of many undegraded PGA fibers (Fig.8 A, D, G). At 8 weeks, there was an obvious increase in both cartilage ECM deposition and the number of mature lacuna, although a few PGA fibers remained observable (Fig.8 B, E, H). At 12 weeks, the constructs had completely transformed into cartilage-like tissues with no visible residual PGA (Fig.8 C, F, I), and abundant cartilage ECM and mature lacuna were observed. Furthermore, the ear-shaped neo-cartilage showed fine elasticity with a certain mechanical strength.

## 4. Discussions

Despite the rapid progress in cartilage engineering, *in vitro* engineering of cartilage with a fine controlled 3D structure, such as a human ear, remains a great challenge due to the lack of appropriate scaffolds. PGA has proven to be one of the most successful scaffolds for cartilage regeneration. However, for *in vitro* engineering of a cartilage with a precise shape, PGA unwoven fibers (the most widely used physical form) still have some drawbacks, such as the difficulties in controlling an accurate shape and in gaining a proper mechanical strength. In the current study, aided by CAD/CAM technique, the PGA fibers were prepared into the accurate shape of a human ear. Furthermore, by coating with PLA, the scaffold could obtain sufficient mechanical strength to retain the original shape during cell culture until the ear-shaped cartilage was finally formed. These results may provide useful information for future external ear reconstructions by *in vitro* engineered cartilage as well as for the engineering of other tissues with complicated 3D structures.

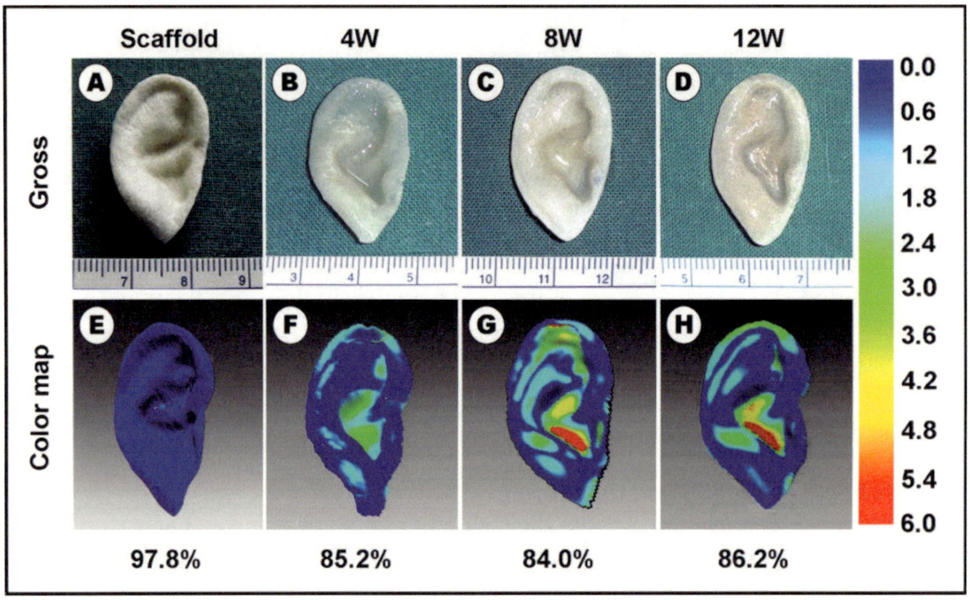


Fig. 7. Shape evaluation of the ear-shaped constructs. The scaffold shows an accurate ear-like structure (A) with a high similarity level compared to the positive mold (E). All the cell-scaffold constructs largely maintain their original ear-like structures at 4weeks (B), 8 weeks (C), and 12 weeks (D). Quantitative analysis show over 84% shape similarity in all the samples (E-H) compared to the positive mold

Preparation of the PGA fibers into an accurate ear structure is the first important step to determine the final shape of the engineered cartilage. To achieve this, a negative mold corresponding to the desired shape is required. Traditionally, the negative mold was fabricated by casting impression materials onto a patient's normal ear(Cao et al., 1997; Isogai et al., 2004), so that the shape of the PGA scaffold prepared by this mold exactly replicated the shape of the ear being casted. However, clinically, the ear aiming to reconstruct should be mirror-symmetrical to the contralateral normal ear. CAD/CAM, as a novel technique, has been widely used for the fabrication of anatomically accurate 3D models(Bill et al., 1995; Ciocca et al., 2007; Erickson et al., 1999; Subburaj et al., 2007). Particularly, this method can accurately perform complicated manipulations of the original 3D data, including Boolean operations, mirror imaging, and scaling(Al et al., 2005; Ciocca, Scotti, 2004; Karayazgan-Saracoglu et al., 2009). CAD/CAM technique was therefore used in the current study for the fabrication of the mirror-image negative mold for a human ear in half size. Using this mold, PGA fibers were able to be accurately prepared into the ear-shaped scaffold that was mirror-symmetrical to the normal ear in half size.

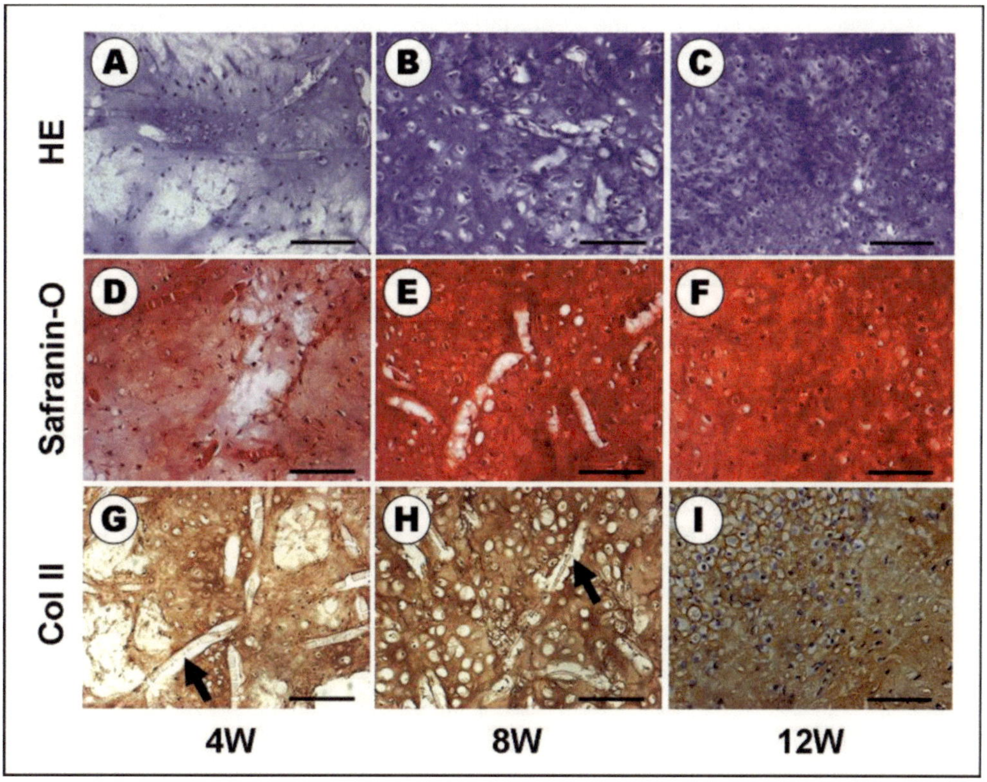


Fig. 8. Histological examinations of the *in vitro* ear-shaped constructs. At 4 weeks, the constructs form heterogeneous cartilage-like tissue along with undegraded PGA fibers (A, D, G). With prolonged culture time, the histological structure of the constructs gradually become compact, accompanied with increased numbers of lacuna structures at 8 weeks (B, E, H). Homogeneous cartilage with abundant ECM and mature lacuna are observed at 12 weeks (C, F, I) with no visible scaffold residuals in the constructs. The black arrows indicate the undegraded PGA fibers. Scale bar = 100μm

After the preparation of the ear-shaped PGA scaffold, the issue of shape retention during *in vitro* chondrogenesis becomes important. The shape maintenance of the cell-scaffold constructs mainly depends on the mechanical strength and degradation rate of the scaffold(Kim et al., 1994). The mechanical strength of PGA scaffold alone is not sufficient for the shape maintenance, and thus PLA coating was used to strengthen its mechanical properties as reported(Cui et al., 2009; Frenkel, Di, 2004; Yang et al., 2001). However, a high amount of PLA in the scaffold would negatively affect cartilage formation because of poor cell compatibility(Moran et al., 2003). Therefore, an appropriate PLA content in the scaffold is important for both cartilage formation and shape maintenance. In the current study, we evaluated the effects of four PLA contents on the scaffolds' mechanical properties and cartilage formation. According to the current results, the mechanical strength of the scaffolds increased with increasing PLA content. However, homogeneous cartilage was only observed in groups with PLA contents of 20% or less. Fortunately, the scaffold with 20% PLA was strong enough to retain the original shape of the cell-scaffold construct until the ear-shaped cartilage was finally formed after 12 weeks.

Besides the mechanical strength, the degradation rate of the scaffold is also an important factor that determines the final shape of engineered tissue. The ideal degradation rate should match the rate of ECM deposition. If the degradation rate of the scaffold is much faster than deposition rate of ECM, the engineered tissue would gradually collapse due to insufficient support, and thus the shape cannot be maintained. According to the histological findings at 8 weeks (Fig.4), the constructs in both 0% and 10% PLA groups had some void regions and lower amounts of residual scaffold, indicating that the degradation rate of the scaffolds in these two groups might be faster than the deposition rate of ECM. In contrast, the constructs in 20% PLA group showed a relatively compact histological structure with more scaffold fibers, indicating the scaffold with 20% PLA has an appropriate degradation rate matching the ECM formation.

In addition, for engineering a complicated structure like a human ear, it is necessary to establish a method to quantitively evaluate the shape exactness of the scaffold as well as to trace the deformation of the constructs during *in vitro* chondrogenesis. 3D laser surface scanning is one of the most popular data acquisition techniques, and has been successfully applied to quantify facial dimensions(Kau, Richmond, 2008; Kau et al., 2005; Toma et al., 2009). It has also been introduced to determine the dimensions of the ear(Coward et al., 2000; Sforza et al., 2005). However, no studies have applied this method to analyze the shape of tissue engineered ear grafts. In the current study, the introduction of 3D laser scanning system provided an effective tool for quantitively evaluating the shape exactness of the ear graft as well as tracing its shape change during *in vitro* engineering.

## 5. Conclusions

In summary, this study established a method to precisely engineer a cartilage *in vitro* with a shape that is mirror-symmetrical to the normal ear. Additionally, a quantitative system for evaluating the shape exactness of the constructs was established as well. These strategies may provide useful tools for future external ear reconstructions by *in vitro* engineered cartilage as well as for engineering of other tissues with complicated 3D structures. Moreover, the *in vitro* engineering system established in this study may also offer useful references for ear-shaped cartilage construction based on stem cells, since the ectopic chondrogenesis of stem cells requires a long-term induction *in vitro*(Liu et al., 2008). In future studies, we will also investigate the fate of this ear-shaped cartilage after subcutaneous implantation, especially in an immunocompetent animal model.

## 6. References

Al Mardini M, Ercoli C, Graser GN. 2005. A technique to produce a mirror-image wax pattern of an ear using rapid prototyping technology. *J Prosthet Dent.* 94(2):195-8.

Aufderheide AC, Athanasiou KA. 2005. Comparison of scaffolds and culture conditions for tissue engineering of the knee meniscus. *Tissue Eng.* 11(7-8):1095-104.

Beahm EK, Walton RL. 2002. Auricular reconstruction for microtia: part I. Anatomy, embryology, and clinical evaluation. *Plast Reconstr Surg.* 109(7):2473-82; quiz following 2482.

Bill JS, Reuther JF, Dittmann W, et al. 1995. Stereolithography in oral and maxillofacial operation planning. *Int J Oral Maxillofac Surg.* 24(1 Pt 2):98-103.

Brent B. 1999. Technical advances in ear reconstruction with autogenous rib cartilage grafts: personal experience with 1200 cases. *Plast Reconstr Surg.* 104(2):319-34; discussion 335-8.

Cao Y, Vacanti JP, Paige KT, et al. 1997. Transplantation of chondrocytes utilizing a polymer-cell construct to produce tissue-engineered cartilage in the shape of a human ear. *Plast Reconstr Surg.* 100(2):297-302; discussion 303-4.

Chung C, Burdick JA. 2008. Engineering cartilage tissue. Adv Drug Deliv Rev. 60(2):243-62.

Ciocca L, Mingucci R, Gassino G, et al. 2007. CAD/CAM ear model and virtual construction of the mold. *J Prosthet Dent.* 98(5):339-43.

Ciocca L, Scotti R. 2004. CAD-CAM generated ear cast by means of a laser scanner and rapid prototyping machine. *J Prosthet Dent.* 92(6):591-5.

Ciorba A, Martini A. 2006. Tissue engineering and cartilage regeneration for auricular reconstruction. *Int J Pediatr Otorhinolaryngol.* 70(9):1507-15.

Coward TJ, Scott BJ, Watson RM, et al. 2000. Laser scanning of the ear identifying the shape and position in subjects with normal facial symmetry. *Int J Oral Maxillofac Surg.* 29(1):18-23.

Cui L, Wu Y, Cen L, et al. 2009. Repair of articular cartilage defect in non-weight bearing areas using adipose derived stem cells loaded polyglycolic acid mesh. *Biomaterials.* 30(14):2683-93.

Erickson DM, Chance D, Schmitt S, et al. 1999. An opinion survey of reported benefits from the use of stereolithographic models. *J Oral Maxillofac Surg.* 57(9):1040-3.

Frenkel SR, Di Cesare PE. 2004. Scaffolds for articular cartilage repair. *Ann Biomed Eng.* 32(1):26-34.

Gunatillake PA, Adhikari R. 2003. Biodegradable synthetic polymers for tissue engineering. *Eur Cell Mater.* 5:1-16; discussion 16.

Haisch A, Klaring S, Groger A, et al. 2002. A tissue-engineering model for the manufacture of auricular-shaped cartilage implants. *Eur Arch Otorhinolaryngol.* 259(6):316-21.

Heath CA, Magari SR. 1996. Mini-review: Mechanical factors affecting cartilage regeneration in vitro. *Biotechnol Bioeng.* 50(4):430-7.

Isogai N, Asamura S, Higashi T, et al. 2004. Tissue engineering of an auricular cartilage model utilizing cultured chondrocyte-poly(L-lactide-epsilon-caprolactone) scaffolds. *Tissue Eng.* 10(5-6):673-87.

Kamil SH, Kojima K, Vacanti MP, et al. 2003. In vitro tissue engineering to generate a human-sized auricle and nasal tip. *Laryngoscope.* 113(1):90-4.

Kamil SH, Vacanti MP, Aminuddin BS, et al. 2004. Tissue engineering of a human sized and shaped auricle using a mold. *Laryngoscope.* 114(5):867-70.

Karayazgan-Saracoglu B, Gunay Y, Atay A. 2009. Fabrication of an auricular prosthesis using computed tomography and rapid prototyping technique. *J Craniofac Surg.* 20(4):1169-72.

Kau CH, Richmond S. 2008. Three-dimensional analysis of facial morphology surface changes in untreated children from 12 to 14 years of age. *Am J Orthod Dentofacial Orthop.* 134(6):751-60.

Kau CH, Richmond S, Zhurov AI, et al. 2005. Reliability of measuring facial morphology with a 3-dimensional laser scanning system. *Am J Orthod Dentofacial Orthop.* 128(4):424-30.

Kim BS, Mooney DJ. 1998. Engineering smooth muscle tissue with a predefined structure. *J Biomed Mater Res.* 41(2):322-32.

Kim WS, Vacanti JP, Cima L, et al. 1994. Cartilage engineered in predetermined shapes employing cell transplantation on synthetic biodegradable polymers. *Plast Reconstr Surg.* 94(2):233-7; discussion 238-40.

Langer R, Vacanti JP. 1993. Tissue engineering. Science. 260(5110):920-6.
Liu K, Zhou GD, Liu W, et al. 2008. The dependence of in vivo stable ectopic chondrogenesis by human mesenchymal stem cells on chondrogenic differentiation in vitro. *Biomaterials.* 29(14):2183-92.
Liu Y, Zhang L, Zhou G, et al. 2010. In vitro engineering of human ear-shaped cartilage assisted with CAD/CAM technology. *Biomaterials.* 31(8):2176-83.
Moran JM, Pazzano D, Bonassar LJ. 2003. Characterization of polylactic acid-polyglycolic acid composites for cartilage tissue engineering. *Tissue Eng.* 9(1):63-70.
Nagata S. 1993. A new method of total reconstruction of the auricle for microtia. *Plast Reconstr Surg.* 92(2):187-201.
Naumann A, Aigner J, Staudenmaier R, et al. 2003. Clinical aspects and strategy for biomaterial engineering of an auricle based on three-dimensional stereolithography. *Eur Arch Otorhinolaryngol.* 260(10):568-75.
Neumeister MW, Wu T, Chambers C. 2006. Vascularized tissue-engineered ears. *Plast Reconstr Surg.* 117(1):116-22.
Rodriguez A, Cao YL, Ibarra C, et al. 1999. Characteristics of cartilage engineered from human pediatric auricular cartilage. *Plast Reconstr Surg.* 103(4):1111-9.
Sforza C, Dellavia C, Tartaglia GM, et al. 2005. Morphometry of the ear in Down's syndrome subjects. A three-dimensional computerized assessment. *Int J Oral Maxillofac Surg.* 34(5):480-6.
Shieh SJ, Terada S, Vacanti JP. 2004. Tissue engineering auricular reconstruction: in vitro and in vivo studies. *Biomaterials.* 25(9):1545-57.
Sterodimas A, de Faria J, Correa WE, et al. 2009. Tissue engineering and auricular reconstruction: a review. *J Plast Reconstr Aesthet Surg.* 62(4):447-52.
Subburaj K, Nair C, Rajesh S, et al. 2007. Rapid development of auricular prosthesis using CAD and rapid prototyping technologies. *Int J Oral Maxillofac Surg.* 36(10):938-43.
TANZER RC. 1959. Total reconstruction of the external ear. Plast Reconstr Surg. 23(1):1-15.
Toma AM, Zhurov A, Playle R, et al. 2009. Reproducibility of facial soft tissue landmarks on 3D laser-scanned facial images. *Orthod Craniofac Res.* 12(1):33-42.
Walton RL, Beahm EK. 2002. Auricular reconstruction for microtia: Part II. Surgical techniques. *Plast Reconstr Surg.* 110(1):234-49; quiz 250-1, 387.
Xu JW, Johnson TS, Motarjem PM, et al. 2005. Tissue-engineered flexible ear-shaped cartilage. *Plast Reconstr Surg.* 115(6):1633-41.
Yan D, Zhou G, Zhou X, et al. 2009. The impact of low levels of collagen IX and pyridinoline on the mechanical properties of in vitro engineered cartilage. *Biomaterials.* 30(5):814-21.
Yang S, Leong KF, Du Z, et al. 2001. The design of scaffolds for use in tissue engineering. Part I. Traditional factors. *Tissue Eng.* 7(6):679-89.
Yu Z, Mu X, Feng S, et al. 2009. Flip-registration procedure of three-dimensional laser surface scanning images on quantitative evaluation of facial asymmetries. *J Craniofac Surg.* 20(1):157-60.
Zhang L. 2010. It is time to reconstruct human auricle more precisely and microinvasively. *Plast Reconstr Surg.* 125(4):155e-156e.
Zhang L, Spector M. 2009. Comparison of three types of chondrocytes in collagen scaffolds for cartilage tissue engineering. *Biomed Mater.* 4(4):45012.

# Part 7

## Central Nervous System

18

# Advances in the Combined Use of Adult Cell Therapy and Scaffolds for Brain Tissue Engineering

Elisa Garbayo[1], Gaëtan J.-R. Delcroix[2], Paul C. Schiller[2] and Claudia N. Montero-Menei[3,4]
[1]*Pharmacy and Pharmaceutical Technology Department, University of Navarra, Pamplona,*
[2]*University of Miami, UMTB, Florida,*
[3]*LUNAM Université, Ingénierie de la Vectorisation Particulaire, F-49933 Angers,*
[4]*INSERM, U646, F-49933 Angers*
[1]*Spain*
[2]*USA*
[3,4]*France*

## 1. Introduction

No long-term effective treatments are currently available for brain neurological disorders such as focal and global cerebral ischemia, traumatic brain injury (TBI) and neurodegenerative disorders. Current treatments are primarily focused on limiting the damage and slowing the degeneration and, most of them are only symptomatic while function restoration is rarely achieved. This underlies the need for alternative therapies such as brain cell therapy that allows functional replacement of missing or damaged neurons by transplanting cells that may differentiate into the desired phenotype and integrate the host parenchyma, or alternatively rescue the affected neuronal population. Due to the high cell death that occurs during neurodegenerative processes, brain neurological disorders are an ideal target for cell-based therapies. Several teams tried to prevent the loss of neurons or to replace them, using neuronal precursors and lately, since stem cell discovery, embryonic stem (ES) cells, adult stem cells, tissue-derived stem cells or more recently induced pluripotent stem (iPS) cells (for review see (Abeliovich & Doege, 2009; Joyce et al., 2010; Chen & Xiao, 2011; Loewenbruck & Storch, 2011). These studies demonstrated the potential of cell therapy to repair the injured brain even if poor survival of grafted cells was reported (Brundin et al., 1985; Schierle et al., 1999; Brundin et al., 2000; Isacson et al., 2003; Olanow et al., 2003). In this book chapter, brain cell therapy studies performed with adult non-transformed cells that, by nature, allow the use of autologous tissues for transplantation and overcome the immunological, availability, as well as ethical concerns will be reported.
It is important to bear in mind that, for organs with a higher level of complexity such as the brain, cell therapy remains a challenging task. Fortunately, relatively recent medical, biological and technological advances in tissue engineering approaches allow functional

tissue and organ recovery by using the appropriate combination of three fundamental "tools" known as the tissue engineering triad: cells, engineered materials and signalling molecules (essentially growth factors). In practice, this is translated into a great versatility of delivery systems, which mimic the natural repairing environment of the brain tissue, creating tuneable and customized spatio-temporal gradients of signals guiding tissue regeneration. Bioactive scaffolds are likely to reinforce the success of cell replacement therapies by providing a microenvironment that facilitates the survival, proliferation, differentiation, and connectivity of transplanted and/or endogenous cells (Pettikiriarachchi et al., 2010). However, delivery of cells with scaffolds to the damaged brain still remains challenging due to practical limitations of delivery. The second part of this book chapter will provide an overview of what solutions tissue engineering may provide for adult cell therapy of the brain. Since the brain is a functionally complex organ, cell growth and differentiation alone are not enough in order to achieve its functional recovery. The correct re-establishment of the axonal connections and neuronal circuits is also necessary. In general, scaffolds for brain therapy should meet several essential requirements like being biocompatible, biodegradable, immunologically inert and able to support neurite outgrowth. The most common materials, fabrication methods and desirable properties of biomaterials used for brain protection repair and regeneration will be presented.

Brain tissue engineering has several limitations and many unanswered questions or concerns should be addressed before reaching the clinic. Moreover, with the current workflow, it takes a long time to select an effective brain tissue engineering strategy to translate into clinical studies. Additionally, recent developments in understanding the basic biology of brain tissue formation in physiological and pathological conditions have resulted in an explosion in the numbers of tissue engineering products that could be potential candidates for treating brain disorders. Screening platforms that bridge the gap between conventional tissue culture and animal models would help to improve understanding of cell-based therapies and optimize central nervous system (CNS) tissue engineering. In this regard, the usefulness of 3D brain organotypic cultures in CNS research as well as in the drug discovery process will be discussed in this book chapter.

## 2. Adult cell therapy for brain neuronal damages

### 2.1 Adult cells for cerebral ischemia and traumatic brain injury therapy

#### 2.1.1 Adult stem cell therapy

Mesenchymal stem cells (MSCs) are the most widely investigated adult stem cells for brain cell therapy of cerebral ischemia and TBI (Table 1). Various animal models of ischemia have been used to investigate the therapeutic effects of MSCs on the lesioned brain and in all these models, bone marrow-derived MSCs resulted in an increased survival of neurons and most of the time in an improved cognitive function of the animals (reviewed in (van Velthoven et al. 2009)). In addition, MSCs have been described to reduce the thickness of the scar walls (Li et al., 2005) and may also favour angiogenesis (Chen et al., 2003; Chopp et al., 2008) as well as synaptogenesis (Chen & Chopp, 2006). Functional improvements may be observed whatever the implantation route, intravenously or intracerebrally, even if reduction in infarct volume is not always observed (Li et al., 2000). This underlines the capacity of MSCs to migrate towards lesions, as was already observed by other groups

(Chen et al., 2001) after IV injection, or after intracerebral graft in a lesioned rat brain (Sykova & Jendelova, 2007; Delcroix et al., 2009). Noteworthy, *in situ* neural or neuronal differentiation of transplanted MSCs has been described in several studies. For example, systemic or intracarotid artery administration of MSCs in rat models of ischemia improve neurogenesis and functional recovery, with the detection of neuronal or glial markers expressed by a fraction of MSCs in the brain (van Velthoven et al., 2009; Li et al., 2002; Esneault et al., 2008; Perasso et al., 2011). However, due to the low grafted cell survival and neural/neuronal differentiation, neurological benefits are often assumed to mainly derive from the increased production of growth factors and other paracrine factors from MSCs in the ischemic tissue (England, 2009; Yarygin et al., 2009). Factors secreted by the ischemic brain itself, such as vascular endothelial growth factor (VEGF) and erythropoietin (EPO), are also thought to play a major role in brain protection from ischemia (Tang et al., 2006). MSCs combined with growth factors also provided functional effects and in some cases reduced the lesion volume, the number of apoptotic cells within the ischemic lesion and stimulated host repair responses as recruitment of host progenitor cells (Esneault et al., 2008; Rahnemai-Azar et al., 2011; van Velthoven et al., 2009). Modulation of inflammatory and immune response or production of neuroprotective chemokines such as fractalkine and monocyte chemoattractant protein-1 (MCP-1) by MSCs are others mechanisms that may be involved in neuronal protection during ischemia (Re & Przedborski, 2006; Ohtaki et al., 2008; Madrigal et al., 2009; Garbayo et al., 2011). Recruitment of host progenitor cells may also contribute to MSC-induced repair processes in response to chemokines [growth-related oncogene (GRO), MCP-1] secreted by the implanted cells (Rahnemai-Azar et al., 2011; Gordon et al., 2009). After a first clinical trial that underwent criticisms a few years ago (Bang et al., 2005; De Keyser, 2005), there is a growing number of phase I-II clinical trials to evaluate the safety and efficacy of MSCs or CD34+ cells delivered either by arterial infusion, IV or intracranial injection in the context of stroke (see clinicaltrials.gov for up to date informations). Moreover, two consecutive meetings have been held recently named the Stem cell Therapies as an Emerging Paradigm in Stroke (STEPS I & II), in order to provide guidelines for preclinical and clinical studies leading to the successful development of cell therapy for stroke (Broderick, 2009; Savitz et al. 2011).

In the context of TBI, MSCs may also improve functional benefits whatever the injection route. However, a maximum of 10 % implanted MSCs may transdifferentiate into neuronal cells *in vivo* so that, again, this mechanism was assumed not to be solely responsible for the functional benefits observed in TBI animal models (Li & Chopp, 2009). Growth factor production by MSCs stimulated by the lesioned brain (Chen et al., 2002) was certainly the main mechanism leading to functional recovery by promoting glial, neuronal and blood vascular remodelling (Li & Chopp, 2009; Richardson et al., 2007). However, despite the benefits obtained using this approach, no significant changes in the lesion volume were observed, a problem that tissue engineering alternatives may resolve (see section 3 of this chapter). Concerning adult neural stem cells (NSCs), very few studies were performed in the context of stroke, and none for TBI. It is however interesting to note that a fraction of adult NSCs, either derived from rat hippocampus (Zhang et al., 2011) or from human wisdom teeth (Yang, Chen et al., 2009), survived within the brain after transplantation and led to significant improvements in stroke. Another population of CD31-/CD146- cells isolated from dental pulp has also been used in stroke rat models, leading to migration and differentiation of the endogenous neuronal progenitor cells as well as vasculogenesis (Sugiyama et al., 2011).

| Cells | Origins | Recipients | Scaffolds | Benefits | References |
|---|---|---|---|---|---|
| MSC | Mouse | MCAO mouse model | | Functional benefits but no reduction in infarct volume | Li, Chopp et al. 2000 |
| | Rat | MCAO rat model | | IV transplantation: migration toward the lesion | Chen, Li et al. 2001 |
| | Human | MCAO rat model | | IV transplantation: importance of growth factor increase in the ischemic tissue | Li, Chen et al. 2002 |
| | Rat | MCAO rat model | | IV transplantation in combination with EPO: functional recovery and neurogenesis<br>A fraction of MSCs with neuronal and glial markers | Esneault, Pacary et al. 2008 |
| | Rat | MCAO rat model | | IV transplantation: reduced thickness of the scar walls | Li, Chen et al. 2005 |
| | Human | MCAO rat model | | IV transplantation: promotion of angiogenesis | Chen, Zhang et al. 2003 |
| | Human | tCCAO mouse model | | Modulation of inflammatory and immune responses | Ohtaki, Ylostalo et al. 2008 |
| | Mouse | 2VO rat model | | IV transplantation: increased number of pyramidal CA1 neurons | Perasso, Cogo et al. 2011 |
| | Human | ACA rat model | FN-coated PLGA microspheres | Neuroprotection of CA1 hippocampal neurons with MIAMI cells: neuroprotection enhanced with microspheres | Garbayo, Raval et al. 2011 |
| | Human | TBI rat model | Collagen cylinder | Scaffolds improved spatial learning, sensorimotor function & reduced the lesion volume | Lu, Mahmood et al. 2007 |
| | Human | TBI rat model | Collagen cylinder | Delayed transplantation: increased angiogenesis in the injured cortex & transcallosal fiber length | Xiong, Qu et al. 2009 |
| | Human | TBI mouse model | Collagen cylinder | Scaffolds improved spatial learning, reduced lesion volume & increased vascular density | Qu, Xiong et al. 2009 |
| | Human | Human (autologous, stroke) | | Proof of safety for MSCs use in human | Bang, Lee et al. 2005 |
| NSC | Human From teeth | MCAO rat model | | Functional benefits | Yang, Chen et al. 2009 |
| | Rat hippocampi | MCAO rat model | | Amelioration of neurological deficits<br>Reduced total infarct volume | Zhang, Jin et al. 2011 |
| CD31-/CD146- progenitor cells | Pig Teeth-derived | MCAO rat model | | Migration and differentiation of the endogenous neuronal progenitor cells<br>Vasculogenesis | Sugiyama, Iohara et al. 2011 |
| iPS cells | Mouse Embryonic fibroblast-derived | MCAO rat model | Fibrin glue | Sub-dural injection reduced the infarct size compared to iPS cells directly injected in the parenchyma<br>Functional improvements and no teratoma when sub-dural injection with fibrin glue is used | Chen, Chang et al. 2011 |
| | Human Fibroblast-derived | MCAO rat model | | Migration of the cells towards injured brain area<br>*In situ* neuronal differentiation of iPS cells<br>Improved sensorimotor function 4-16 days after grafting | Jiang, Lv et al. 2011 |

***Abbreviations:*** ACA: asphyxial cardiac arrest; EPO: erythropoietin; FN: fibronectin; iPS: induced pluripotent stem; IV: intravenous; MCAO: middle cerebral artery occlusion; MIAMI: marrow isolated adult multilineage inducible; MSC: mesenchymal stem cell; NSC: neural stem cell; tCCAO: transient common carotid artery occlusion, UPDRS: unified Parkinson's disease rating scale.

Table 1. Adult and iPS cell therapies for cerebral ischemia and TBI

### 2.1.2 Adult derived pluripotent stem cell therapy

A growing interest is now observed in the scientific community for the iPS cells. These "embryonic-like" cells were primarily derived from adult fibroblasts by the expression of transcription factors (a combination of 4 factors within c-Myc, KLF4, LIN28, Nanog, Oct3/4 and Sox2) (Takahashi et al., 2007; Yu et al., 2007). Several applications are currently envisioned for these pluripotent cells, which share the same potential than ES cells, without the associated ethical problems. In addition to their applications for drug discovery and toxicity testing (Laustriat et al., 2010) these cells may be helpful for complex disease understanding. They are also attractive for cell therapy studies, due to the possibility to obtain a large amount of cells from various lineages and to perform autologous grafts. iPS have now been derived from several types of adult tissues, with variable efficiency, and always using refined protocols to avoid permanent genomic integration and the use of lentiviruses for transducing the cells (Okita et al., 2011; Narsinh, et al., 2011; Lee et al., 2009; Gonzalez et al., 2009; Zhou et al., 2009; Page et al., 2009; Yu et al., 2009). Indeed, uses of viral-based protocols may be at the origins of tumour formation and may also lead to further difficulties to differentiate the cells toward a given phenotype (Yu et al., 2009). The propensity of iPS to form teratomas may also be related to the tissue of origin of the cells (Miura et al., 2009). Despite their great interest, one has to keep in mind that further developments are still required to fully understand their reprogramming process (Nakagawa & Yamanaka, 2011) in order to provide a safe cell therapeutic product for the clinic (Pera, 2011). In the context of ischemic stroke, an interesting study describes the use of iPS-derived neurons and astrocytes directly injected in the damaged cortex of a stroke animal model (Chen et al., 2011), which resulted in a reduced ischemic size with functional improvements. Another study describes the *in situ* migration and differentiation of human iPS cells in a similar context, with an improved sensorimotor function of the animals 4-16 days after grafting (Jiang et al. 2011).

### 2.2 Adult cells for neurodegenerative disorder therapy

Huntington's and Parkinson's diseases (HD and PD respectively) have been widely studied in cell therapy programs due to the relatively small area affected by the diseases, particularly at early stages (see for reviews (Lindvall & Kokaia, 2009; Loewenbruck & Storch, 2011)). In opposition, Alzheimer's disease (AD) would be more complex to treat due to the multiple sites of the brain affected in this disease. Moreover, no tissue engineering strategies have yet been described for AD so that we will only focus on HD and PD in the following (Table 2).

#### 2.2.1 Adult cell therapy

An increasing number of studies performed with foetal tissue, ES cells or NSC grafts, mostly in quinolinic acid (QA)-induced animal models of the disease (Clelland et al., 2008) gave successful results leading to clinical trials with foetal-derived cells. However, cell therapy for HD, an incurable disease, is still not widely available in clinic due to ethical, logistical or safety concerns (Kelly et al., 2009). Adult cells have only been used in early studies with Sertoli cell grafts that protected the lesioned area in an animal model of HD; effects that may result from Sertoli cells trophic and anti-inflammatory potentials (Emerich, 2004).

One of the first cell therapy strategies to treat PD has been performed with adult cell types synthesizing dopamine or its precursor L-DOPA, in order to replenish the striatum level of dopamine (for review (Drucker-Colin & Verdugo-Diaz, 2004; Fernandez-Espejo et al., 2005). Autologous adrenal medulla tissue were grafted in the caudate nucleus of 2 young PD patients, but observed functional improvements were transitory and mainly due to trophic effects as only 1 % of the cells synthesized dopamine (Madrazo et al., 1987). Most importantly, very few cells were detected 1 or 2 years after transplantation (Hurtig et al., 1989). Other studies were performed using cultured chromaffin cell suspensions, but few surviving cells in long term studies did not validate this approach (Drucker-Colin et al., 1999). Human retinal pigment epithelium (hRPE) cells derived from the inner layer of the neural retina are isolated from human eye bank for transplantation purposes (reviewed in (Stover & Watts, 2008)). hRPE cells may be expanded, present a tyroxine hydroxylase activity (Pawelek & Korner, 1982), and produce L-DOPA, which is synthesized as a melanin precursor, so that these cells may supplement the oral administration of L-DOPA *in situ*. Moreover, it has been proposed that hRPE cells may be immune-privileged after transplantation due to their expression of Fas-ligand (Griffith et al., 1995; Jorgensen et al., 1998).

#### 2.2.2 Adult stem cell therapy

Adult stem cells are expected to better differentiate and integrate the host brain compared to adult non-stem cells and are therefore widely investigated for their therapeutic potential in the context of neurodegenerative disorders. Whole bone marrow cells [i.e. hematopoietic stem cells (HSCs) & MSCs] implanted into the bilateral lesioned striatum of HD rat models, reversed functional deficits such as working memory (Lescaudron et al., 2003) even if the cell population responsible for the beneficial effects remained unknown. Noteworthy, autologous grafts may not be an appropriate strategy for the treatment of HD as transplanted cells would also carry the mutant huntingtin gene responsible for the disease. More recently, transplantation of MSCs, either intracerebrally or intravenously, resulted in a decreased atrophy of lesioned rat striatum (Amin et al., 2008) and in some functional benefits (Edalatmanesh et al., 2009) even though only a fraction of cells (1 %) expressed neural phenotypes. Thus, it was suggested that MSCs, by producing growth factors, allowed surviving cells within the caudate nucleus to function more efficiently and to facilitate other compensatory responses (Dunbar et al., 2006). In this regard, another study demonstrated the importance of factors such as stem cell factor (SCF), produced *in situ* in the lesioned striatum, to promote the migration and engraftment of MSCs *via* SCF receptor c-kit (Bantubungi et al., 2008).

In the context of PD, several teams, including ours, reported the neuronal differentiation of human MSCs toward a dopaminergic phenotype *in vitro*, indicating that these cells may constitute an alternative dopamine secreting source of cells (Trzaska & Rameshwar, 2011; Barzilay et al., 2008; Trzaska et al., 2007; Tatard et al., 2007). An interesting study also reported no major differences between MSCs from normal patients compared to MSCs isolated from parkinsonian patients, which may be induced to produce up to 30 % of dopaminergic neurons *in vitro* (Zhang et al., 2008). Human MSCs, pre-induced towards a neuronal phenotype and transplanted in the totally dopaminergic deafferented striatum of rats, led to an improved functional recovery for up to 4 months compared to naïve hMSCs (Levy et al., 2008). A similar recovery has been observed in a rat partial lesion model of PD, in which dopaminergic fibers are spared in the striatum (Bouchez et al., 2008). Moreover, microdialysis demonstrated that part of the striatal pool of dopamine was restored upon MSCs transplantation. Naïve hMSCs or a subpopulation of hMSC pre-treated with epidermal growth factor (EGF) and basic fibroblast growth factor (bFGF) also led to an efficient recovery after partial lesions, suggesting a paracrine restorative effect by MSCs. Indeed, hMSCs, that migrate toward lesions (Hellmann et al., 2006; Sadan et al., 2009; Delcroix et al., 2011) may induce a protective or restorative effect on the remaining neurons within the host brain, due to the secretion of a large panel of neurotrophic factors, including brain-derived neurotrophic factor (BDNF) and glial cell line-derived neurotrophic factor (GDNF), which may also enhance endogenous neurogenesis (Levy et al., 2008; McCoy et al., 2008; Bahat-Stroomza et al., 2009; Sadan et al., 2009). MSCs may also modulate the host response to the lesion (Kim et al., 2008) and probably ultimately replace functional cells within the host brain as a few MSCs with neuronal-like morphology and markers were observed in a total lesion model (Levy et al., 2008). We are therefore starting to better understand the mechanisms of action of these cells in the context of PD, while other teams try, with interesting results, to find new ways to deliver the cells to the brain. Indeed recently, a non invasive intranasal delivery of MSC in an animal model of PD, with significant improvements in the functional outcome of the animals was reported (Danielyan et al., 2011). The growing amount of MSC therapy studies for PD is starting to set the ground for pre-clinical studies, and a trial is already underway to evaluate the safety and efficiency of intrastriatal grafts of autologous MSCs to treat PD (Jaslok Hospital and Research Centre, India, sources clinicaltrials.gov).

NSCs isolated from the adult brain have been evaluated for the treatment of PD and HD due to their potential to differentiate into neurons *in vivo* (Lois & Alvarez-Buylla, 1993), despite the problems encountered for their isolation. A study reported the transplantation of NSCs from adult rat SVZ in a rat model of HD, with NSC survival up to 8 weeks after grafting and migration throughout the brain. To a larger extent than MSCs, up to 15 % of adult NSCs differentiated into mature neurons with specific markers of striatal medium spiny projection neurons and interneurons (Vazey et al., 2006). In addition, priming these cells with lithium chloride resulted in an even further improved functional outcome of the grafted animals (Vazey & Connor, 2011). To treat PD, a clinical trial has been performed with adult NSCs (reviewed in (Arias-Carrion & Yuan, 2009)) and gave interesting proof of concept for the autologous use of these cells. NSCs were isolated from the patient's brain during the insertion of a thalamic stimulator and, after expansion and differentiation, a total of 6 millions cells, and among them GABAergic and dopaminergic cells, were grafted 9 months later into the patient's post-commissural putamen. An improvement in the Unified Parkinson's Disease Rating Scale (UPDRS) score was observed over the next 36 months, although results returned to baseline at 5 years post-operation (Levesque, 2009).

### 2.2.3 Adult cell-derived pluripotent stem cell therapy

As for several other disorders, iPS cell have been evaluated for PD cell therapy (Table 2). Dopaminergic neurons have efficiently been derived from iPS cells, and those cells have led to functional recovery in an animal model of PD (Wernig et al., 2008; Swistowski et al., 2010). In addition, it was recently demonstrated that iPS derived from patients suffering from idiopatic PD, after dopaminergic differentiation *in vitro*, led to an efficient functional recovery of the grafted animals (Hargus et al., 2010).

To conclude this section, it appears clearly that MSCs were, and still are, the most widely investigated adult cells for brain cell therapy. However, their neuronal differentiation potential remains very low or uncertain after transplantation, explaining the scientific interest of the recently discovered iPS cells. In addition, the poor cell survival and engraftment observed when using chromaffin cells, hRPE cells, MSCs and in general all kinds of transplanted cells, has called into question the efficacy of a cell therapy procedure. These issues may now be acknowledged by tissue engineering approaches, discussed in the following section.

## 3. Combined use of adult stem cells and scaffolds for cell delivery and regeneration of CNS disorders

It is now widely admitted that cell survival, differentiation, and more generally behaviour of cells *in vivo* may be greatly enhanced using adequate biomaterial supports. These supportive elements, called scaffolds, may be of various compositions and shapes and may improve cell behaviour due to the 3D environment as well as to the mechanical and signalling cues they provide to transplanted cells. In this regard, scaffolds may for example stimulate cell survival. These types of tissue engineering strategies for brain cell therapy have been primarily developed with neuronal cell lines (PC12 cells) or with cultured dorsal root ganglion neurons and foetal NSCs due to their availability, ease of expansion and their natural ability to integrate and differentiate within the brain. Benefits gained using these tissue engineering approaches are now being translated to adult cells such as MSCs in order to improve their survival, to guide their differentiation and integration within the host brain. In the following section, we describe the properties required for a scaffold in tissue engineering applications within the brain, before reviewing the strategies used to improve brain cell therapy by combination with a tissue engineering approach. Improvement of classical scaffolds by means of a biomimetic approach as well as a novel strategy currently developed in our laboratory, the Pharmacologically Active Microcarriers (PAMs), will be presented.

| | Cells | Origins | Recipients | Scaffolds | Benefits | References |
|---|---|---|---|---|---|---|
| HD | Sertoli | | 3-NP model | | Possible modulation of local inflammation | Emerich 2004 |
| HD | Choroid plexus | Rat | QA rat model | Alginate microcapsules | Selective neuroprotection, no evidence for sparing of striatal neurons | Borlongan, Thanos et al. 2008 |
| HD | HSC+MSC | Rat | QA rat model | | Reduction of working memory deficits | Lescaudron, Unni et al. 2003 |
| HD | MSC | Rat | QA rat model | | Decreased striatal atrophy | Amin, Reza et al. 2008 |
| HD | MSC | Rat | QA rat model | | Attraction of MSCs by SCF production in the striatum | Bantubungi, Blum et al. 2008 |
| HD | MSC | Rat | QA rat model | | IV transplantation: improved motor & cognitive performance | Edalatmanesh, Matin et al. 2009 |
| HD | Adult NSC (SVZ) | Rat | QA rat model | | Survival at 8 weeks, migration & 15 % of cells differentiated in mature neurons with marker of striatal medium spiny neurons | Vazey, Chen et al. 2006 |
| HD | Adult NSC (SVZ) | Rat | QA rat model | | Lithium chloride priming of the NSCs resulted in accelerated functional outcome of the grafted animals<br>Increased formation of projections from newly formed neurons in the damaged host striatum to the globus pallidus | Vazey and Connor 2011 |
| PD | Adrenal chromaffin | Human | Human (autologous) | | Functional improvements observed up to 10 months | Madrazo, Drucker-Colin et al. 1987 |
| PD | Adrenal chromaffin | Human | Human (autologous) | | No cell detected after 1-2 years | Hurtig, Joyce et al. 1989 |
| PD | Adrenal chromaffin | Rat | 6-OHDA rat model (unilateral) | Cytodex® Glass beads | Increased survival and functional benefits (8 months) | Cherksey, Sapirstein et al. 1996 |
| PD | Adrenal chromaffin | Rat | 6-OHDA rat model (unilateral) | Cytodex® Glass beads | Increased survival and functional benefits (12 months) | Borlongan, Saporta et al. 1998 |
| PD | Adrenal chromaffin | Human | Human | | Improved dopamine production with in vitro differentiation | Drucker-Colin, Verdugo-Diaz et al. 1999 |
| PD | hRPE | Human | 6-OHDA rat model (uni & bilateral) | Gelatine microcarriers (Spheramine®) | Functional recovery with chronic inflammation at late time-point | Flores, Cepeda et al. 2007 |
| PD | hRPE | Human | MPTP primate model | Gelatine microcarriers (Spheramine®) | Long term functional improvements an cell survival (18 months)<br>No immunosuppression required | Doudet, Cornfeldt et al. 2004 |
| PD | hRPE | Human | Human | Gelatine microcarriers (Spheramine®) | Open-label study: good tolerability, sustained motor clinical improvement up to 6 months after grafting | Bakay, Raiser et al. 2004 |
| PD | hRPE | Human | Human | Gelatine microcarriers (Spheramine®) | Analysis of the brain of a patient who underwent the last trial performed with Spheramine® (phase II double-blind, randomized, multicenter, placebo-controlled clinical) with no significant functional improvements<br>6 months after grafting, only 0.036 % of surviving cells | Farag, Vinters et al. 2009 |
| PD | MSC | Rat | 6-OHDA rat model (unilateral & partial) | | Adipose tissue-derived MSCs secreting BDNF, GDNF & NGF<br>Functional recovery after transplantation in SN but no neuronal differentiation *in vivo* | McCoy, Martinez et al. 2008 |
| PD | MSC | Human | 6-OHDA rat model (unilateral & partial) | | Induction of MSCs to secrete GDNF & BDNF<br>Transplanted on the day of lesion, MSCs migrated toward the lesioned striatum and had a regenerative effect | Sadan, Bahat-Stromza et al. 2009 |
| PD | MSC | Human | 6-OHDA rat model (unilateral & total) | | Improved functional recovery with differentiated cells (TH expression and DOPA secretion) | Levy, Bahat-Stroomza et al. 2008 |
| PD | MSC | Human | 6-OHDA rat model (unilateral & partial) | | Trophic, restorative effect of MSCs | Bouchez, Sensebe et al. 2008 |
| PD | MSC | Human | 6-OHDA rat model (unilateral & partial) | PAMs with LM surface & releasing NT3 | With PAMs:<br>Increased MIAMI cells survival and differentiation *in vivo*<br>Neuroprotection, neuroreparation of the nigrostriatal pathway<br>Functional recovery | Delcroix, Garbayo et al. 2011 |
| PD | MSC | Rat | 6-OHDA rat model (unilateral & total) | | Innovative intranasal delivery<br>Migration of the cells throughout the brain<br>24 % of cells survival 4.5 months after grafting<br>Improvement of motor function of the Parkinsonian forepaw | Danielyan, Schafer et al. 2011 |
| PD | Adult NSC | Human | Human (autologous)<br>Cortical and sub-cortical tissue-derived | | After expansion and differentiation *in vitro*, NSCs improved UPDRS score over 36 months after grafting<br>Back to baseline 5 years post-operation | Levesque 2009 |
| PD | iPS cells | Mouse Fibroblast-derived | 6-OHDA rat model (unilateral & partial) | | Neuronal differentiation *in vivo* (TH-positive cells)<br>Functional improvements 4 weeks after grafting | Wernig, Zhao et al. 2008 |
| PD | iPS cells | Human Foetal-lung fibroblast-derived | 6-OHDA rat model (unilateral & total) | | Survival of iPSC-derived dopaminergic neurons *in vivo*<br>Functional improvements 12 weeks after grafting | Swistowski, Peng et al. 2010 |
| PD | iPS cells | PD patient Fibroblast-derived | 6-OHDA rat model | | Survival of iPS cell-derived DA neurons, with arborization<br>Functional improvements 16 weeks after grafting<br>Proof of concept for the autologous use of iPS cells | Hargus, Cooper et al. 2010 |

***Abbreviations:*** AD: Alzheimer's disease; BDNF: brain-derived neurotrophic factor; GDNF: glial cell line derived neurotrophic factor; HD: Huntington's disease; hRPE: human retinal pigment epithelium; HSC: hematopoietic stem cell; iPS: induced pluripotent stem; IV: intravenous; LM: laminin; MIAMI: marrow isolated adult multilineage inducible; MPTP: 1-methyl-4-phenyl-1,2,3,6-tetrahydropyridine; MSC: mesenchymal stem cell; NP: nitropropionic acid; NSC: neural stem cell; NT3: neurotrophin 3; OHDA: hydroxydopamine; PAMs: pharmacollogically active microcarriers; PD: Parkinson's Disease; QA: quinolinic acid; SCF: stem cell factor; SN: substantia nigra; UPDRS: unified Parkinson's disease rating scale.

Table 2. Adult and iPS cell therapies for neurodegenerative disorders

### 3.1 Scaffold requirements for brain cell therapy

Implementation of tissue engineering in combination with adult cells for brain therapy is an emerging field and many requirements need to be taken into account in order to produce an appropriate tissue engineered product. The first consideration in scaffold tailoring for brain is size, but scaffolds must also be fully biodegradable and biocompatible, minimizing macrophage and microglial reaction, without inducing neurotoxicity. Scaffolds must be small enough to be easily implanted into the skull cavity in discrete and precise areas of the brain without perturbing the brain 3D organisation as observed with 30 µm microspheres (Veziers et al., 2001). Moreover, small-sized scaffolds render repeated implantations possible by stereotaxy, with no need for open-surgery (Menei et al., 2005). In this sense, microstructured and nanostructured scaffolds, produced by various techniques may be used (reviewed in (Seidlits et al., 2008)).

Scaffolds are based either on natural or on synthetic biomaterials used alone or in mixtures, providing scaffolds with different properties (see for review (Potter et al., 2008; Dalton & Mey, 2009). Scaffolds based on biodegradable gels encapsulating various molecules and cells have been studied, with e.g. polyethylene glycol (PEG) (Namba et al., 2009), diblock copolypeptide (Yang, Song et al., 2009) or hyaluronic acid (Wang & Spector, 2009) hydrogels. However, gel-based scaffold strategies most of the time require open-surgery, unless an *in situ* gelling process is used (Kim, 2009; Yang, Song et al., 2009) but no gel swelling must be observed. In opposition to gel-based scaffolds, particulate scaffolds, that may be constituted of aliphatic polyesters, including poly(lactic-co-glycolic acid) (PLGA), overcome this problem of delivery and have been intensively studied with embryonic cells and NSCs (Newman & McBurney, 2004; Bible et al., 2009). Cell-material interactions may be advantageously increased due to a large specific surface as with microsphere-shaped scaffolds or scaffolds based on nanofibrous technology (Valmikinathan et al., 2008; Cao et al., 2009).

The adverse host cell response, such as glial scar and inflammation, occurring after scaffold implantation have to be minimized (see for review (Fournier et al., 2003)) therefore rendering the biocompatibility and biodegradability of the biomaterials a crucial parameter (Vert, 2009; Yang, Song et al., 2009). For example, implantation of PLGA microspheres into the brain does not induce a specific astrocytic or macrophage/microglia reaction, which is similar to the one observed after control fluid injection. Moreover, PLGA microspheres fully degrade into $CO_2$ and $H_2O$ (Menei et al., 2005). Scaffolds should be able to degrade with time, with degradation products that may also be eliminated by the host, allowing a full integration of transplanted cells into the brain. This criteria was not observed for synthetic poly(methylidene malonate 2.1.2) microspheres implanted into rat brains even if biocompatibility of the intact microspheres was satisfactory (Fournier et al., 2006). It is interesting to note that size of particles may also affect the extent of the host response. *In vitro*, small size phagocytable hydroxyapatite particles (1-30 µm) have been shown to induce a strong production of the inflammatory cytokines tumor necrosis factor-alpha (TNFα), interleukin 6 (IL6) and interleukin 10 (IL10) by human monocytes, the first cells recruited to the inflammation site, which may be correlated to a stronger host inflammatory response. On the other hand, this effect decreased for particles of more than 30 µm in diameter. Importantly, shape is also critical for the extent of the response, needle-shape being potentially more detrimental compared to spherical-shaped particles (Laquerriere et al., 2003).

Cell attachment to the biomaterial critically depends on its surface charges, cells being attracted to positive charged surface due to sialic acid residues on the cell membrane which produce a net negative charge on the cell surface. Furthermore, the first step following implantation of a scaffold within the brain is its coverage by a non-specific layer of proteins, which may contribute to the inflammation process and biocompatibility problems (Fournier et al., 2003). Surface characteristics (charge, hydrophilicity and hydrophobicity) are therefore important points to consider when designing a new type of scaffold. Outer but also inner topography of scaffolds may affect cell behaviour once implanted into the brain. For example, the presence of pores and channels on the surface of synthetic poly-ε-caprolactone (PCL) scaffolds may enhance host astrocytic infiltration and affect host cell migration (Wong et al., 2008). Moreover, access to nutrients is a critical parameter for neuronal cells, which require large amounts of nutrients such as glucose. Therefore, if larger implants are used, vascularisation is required for cell survival. In this sense, porous scaffolds, or scaffolds that become porous after implantation during degradation, may alleviate vascularisation problems.

Finally, a very interesting study first described the effects of matrix elasticity to direct MSCs lineage specification. Soft type matrices mimicking brain being neurogenic while stiffer matrices appeared to be myogenic and furthermore osteogenic. The observed phenotypic specification was irreversible after several weeks in culture, therefore reflecting neuronal commitment of the MSCs cultivated on soft matrices (Engler et al., 2006). Noteworthy, new data from this team seem to indicate that stiffness variation, not just stiffness alone, can be an important regulator of MSCs behaviour (Tse & Engler, 2011). Again, this underlines the importance of the choice of the biomaterial for brain tissue engineering.

### 3.2 Scaffold design and manufacture

The following section reviews materials and fabrication methods used in the development of scaffolds to enhance brain tissue regeneration.

#### 3.2.1 Materials

One of the first considerations when designing a scaffold for brain tissue engineering is the choice of the material. Some of the aspects that should be carefully taken into account are: 1) if the material maintains an appropriate shape after implantation, 2) if sterilization of the scaffold prior to implantation is possible, 3) if the scaffold is brain biocompatible to avoid abnormal immune responses, 4) if the material has the appropriate degradation rate for the desired application and 5) the ability to provide a controlled release of the drug, in case of materials encapsulating a therapeutic molecule such as growth factors. Materials used in the development of scaffolds for brain tissue engineering can be broadly divided into biodegradable and non-biodegradable. Due to the transient nature of the biodegradable polymers, they are preferred for a brain application because they do not require surgical removal when the treatment is finished. Depending on their nature they can be classified in natural and synthetic materials. Both of them used alone or in combination have been tested in clinical practice. Natural materials are very interesting because they contain sites for cell adhesion, allowing for cell infiltration. These natural materials also exhibit similar properties to the soft tissues they are replacing. However, since these materials are obtained from natural sources, they must be purified before use and it is difficult to control the homogeneity of product between batches. Synthetic materials in contrast, have a known composition and can be designed to minimize the immune response or the degradation rate.

Materials (natural and synthetic) used in brain tissue engineering applications, with their benefits and drawbacks will be discussed below in this section.

### 3.2.1.1 Natural materials

*Collagen.* Collagen, one of the most common extracellular matrix (ECM) proteins, has been extensively used as a potential scaffold for neural tissue. Collagen is an easily accessible material that can be isolated from mammals, including rats, bovines and humans. One advantage of collagen use is that the scaffold properties can be easily varied by using different concentrations of collagen or by covalently modifying the cell adhesion sites. However, immune response could arise if cross-species transplantation is used. In the CNS collagen has been mainly used as scaffolds for the treatment of TBI among others (Qu et al., 2009; Qu et al. 2011).

*Alginate.* Alginate is a natural anionic polysaccharide composed of D-mannuronic (M) and L-guluronic (G) residues in varying proportions. It is easily obtained from algae and can be cross-linked to form three dimensional scaffolds. Cross-linking and gel formation takes place when divalent cations, such as calcium, ionically bind carboxylic acid groups of blocks of guluronic residues between chains. They must undergo extensive purification to prevent immune responses after implantation. The main use for alginate in the area of brain tissue engineering is in encapsulating cells with the purpose of immune-isolation from the host. Its efficacy has been evaluated in models of CNS diseases like PD, stroke, ALS, spinal cord injury, TBI and HD among others (Orive et al., 2003; Grandoso et al., 2007; Orive et al., 2009; Purcell et al., 2009).

*Fibrin.* Fibrin is a protein involved in the clotting of blood. Similar to collagen, fibrin scaffolds contain sites for cell adhesion and the scaffold properties vary depending on the concentration of fibrin used. Fibrin matrix has favorable features as a scaffold, that would fit well in the fragile CNS tissue, including biocompatibility, biodegradability, binding capacity to the tissue, low risk of foreign body reaction, physiological flexibility, good plasticity. Fibrin scaffolds have been proposed as potentially suitable vehicle for cell transplantation therapy and, combined with bone marrow stroma cells, have been evaluated after cortical injury in rats (Yasuda et al., 2010).

*Chitosan.* Chitosan is a polysaccharide industrially derived from partial deacetylation of chitin, the major compound of exoskeletons in crustaceans. Chitosan has been reported to be suitable for preparation of particulate systems (micro and nanoparticles) for brain application due to its good biodegradability, biocompatibility, stability and low toxicity. Chitosan nanoparticles coated with polysorbate 80 have been proposed for brain targeting (Aktas et al., 2005). However, despite its desirable characteristic, its actual use is limited because of its poor solubility in water.

*Other natural materials.* Other examples of natural materials used in the fabrication of brain tissue scaffolds are dextran, fibronectin (FN), laminin (LM) or hyaluronic acid. An interesting approach used hyaluronic acid and collagen to develop a tridimensional biodegradable porous scaffold which is a sponge with an open porous structure and mechanical behaviour comparable to brain tissue (Wang & Spector, 2009). Dextran hydrogels have been investigated as drug delivery vehicles and as macroporous scaffolds However, the dextran non-cell-adhesive nature has limited its use in tissue engineering. To overcome this limitations, Levesque *et al.*, proposed the use of macroporous scaffolds of methacrylated dextran (Dex-MA) copolymerized with aminoethyl methacrylate (AEMA) introducing primary amine groups for covalent immobilization of extracellular-matrix-derived peptides for axonal guidance (Levesque & Shoichet, 2006). Recently, the use of bioactive scaffolds generated by cryogelation of dextran or gelatine linked to LM to create niche-like structures that promote the differentiation of stem cells was proposed (Jurga et al. 2011). LM and FN scaffolds have been proposed as appropriate extracellular-matrix based scaffolds that can be exploited to improve cell transplantation into the injured brain (Tate et al., 2009).

#### 3.2.1.2 Synthetic materials

*Poly(α-hydroxyacids).* Poly (α-hydroxyacids) were found to be bioabsorbable and biocompatible in the 1960's. They are the most widely known, studied and used polymers for brain drug delivery due to its biodegradable properties, biocompatibility and the absence of significant toxicity (Athanasiou et al., 1996; Shive & Anderson, 1997). They have suitable mechanical properties for its use in scaffold fabrication. Poly (α-hydroxyacids) are approved by the Food and Drug Administration (FDA) for its use in humans and have been used extensively in medicine in a variety of applications. For instance, more than 100 different molecules have been incorporated in PLGA microspheres since their first application (Menei et al., 2005). Poly (α-hydroxyacids) are constituted of lactic and/or glycolic acid units and degrade *in vivo* by nonenzymatic hydrolysis to lactic and glycolic acids. These can be further metabolized or excreted via normal physiological pathways.

*Polycaprolactone (PCL).* PCL is a biodegradable polyester with minimal bio-reactivity that has been widely used in the biomedical field and in CNS applications in particular. PCL degrades very slowly by hydrolysis of ester bonds. PCL based matrices have been used for instances, as scaffolds for tissue regeneration after controlled cortical impact induced-TBI (Wong et al., 2008).

*Others.* Other examples of synthetic materials used in the fabrication of brain tissue scaffolds are poly (glycerol sebacate) (PGS), gelatin-siloxane (GS), poly (ethylene glycol)/poly (ethylene oxide), poly (ethylene-co-vinyl acetate) (EVA), poly (2-hydrixyethil methacrylate) (pHEMA) and poly (2-hydroxyethyl methacrylate-co-methyl methacrylate) (pHEMA-MMA).

### 3.2.2 Scaffold fabrication techniques

As brain tissue engineering progresses, the need of novel scaffold structures and fabrication techniques has become of great importance (See Walker et al., 2009; Subia, 2010). The most common scaffold production methods are revised below. After fabrication, all of these scaffolds are then characterized in terms of morphology, mechanical, bulk and surface properties.

*Phase separation.* Phase separation technique is based on temperature changes to separate a polymeric solution in two phases. When liquid-liquid phase is separated, it is quenched to form two solid phases. Finally, solvent is removed and porous scaffolds are obtained. Biological molecules can be incorporated to the polymeric solution to obtain drug-loaded scaffolds. Phase separation technique can be combined with other fabrication techniques like leaching or prototyping to create 3D-scaffolds with controlled pore morphology.

*Particulate Leaching.* Leaching is a reproducible technique to fabricate porous, sponge-like scaffolds with a desirable cellular structure for tissue engineering applications. Salt, wax or sugars known as porogens are used to create porous scaffolds. The fabrication process involves casting the polymer mixed with a porogen into a desired shape, and then introducing a solvent that would dissolve the salt and leave the polymer intact, leaving a foam-like structure.

*Electrospinning.* Electrospinning is an efficient method to produce nano or micro-scaffolds comparable to the ECM fibers natively found in the tissue. In this fabrication method, a polymer solution is pumped through a syringe connected to a high voltage source. As a droplet forms at the needle tip, electrostatic repulsions form long fibers that are collected onto a grounded metal plate in the form of a nanofibrous mat. Due to the simplicity of this method, it has received considerable attention for use in tissue engineering and more than 200 polymers have been used for electrospinning. However, it is difficult to obtain 3D scaffolds with well defined pore structure and complex geometry using this technique and electrospinning is preferably used to produce thin 2D sheets.

*Rapid prototyping.* Rapid prototyping describes different manufacturing processes that allow automated fabrication using methods of material bonding or deposition. This fabrication method produces objects with geometry difficult to be created using the "traditional" machining methods of milling, turning, or drilling. Generally, files generated using standard computer aided design (CAD) software are used to produce the scaffolds.

*Other techniques.* Stereolithography uses a laser and a photosensitive liquid polymer solution to fabricate scaffolds. Three dimensional printing has been employed to fabricate porous scaffolds by inkjet printing liquid blinder droplets onto particulate matter.

### 3.3 Overview of strategies used to combine adult stem cells and scaffolds

The following section reviews primary studies that investigated the potential of different combinations of cells and scaffolds to improve the cell therapy benefits for PD, HD, cerebral ischemia and TBI. Studies on hRPE cells in combination with gelatine-based scaffolds, a device that underwent a phase II clinical trial for the treatment of PD will also be described. Several strategies are now focusing on improving cell interactions with the biomaterials by modifying its surface using biological molecules, mainly derived from ECM, to better regulate grafted cell behaviour. After briefly describing the molecular mechanisms of cell-ECM molecule interactions and its effect on cell behaviour, we will present some studies using such "biomimetic scaffolds". Finally, the PAMs developed in our laboratory, that constitute a more advanced approach combining a bioactive surface with the controlled delivery of a growth factor will be presented.

#### 3.3.1 Gel-based scaffolds and cell encapsulating technology

By contributing to maintain brain's integrity as well as by favouring the integration of host cells inside the marginal cavities, gel-based scaffolds, without cells, may be advantageous to repair the brain after an ischemic stroke (Yamashita et al., 2009). Moreover, differentiated iPS injected together with a fibrin glue under the dura mater enhanced the effect of these iPS injected alone, which led to a reduced infarct volume as well as to functional recovery after an ischemic stroke; thereby underlining the importance of a support for the cells (Chen & Xiao, 2011).

In the context of TBI, degradable collagen scaffolds did not reduce the lesion size nor did they improve functional recovery, unless if seeded with hMSCs (Lu et al., 2007). Indeed, four days after TBI, transplantation of a cylindrical collagen scaffold seeded with hMSCs in the lesion cavity induced a reduction in the lesion volume, together with an improved spatial learning and sensorimotor function of the animals. Recently, the same team reported that delayed transplantation of these complexes (7 days after TBI) further enhanced spatial learning and sensorimotor function, and induced angiogenesis in the injured cortex as well as transcallosal fiber length (Qu et al., 2009; Xiong et al., 2009). In experimental models of TBI, MSCs are usually injected adjacent to the lesion within the parenchyma to avoid injection within the lesion cavity. An advantage of this gel-based strategy was the possible use of 3 fold more MSCs compared to parenchymal injection, therefore increasing the regenerative potential. However, the major limitation was the need for open-surgery to implant the device. This issue may potentially be addressed using an *in situ* gelling process (Perale et al., 2011), even if the possible expansion and ensuing damage to the brain parenchyma of the solidifying gel has to be taken in consideration. Nevertheless, all these studies demonstrate the benefits of a 3D support which are probably improving cell survival, even if the underlying mechanisms are still not always fully understood.

### 3.3.2 Cell adhesion on particulate scaffolds

As previously described, the major problems encountered when grafting chromaffin cells in the context of PD was the poor cell survival and the absence of long term effects *in vivo* (Drucker-Colin & Verdugo-Diaz, 2004). Two studies gave the proof of concept that particulate scaffolds may enhance survival of cells adhered onto collagen-coated dextran (Cytodex 3®) or glass bead microcarriers into the brain of hemi-parkinsonian rats. The pivotal finding of these studies is that adult rat adrenal chromaffin cells implanted in the brain after attachment to microcarriers retain their ability for a prolonged period (8-12 months) to correct a striatal dopamine deficit as judged by their efficacy in reducing apomorphine-induced rotation (Cherksey et al., 1996; Borlongan et al., 1998). Even if the underlying mechanisms were not studied, it is noteworthy that no inflammation was detected when implanted into the striatum. During the same period, these scaffolds were used for transplanting human FVM cells in similar rat model of PD (Saporta et al., 1997) and the Cytodex® microcarriers allowed an increased cell survival without immunosuppression. This effect was thought to result from the presence of a protective astrocytic cloak around the cell/microcarrier complexes. These data were the first to demonstrate the need for cell attachment to a 3D complex to improve grafted cell survival, and it also underlined the immunomodulatory benefits that may be gained from the use of a tissue engineering strategy.

A similar strategy combines cultured hRPE cells attached to biocompatible, non biodegradable cross-linked porcine gelatine microcarriers, with a mean diameter of 100 μm. Several preliminary studies in parkinsonian rat models (unilateral and bilateral lesions) proved the efficacy of these grafted complexes, named Spheramine® (Watts et al., 2003). An increased survival of hRPE cells, without immunosuppression, and long term functional improvements were observed, although chronic inflammation was reported at later time-points (5 months) (Flores et al., 2007). These microcarriers have also been implanted in the brain of hemi-parkinsonian monkeys and resulted in long term cell survival and functional improvements at 18 months (Doudet et al., 2004). As expected, hRPE cells unattached to microcarriers did not survive well in the brain, and did not produce a lasting therapeutic effect in various PD animal models. These encouraging results led to an open-label clinical study that included 6 patients with advanced PD receiving 325,000 hRPE cells attached to microcarriers and demonstrated a good tolerability to Spheramine®. Moreover, at 6 months post-operation, the mean UPDRS-M (off) score improved to 34 % from the pre-operation baseline. Half of the patients also demonstrated a reduced Dyskinesia Rating Scale scores (Bakay et al., 2004). The success of this strategy finally led to a phase II double-blind, randomized, multicenter, placebo-controlled (sham surgery) study to evaluate safety, tolerability, and efficacy of Spheramine® implanted bilaterally into the postcommissural putamen of patients with advanced PD, even if the mechanisms underlying the improved survival of hRPE cells upon attachment to gelatin are not yet well elucidated (Stover & Watts, 2008). Unfortunately, preliminary results of the phase II clinical trial seem to indicate that the study failed to demonstrate the efficacy of Spheramine® beyond a remarkable placebo effect. A reason for that may be the lack of long term survival of the cells in the human brain, as only around 0.036 % of cells survival was observed in the brain of a patient who died 6 months after surgery (Farag et al., 2009).

Adult rat choroid plexus cells have the potential to secrete a wide range of growth factors, and have been used as candidate cells for HD cell therapy (Borlongan et al., 2008). In this study, choroid plexus cells were encapsulated in alginate beads to improve cell viability and to prevent host rejection. Transplantation of these cells into rat striatum resulted in an encouraging neuroprotection when a QA lesion was performed 3 days after cell transplantation, even if no evidence was provided concerning the specific sparing of GABAergic medium spiny projection neurons, especially sensitive to degeneration in the context of HD.

### 3.3.3 Molecular mechanisms of cell adhesion to scaffolds

The ECM contains adherent glycoproteins, glycosaminoglycans and ions. ECM proteins, such as collagen, FN, LM, tenascin and proteoglycans, interact with each other forming a supportive scaffold for the cells within the tissue (see for review (Bosman & Stamenkovic, 2003)). Composition and proportion of its constituents vary depending on the type of tissue, thereby giving different mechanical, chemical or signalling cues to the surrounding cells. We will here focus on FN and LM because of their astonishing variety of effects on cells as well as because of their wide use for brain tissue engineering applications. FN and LM structures have been described a few decades ago and are constituted of an assembly of large polypeptides, all of them having specific interaction sites with other ECM molecules or cell surfaces (Hynes & Yamada, 1982; Engvall & Wewer, 1996; Powell & Kleinman, 1997). LM and FN interact with cells *via* the integrin family of receptors, therefore allowing cell attachment to the matrix and further signal transduction. Integrins are a family of proteins constituted of an α subunit and a β subunit which assemble into distinct integrin receptors having specific binding affinities with ECM molecules (Barczyk et al., 2009). These interactions result in highly complex downstream signalling pathways that originate at the focal adhesion sites, where a variety of proteins (e.g. Ilk, Fak, Src) interact with the integrin tails on their cytoplasmic ends, thereby regulating cell survival, proliferation and differentiation (Hynes, 2002; Chen, 2010).

Potential roles of ECM molecules for brain repair are now widely admitted. Endogenous levels of FN and LM increase in a TBI context *in vivo*, thereby suggesting a reparative role of these molecules (Tate et al., 2007). Tissue repair benefits have also been obtained by transplanting LM-based hyaluronic gel scaffolds, without cells, in a rat model of cortical lesion (Hou et al., 2005). A recent study demonstrated an enhanced survival *in vitro* of human MSCs within a PEG hydrogel modified with RGD peptides, a motif involved in

ECM-cell interactions. In this study, RGD peptides attached to the scaffold provided the required adhesion sites to maintain MSC survival, while soluble peptides resulted in a strong decrease of cell viability. Noteworthy, the presence of a glycine spacer between the RGD peptides and the gel further improved MSC survival, therefore underlining the importance of the presentation context of the peptide within the gel (Salinas & Anseth, 2008). Another team described the effect of FN and LM-derived peptides (RGDSP and IKVAV, respectively) using a PEG-hydrogel array to screen the effect of these ECM molecules on hMSCs viability. Their strategy allowed to also analyse the consequences of the hydrogel used, for example, the combination of the IKVAV peptides with degradable gels resulted in decreased sustained viability compared to non-degradable gels (Jongpaiboonkit et al., 2008). A recent study described the single-cell encapsulation of hMSCs in FN- and fibrinogen-containing hydrogel capsules that rescue the cells from apoptosis induced by loss of anchorage, while providing an increased cell metabolic activity in culture. Effects that were certainly mediated *via* the MAPK/ERK signalling cascade and upstream integrin/ECM molecule interactions (Karoubi et al., 2009). Using a subpopulation of hMSCs, we recently showed in a rat model of global ischemia, that marrow-isolated adult multilineage inducible (MIAMI) cells (D'Ippolito et al., 2004) adhered onto PLGA microcarriers with a FN biomimetic surface enhanced the neuroprotection effect observed with the cells alone (Garbayo et al., 2011). This effect may be attributed to an increased survival of the cells on these carriers, or the increased production of neuroprotective mediators by these cells or both.

ECM molecules may also affect proliferation as well as life span of cells, this being the rationale for expanding MSCs on ECM molecules in several *in vitro* protocols (see for example (D'Ippolito et al., 2004; Matsubara et al., 2004)). Modification of PLCL (poly(L-lactic acid)-co-poly-(ε-caprolactone)) nanofibrous scaffolds with collagen I resulted in an increased proliferation of human MSCs and neuronal differentiation after exposure to an induction media, compared to standard PLCL scaffolds (Prabhakaran et al., 2009). The benefits of collagen I was assumed to result from its high cell adhesion properties, thought to be required to ensure a proper neuronal differentiation. In our laboratory, we observed an increased proliferation of the human stromal MIAMI cells expanded on FN compared to standard culture substrates (Delcroix et al., 2011). Extensive crosstalk takes place between integrin and growth factor receptor signalling pathways to stimulate progression through the G1 phase of the cell cycle, as mitogenic signalling may be weak and transient in the absence of integrin-mediated cell adhesion for cells with an anchorage dependent growth, such as MSCs (see (Danen & Yamada, 2001) for review).

Neural precursor differentiation, by enhancing neurite outgrowth *in vitro* or *in vivo* is certainly one of the first and most studied effects of LM and its derived bioactive peptides (Rogers et al., 1983; Grimpe et al., 2002). Accumulating evidence support a bioactive signalling role of LM in the morphological and molecular induction of MSCs toward a neuronal lineage (Qian & Saltzman, 2004; Ho et al., 2006; Delcroix et al., 2011). In our laboratory, the enhanced expression of β3-Tubulin and NFM when human mesenchymal stromal MIAMI cells were differentiated on a LM substrate was also accompanied with an increased cell length and a decreased proliferation rate (Delcroix et al., 2011). In this regard, a study reported the increased length of neurite-like extensions of MSCs, when incorporated within a collagen gel containing FN and LM, an effect at least partly mediated *via* the FAK pathway (Lee et al., 2011). Specific fragments of LM, e.g. the peptides IKVAV & YIGSR play crucial signalling roles in neurite outgrowth and cell adhesion. Accordingly, studies also aim at functionalizing scaffolds with similar peptide fragments to reproduce the effect of the native ECM molecules, but also to isolate one of its specific effects (Powell & Kleinman, 1997; Orive et al., 2009). Some observations of how ECM components control MSC behaviour are summarized in Table 3.

| Effects | ECM component or derived peptides | Observations | References |
|---|---|---|---|
| Survival | ***RGD*** | PEG hydrogel<br>RGD improved hMSCs survival *in vitro*<br>Improvement enhanced with the use of a glycine spacer between RGD and the gel | Salinas and Anseth 2008 |
| | ***RGDSP & IKVAV*** | PEG-hydrogel<br>Both peptides improved hMSCs survival<br>Combination of IKVAV with degradable gels decreased the benefits compared to non-degradable gels | Jongpaiboonkit, King et al. 2008 |
| | ***FN & fibrinogen*** | Single-cell encapsulation of hMSCs in FN- or fibrinogen-containing hydrogel capsules<br>Enhanced survival of cells + increased cell metabolic activity in vitro<br>Validation of the concept in vivo (uninjured hindlimb model in rats) | Karoubi, Ormiston et al. 2009 |
| Proliferation | ***Collagen I*** | PLCL/collagen I nanofibrous scaffolds<br>Increased proliferation of hMSCs *vs* standard PLCL scaffolds<br>Effect due to adequate cell adhesion properties of collagen I<br>Allow neuronal differentiation of hMSCs when exposed to the appropriate inducers | Prabhakaran, Venugopal et al. 2009 |
| | ***FN*** | FN substrate<br>Increased proliferation of MIAMI cells *in vitro* | Delcroix, Garbayo et al. 2011 |
| Neuronal differentiation | ***LM*** | LM substrate<br>Induction of nestin expression by MSCs *in vitro* | Ho, Yu et al. 2006 |
| | | LM substrate<br>Morphological changes in hMSCs during neuronal differentiation *in vitro*<br>Increased percentage of cells with secondary and tertiary branching on LM *vs* FN or PDL | Qian and Saltzman 2004 |
| | | LM substrate<br>Increased cell length and expression of neuronal proteins (β3-Tubulin and NFM) during differentiation of MIAMI cells *in vitro vs* FN or glass | Delcroix, Garbayo et al. 2011 |
| | | Collagen gel enriched with FN and LM<br>Increased neurite length of MSCs incorporated within the gel<br>Effect mediated *via* the FAK pathway | Lee, Yu et al. 2011 |

***Abbreviations:*** ECM: extracellular matrix molecules; FN: fibronectin; LM: laminin; MIAMI: marrow isolated adult multilineage inducible; MSC: mesenchymal stem cell; NFM: neurofilament medium; PDL: poly-D-lysine; PEG: polyethyleneglycol; PLCL: poly(L-lactic acid)-co-poly-(e-caprolactone).

Table 3. Control of MSC behaviour by ECM components

### 3.3.4 The pharmacologically active microcarriers (PAMs): a tool to combine the biomimetic approach and the controlled release of a growth factor

In addition to the biomimetic approach, another way to improve the efficiency of cell grafts is to deliver a growth factor by the transplanted scaffolds, further affecting the fate of both transplanted and host cells. Interestingly, the potential of microsphere-hydrogel scaffolds to deliver 2 growth factors at specific rates has been described (Burdick et al., 2006), but has not yet led to *in vivo* studies. For reviews, see (Tatard et al., 2005; Delcroix, Schiller et al., 2010). Our group has formulated PLGA microspheres which deliver therapeutic proteins in a sustained and controlled manner. The use of these growth factor delivery vectors for neuroprotection or for the repair of the nigro-striatal dopaminergic system has been successfully validated in animal models of AD or of PD, respectively (Pean et al., 2000; Jollivet et al., 2004; Menei et al., 2005). In this sense, we developed the PAMs that combine these two approaches and may be easily injected in the desired tissue (Figure 1). These PAMs are biodegradable and biocompatible PLGA microspheres conveying cells on their surface, therefore providing an adequate 3D microenvironment *in vivo*. Moreover, the controlled delivery of a trophic factor in combination with a biomimetic surface act synergistically to stimulate the survival and/or differentiation of the grafted cells toward a specific phenotype, therefore enhancing their engraftment after their complete degradation (Tatard et al., 2005). Finally, it should be noted that the delivered molecule may also affect the host microenvironment allowing the integration of the grafted cells and/or stimulating the lesioned brain repair capacities.

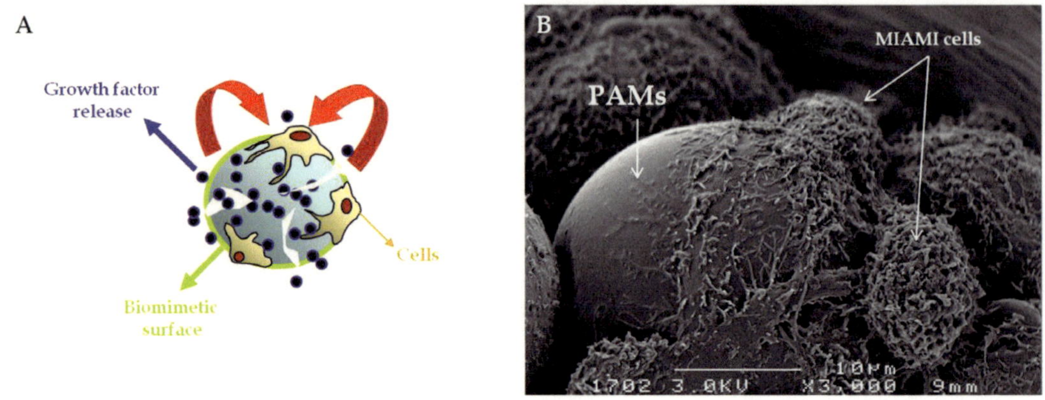


Fig. 1. A) Schematic representation of PAMs. These PAMs are PLGA microspheres with a biomimetic surface that convey cells and release a growth factor in a controlled manner. B) Scanning electron microscopy image of PAMs conveying MIAMI cells on their surface

The proof of concept of this unique and simple device of cell and protein-delivery in neuroprotection and tissue repair for the treatment of neurological disorders has first been validated in a PD rat model using a neuronal cell line (PC12 cells) transported by nerve growth factor (NGF)-releasing PAMs. (Tatard et al., 2004). The efficacy of PAMs for cell therapy of PD in a clinical paradigm was then demonstrated using GDNF-releasing PAMs, conveying a small number of embryonic ventral mesencephalon dopaminergic cells (Tatard et al., 2007). We next implemented the PAMs transporting MIAMI cells for PD adult stem cell therapy. MIAMI cells, may differentiate toward neuronal-like cells in a neurotrophin 3 (NT3) dependent manner (Tatard et al., 2007), especially when pre-treated with EGF and bFGF (Delcroix, Curtis et al., 2010). Moreover, a LM surface enhances the neuronal differentiation (Delcroix et al., 2011). We designed PAMs made of 60 µm PLGA microspheres encapsulating NT3 and covered with a LM biomimetic surface. After adhesion of dopaminergic-induced (DI)-MIAMI cells, the PAM/cell complexes were grafted in the partially dopaminergic-deafferented striatum of rats and led to a strong reduction of the amphetamine-induced rotational behaviour together with the protection/repair of the nigrostriatal pathway. These effects were correlated with the increased survival of DI-MIAMI cells which differentiated towards dopaminergic-like cells and may also secrete growth factors (Delcroix et al., 2011). We thus showed that combining growth factors, cell adhesion molecules and an adapted 3D structure in the same polymeric scaffold allows the synthesis of an adaptable and very efficient system that can deliver stem cells and give them appropriate cues allowing better stem cell survival, differentiation and integration into the host tissues after implantation. After deeper characterization of the underlying mechanisms, this tissue engineering strategy may ultimately set the ground for pre-clinical studies with non-human primates to increase the efficiency of MSC therapy of the brain.

## 4. Unanswered questions in tissue engineering for targeting CNS disorders using conventional tissue culture and animal models

Although brain tissue engineering has shown a certain level of therapeutic benefits, there are still many unanswered questions and concerns that need to be addressed. For instance, one general problem found in cell therapy is the difficulty to reproduce results between laboratories. Differences on the source, method of preparation, differentiation status and age of the stem cells used may be the cause. Moreover, with the current workflow it takes a long time to select an effective stem cell strategy to translate into clinical studies. The main steps to follow include the selection of the source of cells (MSC, NSC, ESC, iPS), the choice of the culture media, the state of stem cell cycle (undifferentiated *vs* differentiated), the mode of culture, the form of growth (neurosphere *vs* monolayer for instance), the study of substrate variations (FN, LN, matrigel among others) and the use of different animal models to test each therapeutic variation. Moreover, very little is known about the exact mechanisms by which stem cells may repair damaged tissues. It is not yet totally understood if stem cells may directly replace lost cells *via* their differentiation potential or if the beneficial effect could be due to their paracrine secretions or their immune regulatory functions, or if it is due to a combination of these effects. An additional and very troublesome problem is the difficulty to track the location and activity of stem cells once they are grafted into the animal. Several strategies have been proposed to follow the grafted cells *in vivo*. One of the most commonly used is to label the cells with fluorescent markers like quantum dots, organic dyes or fluorescent proteins among others. Other authors have grafted sex mismatched stem cells to identify them by fluorescent *in situ* hibridation (FISH) on the X and Y chromosomes. The injection of human cells into rodent animals or *vice versa* to use species-specific antibodies for its identification is another approach.

In general, it is difficult to understand stem cell mechanisms of action and fate using conventional tissue culture or animal models due to the limitations of both systems. *In vitro* models are a good high throughput screening tool but they do not realistically mimic the *in vivo* situation. Moreover, the 3D integration of grafted cells into the host tissue cannot be studied using tissue culture models. On the other hand, it is clear that animal models are essential in proof-of-concept principle experiments and in establishing the preclinical safety and efficacy data required before human clinical trials. However, most of them are expensive, laborious and time consuming to be useful as screening tests. Generally, it can be stated that as a system becomes more complex, the throughput in terms of screening capacity decreases (Sundstrom et al., 2005). Moreover, the real-time monitoring of grafted cells using animal models is not easy. A screening platform in-between *in vitro* and *in vivo* models is required to improve understanding of stem cells. A possible option would be the use of brain organotypic slice culture for optimization of CNS cell therapies. These organotypic cultures mimic *in vivo* models of brain diseases better than cell cultures, they are cost-effective and easier to optimize, use and manipulate than rodent and primate CNS disease models. The usefulness of brain organotypic slices in CNS research as well as in the drug discovery process has been increasing in recent years and will be discussed in the next section.

## 5. Organotypic 3D culture models: novel platforms for optimization of CNS cell therapy and tissue engineering

Since their introduction, organotypic cultures of rat brain slices have become a useful tool to study drug effects. Brain organotypic slices are *ex-vivo* cultures that bridge the gap between *in vitro* and *in vivo* models. Slices are easy to prepare and they preserve the tissue architecture of the brain regions that they originated from, allowing interaction of multiple cell types like neurons, astrocytes and microglia and maintaining neuronal activities. This 3D environment has tremendous importance to evaluate the efficacy of tissue engineering approaches since most of the cells require cues from a truly 3D environment to form relevant physiological tissue structures *in vitro*. This 3D provides external mechanical inputs and cell adhesion parameters, which affects intracellular signalling. Importantly, these slices permit direct treatment or injection of drugs, virus, cells, microscaffolds... making them ideal for screening (Cho et al., 2007).

### 5.1 Preparation of organotypic cultures of CNS tissue

Organotypic brain slice culture was developed by inspiration on explant cultures techniques of various anatomical origins. Since then, several methods have been developed to prepare and maintain slices alive in long-term culture. The roller-tube technique based on the use of roller tubes was first described in 1981 (Gahwiler, 1981). However, most studies now use the membrane interface method described in 1991 (Stoppini et al., 1991) which provides an easier access to the slice culture. The principle of the membrane interface method is to maintain brain slices on a porous membrane filter at the interface between medium and a humidified atmosphere. The medium provides adequate nutrition to tissues through the membrane *via* capillary action. Rats and mice are the most common donor sources, but also rabbits, pigs and human fetuses have been used. Most organotypic brain slice cultures have been derived from neonatal (P0-P10) animals, but recently also adolescent or adult rats have been used and even human postmortem brain tissue slices have been kept alive for a few weeks in culture (Noraberg et al., 2005). Brain slices are relatively easy to prepare (Figure 2).

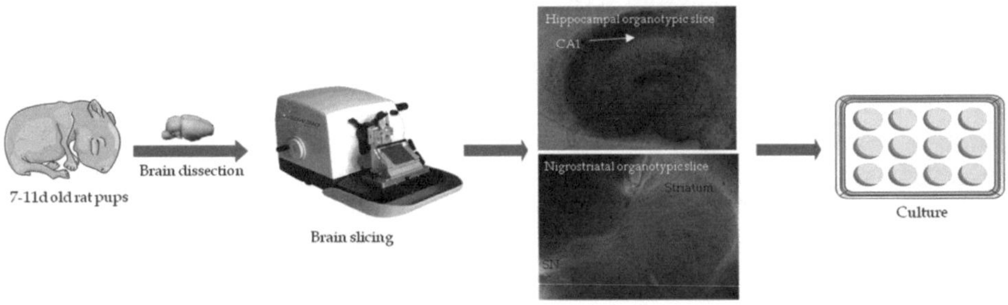


Fig. 2. Organotypic preparation; 7 to 11 days old rats are commonly used to prepare brain organotypic cultures. First the brain is dissected free of skull. Second, slicing is performed with a vibrating microtome. Finally, slices are maintained in cultured using the membrane interface method. Representative bright field images of hippocampal and nigrostriatal organotypics are shown in the figure

Briefly, the brain is dissected free of the skull and placed into a saline solution whose composition resembles that of the cerebro-spinal fluid and is thus often referred to as artificial cerebro-spinal fluid (ACSF). Slicing is performed with a vibrating microtome at high amplitude and very slow speed. Slice thickness varies according to specimen and the type of experiment, from 150 to 400 µm (Lossi et al., 2009). Over the years, slice culture systems have been successfully established from a variety of brain regions including hippocampus (far more frequently), striatum, cortex, spinal cord and cerebellum. Furthermore, a number of tissue slice co-cultures have been developed, which allow the assessment of inter-neural responses across brain regions. Moreover, when organotypic slices are exposed to certain toxic conditions (oxygen-glucose deprivation, neurotoxins, glutamate-mediated excitotoxicity) they develop many pathophysiological features found in brain disorders and consequently, brain organotypic slices can be used as ex-vivo models of CNS diseases. Nowadays, organotypic models for global cerebral ischemia, ischemic stroke, Alzheimer's disease, PD, HD, TBI, epilepisa and amyotrophic lateral sclerosis have been described (For review see (Noraberg et al., 2005; Sundstrom et al., 2005; Cho et al., 2007; Cimarosti & Henley, 2008; Lossi et al., 2009)). Among them, hippocampal organotypic slices exposed to oxygen-glucose deprivation, a model for global cerebral ischemia is the most commonly used.

### 5.2 Applications

Organotypic slices have been commonly used as models to investigate mechanism and treatment strategies for neurodegenerative disorders. Interestingly, the utility of brain slices to test CNS cell therapy efficacy has recently started to be investigated. Organotypic cultures are a very useful tool for screening of candidate stem cells for a specific pathology. They also could be used to track survival, differentiation, proliferation and migration of the transplanted cells with or without microscaffolds or to study graft and host interactions (Figure 3). Stem cell mechanisms of neuroprotection or paracrine secretions of MSCs could be also studied using brain slices. Charriere *et al.*,(Charriere et al. 2010) analyzed the interactions between bone marrow stromal cells and hippocampal slice cultures to clarify putative cross-interactions between MSCs and the CNS that could explain the molecular mechanism of stem cells. The dopaminergic differentiation of ES in a PD organotypic model and the effect of morphogenetic proteins such as LM on the differentiation of the cells has also been studied (Kearns et al., 2006). It was also shown that olfactory ensheathing cells (OEC) when co-cultured with the auditory brain stem slice culture not only promoted neurite outgrowth from the cochlear nucleus region of the brain stem slice but also supported the OEC indicating positive interactions between both (Jiao et al. 2010). Recently, our group examined the potential of MIAMI cells injected into the hippocampus to prevent neuronal damage induced by global ischemia using rat hippocampal slices exposed to oxygen-glucose deprivation. We showed that MIAMI cells prevented neuronal damage. MIAMI cell therapeutic value was significantly increased when delivering the cells complexed with FN-coated biomimetic microcarriers probably by increasing stem cell survival and paracrine secretion of pro-survival and/or anti-inflammatory molecules as concluded from survival, differentiation and gene expression analysis (Garbayo et al., 2011). Furthermore, brain organotypic slices can be used to perform tissue biocompatiblity studies of scaffolds prepared for brain tissue engineering application and to determine how changes in the composition or in the functionalization of the scaffold could compromise their brain biocompatibility. In this context, Kristensen *et al.*, have used organotypic brain slice cultures to assess silicon-based arrays biocompatibility (Kristensen et al., 2001). Other exciting application could be to use them to determine what scaffold composition best promotes proper adhesion and proliferation of the stem cells. Recently, in various pathologies it has been reported that multiple growth factor delivery is more effective than single growth factor administration since it mimics better the natural microenvironments of tissue formation and repair (Richardson et al., 2001; Barrientos et al., 2008). The difficulties of this approach are principally to find the optimised growth factor ratio, each factor at a physiological dose and in a specific spatiotemporal pattern. Brain organotypic slices could be used to determine proper growth factor cocktail for a specific brain damage and to control the relationship between growth factor concentration gradient and timing.

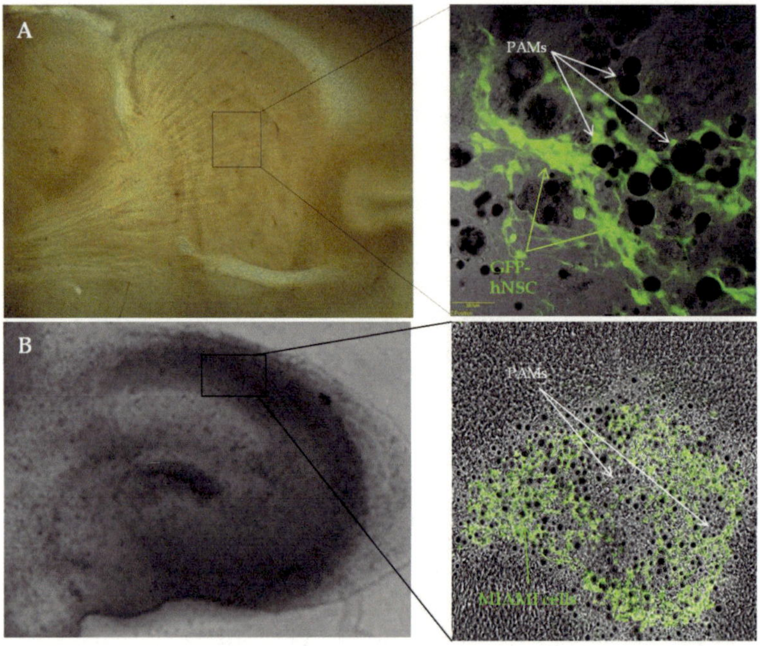


Fig. 3. A) Nigrostriatal organotypic culture with GFP-hNSC forming complexes with PAMs injected into the striatum. GFP-hNSC cells were kindly provided by Dr Martinez-Serrano. B) Hippocampal organotypic culture with MIAMI cells forming complexes with PAMs injected into the CA1 hippocampus

An innovative approach using modified organotypic slices derived from a transgenic, mutant, null, gain-of-function, loss-of function, knock-in, or knockout animals was recently proposed (Li & Loudon, 2008). They discuss the possibility of using modified organotypic slice cultures to understand how implanted cells interact with resident cellular matrix and injured residential cells to predict their *in vivo* behaviour (Li & Loudon, 2008). Organotypic cultures could also be used to evaluate how temporal expression levels of stem cell chemokine receptors can be quantitatively related with their migration capacity toward brain tumor- or lesion producing -signal ligand SDF-1. The foundation of this experimental platform is to establish a system that mimics *in vivo* properties, first to maintain stem cells in a quiescent state, and then induce stem cells to produce targeting molecule cytokine receptors and matrix remodeling enzymes. Li & Loudon showed that migration of stem cells was enhanced by an intermediate concentration of SDF-1 gradient but inhibited by higher concentrations, with no stimulation at low concentrations (Li & Loudon, 2008).

### 5.3 Limitations of organotypics

Organotypic slices are not without limitations. The most limiting features are: 1) currently brain slices are only produced from juvenile donor animals as it is known that young tissue has more neuronal plasticity and are more resistant than older tissue. Recent reports have used older donors to prepare brain organotypic slices but although it is possible, a low yield was observed and slices remained viable for only 3 to 4 weeks. 2) Automation is a challenge since a skilled operator is required for the production of the slices. 3) Not all brain areas are amenable to culture being the most appropriate the regions with a lamellar structure that can be aligned parallel to the plan of slicing. 4) Organotypics do not have a functional vascular compartment (for review see Sundstrom et al., 2005).

## 6. Conclusions

Although the ideal brain scaffold that satisfies all the requirements does not exist yet, the past several years have seen considerable progress in this field of study. For instance, materials of many types have been used to create brain scaffolds capable of providing sustained delivery of signalling molecules and an adequate 3D support for transplanted cells, thereby increasing cell survival and even guiding cell differentiation and fate *in vivo* Scaffold surface characteristics have also been modified to better mimic the natural brain environment, to control cell attachment, growth and differentiation or for specific uses. A key-point to remember is that most regenerative technologies in the future will probably be combinatorial, including biocompatible scaffolds, stem cells and the various factors necessary for their survival and function. This might be especially true in order to implement iPS cells therapy in the future, to control their proliferation and differentiation potential. Continual progress in the design and fabrication of future scaffolds is required to improve current delivery platforms since there is a need for development of custom matrices either tailored for purpose, or for the individual patient. Recent developments in understanding the basic biology of brain tissue formation in physiological and pathologic conditions have resulted in an explosion in the numbers of bio-engineered and tissue engineering products that could be potential candidates for treating brain disorders. A major challenge for the pharmaceutical industry is to find useful tools for screening since a bottleneck exists between the number of compounds that are interesting and the relatively limited *in vitro* and *in vivo* existing methods. Animal models are still the main choice for such studies but over the past years, brain organotypic cultures have begun to emerge as useful tools for screening of new neuroprotective and neuroregenerative approaches. In the current chapter, the combination of adult stem cells with microscopic scaffolds has been revised. Recently, research on the application of nanotechnology in stem cell and tissue engineering research has gained much attention in the scientific community. A significant and exciting area of research is the use of nanoparticles to engineer different patterned topographies of scaffolds that mimic the ECM, to study their effects on stem cells.

## 7. References

Abeliovich, A. & Doege, C. A. (2009). Reprogramming therapeutics: iPS cell prospects for neurodegenerative disease. *Neuron,* Vol.61, No.3, pp. 337-339.

Aktas, Y., Yemisci, M., Andrieux, K., Gursoy, R. N., Alonso, M. J., Fernandez-Megia, E., Novoa-Carballal, R., Quinoa, E., Riguera, R., Sargon, M. F., Celik, H. H., Demir, A. S., Hincal, A. A., Dalkara, T., Capan, Y. & Couvreur, P. (2005). Development and brain delivery of chitosan-PEG nanoparticles functionalized with the monoclonal antibody OX26. *Bioconjug Chem,* Vol.16, No.6, pp. 1503-1511.

Amin, E. M., Reza, B. A., Morteza, B. R., Maryam, M. M., Ali, M. & Zeinab, N. (2008). Microanatomical evidences for potential of mesenchymal stem cells in amelioration of striatal degeneration. *Neurol Res,* Vol.30, No.10, pp. 1086-1090.

Arias-Carrion, O. & Yuan, T. F. (2009). Autologous neural stem cell transplantation: A new treatment option for Parkinson's disease? *Med Hypotheses,* Vol.73, No.5, pp. 757-759.

Athanasiou, K. A., Niederauer, G. G. & Agrawal, C. M. (1996). Sterilization, toxicity, biocompatibility and clinical applications of polylactic acid/polyglycolic acid copolymers. *Biomaterials,* Vol.17, No.2, pp. 93-102.

Bahat-Stroomza, M., Barhum, Y., Levy, Y. S., Karpov, O., Bulvik, S., Melamed, E. & Offen, D. (2009). Induction of adult human bone marrow mesenchymal stromal cells into functional astrocyte-like cells: potential for restorative treatment in Parkinson's disease. *J Mol Neurosci,* Vol.39, No.1-2, pp. 199-210.

Bakay, R. A., Raiser, C. D., Stover, N. P., Subramanian, T., Cornfeldt, M. L., Schweikert, A. W., Allen, R. C. & Watts, R. (2004). Implantation of Spheramine in advanced Parkinson's disease (PD). *Front Biosci,* Vol.9, pp. 592-602.

Bang, O. Y., Lee, J. S., Lee, P. H. & Lee, G. (2005). Autologous mesenchymal stem cell transplantation in stroke patients. *Ann Neurol,* Vol.57, No.6, pp. 874-882.

Bantubungi, K., Blum, D., Cuvelier, L., Wislet-Gendebien, S., Rogister, B., Brouillet, E. & Schiffmann, S. N. (2008). Stem cell factor and mesenchymal and neural stem cell transplantation in a rat model of Huntington's disease. *Mol Cell Neurosci,* Vol.37, No.3, pp. 454-470.

Barczyk, M., Carracedo, S. & Gullberg, D. (2009). Integrins. *Cell Tissue Res,* Vol.339, No.1, pp. 269-280.

Barrientos, S., Stojadinovic, O., Golinko, M. S., Brem, H. & Tomic-Canic, M. (2008). Growth factors and cytokines in wound healing. *Wound Repair Regen,* Vol.16, No.5, pp. 585-601.

Barzilay, R., Kan, I., Ben-Zur, T., Bulvik, S., Melamed, E. & Offen, D. (2008). Induction of human mesenchymal stem cells into dopamine-producing cells with different differentiation protocols. *Stem Cells Dev,* Vol.17, No.3, pp. 547-554.

Bible, E., Chau, D. Y., Alexander, M. R., Price, J., Shakesheff, K. M. & Modo, M. (2009). The support of neural stem cells transplanted into stroke-induced brain cavities by PLGA particles. *Biomaterials,* Vol.30, No.16, pp. 2985-2994.

Borlongan, C. V., Saporta, S. & Sanberg, P. R. (1998). Intrastriatal transplantation of rat adrenal chromaffin cells seeded on microcarrier beads promote long-term functional recovery in hemiparkinsonian rats. *Exp Neurol,* Vol.151, No.2, pp. 203-214.

Borlongan, C. V., Thanos, C. G., Skinner, S. J., Geaney, M. & Emerich, D. F. (2008). Transplants of encapsulated rat choroid plexus cells exert neuroprotection in a rodent model of Huntington's disease. *Cell Transplant,* Vol.16, No.10, pp. 987-992.

Bosman, F. T. & Stamenkovic, I. (2003). Functional structure and composition of the extracellular matrix. *J Pathol,* Vol.200, No.4, pp. 423-428.

Bouchez, G., Sensebe, L., Vourc'h, P., Garreau, L., Bodard, S., Rico, A., Guilloteau, D., Charbord, P., Besnard, J. C. & Chalon, S. (2008). Partial recovery of dopaminergic pathway after graft of adult mesenchymal stem cells in a rat model of Parkinson's disease. *Neurochem Int,* Vol.52, No.7, pp. 1332-1342.

Broderick, J. P. (2009). Endovascular therapy for acute ischemic stroke. *Stroke,* Vol.40, No.3 Suppl, pp. S103-106.

Brundin, P., Barbin, G., Isacson, O., Mallat, M., Chamak, B., Prochiantz, A., Gage, F. H. & Bjorklund, A. (1985). Survival of intracerebrally grafted rat dopamine neurons previously cultured in vitro. *Neurosci Lett,* Vol.61, No.1-2, pp. 79-84.

Brundin, P., Karlsson, J., Emgard, M., Schierle, G. S., Hansson, O., Petersen, A. & Castilho, R. F. (2000). Improving the survival of grafted dopaminergic neurons: a review over current approaches. *Cell Transplant,* Vol.9, No.2, pp. 179-195.

Burdick, J. A., Ward, M., Liang, E., Young, M. J. & Langer, R. (2006). Stimulation of neurite outgrowth by neurotrophins delivered from degradable hydrogels. *Biomaterials,* Vol.27, No.3, pp. 452-459.

Cao, H., Liu, T. & Chew, S. Y. (2009). The application of nanofibrous scaffolds in neural tissue engineering. *Adv Drug Deliv Rev,* Vol.61, No.12, pp. 1055-1064.

Cimarosti, H. & Henley, J. M. (2008). Investigating the mechanisms underlying neuronal death in ischemia using in vitro oxygen-glucose deprivation: potential involvement of protein SUMOylation. *Neuroscientist,* Vol.14, No.6, pp. 626-636.

Clelland, C. D., Barker, R. A. & Watts, C. (2008). Cell therapy in Huntington disease. *Neurosurg Focus,* Vol.24, No.3-4, pp. E9.

Charriere, K., Risold, P. Y. & Fellmann, D. (2010). In vitro interactions between bone marrow stromal cells and hippocampal slice cultures. *C R Biol,* Vol.333, No.8, pp. 582-590.

Chen, C. & Xiao, S. F. (2011). Induced pluripotent stem cells and neurodegenerative diseases. *Neurosci Bull,* Vol.27, No.2, pp. 107-114.

Chen, J. & Chopp, M. (2006). Neurorestorative treatment of stroke: cell and pharmacological approaches. *NeuroRx,* Vol.3, No.4, pp. 466-473.

Chen, J., Li, Y., Wang, L., Lu, M., Zhang, X. & Chopp, M. (2001). Therapeutic benefit of intracerebral transplantation of bone marrow stromal cells after cerebral ischemia in rats. *J Neurol Sci,* Vol.189, No.1-2, pp. 49-57.

Chen, J., Zhang, Z. G., Li, Y., Wang, L., Xu, Y. X., Gautam, S. C., Lu, M., Zhu, Z. & Chopp, M. (2003). Intravenous administration of human bone marrow stromal cells induces angiogenesis in the ischemic boundary zone after stroke in rats. *Circ Res,* Vol.92, No.6, pp. 692-699.

Chen, S. J., Chang, C. M., Tsai, S. K., Chang, Y. L., Chou, S. J., Huang, S. S., Tai, L. K., Chen, Y. C., Ku, H. H., Li, H. Y. & Chiou, S. H. (2011). Functional improvement of focal cerebral ischemia injury by subdural transplantation of induced pluripotent stem cells with fibrin glue. *Stem Cells Dev,* Vol.19, No.11, pp. 1757-1767.

Chen, X., Katakowski, M., Li, Y., Lu, D., Wang, L., Zhang, L., Chen, J., Xu, Y., Gautam, S., Mahmood, A. & Chopp, M. (2002). Human bone marrow stromal cell cultures conditioned by traumatic brain tissue extracts: growth factor production. *J Neurosci Res,* Vol.69, No.5, pp. 687-691.

Chen, X. D. (2010). Extracellular matrix provides an optimal niche for the maintenance and propagation of mesenchymal stem cells. *Birth Defects Res C Embryo Today,* Vol.90, No.1, pp. 45-54.

Cherksey, B. D., Sapirstein, V. S. & Geraci, A. L. (1996). Adrenal chromaffin cells on microcarriers exhibit enhanced long-term functional effects when implanted into the mammalian brain. *Neuroscience,* Vol.75, No.2, pp. 657-664.

Cho, S., Wood, A. & Bowlby, M. R. (2007). Brain slices as models for neurodegenerative disease and screening platforms to identify novel therapeutics. *Curr Neuropharmacol,* Vol.5, No.1, pp. 19-33.

Chopp, M., Li, Y. & Zhang, J. (2008). Plasticity and remodeling of brain. *J Neurol Sci,* Vol.265, No.1-2, pp. 97-101.

D'Ippolito, G., Diabira, S., Howard, G. A., Menei, P., Roos, B. A. & Schiller, P. C. (2004). Marrow-isolated adult multilineage inducible (MIAMI) cells, a unique population of postnatal young and old human cells with extensive expansion and differentiation potential. *J Cell Sci,* Vol.117, No.Pt 14, pp. 2971-2981.

Dalton, P. D. & Mey, J. (2009). Neural interactions with materials. *Front Biosci,* Vol.14, pp. 769-795.

Danen, E. H. & Yamada, K. M. (2001). Fibronectin, integrins, and growth control. *J Cell Physiol,* Vol.189, No.1, pp. 1-13.

Danielyan, L., Schafer, R., von Ameln-Mayerhofer, A., Bernhard, F., Verleysdonk, S., Buadze, M., Lourhmati, A., Klopfer, T., Schaumann, F., Schmid, B., Koehle, C., Proksch, B., Weissert, R., Reichardt, H. M., van den Brandt, J., Buniatian, G. H., Schwab, M., Gleiter, C. H. & Frey Ii, W. H. (2011). Therapeutic Efficacy of Intranasally Delivered Mesenchymal Stem Cells in a Rat Model of Parkinson Disease. *Rejuvenation Res,* Vol.14, No.1, pp. 3-16.

De Keyser, J. (2005). Autologous mesenchymal stem cell transplantation in stroke patients. *Ann Neurol,* Vol.58, No.4, pp. 653-654; author reply 654-655.

Delcroix, G. J., Curtis, K. M., Schiller, P. C. & Montero-Menei, C. N. (2010). EGF and bFGF pre-treatment enhances neural specification and the response to neuronal commitment of MIAMI cells. *Differentiation,* Vol.80, No.4-5, pp. 213-227.

Delcroix, G. J., Garbayo, E., Sindji, L., Thomas, O., Vanpouille-Box, C., Schiller, P. C. & Montero-Menei, C. N. (2011). The therapeutic potential of human multipotent mesenchymal stromal cells combined with pharmacologically active microcarriers transplanted in hemi-parkinsonian rats. *Biomaterials,* Vol.32, No.6, pp. 1560-1573.

Delcroix, G. J., Jacquart, M., Lemaire, L., Sindji, L., Franconi, F., Le Jeune, J. J. & Montero-Menei, C. N. (2009). Mesenchymal and neural stem cells labeled with HEDP-coated SPIO nanoparticles: in vitro characterization and migration potential in rat brain. *Brain Res,* Vol.1255, pp. 18-31.

Delcroix, G. J. R., Schiller, P. C., Benoit, J.-P. & Montero-Menei, C. N. (2010). Adult cell therapy for brain neuronal damages and the role of tissue engineering. *Biomaterials,* Vol.31, No.8, pp. 2105-2120.

Doudet, D. J., Cornfeldt, M. L., Honey, C. R., Schweikert, A. W. & Allen, R. C. (2004). PET imaging of implanted human retinal pigment epithelial cells in the MPTP-induced primate model of Parkinson's disease. *Exp Neurol,* Vol.189, No.2, pp. 361-368.

Drucker-Colin, R. & Verdugo-Diaz, L. (2004). Cell transplantation for Parkinson's disease: present status. *Cell Mol Neurobiol,* Vol.24, No.3, pp. 301-316.

Drucker-Colin, R., Verdugo-Diaz, L., Morgado-Valle, C., Solis-Maldonado, G., Ondarza, R., Boll, C., Miranda, G., Wang, G. J. & Volkow, N. (1999). Transplant of cultured neuron-like differentiated chromaffin cells in a Parkinson's disease patient. A preliminary report. *Arch Med Res,* Vol.30, No.1, pp. 33-39.

Dunbar, G. L., Sandstrom, M. I., Rossignol, J. & Lescaudron, L. (2006). Neurotrophic enhancers as therapy for behavioral deficits in rodent models of Huntington's disease: use of gangliosides, substituted pyrimidines, and mesenchymal stem cells. *Behav Cogn Neurosci Rev,* Vol.5, No.2, pp. 63-79.

Edalatmanesh, M. A., Matin, M. M., Neshati, Z., Bahrami, A. R. & Kheirabadi, M. (2009). Systemic transplantation of mesenchymal stem cells can reduce cognitive and motor deficits in rats with unilateral lesions of the neostriatum. *Neurol Res,* Vol. 32, No.2, pp. 166-172.

Emerich, D. F. (2004). Sertoli cell grafts for Huntington's disease. An opinion. *Neurotox Res,* Vol.5, No.8, pp. 567.

England, T. (2009). Stem cells for enhancing recovery after stroke: a review. *International journal of stroke,* Vol.4, pp. 101-110.

Engler, A. J., Sen, S., Sweeney, H. L. & Discher, D. E. (2006). Matrix elasticity directs stem cell lineage specification. *Cell,* Vol.126, No.4, pp. 677-689.

Engvall, E. & Wewer, U. M. (1996). Domains of laminin. *J Cell Biochem,* Vol.61, No.4, pp. 493-501.

Esneault, E., Pacary, E., Eddi, D., Freret, T., Tixier, E., Toutain, J., Touzani, O., Schumann-Bard, P., Petit, E., Roussel, S. & Bernaudin, M. (2008). Combined therapeutic strategy using erythropoietin and mesenchymal stem cells potentiates neurogenesis after transient focal cerebral ischemia in rats. *J Cereb Blood Flow Metab,* Vol. 28, No.9, pp. 1552-1563.

Farag, E. S., Vinters, H. V. & Bronstein, J. (2009). Pathologic findings in retinal pigment epithelial cell implantation for Parkinson disease. *Neurology,* Vol.73, No.14, pp. 1095-1102.

Fernandez-Espejo, E., Armengol, J. A., Flores, J. A., Galan-Rodriguez, B. & Ramiro, S. (2005). Cells of the sympathoadrenal lineage: biological properties as donor tissue for cell-replacement therapies for Parkinson's disease. *Brain Res Brain Res Rev,* Vol.49, No.2, pp. 343-354.

Flores, J., Cepeda, I. L., Cornfeldt, M. L., O'Kusky, J. R. & Doudet, D. J. (2007). Characterization and survival of long-term implants of human retinal pigment epithelial cells attached to gelatin microcarriers in a model of Parkinson disease. *J Neuropathol Exp Neurol,* Vol.66, No.7, pp. 585-596.

Fournier, E., Passirani, C., Colin, N., Sagodira, S., Menei, P., Benoit, J. P. & Montero-Menei, C. N. (2006). The brain tissue response to biodegradable poly(methylidene malonate 2.1.2)-based microspheres in the rat. *Biomaterials,* Vol.27, No.28, pp. 4963-4974.

Fournier, E., Passirani, C., Montero-Menei, C. N. & Benoit, J. P. (2003). Biocompatibility of implantable synthetic polymeric drug carriers: focus on brain biocompatibility. *Biomaterials,* Vol.24, No.19, pp. 3311-3331.

Gahwiler, B. H. (1981). Organotypic monolayer cultures of nervous tissue. *J Neurosci Methods,* Vol.4, No.4, pp. 329-342.

Garbayo, E., Raval, A. P., Curtis, K. M., Della-Morte, D., Gomez, L. A., D'Ippolito, G., Reiner, T., Perez-Stable, C., Howard, G. A., Perez-Pinzon, M. A., Montero-Menei, C. N. & Schiller, P. C. (2011). Neuroprotective Properties of Marrow-Isolated Adult Multilineage Inducible Cells in Rat Hippocampus Following Global Cerebral Ischemia Are Enhanced When Complexed to Biomimetic Microcarriers. *J Neurochem,* In press

Gonzalez, F., Barragan Monasterio, M., Tiscornia, G., Montserrat Pulido, N., Vassena, R., Batlle Morera, L., Rodriguez Piza, I. & Izpisua Belmonte, J. C. (2009). Generation of mouse-induced pluripotent stem cells by transient expression of a single nonviral polycistronic vector. *Proc Natl Acad Sci U S A,* Vol.106, No.22, pp. 8918-8922.

Gordon, R. J., McGregor, A. L. & Connor, B. (2009). Chemokines direct neural progenitor cell migration following striatal cell loss. *Mol Cell Neurosci,* Vol.41, No.2, pp. 219-232.

Grandoso, L., Ponce, S., Manuel, I., Arrue, A., Ruiz-Ortega, J. A., Ulibarri, I., Orive, G., Hernandez, R. M., Rodriguez, A., Rodriguez-Puertas, R., Zumarraga, M., Linazasoro, G., Pedraz, J. L. & Ugedo, L. (2007). Long-term survival of encapsulated GDNF secreting cells implanted within the striatum of parkinsonized rats. *Int J Pharm,* Vol.343, No.1-2, pp. 69-78.

Griffith, T. S., Brunner, T., Fletcher, S. M., Green, D. R. & Ferguson, T. A. (1995). Fas ligand-induced apoptosis as a mechanism of immune privilege. *Science,* Vol.270, No.5239, pp. 1189-1192.

Grimpe, B., Dong, S., Doller, C., Temple, K., Malouf, A.T. & Silver, J. (2002). The critical role of basement membrane-independent laminin gamma 1 chain during axon regeneration in the CNS. *J Neurosci,* Vol.22, No.8, pp. 3144-60.

Hargus, G., Cooper, O., Deleidi, M., Levy, A., Lee, K., Marlow, E., Yow, A., Soldner, F., Hockemeyer, D., Hallett, P. J., Osborn, T., Jaenisch, R. & Isacson, O. (2010). Differentiated Parkinson patient-derived induced pluripotent stem cells grow in the adult rodent brain and reduce motor asymmetry in Parkinsonian rats. *Proc Natl Acad Sci U S A,* Vol. 107, No.36, pp. 15921-15926.

Hellmann, M. A., Panet, H., Barhum, Y., Melamed, E. & Offen, D. (2006). Increased survival and migration of engrafted mesenchymal bone marrow stem cells in 6-hydroxydopamine-lesioned rodents. *Neurosci Lett,* Vol.395, No.2, pp. 124-128.

Ho, M., Yu, D., Davidsion, M. C. & Silva, G. A. (2006). Comparison of standard surface chemistries for culturing mesenchymal stem cells prior to neural differentiation. *Biomaterials,* Vol.27, No.24, pp. 4333-4339.

Hou, S., Xu, Q., Tian, W., Cui, F., Cai, Q., Ma, J. & Lee, I. S. (2005). The repair of brain lesion by implantation of hyaluronic acid hydrogels modified with laminin. *J Neurosci Methods,* Vol.148, No.1, pp. 60-70.

Hurtig, H., Joyce, J., Sladek, J. R., Jr. & Trojanowski, J. Q. (1989). Postmortem analysis of adrenal-medulla-to-caudate autograft in a patient with Parkinson's disease. *Ann Neurol,* Vol.25, No.6, pp. 607-614.

Hynes, R. O. (2002). Integrins: bidirectional, allosteric signaling machines. *Cell,* Vol.110, No.6, pp. 673-687.

Hynes, R. O. & Yamada, K. M. (1982). Fibronectins: multifunctional modular glycoproteins. *J Cell Biol,* Vol.95, No.2 Pt 1, pp. 369-377.

Isacson, O., Bjorklund, L. M. & Schumacher, J. M. (2003). Toward full restoration of synaptic and terminal function of the dopaminergic system in Parkinson's disease by stem cells. *Ann Neurol,* Vol.53 Suppl 3, pp. S135-146; discussion S146-138.

Jiang, M., Lv, L., Ji, H., Yang, X., Zhu, W., Cai, L., Gu, X., Chai, C., Huang, S., Sun, J. & Dong, Q. (2011). Induction of pluripotent stem cells transplantation therapy for ischemic stroke. *Mol Cell Biochem,* In press.

Jiao, Y., Novozhilova, E., Karlen, A., Muhr, J. & Olivius, P. (2010). Olfactory ensheathing cells promote neurite outgrowth from co-cultured brain stem slice. *Exp Neurol,* In press.

Jollivet, C., Aubert-Pouessel, A., Clavreul, A., Venier-Julienne, M. C., Remy, S., Montero-Menei, C. N., Benoit, J. P. & Menei, P. (2004). Striatal implantation of GDNF releasing biodegradable microspheres promotes recovery of motor function in a partial model of Parkinson's disease. *Biomaterials,* Vol.25, No.5, pp. 933-942.

Jongpaiboonkit, L., King, W. J. & Murphy, W. L. (2008). Screening for 3D Environments That Support Human Mesenchymal Stem Cell Viability Using Hydrogel Arrays. *Tissue Eng Part A,* Vol. 15, No.2, pp. 343-353.

Jorgensen, A., Wiencke, A. K., la Cour, M., Kaestel, C. G., Madsen, H. O., Hamann, S., Lui, G. M., Scherfig, E., Prause, J. U., Svejgaard, A., Odum, N., Nissen, M. H. & Ropke, C. (1998). Human retinal pigment epithelial cell-induced apoptosis in activated T cells. *Invest Ophthalmol Vis Sci,* Vol.39, No.9, pp. 1590-1599.

Joyce, N., Annett, G., Wirthlin, L., Olson, S., Bauer, G. & Nolta, J. A. (2010). Mesenchymal stem cells for the treatment of neurodegenerative disease. *Regen Med,* Vol.5, No.6, pp. 933-946.

Jurga, M., Dainiak, M. B., Sarnowska, A., Jablonska, A., Tripathi, A., Plieva, F. M., Savina, I. N., Strojek, L., Jungvid, H., Kumar, A., Lukomska, B., Domanska-Janik, K., Forraz, N. & McGuckin, C. P. (2011). The performance of laminin-containing cryogel scaffolds in neural tissue regeneration. *Biomaterials,* Vol.32, No.13, pp. 3423-3434.

Karoubi, G., Ormiston, M. L., Stewart, D. J. & Courtman, D. W. (2009). Single-cell hydrogel encapsulation for enhanced survival of human marrow stromal cells. *Biomaterials,* Vol.30, No.29, pp. 5445-5455.

Kearns, S. M., Scheffler, B., Goetz, A. K., Lin, D. D., Baker, H. D., Roper, S. N., Mandel, R. J. & Steindler, D. A. (2006). A method for a more complete in vitro Parkinson's model: slice culture bioassay for modeling maintenance and repair of the nigrostriatal circuit. *J Neurosci Methods,* Vol.157, No.1, pp. 1-9.

Kelly, C. M., Dunnett, S. B. & Rosser, A. E. (2009). Medium spiny neurons for transplantation in Huntington's disease. *Biochem Soc Trans,* Vol.37, No.Pt 1, pp. 323-328.

Kim, M. (2009). Electrostatic Crosslinked In Situ-Forming In Vivo Scaffold For Rat Bone Marrow Mesenchymal Stem Cells. *Tissue Eng Part A,* Vol.15, No.10, pp. 3201-9

Kim, Y. J., Park, H. J., Lee, G., Bang, O. Y., Ahn, Y. H., Joe, E., Kim, H. O. & Lee, P. H. (2008). Neuroprotective effects of human mesenchymal stem cells on dopaminergic neurons through anti-inflammatory action. *Glia,* Vol. 57, No.1, pp. 13-23.

Kristensen, B. W., Noraberg, J., Thiebaud, P., Koudelka-Hep, M. & Zimmer, J. (2001). Biocompatibility of silicon-based arrays of electrodes coupled to organotypic hippocampal brain slice cultures. *Brain Res,* Vol.896, No.1-2, pp. 1-17.

Laquerriere, P., Grandjean-Laquerriere, A., Jallot, E., Balossier, G., Frayssinet, P. & Guenounou, M. (2003). Importance of hydroxyapatite particles characteristics on cytokines production by human monocytes in vitro. *Biomaterials,* Vol.24, No.16, pp. 2739-2747.

Laustriat, D., Gide, J. & Peschanski, M. (2010). Human pluripotent stem cells in drug discovery and predictive toxicology. *Biochem Soc Trans,* Vol.38, No.4, pp. 1051-1057.

Lee, H., Park, J., Forget, B. G. & Gaines, P. (2009). Induced pluripotent stem cells in regenerative medicine: an argument for continued research on human embryonic stem cells. *Regen Med,* Vol.4, No.5, pp. 759-769.

Lee, J. H., Yu, H. S., Lee, G. S., Ji, A., Hyun, J. K. & Kim, H. W. (2011). Collagen gel three-dimensional matrices combined with adhesive proteins stimulate neuronal differentiation of mesenchymal stem cells. *J R Soc Interface*, In press.

Lescaudron, L., Unni, D. & Dunbar, G. L. (2003). Autologous adult bone marrow stem cell transplantation in an animal model of huntington's disease: behavioral and morphological outcomes. *Int J Neurosci*, Vol.113, No.7, pp. 945-956.

Levesque, M. F. (2009). Therapeutic Microinjection of Autologous Adult Human Neural Stem Cells and Differentiated Neurons for Parkinson's Disease: Five-Year Post-Operative Outcome. *The Open Stem Cell Journal*, Vol.1, pp. 20-29.

Levesque, S. G. & Shoichet, M. S. (2006). Synthesis of cell-adhesive dextran hydrogels and macroporous scaffolds. *Biomaterials*, Vol.27, No.30, pp. 5277-5285.

Levy, Y. S., Bahat-Stroomza, M., Barzilay, R., Burshtein, A., Bulvik, S., Barhum, Y., Panet, H., Melamed, E. & Offen, D. (2008). Regenerative effect of neural-induced human mesenchymal stromal cells in rat models of Parkinson's disease. *Cytotherapy*, Vol.10, No.4, pp. 340-352.

Li, S. C. & Loudon, W. G. (2008). A novel and generalizable organotypic slice platform to evaluate stem cell potential for targeting pediatric brain tumors. *Cancer Cell Int*, Vol.8, pp. 9.

Li, Y., Chen, J., Chen, X. G., Wang, L., Gautam, S. C., Xu, Y. X., Katakowski, M., Zhang, L. J., Lu, M., Janakiraman, N. & Chopp, M. (2002). Human marrow stromal cell therapy for stroke in rat: neurotrophins and functional recovery. *Neurology*, Vol.59, No.4, pp. 514-523.

Li, Y., Chen, J., Zhang, C. L., Wang, L., Lu, D., Katakowski, M., Gao, Q., Shen, L. H., Zhang, J., Lu, M. & Chopp, M. (2005). Gliosis and brain remodeling after treatment of stroke in rats with marrow stromal cells. *Glia*, Vol.49, No.3, pp. 407-417.

Li, Y. & Chopp, M. (2009). Marrow stromal cell transplantation in stroke and traumatic brain injury. *Neurosci Lett*, Vol.456, No.3, pp. 120-123.

Li, Y., Chopp, M., Chen, J., Wang, L., Gautam, S. C., Xu, Y. X. & Zhang, Z. (2000). Intrastriatal transplantation of bone marrow nonhematopoietic cells improves functional recovery after stroke in adult mice. *J Cereb Blood Flow Metab*, Vol.20, No.9, pp. 1311-1319.

Lindvall, O. & Kokaia, Z. (2009). Prospects of stem cell therapy for replacing dopamine neurons in Parkinson's disease. *Trends Pharmacol Sci*, Vol.30, No.5, pp. 260-267.

Loewenbruck, K. & Storch, A. (2011). Stem cell-based therapies in Parkinson's disease: future hope or current treatment option? *J Neurol*, In press.

Lois, C. & Alvarez-Buylla, A. (1993). Proliferating subventricular zone cells in the adult mammalian forebrain can differentiate into neurons and glia. *Proc Natl Acad Sci U S A*, Vol.90, No.5, pp. 2074-2077.

Lossi, L., Alasia, S., Salio, C. & Merighi, A. (2009). Cell death and proliferation in acute slices and organotypic cultures of mammalian CNS. *Prog Neurobiol*, Vol.88, No.4, pp. 221-245.

Lu, D., Mahmood, A., Qu, C., Hong, X., Kaplan, D. & Chopp, M. (2007). Collagen scaffolds populated with human marrow stromal cells reduce lesion volume and improve functional outcome after traumatic brain injury. *Neurosurgery*, Vol.61, No.3, pp. 596-602; discussion 602-593.

Madrazo, I., Drucker-Colin, R., Diaz, V., Martinez-Mata, J., Torres, C. & Becerril, J. J. (1987). Open microsurgical autograft of adrenal medulla to the right caudate nucleus in two patients with intractable Parkinson's disease. *N Engl J Med,* Vol.316, No.14, pp. 831-834.

Madrigal, J. L., Leza, J. C., Polak, P., Kalinin, S. & Feinstein, D. L. (2009). Astrocyte-derived MCP-1 mediates neuroprotective effects of noradrenaline. *J Neurosci,* Vol.29, No.1, pp. 263-267.

Matsubara, T., Tsutsumi, S., Pan, H., Hiraoka, H., Oda, R., Nishimura, M., Kawaguchi, H., Nakamura, K. & Kato, Y. (2004). A new technique to expand human mesenchymal stem cells using basement membrane extracellular matrix. *Biochem Biophys Res Commun,* Vol.313, No.3, pp. 503-508.

McCoy, M. K., Martinez, T. N., Ruhn, K. A., Wrage, P. C., Keefer, E. W., Botterman, B. R., Tansey, K. E. & Tansey, M. G. (2008). Autologous transplants of Adipose-Derived Adult Stromal (ADAS) cells afford dopaminergic neuroprotection in a model of Parkinson's disease. *Exp Neurol,* Vol.210, No.1, pp. 14-29.

Menei, P., Montero-Menei, C., Venier, M. C. & Benoit, J. P. (2005). Drug delivery into the brain using poly(lactide-co-glycolide) microspheres. *Expert Opin Drug Deliv,* Vol.2, No.2, pp. 363-376.

Miura, K., Okada, Y., Aoi, T., Okada, A., Takahashi, K., Okita, K., Nakagawa, M., Koyanagi, M., Tanabe, K., Ohnuki, M., Ogawa, D., Ikeda, E., Okano, H. & Yamanaka, S. (2009). Variation in the safety of induced pluripotent stem cell lines. *Nat Biotechnol,* Vol.7, No.8, pp. 743-5.

Nakagawa, M. & Yamanaka, S. (2011). Reprogramming of somatic cells to pluripotency. *Adv Exp Med Biol,* Vol.695, pp. 215-224.

Namba, R. M., Cole, A. A., Bjugstad, K. B. & Mahoney, M. J. (2009). Development of porous PEG hydrogels that enable efficient, uniform cell-seeding and permit early neural process extension. *Acta Biomater,* Vol.5, No.6, pp. 1884-1897.

Narsinh KH, Jia F, Robbins RC, Kay MA, Longaker MT, Wu JC. (2011) Generation of adult human induced pluripotent stem cells using nonviral minicircle DNA vectors. *Nat Proc.* Vol.6 pp.78-88

Newman, K. D. & McBurney, M. W. (2004). Poly(d,l lactic-co-glycolic acid) microspheres as biodegradable microcarriers for pluripotent stem cells. *Biomaterials,* Vol.25, No.26, pp. 5763-5771.

Noraberg, J., Poulsen, F. R., Blaabjerg, M., Kristensen, B. W., Bonde, C., Montero, M., Meyer, M., Gramsbergen, J. B. & Zimmer, J. (2005). Organotypic hippocampal slice cultures for studies of brain damage, neuroprotection and neurorepair. *Curr Drug Targets CNS Neurol Disord,* Vol.4, No.4, pp. 435-452.

Ohtaki, H., Ylostalo, J. H., Foraker, J. E., Robinson, A. P., Reger, R. L., Shioda, S. & Prockop, D. J. (2008). Stem/progenitor cells from bone marrow decrease neuronal death in global ischemia by modulation of inflammatory/immune responses. *Proc Natl Acad Sci U S A,* Vol.105, No.38, pp. 14638-14643.

Okita, K., Matsumura, Y., Sato, Y., Okada, A., Morizane, A., Okamoto, S., Hong, H., Nakagawa, M., Tanabe, K., Tezuka, K. I., Shibata, T., Kunisada, T., Takahashi, M., Takahashi, J., Saji, H. & Yamanaka, S. (2011). A more efficient method to generate integration-free human iPS cells. *Nat Methods,* In press.

Olanow, C. W., Schapira, A. H. & Agid, Y. (2003). Neuroprotection for Parkinson's disease: prospects and promises. *Ann Neurol,* Vol.53 Suppl 3, pp. S1-2.

Orive, G., De Castro, M., Kong, H. J., Hernandez, R. M., Ponce, S., Mooney, D. J. & Pedraz, J. L. (2009). Bioactive cell-hydrogel microcapsules for cell-based drug delivery. *J Control Release,* Vol.135, No.3, pp. 203-210.

Orive, G., Gascon, A. R., Hernandez, R. M., Igartua, M. & Luis Pedraz, J. (2003). Cell microencapsulation technology for biomedical purposes: novel insights and challenges. *Trends Pharmacol Sci,* Vol.24, No.5, pp. 207-210.

Page, R. L., Ambady, S., Holmes, W. F., Vilner, L., Kole, D., Kashpur, O., Huntress, V., Vojtic, I., Whitton, H. & Dominko, T. (2009). Induction of Stem Cell Gene Expression in Adult Human Fibroblasts without Transgenes. *Cloning Stem Cells,* pp.

Pawelek, J. M. & Korner, A. M. (1982). The biosynthesis of mammalian melanin. *Am Sci,* Vol.70, No.2, pp. 136-145.

Pean, J. M., Menei, P., Morel, O., Montero-Menei, C. N. & Benoit, J. P. (2000). Intraseptal implantation of NGF-releasing microspheres promote the survival of axotomized cholinergic neurons. *Biomaterials,* Vol.21, No.20, pp. 2097-2101.

Pera, M. F. (2011). Stem cells: The dark side of induced pluripotency. *Nature,* Vol.471, No.7336, pp. 46-47.

Perale, G., Giordano, C., Bianco, F., Rossi, F., Tunesi, M., Daniele, F., Crivelli, F., Matteoli, M. & Masi, M. (2011). Hydrogel for cell housing in the brain and in the spinal cord. *Int J Artif Organs,* Vol.34, No.3, pp. 46-7

Perasso, L., Cogo, C. E., Giunti, D., Gandolfo, C., Ruggeri, P., Uccelli, A. & Balestrino, M. (2011). Systemic administration of mesenchymal stem cells increases neuron survival after global cerebral ischemia in vivo (2VO). *Neural Plast,* Vol.2010, pp. 534925.

Pettikiriarachchi, J. T. S., Parish, C.L., Shoichet, M.S., Forsythe, J.S. & Nisbet D.R. (2010). Biomaterials for brain tissue engineering. *Australian Journal of Chemistry,* Vol.63, No.8, pp. 1143-1154.

Potter, W., Kalil, R. E. & Kao, W. J. (2008). Biomimetic material systems for neural progenitor cell-based therapy. *Front Biosci,* Vol.13, pp. 806-821.

Powell, S. K. & Kleinman, H. K. (1997). Neuronal laminins and their cellular receptors. *Int J Biochem Cell Biol,* Vol.29, No.3, pp. 401-414.

Prabhakaran, M. P., Venugopal, J. R. & Ramakrishna, S. (2009). Mesenchymal stem cell differentiation to neuronal cells on electrospun nanofibrous substrates for nerve tissue engineering. *Biomaterials,* Vol.30, No.28, pp. 4996-5003.

Purcell, E. K., Singh, A. & Kipke, D. R. (2009). Alginate composition effects on a neural stem cell-seeded scaffold. *Tissue Eng Part C Methods,* Vol.15, No.4, pp. 541-550.

Qian, L. & Saltzman, W. M. (2004). Improving the expansion and neuronal differentiation of mesenchymal stem cells through culture surface modification. *Biomaterials,* Vol.25, No.7-8, pp. 1331-1337.

Qu, C., Mahmood, A., Liu, X. S., Xiong, Y., Wang, L., Wu, H., Li, B., Zhang, Z. G., Kaplan, D. L. & Chopp, M. (2011). The treatment of TBI with human marrow stromal cells impregnated into collagen scaffold: functional outcome and gene expression profile. *Brain Res,* Vol.1371, pp. 129-139.

Qu, C., Xiong, Y., Mahmood, A., Kaplan, D. L., Goussev, A., Ning, R. & Chopp, M. (2009). Treatment of traumatic brain injury in mice with bone marrow stromal cell-impregnated collagen scaffolds. *J Neurosurg,* Vol. 111, No.4, pp.658-55.

Rahnemai-Azar, A., D'Ippolito, G., Gomez, L. A., Reiner, T., Vazquez-Padron, R. I., Perez-Stable, C., Roos, B. A., Pham, S. M. & Schiller, P. C. (2011). Human marrow-isolated adult multilineage-inducible (MIAMI) cells protect against peripheral vascular ischemia in a mouse model. *Cytotherapy,* Vol.13, No.2, pp. 179-192.

Rogers, S. L., Letourneau, P.C., Palm, S.L., McCarthy, J. & Furcht L.T. (1983). Neurite extension by peripheral and central nervous system neurons in response to substratum-bound fibronectin and laminin. *Dev Biol,* Vol. 98, No.1, pp. 212-220.

Re, D. B. & Przedborski, S. (2006). Fractalkine: moving from chemotaxis to neuroprotection. *Nat Neurosci,* Vol.9, No.7, pp. 859-861.

Richardson, R. M., Sun, D. & Bullock, M. R. (2007). Neurogenesis after traumatic brain injury. *Neurosurg Clin N Am,* Vol.18, No.1, pp. 169-181, xi.

Richardson, T. P., Peters, M. C., Ennett, A. B. & Mooney, D. J. (2001). Polymeric system for dual growth factor delivery. *Nat Biotechnol,* Vol.19, No.11, pp. 1029-1034.

Sadan, O., Bahat-Stromza, M., Barhum, Y., Levy, Y. S., Pisnevsky, A., Peretz, H., Bar Ilan, A., Bulvik, S., Shemesh, N., Krepel, D., Cohen, Y., Melamed, E. & Offen, D. (2009). Protective effects of neurotrophic factors secreting cells in a 6OHDA rat model of Parkinson disease. *Stem Cells Dev,* Vol.18, No.8, pp. 1179-1190

Salinas, C. N. & Anseth, K. S. (2008). The influence of the RGD peptide motif and its contextual presentation in PEG gels on human mesenchymal stem cell viability. *J Tissue Eng Regen Med,* Vol.2, No.5, pp. 296-304

Saporta, S., Borlongan, C., Moore, J., Mejia-Millan, E., Jones, S. L., Bonness, P., Randall, T. S., Allen, R. C., Freeman, T. B. & Sanberg, P. R. (1997). Microcarrier enhanced survival of human and rat fetal ventral mesencephalon cells implanted in the rat striatum. *Cell Transplant,* Vol.6, No.6, pp. 579-584.

Savitz, S. I., Chopp, M., Deans, R., Carmichael, S. T., Phinney, D. & Wechsler, L. (2011). Stem Cell Therapy as an Emerging Paradigm for Stroke (STEPS) II. *Stroke,* Vol.42, No.3, pp. 825-829.

Schierle, G. S., Hansson, O., Leist, M., Nicotera, P., Widner, H. & Brundin, P. (1999). Caspase inhibition reduces apoptosis and increases survival of nigral transplants. *Nat Med,* Vol.5, No.1, pp. 97-100.

Seidlits, S. K., Lee, J. Y. & Schmidt, C. E. (2008). Nanostructured scaffolds for neural applications. *Nanomed,* Vol.3, No.2, pp. 183-199.

Shive, M. S. & Anderson, J. M. (1997). Biodegradation and biocompatibility of PLA and PLGA microspheres. *Adv Drug Deliv Rev,* Vol.28, No.1, pp. 5-24.

Stoppini, L., Buchs, P. A. & Muller, D. (1991). A simple method for organotypic cultures of nervous tissue. *J Neurosci Methods,* Vol.37, No.2, pp. 173-182.

Stover, N. P. & Watts, R. L. (2008). Spheramine for treatment of Parkinson's disease. *Neurotherapeutics,* Vol.5, No.2, pp. 252-259.

Subia, B., Kundu, J. & Kundu, S.C. (2010). Biomaterial scaffold fabrication techniques for potential tissue engineering applications., InTech.

Sugiyama, M., Iohara, K., Wakita, H., Hattori, H., Ueda, M., Matsushita, K. & Nakashima, M. (2011). Dental Pulp Derived CD31-/CD146- Side Population Stem/Progenitor

Cells Enhance Recovery of Focal Cerebral Ischemia in Rats. *Tissue Eng Part A,* In press.

Sundstrom, L., Morrison, B., 3rd, Bradley, M. & Pringle, A. (2005). Organotypic cultures as tools for functional screening in the CNS. *Drug Discov Today,* Vol.10, No.14, pp. 993-1000.

Swistowski, A., Peng, J., Liu, Q., Mali, P., Rao, M. S., Cheng, L. & Zeng, X. (2010). Efficient Generation of Functional Dopaminergic Neurons from Human Induced pluripotent Stem Cells under Defined Conditions. *Stem Cells,* Vol.28, No.10, pp. 1893-1904.

Sykova, E. & Jendelova, P. (2007). In vivo tracking of stem cells in brain and spinal cord injury. *Prog Brain Res,* Vol.161, pp. 367-383.

Takahashi, K., Okita, K., Nakagawa, M. & Yamanaka, S. (2007). Induction of pluripotent stem cells from fibroblast cultures. *Nat Protoc,* Vol.2, No.12, pp. 3081-3089.

Tang, Y., Pacary, E., Freret, T., Divoux, D., Petit, E., Schumann-Bard, P. & Bernaudin, M. (2006). Effect of hypoxic preconditioning on brain genomic response before and following ischemia in the adult mouse: identification of potential neuroprotective candidates for stroke. *Neurobiol Dis,* Vol.21, No.1, pp. 18-28.

Tatard, V. M., D'Ippolito, G., Diabira, S., Valeyev, A., Hackman, J., McCarthy, M., Bouckenooghe, T., Menei, P., Montero-Menei, C. N. & Schiller, P. C. (2007). Neurotrophin-directed differentiation of human adult marrow stromal cells to dopaminergic-like neurons. *Bone,* Vol.40, No.2, pp. 360-373.

Tatard, V. M., Menei, P., Benoit, J. P. & Montero-Menei, C. N. (2005). Combining polymeric devices and stem cells for the treatment of neurological disorders: a promising therapeutic approach. *Curr Drug Targets,* Vol.6, No.1, pp. 81-96.

Tatard, V. M., Sindji, L., Branton, J. G., Aubert-Pouessel, A., Colleau, J., Benoit, J. P. & Montero-Menei, C. N. (2007). Pharmacologically active microcarriers releasing glial cell line - derived neurotrophic factor: Survival and differentiation of embryonic dopaminergic neurons after grafting in hemiparkinsonian rats. *Biomaterials,* Vol.28, No.11, pp. 1978-1988.

Tatard, V. M., Venier-Julienne, M. C., Benoit, J. P., Menei, P. & Montero-Menei, C. N. (2004). In vivo evaluation of pharmacologically active microcarriers releasing nerve growth factor and conveying PC12 cells. *Cell Transplant,* Vol.13, No.5, pp. 573-583.

Tatard, V. M., Venier-Julienne, M. C., Saulnier, P., Prechter, E., Benoit, J. P., Menei, P. & Montero-Menei, C. N. (2005). Pharmacologically active microcarriers: a tool for cell therapy. *Biomaterials,* Vol.26, No.17, pp. 3727-3737.

Tate, C. C., Shear, D. A., Tate, M. C., Archer, D. R., Stein, D. G. & LaPlaca, M. C. (2009). Laminin and fibronectin scaffolds enhance neural stem cell transplantation into the injured brain. *J Tissue Eng Regen Med,* Vol.3, No.3, pp. 208-217.

Tate, C. C., Tate, M. C. & LaPlaca, M. C. (2007). Fibronectin and laminin increase in the mouse brain after controlled cortical impact injury. *J Neurotrauma,* Vol.24, No.1, pp. 226-230.

Trzaska, K. A., Kuzhikandathil, E. V. & Rameshwar, P. (2007). Specification of a dopaminergic phenotype from adult human mesenchymal stem cells. *Stem Cells,* Vol.25, No.11, pp. 2797-2808.

Trzaska, K. A. & Rameshwar, P. (2011). Dopaminergic Neuronal Differentiation Protocol for Human Mesenchymal Stem Cells. *Methods Mol Biol,* Vol.698, pp. 295-303.

Tse, J. R. & Engler, A. J. (2011). Stiffness gradients mimicking in vivo tissue variation regulate mesenchymal stem cell fate. *PLoS ONE,* Vol.6, No.1, pp. e15978.

Valmikinathan, C. M., Tian, J., Wang, J. & Yu, X. (2008). Novel nanofibrous spiral scaffolds for neural tissue engineering. *J Neural Eng,* Vol.5, No.4, pp. 422-432.

van Velthoven, C. T., Kavelaars, A., van Bel, F. & Heijnen, C. J. (2009). Regeneration of the ischemic brain by engineered stem cells: fuelling endogenous repair processes. *Brain Res Rev,* Vol.61, No.1, pp. 1-13.

Vazey, E. M. & Connor, B. (2011). Differential fate and functional outcome of lithium chloride primed adult neural progenitor cell transplants in a rat model of Huntington disease. *Stem Cell Res Ther,* Vol.1, No.5, pp. 41.

Vazey, E. M., Chen, K., Hughes, S. M. & Connor, B. (2006). Transplanted adult neural progenitor cells survive, differentiate and reduce motor function impairment in a rodent model of Huntington's disease. *Exp Neurol,* Vol.199, No.2, pp. 384-396.

Vert, M. (2009). Degradable and bioresorbable polymers in surgery and in pharmacology: beliefs and facts. *J Mater Sci Mater Med,* Vol.20, No.2, pp. 437-446.

Veziers, J., Lesourd, M., Jollivet, C., Montero-Menei, C., Benoit, J. P. & Menei, P. (2001). Analysis of brain biocompatibility of drug-releasing biodegradable microspheres by scanning and transmission electron microscopy. *J Neurosurg,* Vol.95, No.3, pp. 489-494.

Walker, P. A., Aroom, K. R., Jimenez, F., Shah, S. K., Harting, M. T., Gill, B. S. & Cox, C. S., Jr. (2009). Advances in progenitor cell therapy using scaffolding constructs for central nervous system injury. *Stem Cell Rev,* Vol.5, No.3, pp. 283-300.

Wang, T. W. & Spector, M. (2009). Development of hyaluronic acid-based scaffolds for brain tissue engineering. *Acta Biomater,* Vol.5, No.7, pp. 2371-2384.

Watts, R. L., Raiser, C. D., Stover, N. P., Cornfeldt, M. L., Schweikert, A. W., Allen, R. C., Subramanian, T., Doudet, D., Honey, C. R. & Bakay, R. A. (2003). Stereotaxic intrastriatal implantation of human retinal pigment epithelial (hRPE) cells attached to gelatin microcarriers: a potential new cell therapy for Parkinson's disease. *J Neural Transm Suppl,* No.65, pp. 215-227.

Wernig, M., Zhao, J. P., Pruszak, J., Hedlund, E., Fu, D., Soldner, F., Broccoli, V., Constantine-Paton, M., Isacson, O. & Jaenisch, R. (2008). Neurons derived from reprogrammed fibroblasts functionally integrate into the fetal brain and improve symptoms of rats with Parkinson's disease. *Proc Natl Acad Sci U S A,* Vol.105, No.15, pp. 5856-5861.

Wong, D. Y., Krebsbach, P. H. & Hollister, S. J. (2008). Brain cortex regeneration affected by scaffold architectures. *J Neurosurg,* Vol.109, No.4, pp. 715-722.

Xiong, Y., Qu, C., Mahmood, A., Liu, Z., Ning, R., Li, Y., Kaplan, D. L., Schallert, T. & Chopp, M. (2009). Delayed transplantation of human marrow stromal cell-seeded scaffolds increases transcallosal neural fiber length, angiogenesis, and hippocampal neuronal survival and improves functional outcome after traumatic brain injury in rats. *Brain Res,* Vol. 1263, pp. 183-191.

Yamashita, T., Deguchi, K., Sehara, Y., Lukic-Panin, V., Zhang, H., Kamiya, T. & Abe, K. (2009). Therapeutic strategy for ischemic stroke. *Neurochem Res,* Vol.34, No.4, pp. 707-710.

Yang, C. Y., Song, B., Ao, Y., Nowak, A. P., Abelowitz, R. B., Korsak, R. A., Havton, L. A.,Deming, T. J. & Sofroniew, M. V. (2009). Biocompatibility of amphiphilic diblock copolypeptide hydrogels in the central nervous system. *Biomaterials,* Vol. 30, No.15, pp. 2881-2898.

Yang, K. L., Chen, M. F., Liao, C. H., Pang, C. Y. & Lin, P. Y. (2009). A simple and efficient method for generating Nurr1-positive neuronal stem cells from human wisdom teeth (tNSC) and the potential of tNSC for stroke therapy. *Cytotherapy,* Vol. 11, No.5, pp. 1-12.

Yarygin, K. N., Kholodenko, I. V., Konieva, A. A., Burunova, V. V., Tairova, R. T., Gubsky, L. V., Cheglakov, I. B., Pirogov, Y. A., Yarygin, V. N. & Skvortsova, V. I. (2009). Mechanisms of positive effects of transplantation of human placental mesenchymal stem cells on recovery of rats after experimental ischemic stroke. *Bull Exp Biol Med,* Vol.148, No.6, pp. 862-868.

Yasuda, H., Kuroda, S., Shichinohe, H., Kamei, S., Kawamura, R. & Iwasaki, Y. (2010). Effect of biodegradable fibrin scaffold on survival, migration, and differentiation of transplanted bone marrow stromal cells after cortical injury in rats. *J Neurosurg,* Vol.112, No.2, pp. 336-344.

Yu, J., Hu, K., Smuga-Otto, K., Tian, S., Stewart, R., Slukvin, II & Thomson, J. A. (2009). Human induced pluripotent stem cells free of vector and transgene sequences. *Science,* Vol.324, No.5928, pp. 797-801.

Yu, J., Vodyanik, M. A., Smuga-Otto, K., Antosiewicz-Bourget, J., Frane, J. L., Tian, S., Nie, J., Jonsdottir, G. A., Ruotti, V., Stewart, R., Slukvin, II & Thomson, J. A. (2007). Induced pluripotent stem cell lines derived from human somatic cells. *Science,* Vol.318, No.5858, pp. 1917-1920.

Zhang, Z., Jin, D., Yang, Z., Shen, B. & Liu, M. (2011). Effects of 17beta-estradiol pre-treated adult neural stem cells on neuronal differentiation and neurological recovery in rats with cerebral ischemia. *Brain Inj,* Vol.25, No.2, pp. 227-236.

Zhang, Z., Wang, X. & Wang, S. (2008). Isolation and characterization of mesenchymal stem cells derived from bone marrow of patients with Parkinson's disease. *In Vitro Cell Dev Biol Anim,* Vol.44, No.5-6, pp. 169-177.

Zhou, H., Wu, S., Joo, J. Y., Zhu, S., Han, D. W., Lin, T., Trauger, S., Bien, G., Yao, S., Zhu, Y., Siuzdak, G., Scholer, H. R., Duan, L. & Ding, S. (2009). Generation of induced pluripotent stem cells using recombinant proteins. *Cell Stem Cell,* Vol.4, No.5, pp. 381-384.

# Part 8

# Endocrine Organs

# 19

# Regenerative Medicine and Tissue Engineering for the Treatment of Diabetes

Matsumoto S, SoRelle JA and Shimoda M
*Baylor Research Institute / Baylor University / Baylor University Medical Center*
*USA*

## 1. Introduction

Diabetes mellitus is a disease of insulin insufficiency, which causes hyperglycemia and has both acute and chronic complications. Acute complications consist of hyperglycemic ketoacidosis and hypoglycemic episodes. Chronic complications consist of micro- and macro-angiopathies. Micro-angiopathy leads to diabetic nephropathy, neuropathy and retinopathy; macro-angiopathy leads to brain infarction, brain hemorrhage and cardiac infarction. Both acute and chronic complications significantly deteriorate the quality of life of diabetic patients and are sometimes fetal.

More than 23.7 million people suffer from diabetes in the USA today and that number will reach 44.1 million in 2034 (Huang et al., 2009). It has been demonstrated that the loss of beta-cell mass is approximately 95% in type 1 diabetic patients, 65% in type 2 diabetic patients, and even 50% in metabolic syndrome patients (Butler et al., 2003, 2007; Meier et al., 2005, 2008).

From the treatment viewpoint, diabetes is categorized into non-insulin dependent diabetes mellitus (NIDDM) and insulin dependent diabetes mellitus (IDDM). The standard therapy for the IDDM is insulin injection (Table 1). However, in the advanced phase, those patients are not able to control blood glucose levels by insulin injection. Beta cell replacement therapies including whole pancreas transplantation and islet cell transplantation are currently applied clinically for such patients (Matsumoto, 2010a) (Table 1). Unfortunately, there are more than one million IDDM patients in the United States and the number of cadaveric organs available is approximately 7,000 in each year. There is a clear donor shortage and regenerative medicine and/or tissue engineering for creating insulin-producing cells are critical to overcome this issue.

Diabetes is an excellent candidate for regenerative medicine and tissue engineering because only the beta cell with or without alpha cells is necessary to be generated for improving glycemic control. Since islet cell transplantation has already been proven to be an effective treatment for diabetic patients, generating insulin-producing cells is guaranteed for clinical effectiveness.

Indeed, tissue engineering using pig islet cells has already been clinically attempted as a bio-artificial islet transplantation for the treatment of IDDM patients (Table 1). In addition, some approaches including direct signal delivery to pancreas and neural relay of signal from liver to pancreas have been established for beta cell regeneration experimentally (Table 1).

| Category | Treatment | Clinical application | Donor sources | Need for Anti-rejection drugs | Prevention of acute and chronic complications |
|---|---|---|---|---|---|
| Insulin therapy | Insulin injection | Standard therapy | N.A. | No | No |
| Beta cell replacement | Pancreas transplant | Standard therapy | Cadaveric or living donor | Yes | Yes |
| Beta cell replacement | Islet transplant | Semi-standard therapy | Cadaveric or living donor | Yes | Yes |
| Bio-artificial islet | Xeno islet transplant | Under clinical trial | Pig | No | Possible |
| Bio-artificial islet | Generated islet transplant | Not clinically available | Stem cells (ES cells, iPS cells, pancreatic stem cells) | Yes or No | Unknown |
| Beta cell regeneration | Direct signal delivery | Not clinically available | N.A. | Yes for type 1 diabetes | Unknown |
| Beta cell regeneration | Neural relay of signal | Not clinically available | N.A. | Yes for type 1 diabetes | Unknown |

Table 1. Current and future treatments for insulin dependent diabetes mellitus (IDDM). Allogeneic islet transplantation for the treatment of type 1 diabetes is considered as the standard therapy in some countries. Of note, bio-artificial islet transplantation using pig islets has been already initiated for the treatment of type 1 diabetic patients. ES cells: Embryonic stem cells, iPS cells: induced pluripotent stem cells, N.A.: not applicable

Thus this field is one of the most advanced areas for regenerative medicine and tissue engineering.

In this book chapter we describe the current status of regenerative medicine and tissue engineering for creating insulin-producing cells and clinical application or the path to the clinical application of those technologies.

## 2. Regenerative medicine of beta cells

Regenerative medicine of beta cells consists of two major categories. The one is a replacement/implantation of alternative cell sources instead of human islets from cadaver donors. Such cell sources include embryonic stem (ES) cells, induced pluripotent stem (iPS) cells, or other systemic stem cells. They are expanded and differentiated to insulin producing cells *in vitro*, and implanted into a diabetic patient. However, although several straightforward protocols were established, both the efficiency of *in vitro* programming and the function of derived-beta cells remain unsatisfactory. In addition, safety concerns due to inherent risks of neoplasm originating from residual stem cells remain a major hurdle (Borowiak & Melton, 2009; Ricordi & Edlund, 2008; McKnight et al., 2010). To avoid the risks of neoplasm of beta cell generation from stem cells, beta cell transdifferentiation from exocrine tissues has been performed (Minami et al., 2005). Impressively, simple culture of exocrine tissue with EGF and nicotinamide enabled the transdifferentiation of exocrine tissue to beta cells. Additionally, exocrine tissues from type 1 diabetic mice model were able to trans-differentiate into beta cells (Okuno et al., 2007). However, the efficacy was not effective yet to generate enough beta cells to reverse diabetes, and this approach requires human exocrine tissue.

Another medical approach is bona fide regeneration of islet cells/beta cells in a patient. Because there is a slow rate of beta cell turnover in the human pancreas even after injury, regenerative medicine is focusing on stimulating either beta cell replication or neogenesis. Finding a molecular intervention that can be safely used *in vivo* seems challenging but not impossible. There are ways in which neogenesis might be stimulated to expand beta cell mass using agents such as exendin-4, gastrin and epidermal growth factor (Bonner-Weir & Weir, 2005). Another method is differentiation/transdifferentiation in which either existing stem cells or differentiated cells can be programmed/reprogrammed to change their identity. To achieve this goal, most studies have used a gene induction method with a viral vector. It was demonstrated that pancreatic acinar cells might be reprogrammed in mice with injections into the pancreas of adenoviruses expressing three transcription factors, pancreatic duodenal homeobox-1 (PDX-1), musculoaponeurotic fibrosarcoma oncogene homolog A and neurogenin-3 (Zhou et al., 2008). However, gene delivery with viral vectors has shown adverse effects, which have been related to enhancer-mediated mutagenesis of genomic DNA (Hacein-Bey-Abina S et al., 2003) or immunological responses to viral proteins (Manno et al., 2006). Before these permanent or long-term side effects are fully understood and resolved, the safety of using viral vectors must be established.

In this chapter, for bona fide beta cell regeneration, we will introduce two unique methods. The first method is gene delivery using ultrasound targeted micro-bubble destruction (UTMD) technology. UTMD technology allows us to deliver genes specifically into the pancreas by using ultrasound without using viral vectors (Chen et al., 2006, 2010). The other method is activating neural relay mechanism to stimulate beta cell regeneration and insulin secretion in naïve pancreas (Imai et al., 2010). The initial signal is activated in the liver and the signal relay to pancreas via neural system. The unique signal in the liver can actually stimulate beta cell regeneration and insulin secretion in the naïve pancreas.

### 2.1 Ultrasound targeted micro-bubble destruction (UTMD) for gene delivery to regenerate beta cells

#### 2.1.1 Development of UTMD for gene delivery

In order to obtain high gene expression after gene delivery *in vivo* without viral vectors, we have established an ultrasound-mediated gene transfer method named Ultrasound Targeted Microbubble Destruction (UTMD) and achieved efficient gene transfer of plasmid DNA (pDNA) *in vivo* (Chen S et al., 2006, 2010; Korpanty G et al., 2005). Delivery of pDNA does not transport toxic or immunogenic viral protein or polymer particles. UTMD is known to be a novel and potential gene delivery method *in vivo*. The mechanism of gene delivery is as follows: the microbubbles consist of lipid shell and perfluorocarbon gas on the inside. The plasmid gene to be delivered resides in the shell. After infusing the microbubbles with pDNA intravenously, they are detected in the target organ by echography. Under ultrasound exposure, the microbubbles burst and the energy creates transient pores in membranes of surrounding cells, and pDNAs are inserted into the cells (Figure 1).

UTMD has many of the desired characteristics of gene therapy including low toxicity, low immunogenicity, potential for repeated application, organ specificity and broad applicability to acoustically accessible organs.

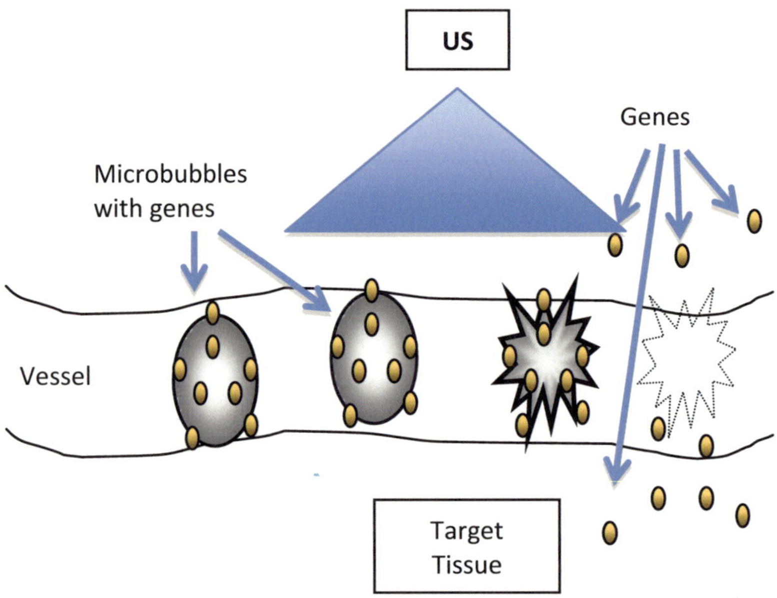


Fig. 1. The mechanism of Ultrasound Targeted Microbubble Destruction (UTMD) for gene delivery. In vessels, microbubbles can be destroyed with high mechanical index ultrasound in the target organ, and the released genes (pDNAs) pass through the vasculature, thus releasing the genes (pDNAs) into the surrounding tissue. The microbubble destruction can only happen under ultrasound

Using this technology, plasmids containing rat insulin 1 promoter (RIP)-human insulin and RIP-hexokinase I were successfully delivered to the islets of adult rats (Chen S et al., 2006). Delivery of RIP-human insulin plasmid resulted in clear increases in circulating human C-peptide and decreased blood glucose levels. Delivery of RIP-hexokinase plasmid resulted in a clear increase in hexokinase I protein expression in islets. Furthermore, delivery of RIP-

NeuroD1 by UTMD technology into streptozotocin induced diabetic rats resulted in promotion of islet regeneration in the naïve pancreas with the return of normal glucose, insulin and C-peptide levels (Chen S et al., 2010). Thus, an exciting new possibility has emerged with this technology. Much work is now underway to determine the potential clinical applications of this *in vivo* gene induction used alone or in combination with other regeneration techniques.

#### 2.1.2 Path to clinical application of beta cell regeneration by UTMD

Even though UTMD for gene delivery is promising for beta cell regeneration, there are several issues that remain to be solved before clinical application. First of all, the safety of this technology needs to be confirmed using a large animal model. Microbubbles have been clinically used as contrast agents for ultrasound; however, the material of microbubbles for UTMD is modified for gene delivery. Therefore, it is necessary to assure the microbubbles for UTMD is safe. Furthermore, since the ability of beta cell regeneration in rodent is much higher in human or large animal (Noguchi et al., 2009b, 2010), it is important to confirm the efficacy of beta cell regeneration in large animal. The next issue is the long-term effect of beta cell regeneration by gene delivery with UTMD. In the rodent model, normoglycemia were maintained for up to 3 months (Chen et al., 2010). Identifying the mechanisms of failure to maintain long-term insulin independence is important. On the other hand, the UTMD technology is relatively easy to be applied. Therefore repeating this technology is feasible. In such case, the effect and safety of repeated UTMD need to be assessed.

When applying the UTMD method for type 1 diabetic patients, it is necessary to prevent autoimmune recurrence. Therefore, immunosuppressive drugs might be necessary to prevent the immunological rejection of regenerated beta cells.

### 2.2 Neural relay for beta cell regeneration in naïve pancreas

The concept of neural relay is very unique. Beta cell proliferative activity changes dynamically to meet systemic needs throughout life. One condition in which beta cell proliferation is enhanced is obesity-related insulin resistance. However, the mechanism underlying this compensatory beta cell response is not well understood.

Recently, Katagiri et al. have identified a neuronal relay, originating in the liver, which enhances both insulin secretion and pancreatic beta cell proliferation for the possible mechanism of obesity-related insulin resistance (Katagiri et al., 2009). Blockade of this neural relay in murine obesity models inhibited pancreatic islet expansion during obesity development, showing this inter-organ communication system to be physiologically involved in compensatory beta cell proliferation. They demonstrated that proliferation of pre-existing beta cells contributes to a beta cell increment by neural relay mechanism. Therefore, this neural relay system is not only for explaining the mechanism of obesity-related insulin resistance but also might be applicable for beta cell regeneration in the naïve pancreas for the treatment of diabetes.

#### 2.2.1 Discovery of neural relay

Metabolism in different organs and tissues works in a coordinated manner. This coordinated metabolism requires inter-organ/tissues communication, therefore the communication among organs and tissues are critical for maintaining normal metabolism (Katagiri et al., 2009). It has been demonstrated that humoral factors including hormones and cytokines play major roles. However, a number of studies have shown that unexpected metabolic phenotypes also make a contribution suggesting the presence of currently unknown metabolic communication systems.

Recently, it was demonstrated that neuronal signaling plays important roles in inter-organ metabolic communication (Yamada et al., 2006). Obesity induces insulin hypersecretion and pancreatic beta cell hyperplasia in response to insulin resistance. These compensatory responses of pancreatic beta cells prevent hyperglycemia; however, this causes hyperinsulinemia which is involved in the pathogenesis of the metabolic syndrome. To elucidate the mechanisms of these compensatory mechanisms, Katagiri et al. activated several proteins which are known to be activated in the livers of obese and lean mice. They discovered that extracellular signal-regulated kinase (ERK) plays an important role in compensatory pancreatic beta cell responses. To activate ERK in the liver, they used adenoviral gene transduction system and discovered that liver-selective ERK activation induced insulin hypersecretion and pancreatic beta cell proliferation (Fujishiro et al., 2003). These pancreatic effects of hepatic ERK activation were inhibited by either splanchnic afferent blockade with pancreatic vagus dissection or midbrain transection. This result indicated that a neuronal relay system from liver to the pancreas consists of the afferent splanchnic nerve, the central nervous system and efferent vagus (Figure 2).

Furthermore, blockage of the neuronal relay at several levels in murine obesity model inhibited pancreatic islet expansion during obesity development indicating that the neural relay played an important role in the inter-organ mechanism in compensatory beta cell responses.

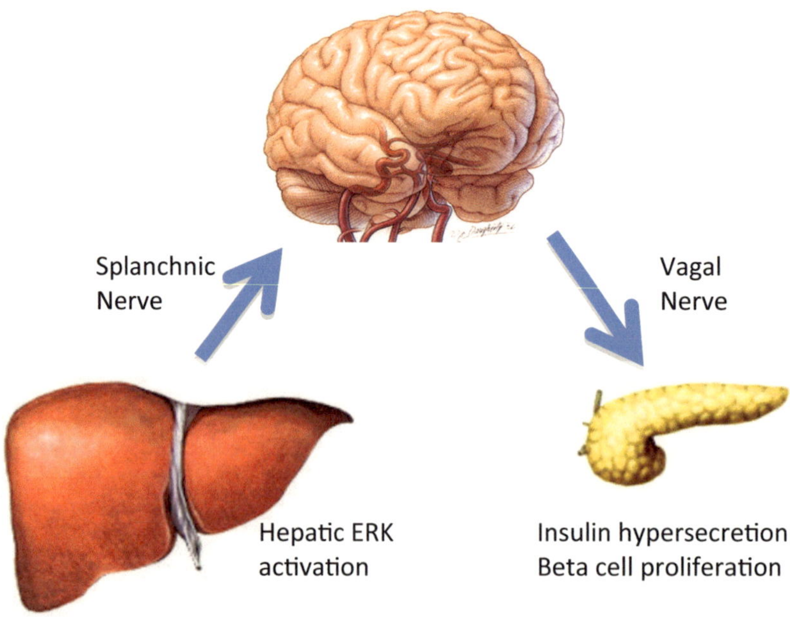


Fig. 2. The concept of neural relay for beta cell regeneration. The signal of hepatic ERK activation reaches the brain via the splanchnic nerve. Then the signal reaches the pancreas via the vagal nerve. This signal stimulates insulin hypersecretion and beta cell proliferation in the pancreas

#### 2.2.2 Potential application of neural relay for beta cell regeneration

Regeneration of beta cells in the original pancreas could be an ideal cure for type 1 diabetic patients. Since the neural relay system can stimulate beta cell proliferation and insulin secretion, this system might be used for beta cell regeneration.

Indeed, it has been demonstrated that liver-selective activation of extracellular signal-regulated kinase (ERK) using an adenoviral gene transduction system resulted in an increase in beta cell mass and normalization of serum glucose levels in streptozotocin (STZ) induced diabetic mice and Akita diabetic mice (Imai et al., 2008). A bromodeoxyuridine (BrdU) staining study demonstrated that beta cell proliferation was the mechanism for increasing beta cell mass by neural relay (Imai et al., 2008). Therefore, stimulating the ERK pathway is a promising idea for beta cell regeneration in the naïve pancreas. However, it should be noted that they used only the STZ model and Akita diabetic mice model. STZ model is a model of insulin dependent diabetes mellitus, however, no autoimmune mechanism is involved. Therefore, although the neural relay works on the STZ model it is still unknown whether this method can be effective on the type 1 diabetic patients with autoimmune disease. In addition, the STZ model cannot completely eliminate beta cells therefore beta cell proliferation was possible. On the contrary, type 1 diabetic patients without insulin secretory ability have completely lost their beta cells. Therefore, it might be impossible to induce beta cell proliferation because no beta cells remain. In order to apply neural relay technology for type 1 diabetes, it is necessary to confirm the efficacy using NOD mice which is an autoimmune induced type 1 diabetes model. Additionally, it is necessary to examine the minimum number of remaining beta cells which will be proliferated to reverse diabetes. If a significant amount of beta cells are necessary, this method can only be applied for the patients who still have insulin secretory ability. Akita diabetic mouse is a model of type 2 diabetes. In general, patients with type 2 diabetes do not require insulin injection; therefore the indication of neural relay for type 2 diabetes should be limited. However, currently, insulin therapy is applied for type 2 diabetic patients in order to save functional beta cell mass. Therefore, neural relay therapy might be an excellent option for type 2 diabetic patients with insulin therapy.

The advantage of this method is that no stem cells such as embryonic stem (ES) cells (Thomson et al., 1998) or induced pluripotent stem (iPS) cells (Takahashi et al., 2007) or pancreatic stem cells are necessary. Therefore, the notorious problem of carcinogenesis of those stem cell derived beta cells is no longer an issue. In addition, *ex vivo* manipulation to create beta cells from stem cells is not necessary.

The possible disadvantage of this method is hepatic injury by activation of hepatic ERK. Also, the adenoviral system may not be appropriate for clinical trials because of the risk of viral infection. UTMD system for activation of hepatic ERK might be useful approach to avoid using viral transfection. Identifying the efficient activation of hepatic ERK should be the key for beta cell regeneration. When ERK stimulation for beta cell regeneration will be applied for type 1 diabetic patients, the prevention autoimmune recurrence and immunosuppressive drugs might be necessary. Furthermore, in advanced type 1 diabetes, all beta cells are destroyed as mentioned above. Therefore, it might be impossible to proliferate islets because no original islets exist. In this case, a combination of beta cell generation from pancreatic stem cells and neural relay might be useful.

## 3. Bio-artificial islets using pig islets

Islet transplantation using the bio-artificial islets created from alternative sources instead of human islets is very attractive to overcome the issue of severe donor shortage for the treatment of diabetes.

Very importantly, two clinical trials using bio-artificial islets consisting of piglet islets have been already clinically performed with promising results. The first trial was performed in Mexico using neonatal pig islets combined with Sertoli cells for the treatment of pediatric type 1 diabetic patients (Valdes-Gonzales et al., 2005a, 2007). Islets and Sertoli cells were put into a chamber. The Sertoli cells protected islets from immunological attacks. The other series were performed by New Zealand group for the treatment of adult type 1 diabetic patients with severe hypoglycemic episodes (Elliott et al., 2011). The New Zealand group used microencapsulated pig islets and those bio-artificial islets were transplanted into the abdominal cavity.

Impressively, both groups achieved insulin independence after transplantation in some cases without use of immunosuppressive drugs.

### 3.1 Bio-artificial islets with Sertoli cells

#### 3.1.1 Preparation of bio-artificial islets with Sertoli cells and transplantation

Valdes-Gonzalez et al. performed bio-artificial islet transplantation into twelve pediatric type 1 diabetic patients (Valdes-Gonzalez et al., 2005a). Their bio-artificial islet consists of collagen-generating devices, islets from neonatal pigs and Sertoli cells. The collagen-generating devices were not considered as immune-isolation devices. Islets were isolated from male 7-10 days old piglets. The animals were bred in New Zealand in a specific pathogen-free environment in accordance with the Association for Assessment and Accreditation of Laboratory Animal Care. After pancreas retrieval, pancreases were digested using collagenase for islet isolation (Elliott et al., 2000; Valdes-Gonzalez et al., 2005b). Islets were placed in RPMI-1640, 2% human serum albumin, 0.12% nicotinamide and 1.5mg/l ciproxine at room temperature and centrifuged at 1000 rpm for 20 min. The average islet yield was 290,730 islet equivalents (IEQ: 1IEQ=1 of 150 μm islet). The purity of islets was assessed with dithizone staining and was greater than 85% in all cases. The viability of islets assessed by acridine orange/propidium iodide staining was more than 85% in all cases. Isolated Sertoli cell-enriched testicular cells were placed in DMEM media with 0.12% nicotinamide and 1.5mg/l ciproxine. All cell preparations underwent full microbiologic screening both in New Zealand and again at the time of transplantation. Cells from ten neonatal pigs were used for each transplant. Sertoli cells and islets were mixed together immediately prior to transplantation.

Two devices were implanted subcutaneously in the upper anterior wall of the patient's abdomen under general anesthesia. The devices were left in place for two months to allow formation of vascularized collagen tissue that completely surrounded and penetrated the device.

Both islets and Sertoli cells were isolated in New Zealand and sent at room temperature in culture media to Mexico. Islets and Sertoli cells were cultured for one day before transplantation. The transplant procedure was carried out by infusing 250,000 islets with 30-100 Sertoli cells per islet. The number of islets per body weight ranged from approximately 14,000 to 21,000IEQ/kg. From 6 to 9 months later, all patients except one, received a second islet and Sertoli cell transplant into previously implanted new devices. No immunosuppressive drug was administered at any point.

#### 3.1.2 Clinical outcomes

Immediately upon entering the study, the patients followed a diet and exercise regimen standard for diabetic patients, with periodic weight and height measurements. The patients were instructed to record blood glucose determination seven times a day (pre- and postprandial, and 3 am). Eleven age and disease matched control group was subjected for 10 months to exactly the same exhaustive endocrine monitoring, and diet and exercise program without receiving a transplant.

In the transplanted patients following the first and more markedly after the second transplant, cluster analysis revealed that two distinct insulin requirement patterns appeared. Half of the patients had a 50% or greater reduction in their insulin requirements, and the other six patients showed a slight increase. This increase in these patients seen corresponded with that that seen in the control group. Half of the patients significantly reduced amount of insulin compared to the control group from the first month post transplant onwards. Two patients achieved insulin independence. The first one was a 15 year old female who had exogenous insulin requirements of 61 U/day before transplant and HbA1c was 13.4%. After the first transplant, she reduced her insulin requirements by 73% and after the second transplant she began to have intermittent period of 3-5 days, alternating between periods of no insulin injections, followed by periods of 1-2 U/day. This pattern last for 3 months, and HbA1c reached 9.6%. The second patient was a 16-year-old female who had exogenous insulin requirement of 55 U/day and HbA1c was 12%. Six months after the first transplant, the patient showed a 6 week reduction to 1-3 U/day and HbA1c was 6.8%. After the second transplant she was totally free of insulin for two consecutive months and her HbA1c was 6.5-7.8%. Interestingly all patients improved glycemic control after transplantation irrespective of the amount of insulin reduction. Long-term follow-up of those patients revealed that all patients have positive porcine C-peptides in urine (Valdes-Gonzalez et al., 2010).

In terms of safety, routine microbiological screening of all patients and close family have been consistently negative, although two patients exhibited transient chimerism as evidenced by porcine DNA thru polymerase chain reaction (PCR). No complication related to the surgery or to the presence of the cells has occurred at any time.

### 3.2 Bio-artificial islets using micro-encapsulation technology

#### 3.2.1 Preparation of bio-artificial islets using micro-encapsulation technology and transplantation

Elliott et al. performed bio-artificial islet transplantation into adult type 1 diabetic patients with severe hypoglycemic episodes (Elliot et al., 2011). Their bio-artificial islets were made of micro-encapsulated islets from neonatal pigs. This group used the same islet isolation procedures using the same herd with Valdes-Gonzalez. The islets were micro-encapsulated using alginate. The capsule allows entering nutrients and glucose inside the capsule and passing insulin outside the capsule (Figure 3). Meanwhile, the capsule blocks antibodies resulting in protecting the islet from immunological attack (Figure 3).

The encapsulation process was a modification of the method described by Calafiore (Calafiore et al., 2006). Encapsulation material was started with raw pharmaceutical grade alginate powder. The alginate powder was dissolved in sterile pyrogen-free deionized water over 24 to 36 hours in the dark at room temperature and 3% NaCl was added. The solution underwent multiple sequential passages through methylcellulose and polyester filters to ensure sterility. The final 1.6% solution was stored in the dark room at 4°C to avoid alginate depolymerization. The islet tissue pellet, usually amounting to a few tenths of a milliliter, was thoroughly mixed with the 1.6% alginate solution. The alginate/islet proportion was adjusted so that one capsule would contain one islet, with fewer than 5% empty capsules. The suspension was extruded through a microdroplet generator, combining air shears with mechanical pressure: the alginate droplets were collected in 1.2% $CaCl_2$ immediately turning into gel micro-beads. They were sequentially overcoated with poly-L-ornithine and an outer alginate layer.

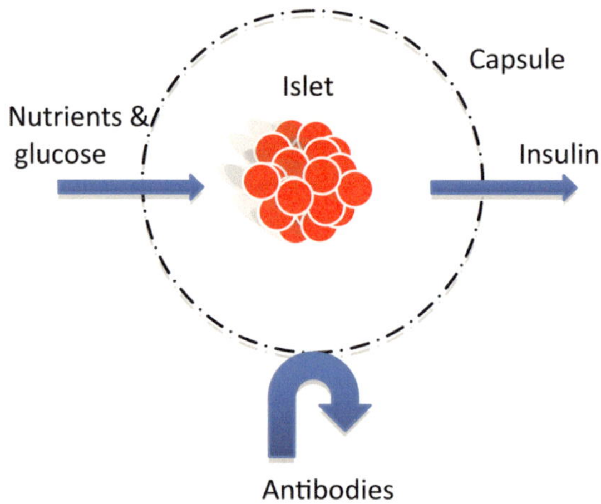


Fig. 3. The concept of microencapsulation of an islet. Nutrients, glucose and insulin can pass through the capsule, but antibodies cannot enter into the capsule

Bio-artificial islets consisted of microencaplusted neonatal pig islets were transplanted into the peritoneal cavity via a laparoscope under general anesthesia. Four patients received 10,000 IEQ/kg body weight islets and the other four patients received 15,000IEQ/kg body weight islets.

### 3.2.2 Clinical outcomes

In order to perform clinical trials, they gained regulatory approval from the relevant authorities after prolonged national and international consultation (Elliott et al., 2011). A national consensus on the bioethical issues was conducted and a separate national consultation on the acceptability of the science was also conducted. Approval from Medsafe the relevant department of the Ministry of Health was obtained. Eight adult patients with longstanding proven type 1 diabetes who met all inclusion and exclusion criteria, were selected on the basis of severe recurrent hypoglycemia usually with hypoglycemic unawareness.

To date, most patients have shown modest reduction in insulin dose commencing about four weeks after transplantation with reduction in HbA1c levels. Most outstanding has been the reduction in severe hypoglycemic episodes and reduction or abolition of unaware hypoglycemia. For example, the first patient had an average of 4 episodes of unaware hypoglycemia per week, which was completely diminished after 8 weeks of transplantation. Transitory insulin independence of several months duration has been seen.

In terms of safety, evidence of xenosis in the xenotransplant recipients has been diligently sought but not found. This is reasonable given the credentials of the source herd used. No serious adverse events related to the surgery or to the presence of the cells have occurred at any time.

### 3.3 Future direction of bio-artificial islets

Since year 2000 after the publication of the Edmonton protocol, allogeneic islet transplantation has become popular as the treatment of type 1 diabetes (Shapiro et al., 2000). Using the Edmonton protocol, type 1 diabetic patients who had severe hypoglycemic episodes became insulin independent and free from hypoglycemic episodes after allogenic islet transplantation (Shapiro et al., 2000). The allogeneic islet transplantation has been expanded using non-heart beating donors (Markmann et al., 2003; Matsumoto et al., 2006b) and even living donor (Matsumoto et al., 2005, 2006a). However, the drawbacks of the Edmonton protocol include necessity of multiple donor organs, unstable islet isolation results, necessity of immunosuppressive drugs, difficulty of maintaining long-term insulin independence and severe shortage of donor organs (Ryan et al., 2005; Shapiro et al., 2006). Currently, we introduced new pancreas preservation (Matsumoto et al., 2002a, 2002b, 2010b) and islet isolation strategies (Noguchi et al., 2009a; Shimoda et al., 2010) and immunosuppressive therapy to improve the efficacy of islet isolation (Matsumoto et al., 2011). Now, we have a very stable islet isolation method, in addition, a single donor pancreas is enough to achieve insulin independence (Matsumoto et al., 2011). Our preliminary data demonstrated that super-high dose islet transplantation could lead to long-term insulin independence after allogeneic islet transplantation (Matsumoto et al., 2010c). However, the severe donor shortage can be never be solved by allogeneic islet transplantation alone. Bio-artificial islets using porcine islets can solve the issue of donor shortage. In addition, both bio-artificial islets with encapsulated islets and islets with Sertoli cells do not require immunosuppressive drugs. This is huge benefit of bio-artificial islet transplantation because one of the major issues of allogeneic islet transplantation is the side effects and cost of immunosuppression (Hatanaka et al., 2010).

Currently, bio-artificial islets were transplanted into the abdominal cavity or under skin. These transplant sites have unique advantages. In the case of allogeneic islet transplantation, islets were transplanted into liver. Multiple infusions of isolated islets into liver cause portal hypertension. Therefore allogenic islet transplantation has the limitation of the numbers of transplantation. In the case of bio-artificial islet transplantation, there is no risk for portal hypertension. Therefore, there is no limitation of number of transplantation. These comparisons were summarized in table 2.

Impressively, pigs in New Zealand have been maintained in a clean, non-pathogenic environment and seem suitable for clinical use. Both clinical trials of bio-artificial islets used these pigs. Expansion of the herd of pigs should be the key to enhancing the bio-artificial islet project. Islet isolation from neonatal piglets is relatively easy and stable; this is important advantage for commercialization.

| | Standard Allogeneic Islet Transplantation | Advanced Allogeneic Islet Transplantation | Bio-artificial Islet Transplantation |
|---|---|---|---|
| Primary Endpoints | Preventing hypoglycemia | Insulin free | Preventing hypoglycemia |
| Secondary Endpoints | Insulin free<br>Preventing diabetic complications | Preventing diabetic complications | Insulin free<br>Preventing diabetic complications |
| Donor numbers | 2 or more human pancreases | 1 human pancreas | 10 piglets pancreases |
| Stability of islet isolation | Not stable | Stable | Stable |
| Transplant site | Liver | Liver | Intra-peritoneum<br>Under skin |
| Immunosuppression | Necessary | Necessary | Not necessary |
| Attaining insulin independence | Most likely | Most likely | Possible |
| Long-term insulin free | Difficult | Possible | Difficult |
| Long-term function | Possible | Most likely | Possible |
| Re-transplant | Up to 3 or 4 times | Up to 3 or 4 times | No limitation |

Table 2. Comparison among the standard allogeneic islet transplantation, the advanced allogeneic islet transplantation and bio-artificial islet transplantation. Of note, bio-artificial islet transplantation has several important advantages including using alternative source, no immunosuppressive drugs and no limitation of re-transplantation

As shown Elliott et al., the bio-artificial islets can eliminate hypoglycemic unawareness, and this is the one of the major goals of allogeneic islet transplantation. Therefore, the patients with hypoglycemic unawareness will be suitable candidates for bio-artificial islet transplantation instead of allogeneic islet transplantation in future.

More importantly, allogeneic islet transplantation can reduce or eliminate diabetic secondary complications such as diabetic nephropathy, retinopathy and neuropathy (Thompson et al., 2011). Especially, if the bio-artificial islet transplantation can also reduce or eliminate such diabetic secondary complications, this treatment will be very valuable because the real problems of diabetes are the secondary complications.

Cost effectiveness is an issue of bio-artificial islets, because maintenance of clean pigs is expensive. The system to maintain cleanness of a huge herd of pigs needs to be developed to overcome the cost issue. The major concern of bio-artificial islet transplantation is zoonosis. Especially, creating a new viral disease by xeno-transplantation must be avoided. Infection of porcine endogenous retrovirus after xeno-transplantation into immune compromised mice demonstrated the risk of the combination of immunosuppression and xeno-transplantation (van der Laan LJ et al., 2000). Therefore current bio-artificial islet transplantations have been performed without immunosuppression.

Acceptance of pig islets by patients is an emotional and highly debated issue for xeno-transplantation. Our survey of type 1 diabetic patients revealed that more than 60% of type 1 diabetic patients were willingly to accept pig islets if the treatment was effective (Hatanaka et al., 2010).

## 4. Conclusions

We introduced two unique gene therapies UMTD and neural relay for beta cell regenerations. Both methods are not clinically applied yet; however due to their relatively safe feature, we believe those methods can be clinically used in near future. We also described the clinical applications of bio-artificial islets using neonatal porcine islets with promising results. Most importantly, so far there are no severe adverse events. Although the results of bio-artificial islet transplantation are not as effective as allogeneic islet transplantation there are many aspects including islet isolation methods (Shimoda et al., 2011a, 2011b), islet culture methods, transplantation sites and patients' treatments, which can be improved.

Currently, diabetes is considered a non-curable disease therefore current treatments are focusing on improving quality of life and preventing diabetic complications (Takita 2011, Hatanaka 2011). However, we believe that the bio-artificial islets and/or gene therapy for beta cell regeneration will cure a majority of both types of diabetes in the future.

## 5. Acknowledgement

This work is partially supported by Juvenile Diabetes Research Foundation (JDRF).

## 6. References

Bonner-Weir S, Taneja M, Weir GC, Tatarkiewickz K, Song KH, Sharma A, & O'Neil JJ. (2000) In vitro cultivation of human islets from expanded ductal tissue. *Proc Natl Acad Sci USA*. 97: 7999-8004.

Bonner-Weir S, & Sharma A. (2002) Pancreatic stem cells. *J Pathol* 197: 519-6.

Bonner-Weir S, & Weir GC. (2005) New sources of pancreatic beta-cells. *Nat Biotechnol.* 23:857-61.

Borowiak M & D.A. & Melton DA. (2009) How to make beta cells?, Curr Opin Cell Biol 21: 727-732.

Butler AE, Janson J, Bonner-Weir S, Ritzel R, Rizza RA, & Butler PC. (2003) Beta-cell deficit and increased beta-cell apoptosis in humans with type 2 diabetes. *Diabetes* 52:102-10.

Butler PC, Meier JJ, Butler AE, Bhushan A. (2007) The replication of β cells in normal physiology, in disease and for therapy. *Nature Clinical Practice Endocrinologu & Metabolism* 3: 758-68.

Chen S, Ding JH, Bekeredjian R, Yang BZ, Shohet RV, Johnston SA, Hohmeir HE, Newgard CB, & Grayburn PA. (2006) Efficient gene delivery to pancreatic islets with ultrasonic microbubble destruction technology. *Proc Natl Acad Sci USA*, 103: 8469-74.

Chen S, Shimoda M, Wang MY, Ding J, Noguchi H, Matsumoto S, & Grayburn PA. (2010). Regeneration of pancreatic islets in vivo by ultrasound-targeted gene therapy. *Gene Ther,* 17: 1411-20.

Elliott RB, Escobar L, Garkavenko O, Croxoson MC, Schroeder BA, McGregor M, Ferguson G, Beck Man N & Ferguson S. (2000) No evidence of infection with porcine endogenous retrovirus in recipients of encaplulated porcine islet xenografts Cell Transplant 9: 895-901.

Fujishiro M, Gotoh Y, Katagiri H, Sakoda H, Ogihara T, Anai M, Onishi Y, Ono H, Abe M, Shojima N, Fukushima Y, Kikuchi M, Oka Y, & Asano T. (2003). Three mitogen-activated protein kinases inhibit insulin signaling by different mechanisms in 3T3-L1 adipocytes. *Mol Endocrinol.* 17:487-97.

Hacein-Bey-Abina S, von Kalle C, Schmidt M, Le Deist F, Wulffraat N, McIntyre E, Radford I, Villeval JL, Fraser CC, Cavazzana-Calvo M, & Fischer A. (2003) A serious adverse event after successful gene therapy for X-linked severe combined immunodeficiency. *N Engl J Med 348:255-6.*

Hatanaka N, Takita M, Yamaguchi T, Kami M, & Matsumoto S. (2010) Interests in beta-cell replacement therapies among Japanese patients with type 1 diabetes. *Diab Res Clin Pract* 89: e5-8.

Hatanaka N, Takita M, Yamaguchi T, Kami M, Matsumoto S. (2011) Development of a novel scale to assess the quality of life in type 1 diabetic patients for beta cell replacement therapy. *Diabetology International* in press

Huang ES., Basu A, O'Grady M, & Capretta JC. (2009) Projecting the Future Diabetes Population Size and Related Costs for the U.S. *Diabetes Care* 32: 2225-9.

Imai J, Oka Y, & Katagiri H. (2010) Identification of a novel mechanism regulating β-cell mass: Neuronal relay from the liver to pancreatic β-cells. *Islets* 1: 75-7.

Imai J, Katagiri H, Yamada T, Ishigaki Y, Suzuki T, Kudo H, Uno K, hasegawa Y, Gao J, Kaneko K, Ishihara H, Niijima A, Nakazato M, Asano T, Minokoshi Y, & Oka Y. (2008) Regulation of pancreatic beta cell mass by neuronal signals from the liver. *Science* 322: 1250-4.

Katagiri H, Imai J, & Oka Y. (2009) Neural relay from the liver induces proliferation of pancreatic beta cells. A path to regenerative medicine using the self-renewal capabilities. *Communicative & Integrative Biology* 2: 425-7.

Korpanty G, Chen S, Shohet RV, Ding J, Yang B, Frenkel PA, & Grayburn PA. (2005) Targeting of VEGF-mediated angiogenesis to rat myocardium using ultrasonic destruction of microbubbles. *Gene Ther. 12:1305-12.*

Manno CS, Pierce GF, Arruda VR, Glader B, Ragni M, Rasko JJ, Ozelo MC, Hoots K, Blatt P, Konkle B, Dake M, Kaye R, Razavi M, Zajko A, Zehnder J, Rustagi PK, Nakai H, Chew A, Leonard D, Wright JF, Lessard RR, Sommer JM, Tigges M, Sabatino D, Luk A, Jiang H, Mingozzi F, Couto L, Ertl HC, High KA, & Kay MA. (2006) Successful transduction of liver in hemophilia by AAV-Factor IX and limitations imposed by the host immune response. *Nat Med 12:342-7.*

Markmann JF, Deng S, Desai NM, Huang X, Velidedeoglu E, Frank A, Liu C, Brayman KL, Lian MM, Wolf B, Bell E, Vitamaniuk M, Doliba N, Matschinsky F, Markmann E, Barker CF, & Naji A. (2003) The use of non-heart-beating donors for isolated pancreatic islet transplantation. Transplantation.75: 1423-9.

Matsumoto S, Rigley T, Qualley S, Kuroda Y, Reems J, & Stevens RB. (2002a) Efficacy of the oxygen-charged static two-layer method for short-term pancreas preservation and islet isolation from nonhuman primate and human pancreata. *Cell Transplant* 11: 769-77.

Matsumoto S, Qualley S, Goel S, Hagman D, Sweet I, Poitout V, Strong DM, Robertson RP, Reems J. (2002b) Effect of the two-layer (University of Wisconsin solution-perfluorochemical plus O2) method of pancreas preservation on human islet isolation, as assessed by the Edmonton Isolation Protocol. *Transplantation* 74: 1414-19

Matsumoto S, Okitsu T, Iwanaga Y, Noguchi H, Nagata H, Yonekawa Y, Yamada Y, Fukuda K, Tsukiyama K, Suzuki H, Kawasaki Y, Shimodaira M, Matsuoka K, Shibata T, Kasai Y, Maekawa T, Shapiro J, & Tanaka K. (2005) Insulin independence after living-donor distal pancreatectomy and islet allotransplantation. *Lancet* 365: 1642-4.

Matsumoto S, Okitsu T, Iwanaga Y, Noguchi H, Nagata H, Yonekawa Y, Liu X, Kamiya H, Ueda M, Hatanaka N, Kobayashi N, Yamada Y, Miyakawa S, Seino Y, Shapiro AM, & Tanaka K. (2006a) Follow-up study of the first successful living donor islet transplantation. Transplantation. 82: 1629-33.

Matsumoto S, Okitsu T, Iwanaga Y, Noguchi H, Nagata H, Yonekawa Y, Yamada Y, Fukuda K, Shibata T, Kasai Y, Maekawa T, Wada H, Nakamura T, & Tanaka K. (2006b) Successful islet transplantation from non-heart-beating donor pancreata using modified Ricordi islet isolation method. *Transplantation* 82: 460-5.

Matsumoto S. (2010a). Islet cell transplantation for type 1 diabetic patients. *J Diabetes*, 2: 16-22.

Matsumoto S, Noguchi H, Shimoda M, Ikemoto, Naziruddin B, Jackson A, Tamura Y, Olsen G, Fujita Y, Chujo D, Takita M, Kobayashi N, Onaca N, & Levy MF. (2010b) Seven consecutive successful clinical islet isolations with pancreatic ductal injection. *Cell Transplant* 19: 291-7.

Matsumoto S, Noguchi H, Takita M, Shimoda M, Tamura Y, Olsen G, Chujo D, Sugimoto K, Itoh T, Naziruddin B, Onaca N, & Levy MF. (2010c) Super high dose islet transplantation is associated with high SUITO index and prolonged insulin independence: A case report. *Transplant Proc* 42: 2156-58.

Matsumoto S, Takita M, Chaussabel D, Noguchi H, Shimoda M, Sugimoto K, Itoh T, Chujo D, SoRelle J, Onaca N, Naziruddin B, & Levy MF. (2011) Improving efficacy of clinical islet transplantation with iodixanol based islet purification, thymoglobulin induction and blockage of IL-1 beta and TNF-alpha. *Cell Transplant* in press

Meier JJ, Bhushan A, Butler AE, Rizza RA, & Butler PC. (2005) Sustained beta cell apoptosis in patients with long-standing type 1 diabetes: indirect evidence for islet regeneration? *Diabetologia* 48:2221-2228

Meier JJ. (2008) Beta cell mass in diabetes: a realistic therapeutic target? *Diabetologia* 51: 703-713

Minami K, Okuno M, Miyawaki K, Okumachi A, Ishizaki K, Oyama K, Kawaguchi M, Ishizuka N, Iwanaga T, & Seino S. (2005) Lineage tracing and characterization of insulin-secreting cells generated from adult pancreatic acinar cells. *Proc Natl Acad Sci U S A*. 102: 15116-21.

Noguchi H, Ikemoto T, Naziruddin B, Jackson A, Shimoda M, Fujita Y, Chujo D, Takita M, Kobayashi N, Onaca N, Levy MF, & Matsumoto S. (2009a) Iodixanol-controlled density gradient during islet purification improves recovery rate in human islet isolation. Transplantation. 87: 1629-35.

Noguchi H, Oishi K, Ueda M, Yukawa H, Hayashi S, Kobayashi N, Levy M, & Matsumoto S. (2009b) Establishment of mouse pancreatic stem cell line. *Cell Transplant* 18: 563-71.

Noguchi H, Naziruddin B, Jackson A, Shimoda M, Ikemoto T, Fujita Y, Chujo D, Takita M, Kobayashi N, Onaca N, Hayashi S, Levy M, & Matsumoto S. (2010) Characterization of human pancreatic progenitor cells. *Cell Transplant* 19; 879-86.

Okuno M, Minami K, Okumachi A, Miyawaki K, Yokoi N, Toyokuni S, & Seino S. (2006) Generation of insulin-secreting cells from pancreatic acinar cells of animal models of type 1 diabetes. *Am J Physiol Endocrinol Metab.* 292: E158-65.

Ricordi R & Edlund H. (2008) Toward a renewable source of pancreatic beta-cells, Nat Biotechnol 26: 397–8.

Ryan EA, Paty BW, Senior PA, Bigam D, Alfadhli E, Kneteman NM, Lakey JR, & Shapiro AM. (2005) Five-year follow-up after clinical islet transplantation. *Diabetes* 54: 2060-9.

Shapiro AM, Lakey JR, Ryan EA, Korbutt GS, Toth E, Warnock GL, Kneteman NM, & Rajotte RV. (2000) Islet transplantation in seven patients with type 1 diabetes mellitus using a glucocorticoid-free immunosuppressive regimen. *N Engl J Med* 343: 230-8.

Shapiro AM, Ricordi C, Hering BJ, Auchincloss H, Lindblad R, Robertson RP, Secchi A, Brendel MD, Berney T, Brennan DC, Cagliero E, Alejandro R, Ryan EA, DiMercurio B, Morel P, Polonsky KS, Reems JA, Bretzel RG, Bertuzzi F, Froud T, Kandaswamy R, Sutherland DE, Eisenbarth G, Segal M, Preiksaitis J, Korbutt GS, Barton FB, Viviano L, Seyfert-Margolis V, Bluestone J, & Lakey JR. (2006) International trial of the Edmonton protocol for islet transplantation. *N Engl J Med* 355: 1313-30.

Shimoda M, Noguchi H, Naziruddin B, Fujita Y, Chujo D, Takita M, Peng H, Tamura Y, Olsen GS, Sugimoto K, Itoh T, Onaca N, Levy MF, Grayburn PA, & Matsumoto S.

(2010) Improved method of human islet isolation for young donors. *Transplant Proc* 42: 2024-6.

Shimoda M, Noguchi H, Fujita Y, Takita M, Ikemoto T, Chujo D, Naziruddin B, Levy MF, Kobayashi N, Grayburn PA, & Matsumoto S. (2011) Improvement of porcine islet isolation by Aralast inhibition of trypsin with pancreatic ductal preservation method. *Cell Transplant* in press

Shimoda M, Noguchi H, Fujita Y, Takita M, Ikemoto T, Chujo D, Naziruddin B, Levy MF, Kobayashi N, Grayburn PA, & Matsumoto S. (2011) Islet purification method using large bottle effectively achieves high islet yield from pig pancreas. *Cell Transplant in press*

Takahashi K, Tanae K, Ohnuki M, Narita M, Ichisaka T, Tomoda K, & Yamanaka S. (2007) Induction of pluripotent stem cells from adult human fibroblasts by defined factors. *Cell* 131: 861-72.

Takita M, Matsumoto S, Qin H, Noguchi H, Shimoda M, Itoh T, Sugimoto K, Naziruddin B, Onaca N, & Levy M. (2011) Secretory Unit of Islet Transplant Objects (SUITO) Index can predict severity of hypoglycemic episodes in clinical islet cell transplantation. *Cell Transplant* in press

Thompson DM, Meloche M, Ao Z, Paty B, Keown P, Shapiro RJ, Ho S, Worsley D, Fung M, Meneilly G, Begg I, Al Mehthel M, Kondi J, Harris C, Fensom B, Kozak SE, Tong SO, Trinh M, & Warnock GL. (2011) Reduced progression of diabetic microvascular complications with islet cell transplantation compared with intensive medical therapy. *Transplantation* 91: 373-8.

Thomson JA, Itskovitz-Eldor J, Shapiro SS, Waknitz MA, Swiergiel JJ, Marshal VS & Jones JM. (1998) Embryonic stem cell lines derived from human blastocytes. *Science* 282: 1145-7.

Valdés-González RA, Dorantes LM, Garibay-Nieto GN, Bracho-Blanchet E, Mendez AJ, Dávila-Pérez R, Elliott RB, Terán L, & White DJ. (2005a). Xenotransplantation of porcine neonatal islets of Lanberhans and Sertoli cells: a 4-year study. *Eur J Endocrinol.* 153: 419-27.

Valdés-González RA, Silva-Torres ML, Ramírez-González B, Terán L, Ormsby CE, & Ayala-Sumuano JT. (2005b) Improved method for isolation of porcine neonatal pancreatic cell clusters. *Xenotransplantation* 12: 240-4.

Valdés-González RA, White DJ, Dorantes LM, Terán L, Garibay-Nieto GN, Bracho-Blanchet E, Dávila-Pérez R, Evia-Viscarra L, Ormsby CE, Ayala-Sumuano JT, Silva-Torres ML, & Ramírez-González B. (2007). Three-yr follow-up of a type 1 diabetes mellitus patient with an islet xenotransplant. *Clin Transplant,* 21: 352-7.

Valdes-Gonzalez R, Rodriguez-Ventura AL, White DJ, Bracho-Blanchet E, Castillo A, Ramírez-González B, López-Santos MG, León-Mancilla BH, & Dorantes LM. (2010) Long-term follow of patients with type 1 diabetes transplanted with neonatal pig islets. *Clin Exp Immunol.* 162: 537-42.

van der Laan LJ, Lockey C, Griffeth BC, Frasier FS, Wilson CA, Onions DE, Hering BJ, Long Z, Otto E, Torbett BE, & Salomon DR. (2000) Infection by porcine endogenous retrovirus after islet xenotransplantation in SCID mice. *Nature.* 407(6800): 90-4.

Yamada T, Katagiri H, Ishigaki Y, Ogihara T, Imai J, Uno K, Hasegawa Y, Gao J, Ishihara H, Niijima A, Mano H, Aburatani H, Asano T, & Oka Y. (2006). Signals from intra-abdominal fat modulate insulin and leptin sensitivity through different mechanisms: neuronal involvement in food-intake regulation. *Cell Metab* 3: 223-9.

Zhou Q, Brown J, Kanarek A, Rajagopal J, & Melton DA. (2008) In vivo reprogramming of adult pancreatic exocrine cells to beta-cells. *Nature* 455: 627-32.

# 20

# Perspectives of Islet Cell Transplantation as a Therapeutic Approach for Diabetes Mellitus

Prabha D. Nair and Neena Aloysious
*Division of Tissue Engineering & Regeneration Technologies, Sree Chitra Tirunal Institute for Medical Sciences & Technology, Trivandrum India*

## 1. Introduction

Diabetes mellitus, the metabolic disorder is rapidly on the rise, becoming one of the main threats to human health and imposing large socio-economic burden on the society in the 21st century (Dall et al.; 2010). International Diabetes Federation in 2011 estimated that over 300 million people around the world have diabetes and is expected to rise to 500 million within next 20 years. The global prevalence of diabetes is shifting significantly from the developed countries to the developing countries. Current therapy for diabetes involves oral antidiabetic drugs and insulin administration; these approaches do not mimic the pulsatile insulin secretory patterns of native β islets for the regulation of glucose in real-time nor provide tight control of blood glucose to avoid late complications of the disease. Whole pancreas transplantation holds promise towards a cure for diabetes, but this procedure requires major surgery and lifelong immunosuppression to prevent graft rejection. Transplantation of islet cells isolated from a donor pancreas has been shown to control glucose levels successfully. Being less invasive, it is a better alternative to pancreas transplantation yet scarcity of donors, maintenance of islet functions such as cell growth and survival *in vitro,* and concern over the adverse effect of life long immunosuppressants used to prevent graft rejection precludes the benefits of islet transplantation from becoming universally acceptable.

One approach to overcome these obstacles of immune rejection is islet encapsulation (Kizilel et al., 2005; Mikos et al; 1994) that uses an immuno-protective biomaterial to create a permselective membrane around a group of islet cells. Transplanted islet cells are separated from the immunological system of host by means of an artificial selectively permeable membrane which allows passage of metabolites and nutrients, while excluding based on size, the larger proteins and cells of the immune system. Thus, encapsulation is designed to limit, and ideally eliminate, an immunological response to the non-host islet cells. Isolation of the islet cells from the human immune system may also make xeno-transplants such as porcine islets, stem cells derived insulin producing cells possible, eliminating the supply problem that exists and the usage of immune suppressive drugs.

Current research is directed towards exploration of alternative sources of pancreatic islet cells. Pancreatic β- cell lines, embryonic stem cells (ESC), adult progenitor cells (APC), and regenerating native islet cells, generation of β cells by therapeutic cloning and differentiation of islets from pancreatic duct cells as well as stem cells are being explored for their potential to serve as β-cell sources. Large numbers of cells can be generated from β cell lines, although their unrestricted proliferation is also a serious concern in the context of cellular therapies. Unlike stem and progenitor cells, β islet cells have limited ability to multiply under normal conditions, although investigations into factors that stimulate β-cell regeneration have yielded promising results. Therapeutic molecules capable of increasing β-cell mass in vivo may eliminate the need for invasive surgical procedures. However, the extent of adult β cell regenerative capacity is unclear. Ex-vivo expansion of islets is indispensable prior to transplantation regardless of the source of islets. Monolayer culture of islets fail to maintain the three dimensional shape and integrity of islets resulting in consequent apoptosis and necrosis. *In vitro* culturing of pancreatic islets result in the loss of extracellular matrix (ECM), basement membrane and introduction of free radicals in isolation procedure, that cause islet cells to loose their three dimensional (3D) structure, function and ultimately undergo apoptosis (Parakevas et al., 1997). These problems could be alleviated by using tissue engineering principles and culturing pancreatic islets in 3 dimensional (3D) scaffold which can help in maintaining cell-cell, and cell-matrix interactions.

## 2. Islet transplantation

Islet cell transplantation is an effectual treatment for improving glycemic condition in diabetic patients thereby reducing the late complications of disease (Shapiro, 2003). The procedure involves isolating islets from a deceased organ donor, purifying, processing and infusing them into diabetic patients.

In the early 1970's, Dr. Clyde Barker at the University of Pennsylvania and the late Paul Lacy at the Washington University in St. Louis were the pioneers in exploring the concept of islet transplantation as a means to cure diabetes. In 1972, Ballinger and Lacy reported amelioration of diabetes in islet recipient rats (Ballinger & Lacy, 1972). In 1973, Reckard and Barker were the first to show that islet transplantation could completely and permanently restore normoglycemia in rodent models of chemically induced diabetes (Reckard & Barker, 1973).

Human islets isolation procedure is more complex than rodent islets isolation (Gray et al., 1984). Ricordi's automated isolation method had given hope for the production of abundant islets for the clinical use (Ricordi et al., 1989). Scharp et al. performed the islets transplantation under immunosuppression in diabetes patients and patients were insulin independent at the period of 22 days (Scharp et al., 1990). This was followed by several other cases, but success rates continued to be low (International Islet Transplant Registry). In 1999, Bretzel et al reported a markedly improved 3-month islet graft function rate of at least 75% in 24 consecutive patients (Bretzel et al., 1999). In the 1-year follow-up of 37 patients, 24% had achieved insulin independence (Bretzel et al., 2000). Between 1999 and 2005 about 650 patients were treated worldwide (Bretzel et al., 2007).Unfortunately, long-term results did not prove that promising.

The first successful islet allograft was reported in 1990 with steroid free immunosuppressant tacrolimus (Tzakis et al., 1990). The success rate of islet transplantation became outstanding after the Edmonton trial in 2000, which described successful intraportal alloislet transplantation, defined as insulin independence, in 7 consecutive patients with hyperlabile diabetes and frequent episodes of hypoglycemia (Shapiro et al., 2000). The success was partly ascribed to the usage of a steroid free immunosuppressive regimen which was a new combination of immunosuppressive drugs, consisting of sirolimus, tacrolimus and daclizumab, excluding the diabetogenic glucocorticoids and large numbers of donor islets (Shapiro et al., 2006).

The short-term results of islet transplantation with Edmonton trial were promising, with insulin independence in approximately 80% of patients at 1 year post-transplant but the proportion of insulin independent recipients declined after the first year post-transplant. Five year follow-up study after transplantation revealed that only 10% of the patients remained insulin independent (Ryan et al., 2005; Shapiro et al., 2006). The suggested reasons attributing for this decline include alloimmune rejection, autoimmune recurrence, and/or toxicity of immunosuppressive medications (Monti et al., 2008). However, about 80% were still C-peptide positive, indicating functioning grafts. Now, a slightly modified Edmonton protocol is used worldwide with reproducible results (Shapiro AM et al., 2006).

Though islet transplantation research has made significant progress, concern over toxicity as well as cost of immunosuppressive therapy still remains. Insulin independence and long term graft survival were achieved for more than three years through a modified immunosuppressive protocol (Bellin et al., 2008) even so the requirement of multiple donors to obtain 10000 islet equivalents per kilogram of patient's weight remains unsolved. Although insulin independence remains the ultimate goal, today, stabilization of glucose levels and avoidance of hypoglycemia are considered to be the main indications for islet transplantation.

## 3. Alternate sources of pancreatic β cells

The scarcity of donor pancreas for islet transplantation is a major obstacle to the widespread use of islet transplantation which urged the focus towards alternate sources of β cells for future transplants. Several alternative means have been suggested which include use of xenogenic islets and immortalized beta cell lines (Narushima et al., 2005). Recent advances in the field of stem cell differentiation and regeneration therapy have focused on new ways to generate insulin-producing beta cells that can be used for transplantation. Several candidate cells have been identified including embryonic stem cells (ESC) and adult stem cells or progenitor cells residing in the pancreas or other organs. The differentiated beta cells have shown to regenerate by replication, which opens the possibility to generate novel beta cells *in vitro* and / or *in vivo* from pre-existing beta cells. Additionally, there are reports that show the successful use of liver cells, endocrine cells from the gut, and bone marrow derived stem cells as source to generate islets by cell transdifferentiation.

### 3.1 Xenogenic islets

In a xenogenic approach, islets from different species are used for transplantation purpose. Porcine islets serve as a potential source in view of the fact that porcine insulin differ from human insulin by 1 amino acid. Neonatal porcine islets were also induced to mature endocrine phenotype under *in vitro* and *in vivo* conditions (Korbutt et al., 1997). Xenogenic tissues induce more vigorous rejection than that of allogenic tissue; hence immunosuppressant dosage should be high enough to prevent graft rejection. Alternately, the cells of xeno origin can be immunoisolated by encapsulation technology to separate the transplanted cells from host immune system which will be discussed later in this review.

Risks associated with transfer of endogenous virus from porcine cells to human genome (Patience et al., 1997; Van der Laan et al., 2000) and the public concerns regarding the use of animal organs for transplantation are major factors impeding the success of this approach in clinical applications.

### 3.2 Stem cells to treat diabetes

Stem cells are non specialized cells which have the ability to self regenerate and differentiate into specialized cell types depending on the niche or external signaling cues (Smith, 2006). Stem cells offer a limitless supply source for islets as well reduces the graft rejection problems (Street et al., 2004). Ideally stem cells used for cell based therapy should meet the following criteria (Gimble, 2003) : It should be available in abundant quantities (millions to billions of cells), harvest procedure should be less invasive, have multilineage differentiation potential and could be efficiently transplanted to the host.

#### 3.2.1 Embryonic Stem Cells (ESC)

Embryonic stem cells which are derived from the inner cell mass of pre-implantation blastocysts have gained the attention of researchers due to its pluripotent nature. Human embryonic stem cells (hES) hold promise for research and clinical applications. hES have some unique abilities as compared to all sources of adult cells: 1) the expansion of ESC in the undifferentiated state is nearly unlimited; and 2) ESC can give rise to all cell types including pancreatic insulin-producing beta cells. Attempts of directed differentiation of hES to cardiomyocytes (Klug et al., 1996), and neurons (Li et al., 1998) have been reported. Many studies have reported the differentiation of mouse and human embryonic stem cells to islet like clusters (Segev et al., 2004; Vaca et al., 2006) either by modifying the culture conditions or by genetic manipulation. Lineage tracing experiments revealed that beta cells are derived from the embryonic endoderm followed by a sequential and transient activation of specific transcription factors like Pdx1, NeuroD / Beta 2, Isl1, Nkx6.1, Nkx2.2, MafA, Pax4, and Pax6. Most published protocols aim to mimic these differentiation factors for the stepwise development of the endo and exocrine pancreas during the embryonic phase. In one approach ESC were induced to generate embryoid bodies under *in vitro* culture and the nestin positive cells were selected to differentiate towards $\beta$ cell lineage by culturing in serum free conditions (Lumelsky et al., 2001). Manipulation of the culture conditions with various growth supplements like insulin, transferrin, selenium and fibronectin (ITSFn), B27,bFGF and nicotinamide resulted in regulated secretion of insulin. Phosphoinositide kinase inhibitors have been reported to promote the differentiation of larger numbers of ESC towards functional $\beta$ cells (Hori et al., 2002). Genetic approach involves the incorporation of a reporter selector gene whose expression controlled by an insulin promoter. Here mouse embryonic stem cells were cultured in manipulated culture conditions to develop into embryoid bodies which were then differentiated to insulin producing cells. Insulin containing population which exhibited *in vitro* regulated hormone secretion was selected for transplantation into diabetic mice. Genetically engineered insulin producing cells maintained glucose homeostasis *in vivo* in mouse models but 40% of animals reversed to hyperglycemic condition after 12 weeks (Soria et al., 2000). ESC culture under serum-free conditions or only low-level fetal calf serum together with stage specific differentiation factors results in temporal expression of pancreatic lineage genes. The final differentiated cells do express insulin secretory granules / C peptide in about 2 - 8 % of the

total cell population and secrete insulin / C-peptide after glucose challenge (D'Amour et al., 2006; Baharvand et al., 2006; Jiang et al., 2007). Feeder free conditions were developed for the differentiation of embryonic stem cells to insulin like clusters which allowed single species propagation of ESC thus avoiding possible zoonotic infection of cells evident by the increased expression of Pdx 1 and reduced expression of Oct-3/4 (Assady et al., 2001). The activation of Pax-4 and Pdx 1 gene expression in embryonic stem cells upregulated the genes involved in the differentiation towards pancreatic endocrine lineage. Pax-4 expression led to an increase in nestin positive cells and insulin producing beta cells but not glucagon producing alpha cells (Moritoh et al., 2003). The development pathway for induction of ESC to insulin producing cells involves series of events which include generation of endoderm lineage cells followed by precursors of pancreatic endocrine lineage cells, and finally the insulin-secreting cells. D'Amour et al., developed a five stage protocol for differentiation of hES to pancreatic endocrine hormone expressing cells through a series of endodermal intermediates resembling those that occur during pancreatic development *in vivo* (D'Amour et al., 2006). Controversial findings also have been reported regarding the differentiation of ESC to pancreatic beta cells.
Insulin immunoreactivity was demonstrated to occur as the consequence of insulin uptake from medium (Rajagopal, J., 2003). Few authors proved that nestin positive cells tend to differentiate towards neuronal lineage rather than pancreatic lineage (Sipione et al., 2004). Kania et al explained the cause for the controversial results in generation of pancreatic cells. It was suggested that insulin producing cells derived from embryonic stem cells without applying lineage selection is dependent on the differentiation factors and protocols (Kania et al., 2004).

Though hES are versatile cells, ethical concerns on the use of human hES, and chances of teratoma formation (Fujikawa et al., 2005) limits their usage. Direct transplantation of embryonic stem cells has reported to culminate in teratoma formation (Nussbaum et al., 2007) from contaminating undifferentiated ESCs. Safe transplantation of hES could be attempted by viral vector mediated transfection *in vitro*, yet the risks associated with cytomegalovirus promoters in transfection cannot be ruled out.

### 3.2.2 Adult Stem Cells

The potential use of adult stem cells offers the advantage of an autologous model whereby a patient's own cells can be used, thereby circumventing immune rejection. Adult stem cells (ASC) are multipotent cells capable of self renewal. They have been reported to be present in almost every tissue like bone marrow, blood, heart, liver, pancreas, adipose tissue and could be transplanted directly without genetic modification or pre-treatments. They exhibit high degree of genomic stability during culture conditions. ASC lack tissue specific characteristics but it could be differentiated to specialized cell types under the influence of appropriate signaling cues (Barry & Murphy, 2004). The stem cell microenvironment plays an important role in its differentiation to committed cells (Galli et al., 2000; Zhao et al., 2002). The potential of adult human stem cells from various sources to differentiate to insulin producing cells have been explored by various research groups. The relative ease of isolating adult stem cells and their expansion makes it an ideal source for cell based therapy.

#### 3.2.2.1 Pancreatic stem cells

Pancreatic progenitor/stem cells which are closely related with beta cell lineage represent an attractive source for generation of beta cells (Serup et al., 2001). Human pancreatic ductal cells and islet stem cells have been expanded and differentiated to islet like clusters capable of producing insulin *in vitro* which were capable of reversing of diabetes in non obese diabetic mice thus normalizing blood glucose levels for more than 3 months (Ramiya et al., 2000). Nestin positive cell derived islet cell clusters expressed pancreatic endocrine markers like Glut2, glucagon, and homeodomain transcription factor PDX-1 as well as pancreatic exocrine genes (Zulewski et al., 2001). Glucagon like peptide -1, an intestinal hormone was shown to stimulate the neogenesis of beta cells by inducing the expression of various transcription factors involved in beta cell development (Abraham et al., 2002). Exocrine pancreatic tissue (Baeyens et al., 2005) and neurogenin 3 positive cells (Gu et al., 2002) could also serve as alternative source for beta cells. Even though pancreatic cells seem to be the better source than embryonic stem cells, the fraction of precursor cells isolated from pancreas is very less and heterogenous. Furthermore the harvest procedure from pancreas is also invasive thus limiting this source being applicable in clinical purposes.

#### 3.2.2.2 Bone marrow stem cells (BMSC)

Bone marrow stem cells were induced to differentiate to mature endocrine pancreatic lineage *in vitro* (Y. Sun et al., 2007). The *in vitro* differentiation of human bone marrow stem cells (hBMSC) to endocrine pancreatic cell types were investigated by genetic manipulation using adenovirus coding for mouse transcription factors involved in the early phase of endocrine developmental pathway (Karnielli et al., 2007). The results suggested that bone marrow stem cells shifted towards pancreatic endocrine phenotype with expression of insulin and other transcription factors involved in β cell development. Enhanced green fluorescent protein (GFP) system based genetic approach was utilized to study the differentiation of BMSC to islet like cells. BMSC from transgenic (GFP) male mice were transplanted into sublethally irradiated female mice. After 4 weeks 1.7-3% bone marrow derived GFP positive cells were found only in the pancreatic islets which ruled out the *in vivo* occurrence of cell fusion. The results indicated that bone marrow derived cells activated insulin gene expression after entering pancreatic islet niche (Ianus et al., 2003). However controversial finding was also reported suggesting that bone marrow derived stem cells cannot differentiate to beta cells without the influence of differentiation factors (Lechner et al., 2004) and favorable microenvironment (Moriscot et al., 2005). Recently Phadnis et al evaluated the fate of transplanted epithelialised human bone marrow stem cells under the kidney capsule of pancreatomized NOD/SCID mice (Phadnis et al., 2011).The results suggested that regenerating pancreas secreted paracrine factors and provided the niche which induced the differentiation of hBMSC to mature islet like aggregates capable of secreting insulin.

#### 3.2.2.3 Adipose stem cells

Human subcutaneous adipose tissue, abundant and easily accessible serves as a potential source of adult mesenchymal stem cells. The harvest procedure by lipoaspiration / liposuction is less invasive. Adipose stem cells have been reported to exhibit an increased *in vitro* proliferative potential than bone marrow stromal cells (De Ugarte et al., 2003). Adipose stem cells release cytokines TGF-β and IL-10 which are responsible for its immunomodulatory properties (Puissant et al., 2005). The immunosuppressive property of adipose stem cell has been exploited for the treatment of severe graft versus host disease (Yanez et al., 2006). The differentiation potential of these cells to pancreatic endocrine cells have been investigated by several research groups. Human adipose stem cells induced to islet like cells in serum free differentiation medium for 3 days exhibited an upregulation of pancreatic developmental transcription factors like Isl-1, Ngn3 along with islet hormones such as insulin, glucagon and somatostatin (Timper et al., 2006). A novel protocol using taurine designed for islet differentiation generated 47-51% C- peptide positive cells when compared to reports where the yield was only 2-8% (Jiang et al., 2007).

#### 3.2.2.4 Progenitor cells and stem cells from other tissues

The mechanisms involved in the generation of new beta cells in postnatal life remains controversial. Lineage tracing experiments suggest that after partial pancreatectomy, the majority of novel beta cells derive from pre-existing beta cells by beta cell proliferation rather than stem cell differentiation (Dor et al., 2004; Georgia and Bhushan, 2004). These findings raise new hope that it may be possible to use beta cells as source for the *in vitro* cell expansion in cell therapy. For example, it has been shown that human beta cells can transiently dedifferentiate to fibroblast-like cells, which can proliferate and redifferentiate into insulin-positive cells (Gershengorn et al., 2004; Lechner et al., 2005). However, direct evidence that this process can be used to produce fully mature beta cells for transplantation is missing, thus far.

It is important to note that the above mentioned experiments do not exclude the existence of pancreatic stem cells. There is convincing evidence that under specific experimental conditions adult stem cells, which reside in the pancreatic ducts, within islets, or exocrine tissue, can develop into beta cells. Bonner-Weir and colleagues as well as other groups provide evidence in several studies that murine and human stem cells or progenitor cells reside in the epithelium of the small pancreatic ducts, expressing the marker cytokeratin-19 (CK19). The ductal cells can expand and give rise to novel beta cells after 90% pancreatectomy or treatment with streptozotocin (Bonner-Weir et al., 1993; Wang et al., 1995; Bonner-Weir S et al., 2003; Gao et al., 2003). This suggests that still unknown cellular or soluble factors are needed to induce terminal differentiation into the endocrine fate. The identification of these critical factors is one focus of current research.

Amnion epithelial cells (Hou et al., 2008; Kadam et al., 2010a), and stem cells from liver (Yang et al., 2002), wharton's jelly of umbilical cord (Chao et al., 2008), umbilical cord blood (Phuc et al., 2010), placenta (Kadam et al., 2010b), gall bladder (Sahu et al., 2009) etc were also differentiated to insulin producing cells. In every case the differentiated islets expressed islet specific proteins and were capable of secreting insulin in response to glucose stimulation, albeit less than native islets. A better understanding of the events that control stem cell commitment and the signaling pathways for differentiation to pancreatic lineage is required to improve the culture conditions for *in vitro* generation of islet like clusters.

## 4. Tissue engineering based strategies

Tissue engineering applies the principles of cell transplantation, material science, and engineering towards the development of biologic substitutes that can restore and maintain normal function. The success of tissue engineering relies on the development of a suitable scaffold, cell source and defined culture conditions. Tissue engineering strategies employed for islet transplantation could be categorized into use of scaffolds for simulation of the native ECM and immunoisolation via encapsulation of islets.

The biomaterial chosen to synthesize the scaffold should be biocompatible and biodegradable (Mohan & Nair, 2005). The scaffold should reproduce an extracellular environment for supporting cell function (Dufour et al., 2005). Neither the scaffold nor its degradation products should be toxic to host. Biocompatibility of chosen biomaterial plays an important role in controlling the rate of protein adsorption and fibrosis. The scaffold should provide three dimensional space for neotisssue formation and the mechanical strength should match the native tissue to withstand *in vivo* forces. It should be highly porous and pores should have interconnectivity to facilitate tissue ingrowth, diffusion of nutrients, oxygen and metabolites (Blomeier et al., 2006). Scaffolds should provide an initial support for the islet grafts during the early post transplant period, enabling the development of a capillary bed and recovery of extracellular matrix interactions (Stock & Vacanti, 2009). Polymers such as polylactic acid, polyglycolic acid, polylactide-co-polyglycolide (PLGA) (Mao et al., 2009), polyvinyl alcohol – polycaprolactone (Mohan et al., 2010) were extensively used for scaffold fabrication purpose for the tissue engineering of various tissues.

### 4.1 Extracellular matrix - Mimicking nature

A major limitation with two dimensional cultures is the lack of microenvironment indispensable for appropriate spatial organization and function of cells which in turn is provided by extracellular matrix (ECM) in native tissue. The ECM of native islets is mainly constituted by type IV collagen, laminin although fibronectin, collagen I, collagen V also have been detected (Stendahl et al., 2009). The effect of ECM proteins on adhesion and proliferation of rat islets have been studied. The results indicated that there were strong interactions between islets and ECM proteins via integrins (Chen et al., 2008). ECM protein coated scaffolds have shown improved graft efficacy at implanted site (Salvay et al., 2008).

Earlier reports of islets cultured under two dimensional conditions exhibited low survival rate and reduced insulin secretion (Paraskevas et al., 1997). Progress in survivability of islets and increased insulin secretion has been achieved by adopting 3D culture conditions. Polyglycolic acid scaffolds coated with lysine were shown to promote islet cell adhesion and survival. The control cells grown in 2D culture underwent apoptosis by day 7 due to accumulation of metabolites and shortage of nutrients (Chun et al., 2008).

Pancreatic islets cultured on agarose cryogel sponge (Bloch et al., 2005) were reported to secrete 15 fold higher insulin at 3mM glucose than islets cultured on polystyrene cell culture dish but failed to respond to stimulation at 16.7mM glucose. The failure was due to limited oxygen and nutrient diffusion across agarose cryogels. Adequate oxygen is a critical parameter to islet cell function and survival (Sweet et al., 2002). The decreased insulin response of pancreatic islets cultured on scaffold for prolonged period due to inefficient oxygen transfer were also reported (Cui et al., 2001).

A novel 3D culture system comprising a semi-interpenetrating polymer network of gelatin and polyvinyl pyrrolidone was designed in our laboratory for pancreatic islets and stem cells to promote their survival and function. The porous nature of the scaffold facilitated efficient nutrient and metabolite exchange. Islets seeded on this scaffold maintained their morphology for more than 30 days whereas control islets cultured on cell culture dish underwent apoptosis by 7th day. The test islets secreted insulin on stimulation with glucose which was comparable to that of freshly isolated mouse islets (Muthyala et al.,2010).

Zhao et al demonstrated the use of three dimensional self assembling peptide nanofiber hydrogel scaffold for islet culture. The peptides formed two beta sheet structures with hydrophilic and hydrophobic surfaces in aqueous solution. The hydrophobic moiety facilitated its self assembly in water and the nanofiber structures were flexible for fabrication to different geometrical shapes that allowed for efficient nutrient and metabolite transfer. The nanofiber scaffold simulated the microenvironment *in vitro* as in native condition which accounted for improved islet viability and function (Zhao et al., 2010).

### 4.2 Immunoisolation strategies for islet transplantation

The principle behind immunoisolation is protection of islets from host immune system using a selectively permeable membrane as a barrier. Low molecular weight substances which include nutrients, oxygen, secretory molecules and cell signaling molecules freely diffuse through the membrane, but passage of immune cells and its products which have high molecular weight is prevented. Immunoisolation mechanism encourages the use of allogenic/ xenogenic sources of islets for transplantation and holds promise towards use of autologous stem cell derived islets in type I diabetic patients. Immunoisolation mechanism includes macroencapsulation and microencapsulation (Narang & Mahato, 2006) of cells.

#### 4.2.1 Microencapsulation

Microencapsulation is the encapsulation of single islets or small groups of islets. These capsules are usually spherical in shape (Chang, 1964). Microcapsules offer the advantage of increased oxygen and nutrient transport due to the large surface area to volume ratio. Microcapsules are advantageous due to several reasons like greater surface to volume ratio, and ease of implantation. The spherical shapes owe to better diffusion capacity and are mechanically stable. The primary drawback of microencapsulation is the difficulty in removing the implants if necessary. Moreover the implantation could be achieved by simple injection procedure (De Vos et al., 2002). Porcine islets microencapsulated in alginate-polylysine-alginate transplanted to diabetic monkeys could achieve normoglycemia without immunosuppression for more than 800 days (Y. Sun et al., 1996). Human and rat islets encapsulated in alginate gels when transplanted in mice survived for 7 months (Schneider et al., 2005). Xenogenic islets immobilized in microcapsules fabricated from alginate -PLL when implanted into peritoneum of non immunosuppressed diabetic rats remained in excellent condition for more than 40 weeks (Lanza et al., 1999). Despite of these advantages some authors have reported reduced functionality of microencapsulated islets in response to glucose challenge (Sandler et al., 1997).

#### 4.2.2 Macroencapsulation

Macroencapsules contain a large mass of islet cells within a diffusion chamber, which are usually formed from spun coat membranes or spun drawn hollow fibers. The advantages of macrocapsules are they could be easily retrieved when required and can be shaped in required geometries such as tubes or discs. Two approaches such as intravascular and extravascular have been tried out in macroencapsulation. Intravascular approach utilizes the principle of perfusion chambers which consists of microporous tubular structures perfused with blood and enclosed within another tube. Islets were seeded in the space between the hollow fibers and the device is anastomised to the host vasculature (Chick et al, 1975). Polyacrylonitrile and polyvinylchloride copolymers have been chosen as materials for creating artificial microcapillaries. Results from implantation of intravascular macrocapsules of islets have shown restoration of normoglycemia in various animal models (AM. Sun et al., 1977). Due to the direct contact of device with the blood, intense anticoagulation is required to prevent thrombus formation, consequently the material chosen should be highly blood compatible and thromboresistant. These concerns have shifted the attention towards extravascular devices.

Extravascular devices are based on the principle of diffusion chambers which does not require anastomosis to host vasculature. The geometry could be planar in the form of flat or hollow fiber model (Scharp et al., 1984). This approach does not pose severe biocompatibility issues and risks to the patient as that of intravascular devices. Extravascular devices can be implanted to different sites such as peritoneal cavity (Lanza et al., 1999), subcutaneously or under kidney capsules (Siebers et al., 1990) with minimal surgical risks. Most commonly used biomaterials for macrocapsule fabrication are nitrocellulose acetate, 2-hydroxyethyl methacrylate (HEMA), acrylonitrile, polyacrylonitrile and polyvinylchloride copolymer, and alginate.

The biocompatibility of immunoisolation membrane depends on several factors like geometry of the device, implantation site and material chosen. Hollow fiber geometry is preferred because of its reduced surface area of contact with the host per islet and reduced foreign body response. Higher density of islet cells often results in reduced viability and necrosis at the center due to nutrient limitation. Smooth outer surface and hydrogels have been reported to improve the biocompatibility by the absence of interfacial tension, thus reducing protein adsorption, cell adhesion and fibrosis (Burczak et al., 1996). Nair et al studied the effect of degree of hydrophilicity on tissue response of polyurethane interpenetrating networks (IPN) (Nair et al., 1992).The results indicated that an increase in hydrophilicity of polyurethane -polyvinyl pyrrolidone IPN's elicited an inert tissue response.

George et al., (2002), Nair (2009, Indian Patent 230740) demonstrated the use of non porous polyurethane membranes and porous polyurethane IPN macrocapsules as islet immunoisolative matrices. Islet cell morphology remained intact and insulin secretion ability was also retained within the immunoislation membranes. Membranes allowed diffusion of glucose and insulin while retained transplant rejection factors like antibodies, immunoglobulins and immune cells. Reduced protein adsorption and cell adhesion on polyurethane membranes contributed to improve the biocompatibility which made them ideal for immunoisolation. The IPN macrocapsules also served as an *in vivo* bioreactor cum immunoisolation device permitting immature islet like clusters derived from a variety of stem cell sources to mature completely and control glycemic levels of streptozotocin induced diabetic animal models without immunosuppression for periods upto 3 months. (Kadam, 2010a; Phadnis, 2011; Muthyala, 2011;; Kadam, 2010 b).Hybrid systems involving macro and microencapsulation have also been fabricated and analyzed for its efficiency in immunoisolation. Chitosan/gelatin hydrogel system was used as an immunoisolative matrix to protect the microencapsulated islet cells from recipient's immune system in xenotransplantation. Mouse insulinoma /agarose microspheres macroencapsulated in chitosan/gelatin hydrogel reversed diabetes in rats. The study suggests that this could be applied as a cell carrier for injectable bioartificial pancreas after certain modifications (Yang et al., 2008).

### 4.3 Combined approach of tissue engineering and immunoisolation

Muthyala et al (2011) employed a combination approach utilising the properties of scaffold to mimic the native ECM and macroencapsulation for immunoisolation to protect the islets from immune cell responses. Pancreatic progenitor derived islets were seeded on gelatin-polyvinylpyrrolidone scaffolds and further macroencapsulated in a polyurethane-polyvinylpyrrolidone semi IPN macrocapsule. The construct when implanted into peritoneal cavity of diabetic rats normalized glycemic condition all through the study period of 3 months. Animals implanted with tissue engineered islets without macroencapsulation showed no reversal of hyperglycemia and died within 15-20 days due to infiltration of host immune cells. Hence the combination approach was found to be very effective in achieving euglycemia by maintaining islet survival and functionality as well as protecting the cells from host immune attack.

## 5. Site for transplantation

The optimal site should be chosen for transplanting islets inorder to meet its high energy requirement and metabolic rate (Hardy et al., 2010). The implantation site has effect on hypoxic conditions which determines islet survival. Safety considerations have been raised regarding the optimal site for transplantation so as to improve islet engraftment and survival (Dufrane et al., 2006; Pillegi et al., 2001). Graft vascularization is an important criterion in islet survival and function (Jansson & Carlsson, 2002). Although immunoisolation prevents the integration of host blood vessels with transplanted islets effective diffusion of nutrients and oxygen can occur within 200μm distances hence highly vascularised sites should be chosen for transplantation. Islet transplantation into prevascularized sites dramatically improves graft survival and function relative to transplantation into non-modified tissue (Balamurugan et al., 2003). Vascularization can be introduced in graft by incorporation of angiogenic growth factors like VEGF (Stendahl et al., 2008) or endothelial cells (Miki et al., 2006). Insulin independence have been achieved by intra-portal islet transplantation in diabetic patients (Shapiro et al., 2005), however liver could not be considered as an optimal site since islets in liver will be exposed to high nutrient concentration and other factors that are toxic and may result in impairment of beta cells(Hiller et al., 1991; Wilson & Chaikof, 2008). Peritoneal cavity has also been tried for implantation of islets which requires 200%-400% more islets (Siebers et al., 1993). Subcutaneous site (Pillegi et al., 2006) have been chosen in diabetic athymic mice for transplantation of islets cultured on plasma –fibroblast gel scaffold (Perez-Basterrechea et al., 2009). Normoglycemia was achieved over 60 day period and vascularization was observed in and around islets. Kidney subcapsular spaces have also been chosen as implantation site to improve biocompatibility of tissue engineered constructs. Islets cultured on biodegradable polymer scaffold transplanted to omental pouch of diabetic dogs resulted in achievement of euglycemia upto 152 days till graft was taken out (Kin et al., 2008).

## 6. Conclusions

Curative therapy for diabetes mellitus mainly implies replacement of functional insulin-producing pancreatic cells, with pancreas or islet-cell transplants. Shortage of donor organs spurs research into alternative means of generating cells from islet expansion, encapsulated islet xenografts, human islet cell-lines, and stem cells. Embryonic and adult stem cells are potential sources for cell replacement and merit further scientific investigation. The expense of the benefit of cell transplantation is the need for immunosuppressive treatment of the recipient, with all its potential risks. Biocompatible macrocapsules for transplantation of islets and islet-like cell clusters differentiated from stem cells help overcome the immune rejection without Immunosuppressive drug therapy. A tissue engineering approach aims to mimic the natural extracellular matrix environment for supporting the transplanted islet cells without sacrificing form and function. A combination approach of tissue engineering, immunoisolation and most appropriately differentiated islet may propel clinical trials involving engineered strategies for cell replacement in diabetic patients, in the not too distant future.

## 7. Acknowledgment

Authors are grateful to Director, SCTIMST and Head BMT Wing for permissions to publish this work and funding from the DBT, Govt. of India. N. Aloysious also acknowledges the CSIR for the JRF fellowship.

## 8. References

Abraham, EJ.; Leech, CA.; Lin, JC.; Zulewski, H. & Habener, JF. (2002). Insulinotropic Hormone Glucagon-Like Peptide-1 Differentiation of Human Pancreatic Islet-Derived Progenitor Cells into Insulin-Producing Cells. *Endocrinology*, Vol.143, No.8, pp. 3152-3161

Assady, S.; Maor, G.; Amit, M.; Itskovitz-Eldor, J.; Skorecki, KL. & Tzukerman, M. (2001). Insulin production by human embryonic stem cells. *Diabetes*, Vol.50, No. 8, pp. 1691–1697

Baeyens, L.; De Breuck, S.; Lardon, J.; Mfopou, JK.; Rooman, I. & Bouwens, L. (2005). *In vitro* generation of insulin-producing beta cells from adult exocrine pancreatic cells. *Diabetologia*, Vol.48, No.1, pp. 49–57

Baharvand, H.; Jafary, H.; Massumi, M. & Ashtiani, SK. (2006). Generation of insulin secreting cells from human embryonic stem cells. *Development, Growth & Differentiation*, Vol.48, No.5, pp. 323-32

Balamurugan, AN.; Gu, Y.; Tabata, Y.; Miyamoto, M.; Cui, W.; Hori, H.; Satake, A.; Nagata, N.; Wang, W. & Inoue, K. (2003). Bioartificial pancreas transplantation at prevascularized intermuscular space: effect of angio-genesis induction on islet survival. *Pancreas*, Vol.26, No.3, pp. 279–285

Ballilnger, WF. & Lacy PE. (1972). Transplantation of intact pancreatic islets in rats. *Surgery*,Vol. 72, No.2, pp. 175-186

Barry FP. & Murphy, JM. (2004). Mesenchymal stem cells: clinical applications and biological characterization. *The International Journal of Biochemistry & Cell Biology*, Vol.36, No.4, pp. 568–584

Bellin, MD.; Kandaswamy, R.; Parkey, J.; Zhang, HJ.; Liu, B.; Ihm, SH.; Ansite, JD.; Witson, J.; Bansal-Pakala, P.; Balamurugan, AN.; Papas, K.; Sutherland, DE.; Moran, A. & Hering, BJ. (2008). Prolonged insulin independence after islet allotransplants in recipients with type 1 diabetes. *American Journal of Transplantation*, Vol. 8, No.11, pp. 2463-2470

Bloch, K.; Lozinsky, VI.; Galaev, IY.; Yavriyanz, K.; Vorobeychik, M.; Azarov, D.; Damshkaln, LG.; Mattiasson, B. & Vardi, P. (2005). Functional activity of insulinoma cells (INS-1E) and pancreatic islets cultured in agarose cryogel sponges. *Journal of Biomedical Materials Research Part A*, Vol.75, No.4, pp. 802–809

Blomeier, H.; Zhang, X.; Rives, C.; Brissova, M.; Hughes, E.; Baker, M.; Powers, AC.; Kaufman, DB.; Shea LD. & Lowe, WL Jr. (2006). Polymer Scaffolds as Synthetic Microenvironments for Extrahepatic Islet Transplantation. *Transplantation*, Vol.82, No.4, pp. 452–459

Bonner-Weir, S.; Baxter, LA.; Schuppin, GT. & Smith, FE. (1993). A second pathway for regeneration of adult exocrine and endocrine pancreas. A possible recapitulation of embryonic development . *Diabetes*, Vol. 42, No.12, pp. 1715 - 1720

Bonner-Weir, S.; Taneja, M.; Weir, GC.; Tatarkiewicz, K.; Song, KH.; Sharma, A. & O ' Neil, JJ. (2000). *In vitro* cultivation of human islets from expanded ductal tissue. *Proceedings of the National Academy of Sciences* USA , Vol.97, Vol.14, pp. 7999 - 8004

Bretzel, RG. (2000). Current status and perspectives in clinical islet transplantation. *Journal of Hepato-Biliary-Pancreatic Surgery*, Vol.7, No.4, pp. 370-373

Bretzel, RG.; Brandhorst, D.; Brandhorst, H.; et al. (1999). Improved survival of intraportal pancreatic islet cell allografts in patients with type-1 diabetes mellitus by refined peritransplant management. *Journal of Molecular Medicine*, Vol.77, No.1, pp. 1432-1440

Bretzel, RG.; Jahr, H.; Eckhard, M.; Martin, I,.; Winter, D. & Brendel, MD. (2007). Islet cell transplantation today. *Langenbeck's Archives of Surgery*, Vol.392, No.3, pp. 239-253

Burczak, K.; Gamian, E. & Kochman, A. (1996). Long-term in vivo performance and biocompatibility of poly(viny1 alcohol) hydrogel macrocapsules for hybrid type artificial pancreas. *Biomaterials*, Vol. 17, No. 24, pp. 2351-2356

Chandra, VGS.; Phadnis, S.; Nair, PD. & Bhonde, RR. (2009). Generation of Pancreatic Hormone-Expressing Islet-Like Cell Aggregates from Murine Adipose Tissue-Derived Stem Cells. *Stem Cells*, Vol.27, No. 8, pp. 1941–1953

Chang TM. Semipermeable microcapsules. (1964). *Science*, Vol.146, pp. 524-525

Chao, KC.; Chao, KF.; Fu, YS. & Liu, SH. (2008). Islet-Like Clusters Derived from Mesenchymal Stem Cells in Wharton's Jelly of the Human Umbilical Cord for Transplantation to Control Type 1 Diabetes. *Plos One*, Vol.3, No.1, e1451

Chen, G.; Kawazoe, N. & Tateishi, T. (2008). Effects of ECM Proteins and Cationic Polymers on the Adhesion and Proliferation of Rat Islet Cells. *The Open Biotechnology Journal*, Vol.2, pp. 133-137

Chick, WL.; Like, AA. & Lauris, V. (1975). Beta cell. culture on synthetic capillaries:an artificial endocrine pancreas. *Science*, Vol.187, No.4179, pp. 847-849

Chun, S.; Huang, Y.; Xie, WJ.; Hou, Y.; Huang, RP.; Song, YM.; Liu, XM.; Zheng, W.; Shi, Y. & Song, CF. (2008). Adhesive Growth of Pancreatic Islet Cells on a Polyglycolic Acid Fibrous Scaffold. *Transplantation Proceedings*, Vol.40, No.5, pp. 1658-1663

Cui, W.; Kim, DH.; Imamura, M.; Hyon, SH. & Inoue, K. (2001). Tissue engineered pancreatic islets: culturing rat islets in the chitosan sponge. *Cell Transplantation*, Vol.10, No.4-5, pp. 499 -502

Dall, TM.; Zhang, Y.; Chen, YJ.; Quick, WW.; Yang, WG. & Fogli J. (2010). The economic burden of diabetes. *Health Affairs*, Vol.29, No.2, pp. 297–303

D'Amour, KA.; Bang, AG.; Eliazer, S.; Kelly, OG.; Agulnick, AD.; Smart, NG.; Moorman, MA.; Kroon, E.; Carpenter, MK. & Baetge, EE. (2006). Production of pancreatic hormone- expressing endocrine cells from human embryonic stem cells. *Nature Biotechnology*, Vol.24, No.11, pp. 1392-1401

De Ugarte, DA.; Morizono, K.; Elbarbary, A.; Alfonso, Z.; Zuk, PA.; Zhu, M.; Dragoo, JL.; Ashjian, P.; Thomas, B.; Benhaim, P.; Chen,I.; Fraser, J. & Hedrick, M.H. (2003).

Comparison of multi-lineage cells from human adipose tissue and bone marrow. *Cells Tissues Organs*, Vol.174, No.3, pp. 101–109

De Vos, P.; Hamel, AF. & Tatarkiewicz, K. (2002). Considerations for successful transplantation of encapsulated pancreatic islets. *Diabetologia*, Vol.45, No.2, pp. 159-173

Dor, Y.; Brown, J.; Martinez, OI. & Melton, DA . (2004). Adult pancreatic beta-cells are formed by self-duplication rather than stem-cell differentiation . *Nature,* Vol.429, No.6987, pp. 41 - 46

Dufour, JM.; Rajotte, RV.; Zimmerman, M.; Rezania, A.; Kin, T.; Dixon, DE. & Korbutt, GS. (2005). Development of an ectopic site for islet transplantation, using biodegradable scaffolds. *Tissue Engineering*, Vol.11, No.9/10, pp. 1323-1331

Dufrane, D.; Steenberghe, M.; Goebbels, RM.; Saliez, A.; Guiot, Y. & Gianello, P. (2006). The influence of implantation site on the biocompatibility and survival of alginate encapsulated pig islets in rats. *Biomaterials*, Vol.27, No.17, pp. 3201-3208

Fujikawa, T.; Oh, SH.; Pi, L.; Hatch, HM.; Shupe, T. & Petersen, BE. (2005). Teratoma formation leads to failure of treatment for type I diabetes using embryonic stem cell-derived insulin-producing cells. *The American Journal of Pathology*, Vol.166, No.6, pp. 1781–1791

Galli, R,.;Borello, U.; Gritti, A.; Minasi, MG.; Bjornson, C.; Coletta, M.; Mora, M.; De Angelis, MG.; Fiocco, R.; Cossu, G. & Vescovi, AL. (2000). Skeletal myogenic potential of human and mouse neural stem cells. *Nature Neuroscience*, Vol.3, No.10, pp. 986–991

Gao, R.; Ustinov, J.; Pulkkinen, MA.; Lundin, K.; Korsgren, O. & Otonkoski, T. (2003). Characterization of endocrine progenitor cells and critical factors for their differentiation in human adult pancreatic cell culture . *Diabetes*, Vol.52, No.8, pp. 2007 - 2015

George, S.; Nair, PD.; Risbud, MV. & Bhonde, RR. (2002). Nonporous Polyurethane Membranes as Islet Immunoisolation Matrices - Biocompatibility Studies. *Journal of Biomaterial Applications*, Vol. 16, No. 4, pp. 327-340

Georgia, S. & Bhushan, A . (2004). Beta cell replication is the primary mechanism for maintaining postnatal beta cell mass . *Journal of Clinical Investigation*, Vol.114, No.7, pp. 963 - 968

Gershengorn, MC.; Hardikar, AA.; Wei, C.; Geras-Raaka, E.; Marcus-Samuels, M. & Raaka, MB. (2004). Epithelial-to-mesenchymal transition generates proliferative human islet precursor cells . *Science*, Vol.306, No.5705, pp. 2261 - 2264

Gimble JM. (2003). Adipose tissue-derived therapeutics. *Expert Opinion on Biological Therapy*, Vol.3, No.5, pp. 705-713

Gray, DW.; McShane, P.; Grant, A. & Morris, PJ. (1984). A method for isolation of islets of Langerhans from the human pancreas. *Diabetes*, Vol.33, No.11, pp. 1055-61

Gu, G.; Dubauskaite, J. & Melton, DA. (2002). Direct evidence for the pancreatic lineage: Ngn3+ cells are islet progenitors and are distinct from duct progenitors. *Development*, Vol.129, No.10, pp. 2447–2457

Hardy, MA.; Witkowski, P.; Sondermeijer, H. & Harris, P. (2010). The Long Road to Pancreatic Islet Transplantation. *World Journal of Surgery*, Vol.34, No.4, pp. 625–627

Hiller, W.F.; Klempnauer, J.; Luck, R. & Steiniger, B. (1991). Progressive deterioration of endocrine function after intraportal but not kidney subcapsular rat islet transplantation. *Diabetes*, Vol.40, No.1, pp. 134-140

Hori, Y.; Rulifson, IC.; Tsai, BC.; Heit, JJ.; Cahoy, JD. & Kim, SK. (2002). Growth inhibitors promote differentiation of insulin-producing tissue from embryonic stem cells. *Proceedings of the National Academy of Sciences of the USA*, Vol.99, No.25, pp. 16105–16110

Hou, Y.; Huang, Q.; Liu, T. & Guo, L. (2008). Human amnion epithelial cells can be induced to differentiate into functional insulin-producing cells. *Acta Biochimica et Biophysica Sinica*, Vol.40, No.9, pp. 830-839

Ianus, A.; Holz, GG.; Theise, ND. & Hussain, MA. (2003). In vivo derivation of glucose-competent pancreatic endocrine cells from bone marrow without evidence of cell fusion. *The Journal of Clinical Investigation*, Vol.111, No.6, pp. 843-850

Jansson, L. & Carlsson, PO. (2002). Graft vascular function after transplantation of pancreatic islets. *Diabetologia*, Vol.45, No.6, pp. 749-763

Jiang, J.; Au, M.; Lu, K.; Eshpeter, A.; Korbutt, G.; Fisk, G. & Majumdar, AS. (2007). Generation of insulin producing islet like clusters from human embryonic stem cells. *Stem cells*, Vol.25, No.8, pp. 1940-1953

Kadam, S.; Muthyala, S.; Nair, P. & Bhonde, R. (2010a). Reversal of experimental diabetes in mice by transplantation of neo-islets generated from human amnion--derived mesenchymal stromal cells using immuno-isolatory macrocapsules. *Cytotherapy*, Vol.12, No.8, pp. 982-991

Kadam, S.; Muthyala, S.; Nair, P. & Bhonde, R. (2010b). Human Placenta-Derived Mesenchymal Stem Cells and Islet-Like Cell Clusters Generated From These Cells as Novel Sources for Stem Cell Therapy in Diabetes. *The Review of Diabetic Studies*, Vol.7, No.2, pp. 168-182

Kania, G.; Blyszczuk, P. & Wobus, AM. (2004). The generation of insulin-producing cells from embryonic stem cells - a discussion of controversial findings. *International Journal of Developmental Biology*, Vol.48, No.10, pp. 1061-1064

Karnieli, O.; Izhar-Prato, Y.; Bulvik, S. & Efrat, S. (2007). Generation of insulin-producing cells from human bone marrow mesenchymal stem cells by genetic manipulation. *Stem cells*, Vol.25, No.11, pp. 2837-2844

Kin, T.; O'Neil, JJ.; Pawlick, R.; Korbutt, GS.; Shapiro, AM. & Lakey, JR. (2008). The Use of an Approved Biodegradable Polymer Scaffold as a Solid Support System for Improvement of Islet Engraftment. *Artificial Organs*, Vol.32, No.12, pp. 990-993

Kizilel, S.; Garfinkel, M. & Opara, E. (2005). The bioartificial pancreas: progress and challenges. *Diabetes Technology & Therapeutics*, Vol.7, No.6, pp. 968-985

Klug, MG.; Soonpaa, MH.; Koh, GY. & Field, LJ. (1996). Genetically selected cardiomyocytes from differentiating embryonic stem cells form stable intracardiac grafts. The *Journal of Clinical Investigation*, Vol.98, No.1, pp. 216–224

Korbutt, GS.; Ao, Z.; Flashner, M. & Rajotte, RV. (1997). Neonatal porcine islets as a possible source of tissue for humans and microencapsulation improves the metabolic response of islet graft post transplantation. *Annals of the New York Academy of Sciences*, Vol. 831, No. 1, pp. 294-303

Lanza, RP.; A M Beyer, AM. & Chick, WL. (1999). Xenogenic humoral responses to islets transplanted in biohybrid diffusion chambers. *Transplantation*, Vol. 57, No.9, pp. 1371-1375

Lechner, A.; Nolan, AL.; Blacken, RA. & Habener, JF. (2005). Redifferentiation of insulin secreting cells after *in vitro* expansion of adult human pancreatic islet tissue. *Biochemical and Biophysical Research Communications*, Vol 327, No.2, pp. 581 – 588

Lechner, A.; Yang, YG.; Blacken, RA.; Wang, L.; Nolan, AL. & Habener, JF. (2004). No evidence for significant transdifferentiation of bone marrow into pancreatic beta-cells in vivo. *Diabetes*, Vol.53, No.3, pp. 616-623

Li, M.; Pevny, L.; Lovell-Badge, R. & Smith, A. (1998). Generation of purified neural precursors from embryonic stem cells by lineage selection. *Current Biology*, Vol.8, No.17, pp. 971–974

Lumelsky, N.; Blondel, O.; Laeng, P.; Velasco, I.; Ravin, R. & McKay, R. (2001). Differentiation of embryonic stem cells to insulin-secreting structures similar to pancreatic islets. *Science*, Vol. 292, No.5520, pp. 1389–1394

Mao, GH.; Chen, GA.; Bai, HY.; Song, TR. & Wang, YX. (2009). The reversal of hyperglycaemia in diabetic mice using PLGA scaffolds seeded with islet-like cells derived from human embryonic stem cells. *Biomaterials*, Vol.30, No.9, pp. 1706-1714

Miki, A.; Rivas-Carrillo, JD.; Navarro-Alvarez, N.; Soto-Gutierrez, A.; Chen, Y.;Tanaka, K.; Narushima, M.; Tabata, Y.; Okitsu, T.; Noguchi, H.; Matsumoto, S.; Tanaka, N. & Kobayashi, N. (2006). Maintenance of neovascularization at the implantation site of an artificial device by bFGF and endothelial cell transplant. *Cell Transplantation*, Vol.15, No.10, pp. 893–901

Mikos, A. G., Papadaki, M. G., Kouvroukoglou, S., Ishaug, S. L. and Thomson, R. C. (1994), Mini-review: Islet transplantation to create a bioartificial pancreas. Biotechnology and Bioengineering, 43: 673–677Lavik, E. & Langer, R. (2004). Tissue engineering: current state and perspectives. *Applied Microbiolology and Biotechnology*, Vol.65, No.1, pp. 1–8

Mohan, N. & Nair, PD. (2005). Novel Porous, Polysaccharide Scaffolds for Tissue Engineering Applications. *Trends* in *Biomaterials* & *Artificial Organs*. Vol.18, No.2, pp. 219-224

Mohan, N.; Nair, PD. & Tabata, Y. (2010). Growth factor-mediated effects on chondrogenic differentiation of mesenchymal stem cells in 3D semi-IPN poly(vinyl alcohol)–poly(caprolactone) scaffolds. *Journal of Biomedical Materials Research Part A*, Vol.94, No.1, pp. 146–159

Monti, P.; Scirpoli, M.; Maffi, P.; Ghidoli, N.; De Taddeo, F.; Bertuzzi, F.; Piemonti, L.; Falcone, M.; Secchi, A. & Bonifacio, E. (2008). Islet transplantation in patients with autoimmune diabetes induces homeostatic cytokines that expand autoreactive memory T cells. *The Journal of Clinical Investigation*, Vol.118, No.5, pp. 1806-1814

Moriscot, C.; de Fraipont, F.; Richard, MJ.; Marchand, M.; Savatier, P.; Bosco, D.; Favrot, M. & Benhamou, PY. (2005). Human bone marrow mesenchymal stem cells can express insulin and key transcription factors of the endocrine pancreas developmental pathway upon genetic and/or microenvironmental manipulation *in vitro*. *Stem Cells*, Vol.23, No.4, pp. 594–603

Moritoh, Y.; Yamato, E.; Yasui, Y.; Miyazaki, S. & Miyazaki, J. (2003). Analysis of Insulin-Producing Cells During *In vitro* Differentiation From Feeder-Free Embryonic Stem Cells. *Diabetes*, Vol.52, No.5, pp. 1163-1168

Muthyala, S.; Bhonde, RR. & Nair, PD. (2010). Cytocompatibility studies of mouse pancreatic islets on gelatin - PVP semi IPN scaffolds *in vitro*: Potential implication towards pancreatic tissue engineering. *Islets*, Vol.2, No.6. pp. 357-366

Nair, PD.; Mohanty, M.; Rathinam, K.; Jayabalan, M. & V.N. Krishnamurthy, VN. (1992). Studies on the effect of degree of hydrophilicity on tissue response of Polyurethane interpenetrating polymer networks. *Biomaterials*, Vol. 13, No.8, pp. 537-542.

Nair, PD. (2009) A process for the preparation of a biocompatible polymeric composition of an inter-penetrating polymeric network (IPN), Indian patent 230740

Narang, AS. & Mahato, RI. (2006). Biological and Biomaterial Approaches for Improved Islet Transplantation. *Pharmacological Reviews*, Vol.58, No.2, pp. 194–243

Narushima, M.; Kobayashi, N.; Okitsu, T.; Tanaka, Y.; Li, SA.; Chen, Y.; Miki, A.; Tanaka, K.; Nakaji, S.; Takei, K.; Gutierrez, AS.; Rivas-Carrillo, JD.; Navarro-Alvarez, N.; Jun, HS.; Westerman, KA,.; Noguchi, H.; Lakey, JR.; Leboulch, P.; Tanaka, N. & Yoon, JW. (2005). A human beta cell line for transplantation therapy to control type 1 diabetes. *Nature Biotechnology*, Vol. 23, No.10, pp. 1274 - 1282

Nussbaum, J.; Minami, E.; Laflamme, MA.; Virag, JA.; Ware, CB.; Masino, A.; Muskheli, V.; Pabon, L.; Reinecke, H. & Murry, CE. (2007). Transplantation of undifferentiated murine embryonic stem cells in the heart: teratoma formation and immune response. *The FASEB Journal*, Vol.21, No.7, pp. 1345–1357

Paraskevas, S.; Duguid, WP.; Maysinger, D.; Feldman, L.; Agapitos, D. & Rosenberg, L. (1997). Apoptosis Occurs in Freshly Isolated Human Islets Under Standard Culture Conditions. *Transplantation Proceedings*, Vol.29, No.1-2, pp.750-752

Patience, C.; Takeuchi, Y. & Weiss, RA. (1997). Infection of human cells by an endogenous retrovirus of pigs. *Nature Medicine*, Vol.3, No.3, pp. 282-286

Perez-Basterrechea, M.; Briones, RM.; Alvarez-Viejo, M.; Garcia-Perez, E.; Esteban. MM.; Garcia, V.; Obaya, AJ.; Barneo, L.; Meana, A. & Otero J. (2009). Plasma–Fibroblast Gel as Scaffold for Islet Transplantation. *Tissue Engineering: Part A*, Vol.15, No. 3, pp. 569-577

Phadnis, SM.; Joglekar, MV.; Dalvi, MP.; Muthyala, S.; Nair, PD.; Ghaskadbi, SM.; Bhonde RR. & Hardikar, AA. (2011). Human bone marrow-derived mesenchymal cells differentiate and mature into endocrine pancreatic lineage in vivo. *Cytotherapy*, Vol.13, No.3, pp. 279-293

Phuc, PV.; Nhung, TH.; Loan, DT.; Chung, DC. & Ngoc, PK. (2010). Differentiating of banked human umbilical cord blood-derived mesenchymal stem cells into insulin-secreting cells. *In vitro Cellular & Developmental Biology*, Vol.47, No.1, pp. 54-63

Pileggi, A.; Molano, RD.; Ricordi, C.; Zahr, E.; Collins, J.; Valdes, R. & Inverardi, L. (2006). Reversal of diabetes by pancreatic islet transplantation into a subcutaneous, neovascularized device.*Transplantation*, No.81, pp. 1318–1324

Pileggi, A.; Ricordi, C.; Alessiani, M. & Inverardi, L. (2001). Factors influencing islet of Langerhans graft function and monitoring. *Clinica Chimica Acta*. Vol.310, No.1, pp. 3-16

Puissant, B.; Barreau, C.; Bourin, P.; Clavel, C.; Corre, J.; Bousquet, C.; Taureau, C.; Cousin, B.; Abbal, M.; Laharrague, P.; Penicaud, L.; Casteilla, L. & Blancher, A. (2005). Immunomodulatory effect of human adipose tissue-derived adult stem cells: Comparison with bone marrow mesenchymal stem cells. *British Journal of Haematology*, Vol.129, No.1, pp. 118-129

Rajagopal, J.; Anderson, WJ.; Kume, S.; Martinez, OI. & Melton, DA. (2003). Insulin staining of ES cell progeny from insulin uptake. *Science*, Vol.299, No.5605, pp. 363

Ramiya, VK.; Maraist, M.; Arfors, KE.; Schatz, DA.; Peck, AB. & Cornelius, JG. (2000). Reversal of insulin-dependent diabetes using islet cells generated *in vitro* from pancreatic stem cells. *Nature Medicine*, Vol.6, No.3, pp. 278 –282

Reckard, CR. & Barker, CF. (1973). Transplantation of isolated pancreatic islets across strong and weak histocompatibility barriers. *Transplantation Proceedings*. Vol.5, No.1, pp. 761-763

Ricordi, C.; Lacy, PE.; Finke, EH.; Olack, BJ. & Scharp, DW. (1989). Automated method for isolation of human pancreatic islets. *Diabetes*, Vol.37, No.4, pp. 140-142

Ryan, EA.; Paty, BW.; Senior, PA.; Bigam, D.; Alfadhli, E.; Kneteman, NM.; Lakey, JR. & Shapiro, AM. (2005). Five-year follow-up after clinical islet transplantation. *Diabetes*, Vol.54, No.7, pp. 2060-2069

Sahu, S.; Joglekar, MV.; Dumbre, R.; Phadnis, SM.; Tosh, D. & Hardikar, AA. (2009). Islet-like cell clusters occur naturally in human gall bladder and are retained in diabetic conditions. *Journal of Cellular and Molecular Medicine*, Vol.13, No.5, pp. 999-1000

Salvay, DM.; Rives, CB.; Zhang, X.; Chen, F.; Kaufman, DB.; Lowe, WL Jr. & Shea, LD. (2008). Extracellular Matrix Protein-Coated Scaffolds Promote the Reversal of Diabetes After Extrahepatic Islet Transplantation. *Transplantation*, Vol.85, No.10, pp. 1456–1464

Sandler, S.; Andersson, A.; Eizirik, DL.; Hellerstrom, C.; Espevik, T.; Kulseng, B.; Thu, B.; Pipeleers, DG. & Skjak-Braek, G. (1997). Assessment of insulin secretion *in vitro* from microencapsulated fetal porcine islet like clusters and rat, mouse and human pancreatic islets. *Transplantation*, Vol.63, No.12, pp. 1712-1718

Scharp, DW.; Lacy PE.; Santiago, J, et al. (1990). Insulin independence after islet transplantation into type I diabetic patient. *Diabetes*, Vol.39, No.4, pp. 515-518

Scharp, DW.; Mason, NS. & Sparks, RE. (1984). Islet Immunoisolation:the use of hybrid artificial organs to prevent islet tissue rejection. *World Journal of Surgery*, Vol. 8, No. 2, pp. 221-229

Schneider, S.; Feilen, PJ.; Brunnenmeier, F.; Minnemann, T.; Zimmermann, H.; Zimmermann, U. & Weber, MM. (2005). Long-term graft function of adult rat and human islets encapsulated in novel alginate-based microcapsules after transplantation in immunocompetent diabetic mice. *Diabetes*, Vol. 54, No. 3, pp. 687–693

Segev, H.; Fishman, B.; Ziskind, A.; Shulman, M. & Itskovitz-Eldor, J. (2004). Differentiation of human embryonic stem cells into insulin-producing clusters. *Stem Cells*, Vol.22, No. 3, pp. 265–274

Serup, P.; Madsen, OD. & Mandrup-Poulsen, T. (2001). Islet and stem cell transplantation for treating diabetes. *British Medical Journal*, Vol.322, No.7277, pp. 29 –32

Shapiro, AM.; Gallant, HL.; Hao, EG.; Lakey, JR.; McCready, T.; Rajotte, RV.; Yatscoff, RW. & Kneteman, NM. (2005). The portal immunosuppressive storm: relevance to islet transplantation? *Therapeutic Drug Monitoring*, Vol.27, No.1, pp. 35-37

Shapiro, AM.; Ricordi, C.; Hering, BJ.; Auchincloss, H.; Lindblad, R.; Robertson, RP.; Secchi, A.; Brendel, MD.; Berney, T.; Brennan, DC.; Cagliero, E.; Alejandro, R.; Ryan, EA.; DiMercurio, B.; Morel, P.; Polonsky, KS.; Reems, JA.; Bretzel, RG.; Bertuzzi, F.; Froud, T.; Kandaswamy, R.; Sutherland, DE.; Eisenbarth, G.; Segal, M.; Preiksaitis, J.; Korbutt, GS.; Barton, FB.; Viviano, L.; Seyfert- Margolis, V.; Bluestone, J. &

Lakey, JR. (2006). International trial of the Edmonton protocol for islet transplantation. *The New England Journal of Medicine*. Vol. 355, No.13, pp. 1318-1330

Shapiro, AMJ. (2003). Islet Transplants and Impact on Secondary Diabetic Complications: Does C-Peptide Protect the Kidney?. *Journal of the American Society of Nephrology*, Vol. 14, No.8, pp. 2214–2216

Shapiro, AMJ.; Lakey, JRT.; Ryan, EA.; Korbutt, GS.; Toth, EL.; Warnock, GL.; Kneteman, NM. & Rajotte, RV. (2000). Islet transplantation in seven patients with type 1 diabetes mellitus using a glucocorticoid-free immunosuppressive regimen. *The New England Journal of Medicine*. Vol.343, No.4, pp. 230-238

Siebers, U.; Horcher, A.; Bretzel, RG.; Klock, G.; Zimmermann, U.; Federlin, K. & Zekorn, T. (1993). Transplantation of free and microencapsulated islets in rats: evidence for the requirement of an increased islet mass for transplantation into the peritoneal site. The *International Journal of Artificial Organs*, Vol.16, No.2, pp. 96-99

Siebers, U.; Zekorn, T. & Bretzel, RG. (1990). Histocompatibility of semipermeable membranes for implantable diffusion devices (bioartificial pancreas). *Transplantation Proceedings*, Vol. 22, No. 2, pp. 834-835

Sipione, S.; Eshpeter, A.; Lyon, JG.; Korbutt, GS. & Bleackley, RC. (2004). Insulin expressing cells from differentiated embryonic stem cells are not beta cells. *Diabetologia*, Vol.47, No.3, pp. 499-508

Smith A. (2006). A glossary for stem-cell biology. *Nature*, Vol.441, No.7097, pp. 1060

Soria, B.; Roche, E.; Berna, G.; Leon-Quinto, T.; Reig, JA. & Martin F. (2000). Insulin secreting cells derived from embryonic stem cells normalize glycemia in streptozotocin-induced diabetic mice. *Diabetes*, Vol.49, No.2, pp. 157–162

Stendahl, JC.; Kaufman, DB. & Stupp, SI. (2009). Extracellular Matrix in Pancreatic Islets:Relevance to Scaffold Design and Transplantation. *Cell Transplantation*, Vol.18, No.1, pp. 1–12

Stendahl, JC.; Wang, LJ.; Chow, LW.; Kaufman, DB. & Stupp, SI. (2008). Growth factor delivery from self-assembling nanofibers to facilitate islet transplantation. *Transplantation*, Vol.86, No.3, pp. 478–481

Stock, UA.; and Vacanti, JP. (2009). Tissue engineering: current state and prospects. *Annual Review of Medicine*, Vol.52, No.1, pp. 443-451

Street, CN.; Sipione, S.; Helms, L.; Binette, T.; Rajotte, RV.; Bleackley, RC. & Korbutt, GS. (2004). Stem cell-based approaches to solving the problem of tissue supply for islet transplantation in type 1 diabetes. *The International Journal of Biochemistry & Cell Biology*, Vol.36, No.4, pp. 667–683

Sun, AM.; Parisius, W.; Healy, GM.; Vacek, I. & Macmorine, HG. (1977). The use, in diabetic rats and monkeys, of artificial capillary units containing cultured islets of Langerhans (artificial endocrine pancreas). *Diabetes*, Vol. 26, No.12, pp. 1136–1139

Sun, Y.; Chen, L.; Hou, XG.; Hou, WK.; Dong, JJ.; Sun,L.; Tang, KX.; Wang, B.; Song, J.; Li, H. & Wang, KX. (2007). Differentiation of bone marrow-derived mesenchymal stem cells from diabetic patients into insulin-producing cells *in vitro*. *Chinese Medical Journal*, Vol.120, No.9, pp. 771–776

Sun, Y.; Ma, X.; Zhou, D.; Vacek, I. & Sun, AM. (1996). Normalization of diabetes in spontaneously diabetic cynomologus monkeys by xenografts of microencapsulated porcine islets without immunosuppression. *The Journal of Clinical Investigation*, Vol. 98, No. 6, pp. 1417–1422

Sweet, IR.; Khalil, G.; Wallen, AR.; Steedman, M.; Schenkman, KA.; Reems, JA.; Kahn, SE. & Callis, JB. (2002). Continuous measurement of oxygen consumption by pancreatic islets. *Diabetes Technology & Therapeutics*, Vol.4, No.5, pp. 661- 672

Timper, K.; Seboek, D.; Eberhardt, M.; Linscheid, P.; Christ-Crain, M.; Keller, U.; Muller, B. & Zulewski H. (2006). Human adipose tissue-derived mesenchymal stem cells differentiate into insulin, somatostatin, and glucagon expressing cells. *Biochemical and Biophysical Research Communications*, Vol.341, No.4, pp. 1135-1140

Tzakis, AG.; Ricordi, C.; Alejandro, R.; Zeng, Y.; Fung, JJ.; Todo, S.; Demetris, AJ.; Mintz, DH. & Starzl, TE. (1990). Pancreatic islet transplantation after upper abdominal exenteration and liver replacement. *Lancet*, Vol.336, No.8712, pp. 402-405

Vaca, P.; Martin, F.; Vegara-Meseguer, JM.; Rovira, JM.; Berna, G. & Soria, B. (2006). Induction of differentiation of embryonic stem cells into insulin-secreting cells by fetal soluble factors. *Stem Cells*, Vol. 24, No.2, pp. 258–265

van der Laan, LJ.; Lockey, C.; Griffeth, BC.; Frasier, FS.; Wilson, CA.; Onions, DE.; Hering, BJ.; Long, Z.; Otto, E.; Torbett, BE. & Salomon, DR. (2000). Infection by porcine endogenous retrovirus after islet xenotransplantation in SCID mice. *Nature*, Vol.407, No.6800, pp. 90-94

Wang, RN.; Kloppel, G & Bouwens, L. (1995). Duct - to islet-cell differentiation and islet growth in the pancreas of duct-ligated adult rats . *Diabetologia*, Vol. 38, No.12, pp. 1405 - 1411

Wilson, JT. & Chaikof, EL. (2008). Thrombosis and Inflammation in Intraportal Islet Transplantation: A Review of Pathophysiology and Emerging Therapeutics. *Journal of Diabetes Science* and *Technology*. Vol.2, No.5, pp. 746-759

Yanez, R.; Lamana, ML.; Garcia-Castro, J.; Colmenero, I.; Ramirez, M. & Bueren, JA. (2006). Adipose tissue-derived mesenchymal stem cells have in vivo immunosuppressive properties applicable for the control of the graft-versus-host disease. *Stem Cells*, Vol.24, No.11, pp. 2582-2591

Yang, KC.; Wu, CC.; Lin, FH.; Qi, Z.; Kuo, TF.; Cheng. YH.; Chen, MP. and Sumi, S. (2008). Chitosan/gelatin hydrogel as immunoisolative matrix for injectable bioartificial pancreas. *Xenotransplantation*, Vol.15, No.6, pp. 407–416

Yang, L.; Li, S.; Hatch, H.; Ahrens, K.; Cornelius, JG.; Petersen, BE. & Peck, AB. (2002). *In vitro* trans-differentiation of adult hepatic stem cells into pancreatic endocrine hormone producing cells. *Proceedings of the National Academy of Sciences*, Vol.99, No.12, pp. 8078–8083

Zhao, LR.; Duran, WM.; Reyes, M.; Keene, CD.; Verfaillie, CM. & Low, WC. (2002). Human bone marrow stem cells exhibit neural phenotypes and ameliorate neurological deficits after grafting into the ischaemic brain of rats. *Experimental Neurology*, Vol.174, No.1, pp. 11–20

Zhao, M.; Song, C.; Zhang, W.; Hou, Y.; Huang, R.; Song, Y.; Xie, W.; Shi, Y. & Song, C. (2010). The three-dimensional nanofiber scaffold culture condition improves viability and function of islets. *Journal of Biomedical Materials Research Part A*, Vol.94, No.3, pp. 667–672

Zulewski, H.; Abraham, EJ.; Gerlach, MJ.; Daniel, PB.; Moritz, W.; Muller, B.; Vallejo, M.; Thomas, MK. & Habener, JF. (2001). Multipotential nestin-positive stem cells isolated from adult pancreatic islets differentiate ex vivo into pancreatic endocrine, exocrine, and hepatic phenotypes. *Diabetes*, Vol.50, No.3, pp. 521–533.